Springer Japan KK

S. Torii (Ed.)

Novel Trends in Electroorganic Synthesis

With 348 Figures

 Springer

Sigeru Torii
Department of Applied Chemistry, Faculty of Engineering,
Okayama University, 3-1-1 Tsushima-naka, Okayama, 700-8530 Japan

ISBN 978-4-431-65926-6   ISBN 978-4-431-65924-2 (eBook)
DOI 10.1007/978-4-431-65924-2

Printed on acid-free paper

© Springer Japan 1998
Originally published by Springer -Verlag Tokyo in 1998
Softcover reprint of the hardcover 1st edition 1998

Typesetting: Camera-ready by authors

SPIN: 10663745

Preface

Electroorganic synthesis has been a subject of broad-based study, particularly during the past three decades. New knowledge is accumulating constantly and the time is now ripe to review the recent trends made in the chemistry of electroorganic synthesis. The international symposis were focused on several topics having to do with recent electrochemical reactions, an advanced and leading technology of materials production. Fine chemical transformation, methodology for developing functional materials, investigation of the theoretical and empirical aspects of reaction mechanisms, and electrogeneration of active species and their reactions have been intensively discussed.

This volume comprises the papers presented at the International Symposium on Electroorganic Synthesis held September 24-27, 1997 in Kurashiki as well as at the pre-symposium on The Role of Electrogenerated Active Species in Organic Synthesis held September 21-22 in Okayama, Japan. The aims of these conferences were to survey all aspects of modern electroorganic synthesis, both academic and industrial. The following topics were covered: Part I: Electrooxidation: (1) Alcohols and Phenols; (2) Olefins and Aromatics; (3) Nitrogen-, Sulfur-, and Selenium-Containing Compounds; (4) Phosphorus- and Silicon-Containing Compounds; (5) Halogenation; (6) Reaction with Mediators; (7) Polymers; (8) Electrodes; (9) Miscellaneous. Part II: Electroreduction: (1) Carbonyl Compounds; (2) Halogen-Containing Compounds; (3) Reaction with EG Bases; (4) Nitrogen-Containing Compounds; (5) Boron-, Tin-, and Germanium-Containing Compounds; (6) Silicon Compounds; (7) Metal Complexes; (8) Bio-Electrochemistry; (9) Miscellaneous.

In these proceedings, the oral and poster presentations have been included in the above categories.

I wish to take this opportunity to express my gratitude to all the authors who contributed to the proceedings. I am also grateful to my collaborators, Dr. Hideo Tanaka, Dr. Manabu Kuroboshi, and Miss Akiko Genba, for their enthusiastic assistance in preparing this book as well as to Miss Noriko Sera, Miss Naomi Miyake, and Miss Naoko Noritake, who listed the papers concisely and carefully input the additional editorial data in the computer.

1997

Sigeru Torii
Department of Applied Chemistry
Faculty of Engineering
Okayama University

Contents

(4) Phosphorus- and Silicon-Containing Compounds

(5) Halogenation

(6) Reaction with Mediators

(7) Polymers

(8) Electrodes

(9) Others

Part II: Electroreduction

(1) Carbonyl Compounds

(2) Halogen-Containing Compounds

(6) Silicon Compounds

(7) Metal Complexes

List of Presentators

Numbers in parentheses refer to the pages on which a presentator's paper begins. This list consists of those individuals who actually presented at the Symposium. The names of co-authors are listed in the Table of Contents

Amatore, Christian Andre .... (409)
   Lab. de Chim., École Normale Supérieure, CNRS UA 1110
   24 Rue Lhomond, 75231 Paris Cédex 05, France
Atobe, Mahito .... (253)
   Dept. of Electronic Chem., Interdis. Grad. School of Sci. and Engin., Tokyo Inst. of Tech.
   4259 Nagatsuta, Midori-ku, Yokohama, Kanagawa 226, Japan
Barba, Fructuoso .... (271)
   Depto. de Quim. Org., Univ. de Alcalá
   Campus Universitario, 28871-Alcalá de Henares(Madrid), Spain
Batanero, Hernán Belén .... (321)
   Depto. de Quim. Org., Univ. de Alcalá
   Campus Universitario, 28871-Alcalá de Henares(Madrid), Spain
Becker, James Y. .... (103)
   Dept. of Chem., Ben-Gurion Univ. of the Negev
   P.O.B. 653, Beer-Sheva 84120, Israel
Berdnikov, Eugene Aleksandrovich .... (255)
   Kazan State Univ.
   Galeeva Str. 8, Apt. 54, Kazan, Russia
Bersier, Pierre Martin .... (455)
   Refined / Consultant for Electrocell AB
   Cagliostrostrasse 38, CH-4125 Riehen, Switzerland
Chiba, Kazuhiro .... (25)
   Dept. of Applied Biol. Sci., Tokyo Univ. of Agriculture and Tech.
   3-5-8 Saiwai-cho, Fuchu, Tokyo183, Japan
Chiba, Toshiro .... (123, 313)
   Dept. of Applied Chem., Kitami Inst. of Tech.
   165 Koen-cho, Kitami, Hokkaido 090, Japan
Choi, Jung Hoon .... (341)
   Dept. of Chem., Hanyang Univ.
   Seoul1 33-791, Korea
Comninellis, Christos .... (165)
   Inst. of Chem. Engin., Swiss Federal Inst. of Tech.
   CH-1015 Lausanne, Switzerland
Cretescu, Igor .... (333)
   Faculty of Ind. Chem., Tech. Univ. "Gh. Asachi" Iassy
   Bulevardul D. Mangeron, Nr. 71A, Iasi 6600, Roumania
Ding, Ping .... (41)
   School of Chem. Engin., The Univ. of Birmingham
   Edgbaston, BirminghamB 15 2TT, UK
Duñach, Elisabet .... (391)
   Lab. de Chim. Moléculaire, Unite Associée au CNRS N° 426, Univ. de Nice-Sophia Antipolis
   Parc Valrose, 06108, Nice Cédex 2, France
Erabi, Tatsuo .... (81)
   Faculty of Engin., Tottori Univ.
   4-101 Koyama-cho-minami, Tottori 680, Japan
Evans, Dennis Hyde .... (309, 317)
   Dept. of Chem. and Biochem., Univ. of Delaware
   Newark, DE 19716, USA
Fry, Albert Joseph .... (109)
   Chem. Dept., Wesleyan Univ.
   Middletown, CT 06459, USA

Kanamura, Kiyoshi .... (181)
   Dept. of Energy and Hydrocarbon Chem., Graduate School of Engin., Kyoto Univ.
   Yoshida-honmachi, Sakyo-ku, Kyoto 606-01, Japan
Kashimura, Shigenori .... (355)
   Dept. of Metallurgy, Faculty of Sci. and Engin., Kinki Univ.
   3-4-1, Kowakae, Higashi-Osaka, Osaka 577, Japan
Kashiwagi, Yoshitomo .... (169, 173)
   Faculty of Pharm. Sci., Tohoku Univ.
   Aobayama, Aoba-ku, Sendai, Miyagi 980-77, Japan
Kato, Tsuyoshi .... (295)
   Dept. of Applied Chem., Faculty of Engin., Okayama Univ.
   3-1-1 Tsushima-naka, Okayama 700, Japan
Kawafuchi, Hiroyuki .... (349)
   Toyama National College of Tech.
   13 Hongo-machi, Toyama 939, Japan
Keese, Reinhart .... (395)
   Dept. of Chem. and Biochem., Univ. of Bern
   Freiestrasse 3, CH-3012 Bern, Switzerland
Kim, Shokaku .... (177)
   Dept. of Applied Biol. Sci., Tokyo Univ. of Agriculture and Tech.
   3-5-8 Saiwai-cho, Fuchu, Tokyo183, Japan
Kim, Byeong Hyo .... (325)
   Dept. of Chem., Kwangwoon Univ.
   447-1 Wolgye-Dong, Nowon-ku, Seoul 139-701, Korea
Kimura, Makoto .... (73, 77)
   Dept. of Applied Chem., Graduate School of Engin., Nagoya Univ.
   Chikusa-ku, Nagoya, Aichi 464-01, Japan
Kise, Naoki .... (217)
   Dept. of Biochem., Faculty of Engin., Tottori Univ.
   4-101 Koyama-cho-minami, Tottori 680, Japan
Konno, Akinori .... (211)
   Faculty of Engin., Shizuoka Univ.
   3-5-1 Johoku, Hamamatsu, Shizuoka 432, Japan
Kowal, Andrzej J. .... (11, 445, 447)
   Inst. of Catalysis and Surface Chem., Polish Academy of Sci.
   UL. Niezapominajek 1, 30-239 Krakow, Poland
Kunai, Atsutaka .... (363)
   Dept. of Applied Chem., Faculty of Engin., Hiroshima Univ.
   1-4-1 Kagamiyama, Higashi-Hiroshima, Hiroshima 739, Japan
Kuroboshi, Manabu .... (87)
   Dept. of Applied Chem., Faculty of Engin., Okayama Univ.
   3-1-1 Tsushima-naka, Okayama 700, Japan
Kuwabata, Susumu .... (205)
   Dept. of Applied Chem., Faculty of Engin., Osaka Univ.
   2-1 Yamadaoka, Suita, Osaka 565, Japan
Lessard, Jean .... (65, 329)
   Dépt. de Chim., Univ. de Sherbrooke
   2500, Boulevard Univ., Sherbrooke (Québec) J1K 2R1, Canada
Lhommet, Gérard .... (61)
   Lab. de Chim. des Heterocycles, Univ. Pierre et Marie Curie
   4 Place Jussieu-Case 43, F-75252 Paris Cédex 05, France
Little, Daniel R. .... (221)
   Dept. of Chem., Univ. of California, Santa Barbara
   Santa Barbara, CA 93106, USA
Lorcy, Dominique .... (147)
   Synth. et Electrosynth. Org., Univ. de Rennes 1
   Campus de Beaulieu, Bar 10, 35402 Rennes Cédex, France

Ludvik, Jiří .... (79)
    J. Heyrovsky Inst. of Phys. Chem., Academy of Sci. of the Czech Republic
    Dolejskova 3, 18223 Prague 8, Czech Republic
Lund, Henning .... (213, 431)
    Dept. of Chem., Univ. of Aarhus
    Kemisk Inst., Langelandsgade 140, DK-8800 Aarhus C, Denmark
Maeda, Hatsuo .... (157)
    Faculty of Pharm. Sci., Osaka Univ.
    1-6 Yamadaoka, Suita, Osaka 565, Japan
Maekawa, Hirofumi .... (343)
    Dept. of Chem., Faculty of Engin., Nagaoka Univ. of Tech.
    1603-1 Kamitomioka-machi, Nagaoka, Niigata 940-21, Japan
Maki, Toshihide .... (135)
    Faculty of Pharm. Sci., Nagasaki Univ.
    1-14 Bunkyo-machi, Nagasaki 852, Japan
Mamedov, Vakhid Abdulla Oglu .... (457)
    Faculty of Environmental Chem. and Material, Okayama Univ.
    3-1-1 Tsushima-naka, Okayama 700, Japan
Markó, István E. .... (7)
    Lab. of Org. Chem., Dept. of Chem., Univ. Catholique de Louvain
    Place L. Pasteur 1, 1348 Louvain-la-Neuve, Belgium
Mattiello, Leonardo .... (209)
    Dept. of ICMMPM-Univ. of Rome "La Sapienza"
    Via Del Castro Laurenziano 7, 00161 Rome, Italy
Mazin, Vladimir Markovich .... (293)
    A. N. Frumkin Inst. of Electrochem. of Russian Academy of Sci.
    Leninsky Prospekt 31, 117071, Moscow V-71, Russia
Mazur, Duane J. .... (225)
    The Electrosynthesis Co., Inc.
    72 Ward Road, Lancaster, NY 14086, USA
Misra, Ram A. .... (153, 275)
    Chem. Dept., Faculty of Sci., Banaras Hindu Univ.
    Varanasi 221005, India
Misumi, Yukihiro .... (251)
    Dept. of Applied Chem., Faculty of Engin., Okayama Univ.
    3-1-1 Tsushima-naka, Okayama 700, Japan
Mizuno, Kazuhiko .... (37)
    College of Engin., Osaka Pref. Univ.
    1-1 Gakuen-cho, Sakai, Osaka 593, Japan
Moeller, Kevin David .... (51)
    Dept. of Chem., Washington Univ.
    Campus Box 1134, One Brookings Drive, St. Louis, MO 63130, USA
Moinet, Claude .... (161)
    Lab. d'Électrochim. et Organomet., Univ. de Rennes 1
    Campus de Beaulieu, F-35042 Rennes Cédex, France
Nédélec, Jean-Yves .... (387)
    Centre National de la Recherche Scientifique, Univ. Paris 12-Val de Marne
    2 Rue Henri Dunant, BP28, F-94320 Thiais, France
Nakai, Hidetaka .... (89)
    Dept. of Applied Chem., Kinki Univ.
    3-4-1, Kowakae, Higashi-Osaka, Osaka 577, Japan
Nishiguchi, Ikuzo .... (183, 351)
    Dept. of Chem., Nagaoka Univ. of Tech.
    1603-1 Kamitomioka-cho, Nagaoka, Niigata 940-21, Japan
Nishiyama, Shigeru .... (23)
    Dept. of Chem., Faculty of Sci. and Tech., Keio Univ.
    3-14-1 Hiyoshi, Kohoku-ku, Yokohama, Kanagawa 223, Japan

Nonaka, Tsutomu .... (437)
  Dept. of Electronic Chem., Interdis. Grad. School of Sci. and Engin., Tokyo Inst. of Tech.
  4259 Nagatsuta, Midori-ku, Yokohama, Kanagawa 226, Japan
Novák, Mihály .... (33)
  Inst. of Phys. Chem., Univ. of Szeged
  H-6701 Szeged, P.O.B.105, Hungary
Ogibin, Yury Nikolaevich .... (29)
  N. D. Zelinsky Inst. of Org. Chem., Russian Academy of Sci.
  Leninsky Prospect 47, Moscow 117913, Russia
Ogura, Kotaro .... (197)
  Dept. of Applied Chem. & Chem. Engin., Faculty of Engin., Yamaguchi Univ.
  2557 Tokiwadai, Ube, Yamaguchi 755, Japan
Ohmori, Hidenobu .... (95)
  Faculty of Pharm. Sci., Osaka Univ.
  1-6 Yamadaoka, Suita, Osaka 565, Japan
Okano, Mitsutoshi .... (345)
  Tokyo Inst. of Polytech.
  1583 Iiyama, Atsugi, Kanagawa 243-02, Japan
Okumura, Noriko .... (435)
  Gifu Pharm. Univ.
  5-6-1 Mitahora-Higashi, Gifu 502, Japan
Olea, Maria .... (131)
  Dept. of Phys. Chem., Faculty of Chem. and Chem. Engin., Univ. "Babes-Bolyai"
  11 Arany Janos St., 3400 Cluj-Napoka, Roumania
Ono, Noboru .... (143, 155)
  Dept. of Chem., Faculty of Sci., Ehime Univ.
  2-5 Bunkyo-cho, Matsuyama, Ehime 790, Japan
Osafune, Masahiro .... (127)
  Dept. of Applied Chem., Faculty of Engin., Okayama Univ. of Sci.
  1-1 Ridai-cho, Okayama 700, Japan
Pedersen, Steen Uttrup .... (283)
  Dept. of Chem., Univ. of Aarhus
  Langelandsgade 140, DK-8000 Aarhus C, Denmark
Peters, Dennis G. .... (373)
  Dept. of Chem., Indiana Univ.
  Bloomington, Indiana 47405, USA
Petrosyan, Vladimir .... (279, 417)
  N. D. Zelinsky Inst. of Org. Chem., Russia Academy of Sci.
  Leninsky Prospect 47, Moscow 117913, Russia
Pierre, Fabrice .... (383)
  Lab. "Electrochim. et Organomet.", Univ. de Rennes 1
  Campus de Beaulieu, F-35042 Rennes Cédex, France
Rossi, Leucio .... (193)
  Dept. of Chem., Chem. Engin. and Materials, Univ. of L'Aquila
  Monteluco di Roio, I-67040 L'Aquila, Italy
Sadakane, Masahiro .... (139)
  Kekulè-Inst. für Org. Chem. und Biochem., Univ. Bonn
  Gerhard-Domagk-Str. 1, 53121 Bonn, Germany
Sainsbury, Malcolm .... (339)
  Dept. of Chem., Univ. of Bath
  Claverton Down, Bath, UK
Sasaki, Akiyoshi .... (245)
  Hokkaido National Ind. Research Inst.
  2-17 Tukisamu-Higashi, Toyohira-ku, Sapporo, Hokkaido 062, Japan
Sawaki, Yasuhiko .... (83)
  Dept. of Applied Chem., Faculty of Engin., Nagoya Univ.
  Furo-cho, Chikusa-ku, Nagoya, Aichi 464-01, Japan

Schäfer, Hans J. .... (187, 229)
Org.-Chem. Inst. der Westfälischen-Wilhelms-Univ. Münster
Corrensstr. 40, D-48149 Münster, Germany
Senboku, Hisanori .... (243)
Division of Molecular Chem., Graduate School of Engin., Hokkaido Univ.
N-13, W-8, Kita-ku, Sapporo, Hokkaido 060, Japan
Shinohara, Hiroaki .... (179)
Dept. of Biosci. and Biotech., Faculty of Engin., Okayama Univ.
3-1-1 Tsushima-naka, Okayama 700, Japan
Silvestri, Giuseppe .... (3, 235)
Dept. of Chem. Engin., Univ. of Palermo
Viale delle Scienze, 90128 Palermo, Italy
Sisido, Masahiko .... (413)
Dept. of Biosci. and Biotech., Faculty of Engin., Okayama Univ.
3-1-1 Tsushima-naka, Okayama 700, Japan
Steckhan, Eberhard .... (55, 367)
Kekulè-Inst. für Org. Chem. und Biochem., Univ. Bonn
Gerhard-Domagk-Str. 1, D-53121 Bonn, Germany
Stepanov, Andrei Aleksandrovich .... (299)
A. N. Nesmeyanov Inst. of Organoelement Compounds
28 Vavilov Str., Moscow1 17813, Russia
Sugawara, Masanobu .... (69)
Dept. of Synth. Chem. & Biol. Chem., Graduate School of Engin., Kyoto Univ.
Yoshidahon-machi, Sakyo-ku, Kyoto 606-01, Japan
Sung, Chang Ho .... (449)
Dept. of Electrical Engin., Chonnam National Univ.
300 Yongbong-Dong, Bukgu, Kwangju 500-757, Korea
Takano, Nobuhiro .... (39)
Muroran Inst. of Tech.
27-1 Mizumoto-cho, Muroran, Hokkaido 050, Japan
Taniguchi, Isao .... (175, 405)
Dept. of Applied Chem.and Biochem., Faculty of Engin., Kumamoto Univ.
2-39-1 Kurogami, Kumamoto 860, Japan
Tokuda, Masao .... (239)
Division of Molecular Chem., Graduate School of Engin., Hokkaido Univ.
N-13, W-8, Kita-ku, Sapporo, Hokkaido 060, Japan
Tokumaru, Yoshihisa .... (71)
Dept. of Applied Chem., Faculty of Engin., Okayama Univ.
3-1-1 Tsushima-naka, Okayama 700, Japan
Tomilov, Andrey Petrovich .... (91)
State Research Inst. of Org. Chem. and Tech.
23, Shosse Entuziastov, Moscow 111024, Russia
Uehara, Kaku .... (151)
Research Inst. for Advanced Sci. & Tech., Osaka Pref. Univ.
1-2 Gakuencho, Sakai, Osaka 599, Japan
Uneyama, Kenji .... (301)
Dept. of Applied Chem., Faculty of Engin., Okayama Univ.
3-1-1 Tsushima-naka, Okayama 700, Japan
Utley, James Henry Paul .... (45, 259)
Dept. of Chem., Queen Mary and Westfield College
Mile End Road, London E1 4NS, UK
Yamamura, Shosuke .... (19)
Dept. of Chem., Faculty of Sci. and Tech., Keio Univ.
3-14-1 Hiyoshi, Yokohama, Kanagawa 223, Japan
Yoneda, Norihiko .... (119)
Division of Molecular Chem., Graduate School of Engin., Hokkaido Univ.
N-13, W-8, Kita-ku, Sapporo, Hokkaido 060, Japan

Yoshida, Jun-ichi .... (99)
    Dept. of Synth. Chem. & Biochem., Graduate School of Engin., Kyoto Univ.
    Yoshidahon-machi, Sakyo-ku, Kyoto 606-01, Japan
Yoshizane, Kenji .... (347)
    Dept. of Applied Chem., Faculty of Engin., Okayama Univ. of Sci.
    1-1 Ridai-cho, Okayama 700, Japan
Zuman, Petr .... (233, 337)
    Dept. of Chem., Box 5810, Clarkson Univ.
    Potsdam, NY 13699-5810, USA

Part I
Electrooxidation

ELECTROACTIVATION OF TRANSITION METAL REDOX COUPLES FOR THE CARBONYLATION OF ALCOHOLS TO DIALKYLCARBONATES

Alessandro Galia, Giuseppe Filardo, Onofrio Scialdone, Massimo Musacco, Giuseppe Silvestri

Università di Palermo, Dipartimento di Ingegneria Chimica dei Processi e dei Materiali, Viale delle Scienze, 90128 Palermo, Italy

ABSTRACT

Electrochemical systems which promote the carbonylation of methanol to dimethylcarbonate, based on the anodic activation of transition metal redox couples, or the Br^-/Br_2 couple, are poorly active for the carbonylation of ethanol to diethylcarbonate at room temperature and atmospheric pressure. Positive results have been obtained by combined addiction of $PdBr_2$ and Bu_4NBr to the system. In this way yields up to 50% in diethylcarbonate were obtained, with very good conductivities in long range electrolyses.

INTRODUCTION

The "phosgene free" synthesis of dimethylcarbonate (DMC) by oxidative carbonylation of methanol has a relevant applicative interest, and is at present object of investigation in several laboratories. Several electrochemical systems have been proposed for the synthesis of DMC, involving the catalytic assistance of redox couples: Br^-/Br_2 [1,2,3], Cu(I)/Cu(II) [4,5] or Pd(0)/Pd(II) [6]. Recently we have shown that the same synthesis can be catalysed by a series of compounds of transition metals, some of which never taken under consideration in previous literature, as inorganic salts or as complexes with various ligands [7,8]. In a further development of the research we have considered the extension of this methodology to the synthesis of higher alkylcarbonates. The results obtained in the carbonylation of ethanol to diethylcarbonate (DEC) appeared rather unexpected, and are presented in this communication.

RESULTS AND DISCUSSION

A rather poor conductivity was observed when ethanol alone was used with electrolytes such as NaF, Bu_4NBF_4, or Bu_4NClO_4. Solubility of the electrolytes and conductivity improve when ethanol was added with acetonitrile (ACN).

The stoichiometry of the reactions expected to take place (1-3) is as follows:

anode	$2\ C_2H_5OH + CO \rightarrow (C_2H_5O)_2CO + 2\ e^- + 2\ H^+$	1)
cathode	$2\ C_2H_5OH + 2e^- \rightarrow 2\ C_2H_5O^- + H_2$	2)
bulk	$2\ C_2H_5O^- + 2H^+ \rightarrow 2\ C_2H_5OH$	3)

Suitable catalysts are necessary for the performance of reaction 1. According to our previous experimental results on the electrocarbonylation of methanol to dimethylcarbonate (DMC)[7,8], we have tested the catalytic activity of the transition metal compounds which gave favourable results in that case[9] : CuCl(bipy), CoCl$_2$, PdCl$_2$.

In table 1 the results obtained in the electrocarbonylation of ethanol in the presence of different metal compounds are summarized. Quite surprisingly, the system Cu(I)/Cu(II) has shown a very poor reactivity by anodic activation. Also with Pd(0)/Pd(II) rather low faradic yields, not higher than 2 % were obtained. CoCl$_2$ gave some conductivity problems, therefore we have tested the neutral carbonyl, Co$_2$(CO)$_8$, which by anodic oxidation affords the metal cations Co^{2+} or Co^{3+}, depending on the anode potential. In this case no detectable amounts of diethylcarbonate were produced.

Table 1. Electrocarbonylation of ethanol to diethylcarbonate with different metal compounds.

Expt.	Metal compound (mmoles)	Ea (V vs. SCE)	Electrolyte	DEC faradic yield, %	Charge passed, C
1	CuCl(bipy) 2.1	1.2-1.5	Bu$_4$NBF$_4$	5	360
2	PdCl$_2$ 2.0	1.4	Bu$_4$NClO$_4$	2	3800
3	Co$_2$(CO)$_8$ 2.0	0.8	Bu$_4$NBF$_4$	---	340

Potentiostatic electrolyses; reference SCE. Anodic potential Ea was determined by polarisation curves for each catalyst. Cathode: copper wire; anode carbon felt. Cells with two compartments separated by anionic membrane. Anolyte and catholyte: 60 ml ethanol 0.1M electrolyte; CO saturated, atm. pressure.

The poor performances of the compounds listed in tab. 1 prompted us to test for the electrosynthesis of DEC the redox system Br$^-$/Br$_2$, already successfully proposed [1-3] in the case of DMC under pressures up to 300 atm, and temperatures up to 60°. In our case, at room temperature and atmospheric pressure, faradic yields not higher than 4% were observed (expt. 1, Tab. 2).

A third approach was then attempted, consisting in the contemporary presence of both the bromide ion and the transition metal compound in the anolyte of the cell. Table 2, expts 2 - 6, summarizes the results obtained by this way with tetrabutylammonium bromide: once more cuprous chloride gave very poor results; performances of cobalt, rhodium and palladium chlorides were better, and with palladium bromide the faradic yield vs. DEC rose up to 50%. Furthermore, systems containing both metal compounds and bromide ions show a long range conductivity even better than that observed in the best performances of the electrosynthesis of DMC catalysed by CuCl(bipy).

The system PdBr$_2$/Bu$_4$NBr was also tested for the sysnthesis of DMC, but yields were rather unsatisfactory (expt. 7, Tab. 2)

The catalytic activity of the Pd/Br system is affected by the ratio of palladium to bromide ions as reported in table 3. As observed in the case of the chemical and electrochemical carbonylation of methanol to DMC, also in this case there is an optimum value of the concentration of halide and at lower or higher values lower yields are obtained.

By products of the carbonylation of ethanol with Pd(II)/Br⁻ are: carbon dioxide, ethylformate, ethylacetate, diethylether, acetaldehyde and diethoxyethane. These products are more or less abundant depending on the composition of the electrolytic system, and are all formed in the anolyte. As an example, in the case of expt. 6 of table 2 the product distribution was as follows: DEC 23.8 mmol, carbon dioxide 5.7 mmol, ethylformate 9.2 mmol, ethylacetate 1.7 mmol, diethylether and acetaldehyde about 0.5 mmol, diethoxyethane absent.

Table 2. Electrocarbonylation of ethanol to diethylcarbonate with different redox metal catalysts and Br⁻ ions.

Expt.	Metal catalyst (mmoles)	Ea (V vs. SCE)	Solvent	Bu₄NBr (mmoles)	DEC faradic yield, %	Charge passed, C
1	---	1.0	EtOH/ACN 20/40	12	4	9200
2	CuCl 2	0.9	EtOH/ACN 10/50	6	1	5500
3	RhCl₃ 2	1.0	EtOH/ACN 20/40	12	2	7600
4	CoCl₂ 2	1.0	EtOH/ACN 20/40	12	4	9200
5	PdCl₂ 2	1.0	EtOH/ACN 20/40	12	30	9500
6	PdBr₂ 2	1.0	EtOH/ACN 20/40	8	50	9200
7	PdBr₂ 2	1.0	MeOH/ACN 20/40	8	24	9500

Potentiostatic electrolyses; reference SCE. Anodic potential Ea was determined by polarisation curves for each catalyst. Cathode: copper wire; anode carbon felt. Cells with two compartments separated by anionic membrane . Volume anolyte and catholyte: 60 ml; CO saturated, atm. pressure.

Table 3. Influence of the Pd/Br ratio in the electrocarbonylation of ethanol to diethylcarbonate.

Expt.	Pd/Br	Bu₄NBr (mmoles)	DEC faradic yield, %	Charge passed, C
1	1/5	6	34	6500
2	1/6	8	50	9200
3	1/10	16	30	9900

Potentiostatic electrolyses: anodic potential Ea = 1 V vs. SCE. Cathode: copper wire; anode carbon felt. Cells with two compartments separated by anionic membrane. Solvent: EtOH/ACN 20/40. Volume anolyte and catholyte: 60 ml; CO saturated, atm. pressure. Metal compound: PdBr₂, 2 mmoles in all experiments.

CONCLUSIONS

The experimental results presented here can not be interpreted according to a single reaction pathway: in fact, the transition metal compounds are known to act through the formation of a complex involving both the alcohol and carbon monoxide, which evolves leading to the reduced metal species and the organic carbonate; on the other hand the reaction mechanism proposed in the literature [1] for the carbonylation of methanol to DMC by the couple Br^-/Br_2 refers to the "in situ" formation of $COBr_2$ which reacts with methanol to form DMC and bromide ions which return to the redox cycle. In our case the substantial failure of all the metal based catalytic systems just by substitution of the methyl with an ethyl moiety, and the confirmed influence of the concentration of the halide ions on the performances of the catalytic system are rather consistent with a reaction pathway involving the formation of the complex, although we can not exclude that the metal cations could act as catalysts in the acylic substitution of ethanol on $COBr_2$.

It is quite interesting to observe that the oxygen based chemical system catalysed by copper complexes is active more or less in the same way for the methanol and ethanol carbonylation. The fact that in our case the carbonylation of ethanol fails confirms that the electrochemical pathway involves some peculiar steps, which are not involved in the chemical one.

ACKNOWLEDGEMENTS

This research is supported by the Italian Ministero dell'Università e della Ricerca Scientifica and by the University of Palermo.

References
1. D. Cipris and I. L. Mador, *J. Electrochem. Soc.* **125**, 1954 (1978).
2. R. A. Dombro Jr., G. A. Prentice and M. A. Mc Hugh, *J. Electrochem. Soc.* **135,** 2219 (1988).
3. R. A. Bell, *U.S. Pat.* 4,310,393 (1982); *Chem. Abs.* 96:094012.
4. F. Carbone, G. Filardo, A. Galia, S. Gambino, P. Giannoccaro and G. Silvestri, in *Novel Trends in Electro-organic Synthesis* (Edited by S. Torii), p. 247. Kodansha, Tokyo (1995).
5. K. Otsuka, T. Yagi, I. Yamanaka, *J. Electrochem. Soc.* **142**, 130 (1995).
6. K. Otsuka, T. Yagi, I. Yamanaka, *Electrochim. Acta* **39**, 2019 (1994).
7. A. Galia, G. Filardo, S. Gambino, R. Mascolino, F. Rivetti and G. Silvestri, *Electrochim. Acta* **41**, 2893 (1996).
8. G. Filardo, A. Galia, F. Rivetti, O. Scialdone and G. Silvestri, *Electrochim. Acta* **42**, 1961 (1997).
9. The best faradic yields obtained in our systems for the synthesis of DMC with different metal species were as follows: CuCl(bipy), 85%; $CoCl_2$, 26%; $PdCl_2$, 22% (see ref. 8).

ELECTROORGANIC CYCLISATION OF OXYGEN-CONTAINING COMPOUNDS

István E MARKÓ

Université catholique de Louvain, Département de Chimie, Laboratoire de Chimie Organique, Place Louis Pasteur 1, B-1348 Louvain-la-Neuve, Belgique

ABSTRACT

Spiroketals can be efficiently produced by electrochemical cyclisation of ω-hydroxy-tetrahydropyrans.

The spiroketal motif **1** is a prominent fragment in a large number of biologically active natural products such as insect pheromones, marine toxins and microbial metabolites.[1] As part of our current interest in the efficient and flexible total synthesis of several such natural products, we needed a rapid preparation of various spiroketal subunits. We have recently reported that the intramolecular silyl-modified Sakurai (ISMS) reaction between the readily available annelating agent **2** and ortholactones **3** provides a concise and general entry to these important fragments (Figure 1).[2]

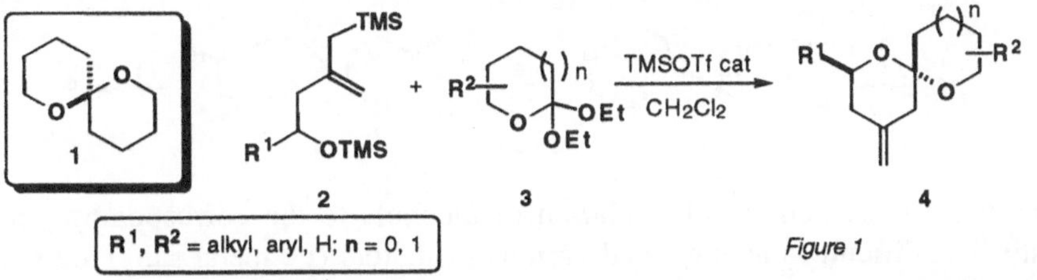

R[1], R[2] = alkyl, aryl, H; n = 0, 1

Figure 1

During our approach to the total synthesis of okadaic acid, we required spiroketal **5** and decided to investigate a novel strategy based on a four-component coupling reaction. Using a modification of the Taddei-Ricci procedure, smooth condensation of aldehyde **6** with crotyltrimethylsilane **7** and ethoxytetrahydropyran **8** occured, affording alcohol **9**. Reductive dechlorination to **10** proceeded in essentially quantitative yield. Oxidative cyclisation of **10** into the requisite spiroketal **5** was effected using an excess of HgO and I_2 under irradition in boiling CCl_4 (Figure 2).[3] Although this approach provided a simple and versatile route to spiroketals such as **5**, it suffered from a number of drawbacks, especially pertaining to the last step of the

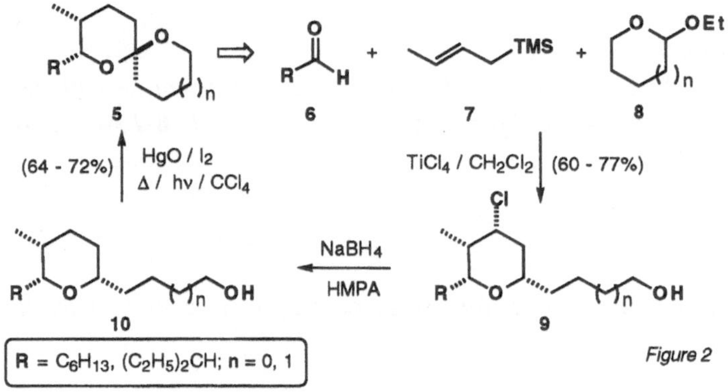

Figure 2

(64 - 72%) | HgO / I₂ — Δ / hv / CCl₄

TiCl₄ / CH₂Cl₂ | (60 - 77%)

NaBH₄ / HMPA

5 ⟸ 6 + 7 (TMS) + 8

10 ← 9

R = C₆H₁₃, (C₂H₅)₂CH; n = 0, 1

sequence for which a large excess of both I$_2$ (4 eqs) and HgO (3 eqs) have to be employed. The high toxicity of the latter reagent precluded scale-up and prompted us to investigate a more ecologically friendly alternative. We envisioned that under electrochemical conditions, loss of an electron and a proton from **10** would generate the alkoxy radical **11**. Ensuing intramolecular hydrogen abstraction would afford a C-centred radical which can lose a second electron to provide the key-oxocarbenium species **12**. Intramolecular cyclisation would then complete the sequence (Figure 4).

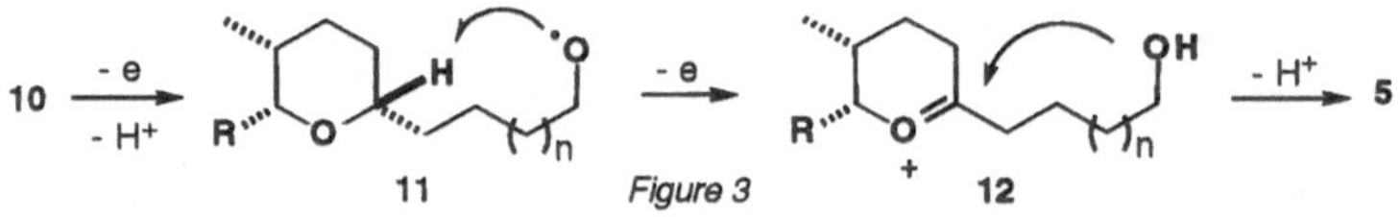

10 $\xrightarrow[- H^+]{- e}$ 11 $\xrightarrow{- e}$ 12 $\xrightarrow{- H^+}$ 5

Figure 3

The direct electrochemical oxidation of alcohols to the corresponding carbonyl derivatives is dfficult to achieve and various mediators (I$_2$, metal salts) are necessary to efficiently promote this transformation.[4] To the best of our knowledge, no example of spiroketalisation under electrochemical conditions has been disclosed so far. Our initial studies were conducted on substrate **13** using various solvents, additives and electrodes. A selection of pertinent conditions is displayed in Table 1.

Preliminary trials using NaI as the additive, with the hope of generating the hypoiodide derivative of **13** *in situ*, in the presence or absence of an electrolyte, failed to give the cyclised material **14** (Table 1, entries 1 and 2). In both cases, the starting material was recovered unaffected. Since α hydroxy radicals can be generated by electrolysis in DMF, we attempted similar experiments on substrate **13**. Once again, alcohol **13** was recovered though some cyclised product **14** could be detected (entry 3). The failure to observe any spiroketal under neutral conditions prompted us to investigate the electrochemical cyclisation of the corresponding alkoxide. Indeed, removal of an electron from an alkoxide to generate an oxygen centred radical is an easier process than electron abstraction from the neutral alcohol function. We were therefore gratified to find that electrolysis of alcohol **13** in CH$_3$CN/^{t}BuOH containing 0.5 eq of ^{t}BuOK afforded the desired spiroketal **14**, though in a modest 13% yield (entry 4). In this reaction, many by products could be observed originating from self-condensation of the CH$_3$CN and Hoffmann degradation of the electrolyte.

Table 1. Testing the conditions for the electrochemical spirocyclisation

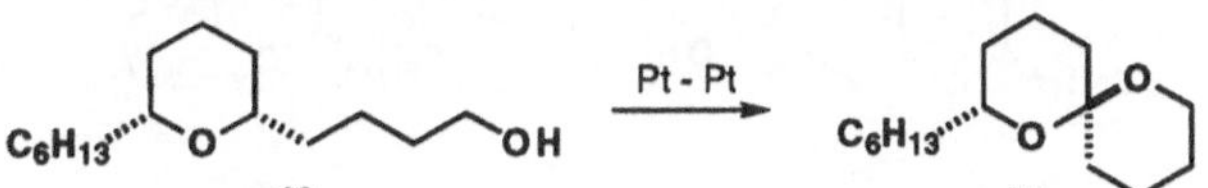

Entry	Solvent	Electrolyte	Additives	Yield
1	MeOH	-	NaI	-
2	MeOH	Et$_4$NBF$_4$	NaI	-
3	DMF	Me$_4$NBF$_4$	-	5%
4	MeCN / tBuOH	Bu$_4$NBF$_4$	tBuOK (0.5 eq)	13%
5	MeOH	Bu$_4$NBF$_4$	NaOMe (1 eq)	15%
6	MeOH	Me$_4$NBF$_4$	NaOMe (1 eq)	30%
7	MeOH	LiBF$_4$	NaOMe (2 eq)	25%
8	**EtOH**	**LiBF$_4$**	**NaOEt (2 eq)**	**61%**
9	iPrOH	LiBF$_4$	NaOPri (2 eq)	10%
10	DMF	-	NaH	5%

Changing to methanol as solvent lowered the amount of by-products (entry 5). Replacement of Bu$_4$NBF$_4$ by Me$_4$NBF$_4$ led to a doubling of the previous yields of **14** (entry 6). Other electrolytes were also tested and LiBF$_4$ was found to be equally efficient and facilitated the purification of **14** (entry 7). The influence the strength of the base was next investigated. Whereas NaOEt in EtOH provided us with an acceptable 60% yield of **14** (entry 8), sodium isopropoxide proved unsuitable (entry 9). Finally, electrochemical oxidation of the *in situ* generated sodium alkoxide (NaH/DMF) failed, affording only trace amounts of the spiroketal **14** (entry 10). Having found suitable conditions for the electrochemical spirocyclisation of **13**, we submitted other precursors to this ring forming process. Some results are collected in Table 2.

Table 2. Electrochemical spirocyclisation

Entry	Substrate	Product	Yield
1	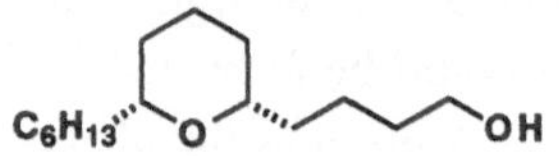	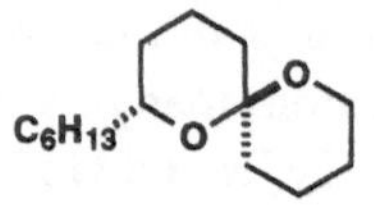	61%
2	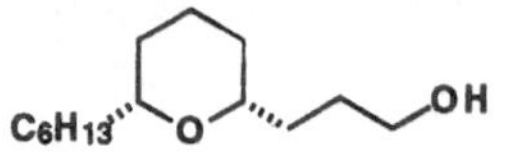	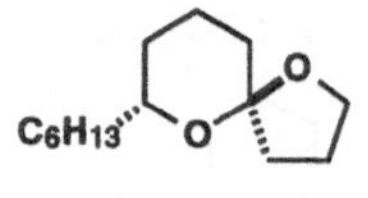	54%
3	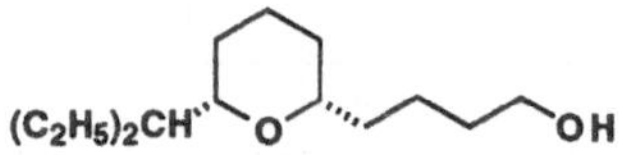	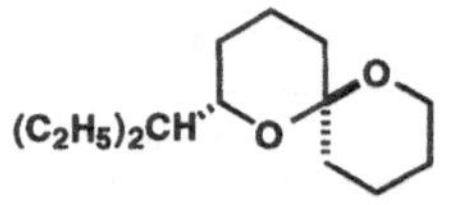	57%
4	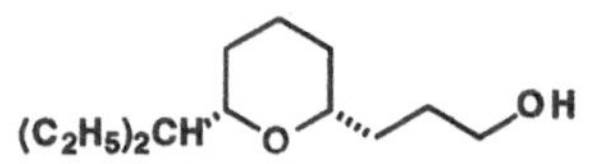	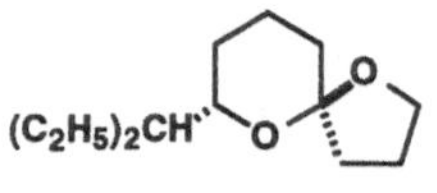	51%

As can be seen from Table 2, both [5,5] and [5,4] spiroketals could be obtained in good yields. The methodology is competitive with those previously reported, with the added benefits that no toxic salts are employed, that the reaction conditions are much milder and that the isolation of the final products is simpler. Further work is aimed at improving the yields of this useful spiroketalisation protocol, exploring the compatibility of the reaction conditions with various protecting and functional groups and understanding the intimate mechanism of this transformation.

AKNOWLEDGEMENTS

IEM is grateful to the Université catholique de Louvain and to Professor S. Torii for his numerous suggestions and continuous encouragements.

REFERENCES

1. F. Perron, and K.M. Albizati, Chem. Rev., 89, 1617 (1989) (b) T.L.B. Boivin, *Tetrahedron*, **1987**, 43, 3309.
2. I. E. Markó, M. Bailey, F. Murphy, J-P. Declercq, B. Tinant, J. Feneau-Dupont, A. Krief and W. Dumont *Synlett*, **1995**, 123 and references cited therein.
3. I.E. Markó, and F. Chellé, F., *Tetrahedron Lett.*, **1997**, 38, 2895.
4. (a) S. Torii Electro-organic Syntheses. Methods and applications, Part I: oxidations, Kodansha-VCH, Tokyo, 1985; (b) T. Shono, Electroorganic Synthesis, Academic Press, 1991.

ELECTROCHEMICAL OXIDATION OF ALCOHOL WITH NICKEL HYDROXIDE ELECTROCATALYST STUDIED BY THE IN SITU TECHNIQUES: INFRARED SPECTROSCOPY, POTENTIAL MODULATED REFLECTANCE SPECTROSCOPY AND ELECTROCHEMICAL METHODS. AN ATTEMPT FOR ELECTROOGANIC SYNTHESIS OF ORGANIC ACIDS

Andrzej KOWAL [a], Richard J. NICHOLS [b], Simon N. PORT [b],
Claudio GUTIERREZ [c], and Jerzy HABER [a]

a. *Institute of Catalysis and Surface Chemistry, Polish Academy of Sciences, Kraków, Poland*
b. *The Department of Chemistry, The University of Liverpool, Liverpool L69 3BX, UK*
c. *Instituto de Quimica Fisica ''Rocasolano'', CSIC, Madrid, Spain*

Both the organic phase and catalyst surface were monitored in situ during the electrochemical synthesis by just using two types of radiation. To monitor the organic phase we use infrared radiation [1], and to examine the catalyst surface we use UV-VIS radiation (320-800 nm) [2].

The electrochemical oxidation of four alcohols (methanol, 1-,2- and tertbutanol) at Ni hydroxide electrodes has been investigated in alkaline electrolytes. In-situ FTIR spectroscopy and electrochemical methods have been used to examine these oxidation reactions. Oxidation of the primary and secondary alcohols commences in the potential region where it is proposed that multilayers NiOOH are formed on the electrode surface, while no reaction occurs with tertiary butanol. Methanol oxidation occurs in two stages, predominantly formate being formed in the potential window 0.36 - 0.44 V (vs. SCE), followed by further oxidation to carbonate at potentials above ca. 0.45 V. Butanoate is the only detected reaction product for 1-butanol electro-oxidation on the potential range 0.36 - 0.5 V. The oxidation of 2-butanol is more complex. In the lower potential range (0.36 - 0.44 V) the major reaction product is butanone, which is further oxidized at higher potentials to either acetate or a mixture of propanoate and formate (or carbonate). In addition, rate constants have been determined for the first stage of the electrochemical oxidation of all the alcohols investigated.

The results for the electrooxidation of the investigated alcohols at Ni/NiOOH in 1M alkaline electrolytes are presented in Fig. 1.

Fig.1 The results for the electrooxidation of the investigated alcohols at Ni/NiOOH in 1M alkaline electrolytes (alcohols with 3 , 2 , 1 , 0 hydrogen in α position) :

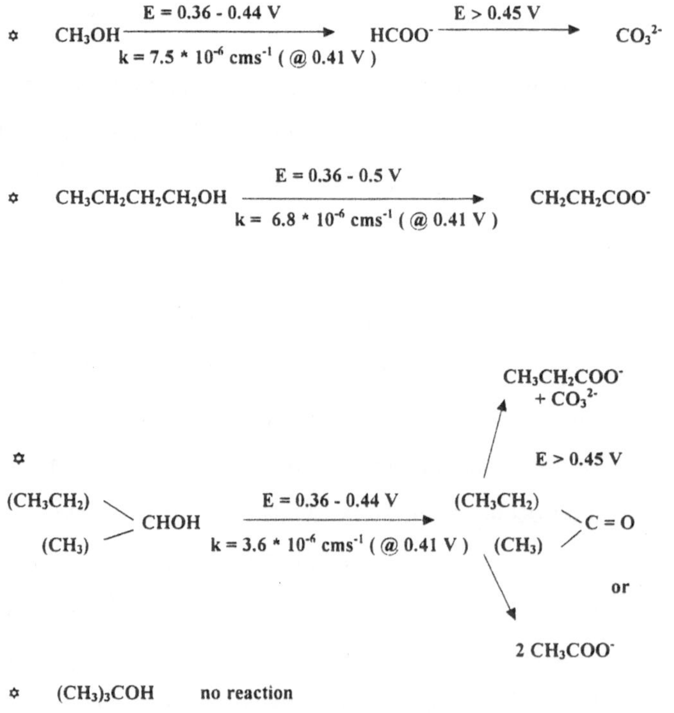

✿ CH_3OH $\xrightarrow[k = 7.5 * 10^{-6} \text{ cms}^{-1} (@ 0.41 \text{ V})]{E = 0.36 - 0.44 \text{ V}}$ $HCOO^-$ $\xrightarrow{E > 0.45 \text{ V}}$ CO_3^{2-}

✿ $CH_3CH_2CH_2CH_2OH$ $\xrightarrow[k = 6.8 * 10^{-6} \text{ cms}^{-1} (@ 0.41 \text{ V})]{E = 0.36 - 0.5 \text{ V}}$ $CH_2CH_2COO^-$

✿

$$\begin{matrix} (CH_3CH_2) \\ \quad \quad \quad \quad > CHOH \\ (CH_3) \end{matrix} \xrightarrow[k = 3.6 * 10^{-6} \text{ cms}^{-1} (@ 0.41 \text{ V})]{E = 0.36 - 0.44 \text{ V}} \begin{matrix} (CH_3CH_2) \\ \quad \quad \quad \quad > C = O \\ (CH_3) \end{matrix}$$

$CH_3CH_2COO^- + CO_3^{2-}$ ($E > 0.45 \text{ V}$)

or

$2 \ CH_3COO^-$

✿ $(CH_3)_3COH$ no reaction

At the next stage, the influence of ethanol (as an example of primary alcohol) on the passivating films on Ni electrode in 1 M NaOH was studied by cyclic voltammetry and potential-modulated reflectance spectroscopy (PMRS) over a wide potential range. The decrease in the PMRS maximum attributed to NiOOH which occurs when ethanol is added, clearly shows that ethanol produces a decrease in the thickness of the NiOOH layer on the Ni electrode; this effect increases in a nearly exponential way with the potential, reaching near saturation at an ethanol concentration of 1 M. The passivating layer of $NiO/Ni(OH)_2$ is not attacked by ethanol.

It is concluded that in order to react with alcohol the anodic oxide of Ni must have hydroxyl group, O^{-2} and Ni on an oxidation level higher than +2; the alcohol must have one hydrogen at least.

In the second part of the presentation, the possibilities of synthesis of organic acids [3-6] like: formic, acetic, propionic, isobutyric, isovaleric and gluconic will be discussed.

In a laboratory scale experiment electrochemical oxidation of alcohols was carried out in an electrolyser of 2 dm^3 capacity, containing two rough Ni anodes of 11 dm^2 surface and three specially prepared Ni + Co plated cathodes, placed vertically in the sequence K, A, K, A, K. Electrolyses were performed using a constant current: the electrolyte was circulated through the system, and the electrolyser was placed in a thermostat. The following parameters were controlled during the synthesis: the anode potential, the potential between anode and cathode, the stability of current, the temperature in the electrolyser and in the thermostat. During the electrolyses the concentration of organic acids, OH^- ions and the amount of CO_3^{-2} were detected by the use of gas and liquid chromatography, and alkacymetric titration with potentiometric detection.

As an example of scaling-up, the process of electrosynthesis of bioassimilable preservative (the mixture of C_2H_5COONa + NaCl) will be discussed. The schematic diagram of technology, pilot scale installation and mass-volume technology diagram will be presented.

References
1) A. Kowal, S. N. Port, R. J. Nichols, Catalysis Today (in press).
2) A. Kowal, C. Guterriez, J. Electroanal. Chem., **1997**, *395*, 243.
3) A. Kowal, J. Haber, E. Fugiel-Mocała, M. Czerwenka, ''Electrode for Electrochemical Processes and Method of Preparation'', Polish Patent No. 148145 **(1990)**.
4) A. Kowal, J. Haber, R. Niewiara, I. Gąsior, ''The Method of Preservative Preparation'', Polish Patent No. 149513 **(1990)**.
5) A. Kowal, J. Haber, M. Czerwenka, M. Adamus, J. Bujak, A. Krawiec, B.Ziobro, ''The Method of Production of 2-aceto-keto-gulonic Acid'', Polish Patent No. 148144 **(1990)**.
6) A. Kowal, J. Haber, ''Electrosynthesis of Bioassimilable Preservative with Nickel Hydroxide Electrode'', 4th World Congress on Chemical Engineering - ''Strategies 2000'', Karlsruhe 1991, p.3.7-32.

REDOX CATALYSIS OF NOVEL HETEROCYCLIC o-QUINONE COENZYMES

Shinobu ITOH, Hirokatsu KAWAKAMI, and Shunichi FUKUZUMI
Department of Applied Chemistry, Faculty of Engineering, Osaka University
Yamada-oka 2-1, Suita, Osaka 565, Japan

Quinones have been widely utilized as a redox catalyst or an electrochemical mediator in a variety of chemical reactions and electrochemical devices. In order to achieve high catalytic efficiency of such systems, developement of a highly reactive and robust catalyst is crucial. In this context, we have recently developed a series of heterocyclic o-quinone compounds that mimic the structures and functions of newly found coenzymes such as pyrroloquinolinequinone (PQQ) and tryptophan tryptophylquinone (TTQ).[1] Here, we demonstrate that a calcium complex of a PQQ analog acts as a very efficient catalyst for the oxidation of simple alcohols such as methanol and ethanol under a very mild condition.[2]

When the trimethyl ester of PQQ (PQQTME) was treated with $Ca(ClO_4)_2$ in anhydrous acetonitrile (CH_3CN), the absorption band at 354 nm due to the quinone shifted to 368 nm and the shoulder around 280 nm decreases with clear isosbestic points at 268, 289, 303, 361, and 422 nm. The 1:1-complex formation between PQQTME and Ca^{2+} with the binding constant K_c of 1900 M^{-1} has been determined by analyzing the spectral change. The following ^{1}H- and ^{13}C-NMR data in CD_3CN indicate that the binding position of Ca^{2+} to PQQTME in solution is the same as that to PQQ in the enzymatic system (**1H** in Scheme 1). In the ^{1}H-NMR spectra, the methyl ester protons at the 7-position move toward down field more than those at the 2- and 9-positions by the complexation with Ca^{2+} ($\Delta\delta$ = +0.14, +0.02, and +0.06, respectively) and the $\Delta\delta$ (down-field shift by the complexation) value of H-8 is also larger than that of H-3 (+0.09 and +0.07, respectively). In the ^{13}C-NMR spectra, C-5 and C-7' (ester carbonyl carbon at the 7-position) shifted down-field (Δppm = +2.0 and +2.8, respectively), while C-4, C-2' (ester carbonyl carbon at the 2-position), and C-9' (ester carbonyl carbon at the 9-position) shifted up-field by the complexation with Ca^{2+} (Δppm = -1.0, -0.2, and -1.0, respectively).

Scheme 1

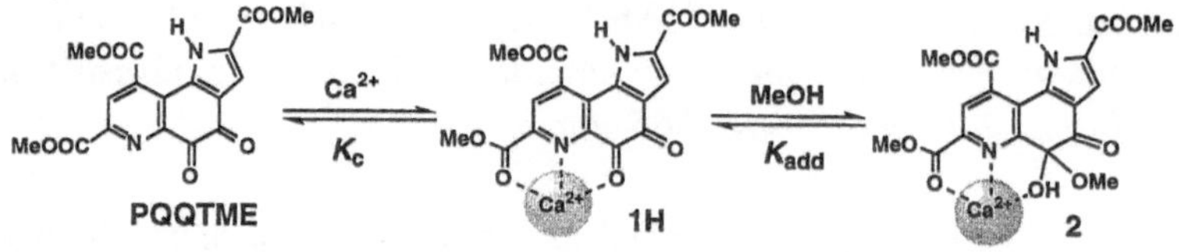

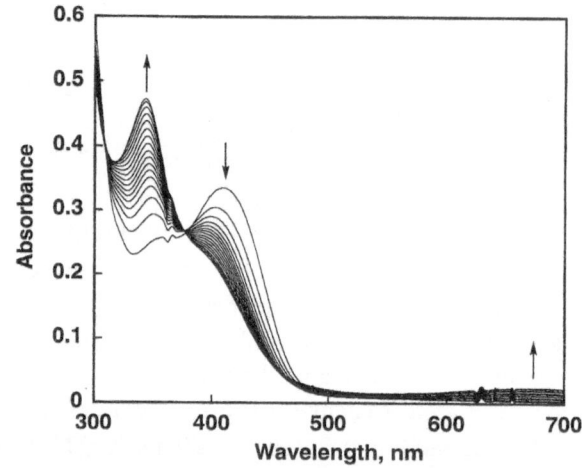

Addition of methanol into a CH$_3$CN solution of Ca^{2+}-complex **1H** resulted in a spectral change corresponding to the C-5 hemiacetal formation (**2** in Scheme 1). The formation constant K_{add} of the Ca^{2+}-complex with methanol was determined as 3.6 M^{-1} which is 6 times larger than that measured in the absence of Ca^{2+} (0.63 M^{-1}),[3] indicating clearly that the complexation with Ca^{2+} enhances the stability of the C-5 hemiacetal. Coordinative interaction of methanol to Ca^{2+} may also enhance the nucleophilicity of methanol by lowering the pK_a value of the -OH group.

To our surprise, addition of methanol into a deaerated CH$_3$CN solution of the Ca^{2+}-complex in the presence of a base such as DBU resulted in formation of *reduced PQQTME* (Figure 1). The final spectrum of the reaction mixture is essentially the same as that obtained by the treatment of authentic PQQTMEH$_2$ (quinol form) with Ca(ClO$_4$)$_2$ and DBU in deaerated CH$_3$CN, and more importantly, *it is also*

Figure 1. Spectral change observed in the oxidation of CH$_3$OH by the calcium complex of PQQTME in deaerated CH$_3$CN at 25°C.

very close to the absorption spectrum of fully reduced MDH.[4] From the reaction mixture in a preparative scale ([PQQTME] = 1.5 x 10^{-4} M, [Ca(ClO$_4$)$_2$] = 8.5 x 10^{-3} M, [DBU] = 2.4 x 10^{-3} M, [CH$_3$OH] = 2.2 M in 100 mL of CH$_3$CN), a reasonable amount of formaldehyde was isolated as the 2,4-dinitrophenylhydrazone derivative. A similar spectral change (quantitative formation of reduced PQQTME) was observed with ethanol as well as methanol, and the oxidation product, acetaldehyde, was obtained quantitatively as 2,4-dinitrophenyl-hydrazone derivative.

The pseudo first-order rate constant (k_{obs}) obtained from the spectral change indicated in Figure 1 shows a Michaelis-Menten type saturation phenomenon when plotted against methanol concentration (Figure 2A). Non-linear curve fitting using the equation of $k_{obs} = kK_{add}[CH_3OH][DBU]/(1 + K_a[DBU] + K_{add}[CH_3OH])$ derived from the reaction mechanism shown in Scheme 2 provided the kinetic parameters as $K_a = 1.4 \times 10^3$ M^{-1}, $K_{add} = 3.7$ M^{-1}, and $k = 0.42$ M^{-1}s^{-1}, where K_a is the deprotonation equilibrium constant (**1⁻** represents deprotonated PQQTME at N-1) and k is the rate constant of the redox reaction from **2** to **3**. Dependence of k_{obs} vs [DBU] (Figure 2B) also afforded a saturation phenomenon as expected from the kinetic equation, from which $K_a = 1.5 \times 10^3$ M^{-1}, $K_{add} = 3.5$ M^{-1}, and $k = 0.42$ M^{-1} s^{-1} were obtained by the computer simulation. The good agreements of those kinetic parameters determined independently from k_{obs} vs [CH$_3$OH] and k_{obs} vs [DBU] together with the agreement of K_{add} determined by the titration (3.6 M^{-1}) and kinetics (3.5 M^{-1} and 3.7 M^{-1}) support the validity of the proposed mechanism and exclude the intermolecular hydride transfer mechanism from methanol to the quinone. A large kinetic isotope effect of 6.4 on the rate constant k was obtained by using CD$_3$OD as a substrate, clearly indicating that the base-catalyzed α-proton abstraction from the substrate is rate-determining in the methanol oxidation reaction (from **2** to **3**).

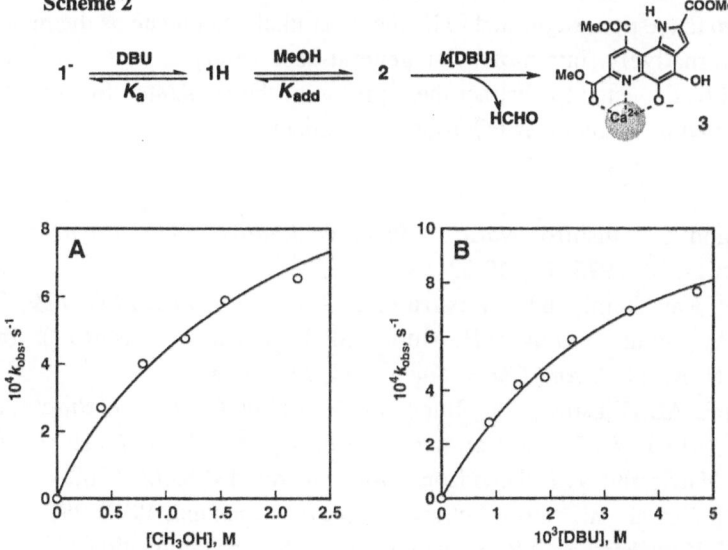

Figure 2. Plot of (A) k_{obs} vs [CH$_3$OH] (at [DBU] = 2.4 x 10^{-3} M) and (B) k_{obs} vs [DBU] (at [CH$_3$OH] = 1.5 M) for the oxidation of CH$_3$OH by the calcium complex of PQQTME (2.5 x 10^{-5} M) in deaerated CH$_3$CN at 25°C. The solid lines are drawn based on the computer simulation using the kinetic equation and the parameters indicated in the text.

Furthermore, the oxidation of ethanol to acetaldehyde proceeded catalytically (1450 % based on PQQTME after 65 h), when the reaction was carried out under aerobic conditions (Scheme 3). We have also found that calcium ion largely accelerates the reoxidation process of reduced PQQ by O_2.[5]

Scheme 3

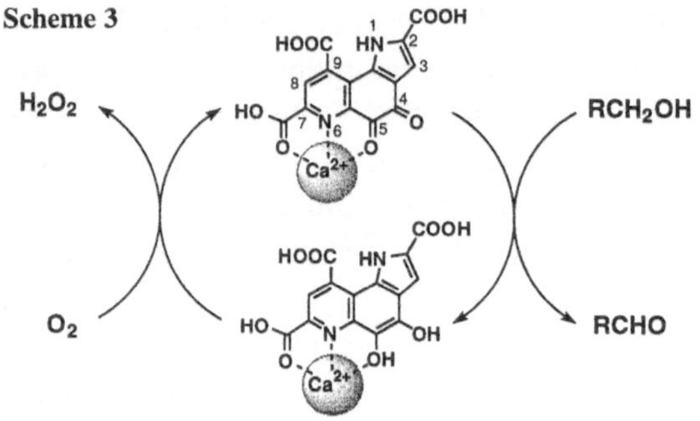

In summary, we have demonstrated here that the Ca^{2+}-complex of PQQTME actually facilitates the alcohol-adduct formation at C-5 and also that Ca^{2+} is required for the base-catalyzed oxidative elimination reaction of the adduct. In the enzymatic system, aspartate (Asp) is suggested to be the most likely candidate as the general-base catalyst. Alternatively, intramolecular general-base catalysis by the C-4 carbonyl oxygen could be expected to abstract the a-proton of the substrate. Such a possibility will be examined by using other PQQ-model compounds.[6]

References

1. S. Itoh and Y. Ohshiro, *Natural Product Reports (The Royal Society of Chemistry, UK)*, **1995**, *12*, 45-53.
2. S. Itoh, H. Kawakami, and S. Fukuzumi, *J. Am. Chem. Soc.*, **1997**, *119*, 439.
3. S. Itoh, M. Ogino, Y. Fukui, H. Murao, M. Komatsu, Y. Ohshiro, T. Inoue, Y. Kai, and N. Kasai, *J. Am. Chem. Soc.*, **1993**, *115*, 9960.
4. (a) J. Frank, M. Dijkstra, J. A. Duine, and C. Balny, C. *Eur. J. Biochem.* **1988**, *174*, 331. (b) I. W. Richardson and C. Anthony, *Biochem. J.* **1992**, *287*, 709. (c) T. K. Harris and V. L. Davidson, *Biochemistry* **1993**, *32*, 4362.
5. S. Itoh, H. Kawakami, and S. Fukuzumi, *Chem. Commun.*, **1997**, 29.
6. S. Itoh, H. Kawakami, and S. Fukuzumi, submitted for publication.

NATURAL PRODUCT SYNTHESES USING PHENOLIC OXIDATION IN THE KEY STEP

Shosuke YAMAMURA

Department of Chemistry, Faculty of Science and Technology, Keio University, Hiyoshi, Yokohama 223, Japan

ABSTRACT

Some bioactive isodityrosine and related oligopeptides have been successfully synthesized by means of anodic or TTN oxidation of o,o'-dihalogenated phenols. Furthermore, electrochemical synthesis of some sesqui- and diterpenes has also been carried out.

INTRODUCTION

Electrochemical studies on phenols and their ethers have been made for a long time. In the case of phenols, however, it is not easy to obtain a desired compound in a regio- and stereo-selective manner, because three unstable species (ArOH+•, ArO•, ArO+) are electrogenerated as a reactive intermediate. Therefore, it is quite important to find the optimum reaction conditions by changing oxidation potential, electrode, and other parameters. Particularly, product selectivity depends on the substituents attached to the aromatic ring and solvents used for anodic oxidation.[1] As compared with electrochemical method, in the case of o,o'-dihalogenated phenols, phenolic oxidation was also carried out using thallium trinitrate (TTN) in MeOH to give the corresponding diaryl ethers.[2]

ANODIC AND TTN OXIDATION OF O,O'-DIHALOGENATED PHENOLS

The glycopeptide antibiotic vancomycin, isolated from *Streptomyces orientalis*, is one of the most challenging molecules for chemists, from the viewpoints of its characteristic heptapeptide core involving diaryl ethers and its potent bactericidal activities of this antibiotic particularly against methicillin resistant *Staphylococcus aureus* (MRSA). In connection with vancomycin, the tyrosine derivative (1) was subjected to anodic oxidation (CCE: 5 mA, +1038 - 1228 mV *vs.* SCE) in MeOH followed by zinc

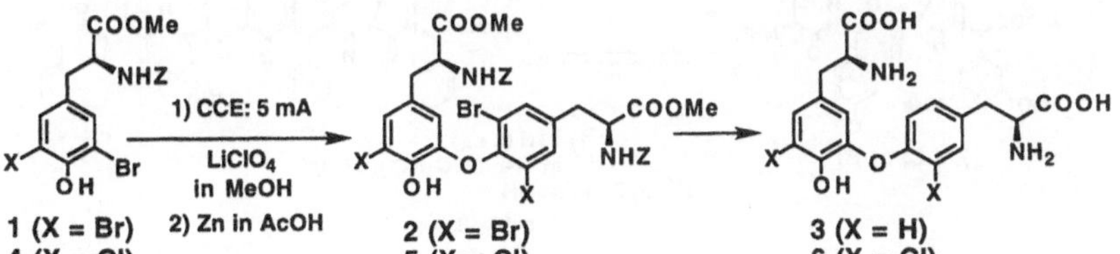

Scheme 1.

19

reduction to afford the corresponding diaryl ether (**2**) in ca. 45% yield,[3] which was quantitatively converted into **3** in 2 steps [1) H_2, 10% Pd-C in MeOH-aq.HCl; 2) 6M HCl]. Furthermore, the 3-bromo-5-chloro-tyrosine derivative (**4**) was also selectively converted into dichloroisodityrosine (**6**) through **5** in ca. 45% overall yield, as shown in Scheme 1. On TTN oxidation in MeOH, we could also obtain isodityrosine (**3**), although its yield (38%) was relatively low. As compared with an

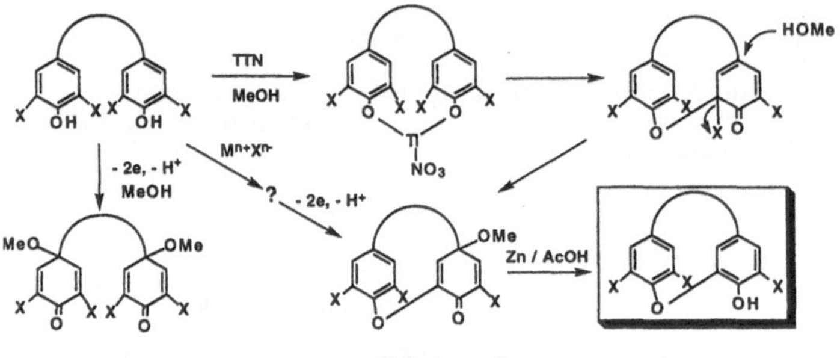
TTN
MeOH
HOMe
Mn+Xn-
- 2e, - H+
MeOH
? - 2e, - H+
NO3
Zn / AcOH

Scheme 2.

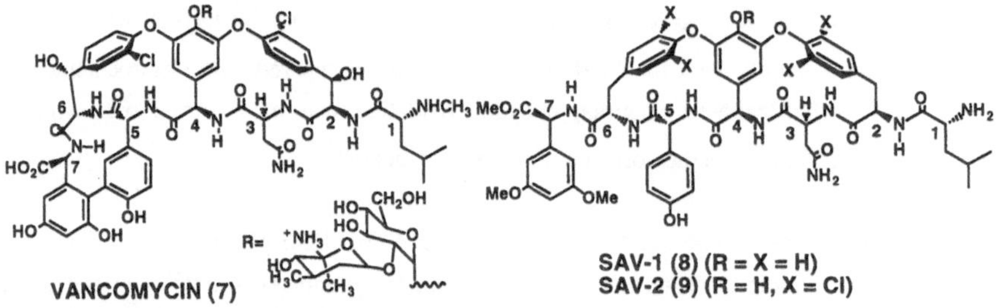
VANCOMYCIN (7)
SAV-1 (8) (R = X = H)
SAV-2 (9) (R = H, X = Cl)

electrochemical method, however, TTN oxidation will be effective for the formation of cyclic diaryl ethers, because of the Tl-O complexation, as shown in Scheme 2, wherein any cyclization products have

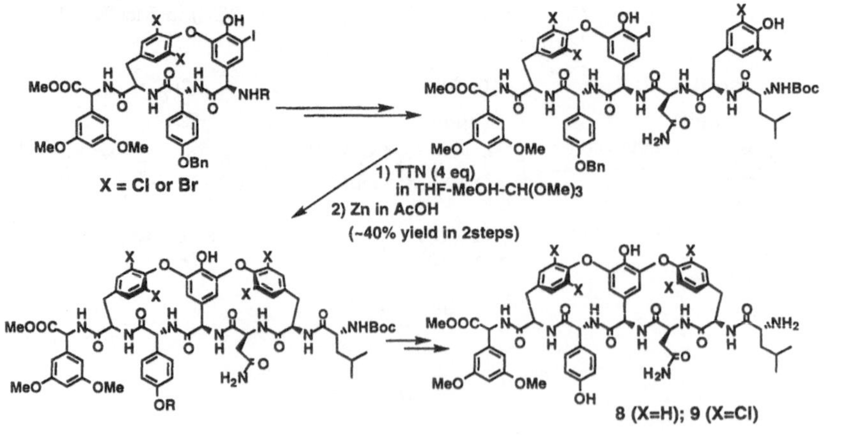
X = Cl or Br
1) TTN (4 eq)
in THF-MeOH-CH(OMe)3
2) Zn in AcOH
(~40% yield in 2steps)
8 (X=H); 9 (X=Cl)

Scheme 3.

not been detected on anodic oxidation.[2]

In connection with vancomycin (**7**), we could synthesize two secoaglucovancomycins (**8** and **9**) using TTN oxidation method, as shown in Scheme 3.[4] These oligopeptides are expected to be a promising agent against vancomycin-resistant MRSA based on a combination of NOE data and molecular dynamics calculation.[5]

ELECTROCHEMICAL FORMATION OF TRICYCLIC COMPOUNDS AND THEIR CONVERSION TO TERPENOIDS

3,4-Dimethoxyphenols [**A**] with a double bond at the side chain are subjected to anodic oxidation under various conditions to give three different types of tricyclic compounds [**B**, **C**, and **D**] in good yields, depending on functional groups attached to the double bond, as shown in Scheme 4. All of these

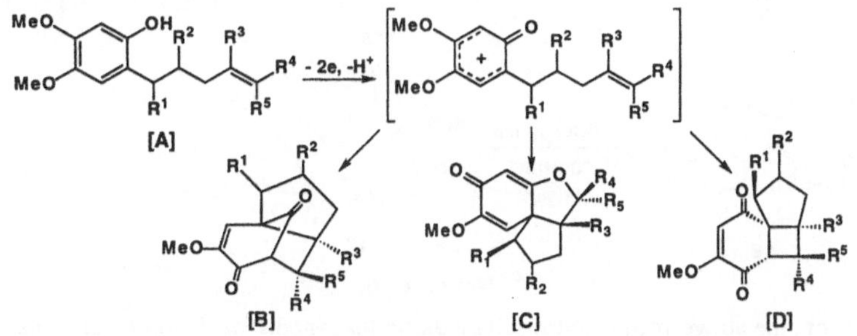

Scheme 4.

cyclization products must be regarded as the promising synthetic intermediates for many sesquiterpenes. From the **B**-type tricyclic intermediates, particularly, 8,14-cedranoxide, silphinene and pentalene have been synthesized.[1, 6]

Furthermore, one of the target compounds is the new 2-epi-cedrene-isoprenologue (**10**), first isolated from *Eremophila georgei* D., which constitutes a new class of diterpenes bearing a tricyclic cedrane-type skeleton in their molecule, whose synthesis is shown in Scheme 5, wherein the requisite key intermediate, 6-acetoxymethyl-2-benzyloxymethyl-6-methyl-9-methoxy-tricyclo-[5.3.1.01,5]undec-9-en-8,11-dione (**11**) was obtained by means of anodic oxidation of the corresponding phenol (**12**). Both α- and β-isomers (**11**) were further converted into the target molecule (**10**).[7] Synthetic study on

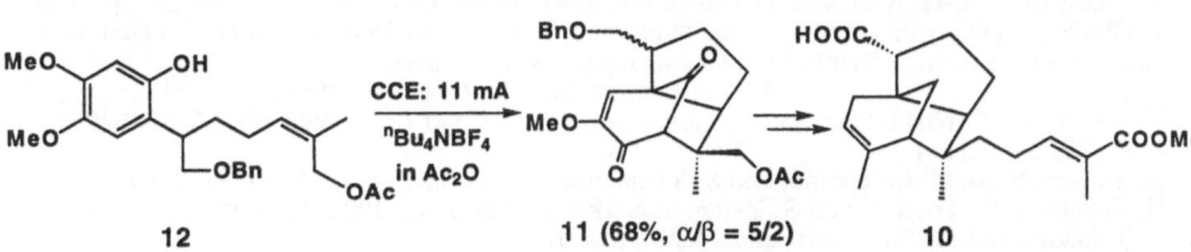

Scheme 5.

another highly oxygenated cedrane-type sesquiterpene (13), one of the constituents of *Acourtia nana*, is in progress starting from the phenol (14) through the corresponding electrochemically formed tricyclo[5.3.1.01,5]undec-9-en-8,11-dione (15) (63%, α/β=3), in which one carbon atom must be introduced, as depicted in Scheme 6. In parallel with this synthetic approach, anodic oxidation of the penta-substituted phenol (16) including all of the carbon atoms of 13 was carried out to afford only the β-isomer (17) in 34% yield,[8] which is also regarded as a promising synthetic intermediate for 13.

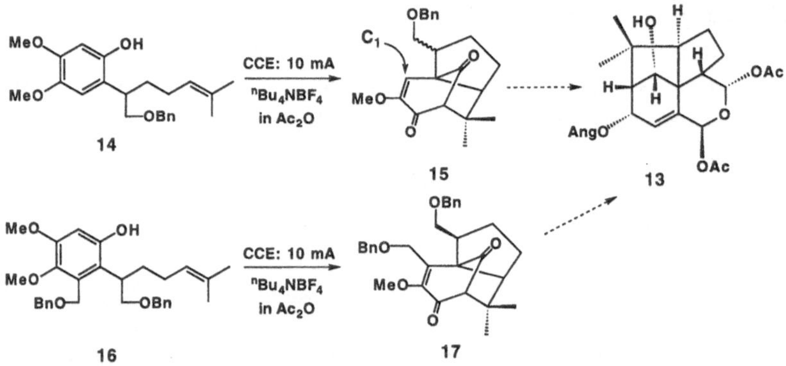

Scheme 6.

In conclusion, the above mentioned results indicate that anodic oxidation of phenols is quite useful and convenient for the synthesis of natural products with a complex structure. And further development of electroorganic synthesis will be able to be expected by controlling each different kind of species (radical cation, radical and cation) in a regio- and stereo-selective manner. Instead of phenols, furthermore, electrochemical studies on aniline derivatives will open a new field.[9]

References

1) S. Yamamura, Natural Products Syntheses by Means of Electrochemical Methodology in "Electroorganic Synthesis: Festschurift for Manuel M. Baiser" (R.D.Little and N.L. Weiberg Ed.), Marcel Dekkar, 307 (1991); S.Yamamura, Y. Shizuri, H. Shigemori, Y. Okuno, and M. Ohkubo, *Tetrahedron*, **1991**, *47*, 635.
2) S. Yamamura and S. Nishiyama, in "Studies in Natural Products Chemistry" (Atta-ur-RahmanEd.), Elsevier, 629 (1992).
3) S. Nishiyama, M.-H. Kim, and S. Yamamura, *Tetrahedron Lett.*, **1994**, *35*, 8397; H. Konishi, T. Okuno, S. Nishiyama, and S. Yamamura, *Tetrahedron Lett.*, **1996**, *37*, 8791; S.Yamamura and S. Nishiyamam, J. Synth. Org. Chem.(Japan), **1997** in press.
4) K. Nakamura, S. Nishiyama, and S. Yamamura, *Tetrahedron Lett.*, **1995**, *36*, 8621.
5) K. Nakamura, S. Nishiyama, and S. Yamamura, *Tetrahedron Lett.*, **1995**, *36*, 8625, 8629; *ibid.*, **1996**, *37*, 191.
6) Y. Shizuri, S. Maki, M. Okuno, and S. Yamamura, *Tetrahedron Lett.*, **1990**, *31*, 7167.
7) H. Takakura, K. Toyoda, and S. Yamamura, *Tetrahedron Lett.*, **1996**, *37*, 4043.
8) H. Takakura and S. Yamamura, unpublished results.
9) S. Nishiyama, Y. Iida, S. Hino, and S. Yamamura, *Electrochimica Acta*, **1997**, *42*, 1943.

REACTIVITY OF O,O'-DIHALOGENATED PHENOLS IN ANODIC OXIDATIONS

Shigeru NISHIYAMA, Hironori KONISHI, Satoshi, IIDA, and Shosuke YAMAMURA

Department of Chemistry, Faculty of Science and Technology, Keio University,

Hiyoshi 3-14-1, Kohoku-ku, Yokohama 223, Japan

ABSTRACT

Anodic oxidations of o,o'-dihalogenated phenols afford diaryl ether or diaryl derivatives as one-electron oxidation products. Halogen substituents control the oxidation mode: the Cl or Br derivatives provide the diaryl ethers, whereas the diaryls are specifically produced from the iodo compounds. The reaction mechanism of these reactions is discussed.

The isodityrosine natural products, such as vancomycin, K-13, OF-4949 and piperazinomycin, share a diaryl ether linkage, which preserves the biologically active stereochemistry of whole molecules. These ether moieties might be constructed by phenolic oxidations of the corresponding phenol precursors in the final step of their biosynthetic pathway. Our biomimetic syntheses of these natural products have been achieved by employing the oxidative coupling of o,o'-dihalogenated

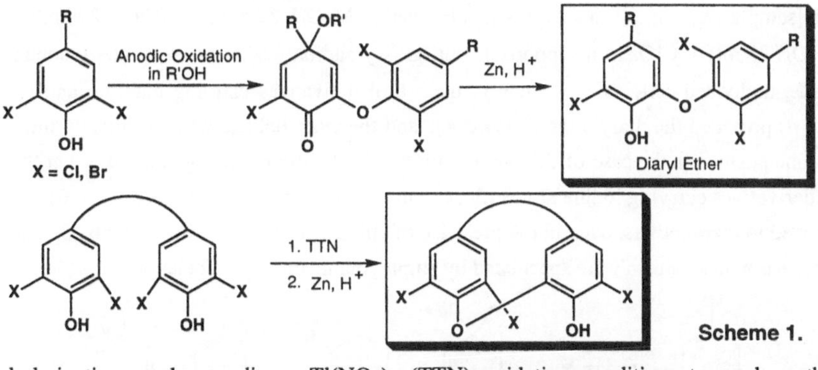

Scheme 1.

phenol derivatives under anodic or Tl(NO$_3$)$_3$ (TTN) oxidation conditions to produce the diaryl ethers,[1] whereas usual synthetic studies have utilized the classical Ullmann reaction. In particular, the anodic oxidation is effective for dimerization, and the TTN oxidation for intramolecular cyclization (Scheme 1). It appears that halogen substituents play a crucial role to determine the oxidation mode.

In the synthesis of isodityrosine itself,[2] anodic oxidation of the dibromo L-tyrosine derivative afforded diaryl ether **1**, contrary to the diiodo derivative which produced the diaryl **2** (Scheme 2). The oxidatin products **1** and **2** were successively submitted to catalytic hydrogenolysis, followed by acid hydrolysis to yield isodityrosine and dityrosine in good yields. Although a direct oxidation of L-tyrosine is known to afford the products in low yields (2 ~ 4%),[3] the o,o'-dihalogenated method could control the oxidation potentials to provide the desired coupling products (40 ~ 50%). This unprecedented chemoselectivity was also observed in chemically unstable phenylglycine derivatives.[2]

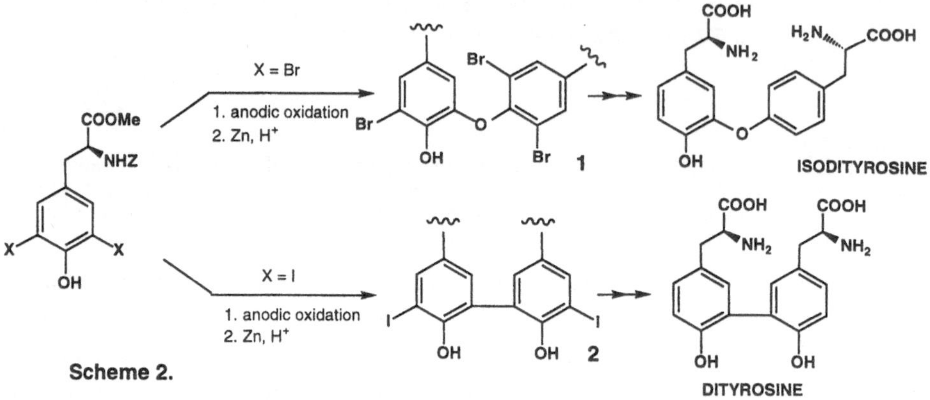

Scheme 2.

To pursue the effect of halogen substituents in the phenolic oxidation, the cresol derivatives carrying various pairs of halogen atoms were oxidized under the constant current electrolysis conditions [sample (2.5 mmol) in solvents (acidic media: MeOH 22.5 ml - HClO$_4$ 2.5 ml; neutral media: MeOH 25 ml), LiClO$_4$ as a supporting salt, glassy carbon beaker as an anode, platinum wire as a cathode, undivided cell]. Consequently, the cresol derivatives carrying chlorine and bromine (**3a, b,** and **d**) provided the diaryl ethers (**type 4**), and the ether linkage was specifically introduced at the bromine position in the case of **3d** possessing a pair of different halogen atoms. On the other hand, the derivatives carrying iodine atoms (**3c, e,** and **f**) provided the diaryls (**type 4 - 6**). Diaryls were preferred to diaryl ethers, even in the presence of chlorine or bromine substituents (**3e** and **3f**).

Their reaction mechanism were speculated by employing theoretical calculations.

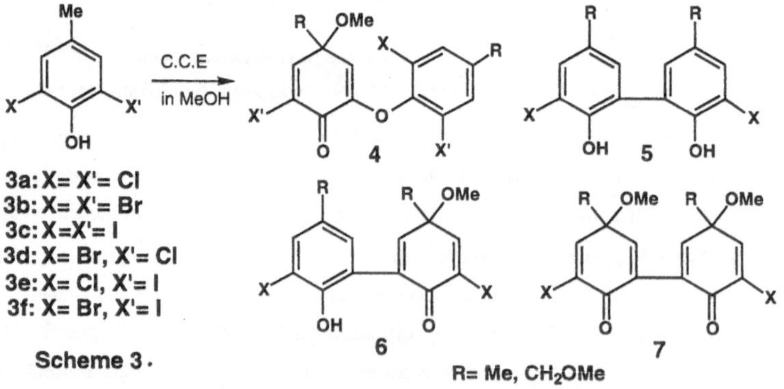

3a: X= X'= Cl
3b: X= X'= Br
3c: X=X'= I
3d: X= Br, X'= Cl
3e: X= Cl, X'= I
3f: X= Br, X'= I

Scheme 3.

R= Me, CH$_2$OMe

1) S. Yamamura and S. Nishiyama, "*Studies in Natural Products Chemistry*", (Atta-ur-Rahman Ed); Elsevier, **1992**, *10F*, 669.
2) S. Nishiyama, M. H. Kim, and S. Yamamura, *Tetrahedron Lett.*, **1994**, *35*, 8397.
3) S. C. Fry, *Meth. Enzymol.*, **1984**, *107*, 388.

ELECTROCHEMICAL SYNTHESIS OF EUGLOBAL SKELETONS VIA BIOMIMIC CYCLOADDITION

Kazuhiro CHIBA

Laboratory of Bio-organic Chemistry, Tokyo University of Agriculture and Technology, 3-5-8 Saiwai-cho, Fuchu, Tokyo 183, Japan

ABSTRACT

Euglobal skeletons were synthesized by electrochemical method and the first total synthesis of natural euglobals were also accomplished by biomimic cycloaddition reaction in electrolytic system.

INTRODUCTION

Euglobals were isolated from *Eucalyptus* spp. as inhibitors of the Epstein-Barr virus activation, or as antimalarial compounds.[1-3] Recently, the construction of these skeletons and derivatives has been widely investigated. These compounds have unique chroman or spirochroman skeletons. These skeletons are suggested to be biosynthesized by hetero Diels-Alder reaction of corresponding *o*-quinone methides with terpenes.

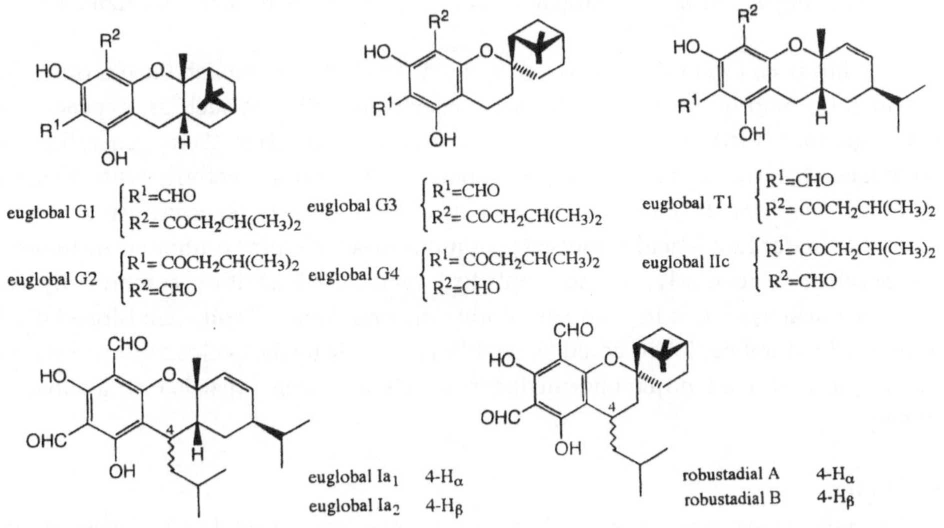

Fig.1 Chemical structures of natural euglobals and robustadials isolated from *Eucalyptus* spp.

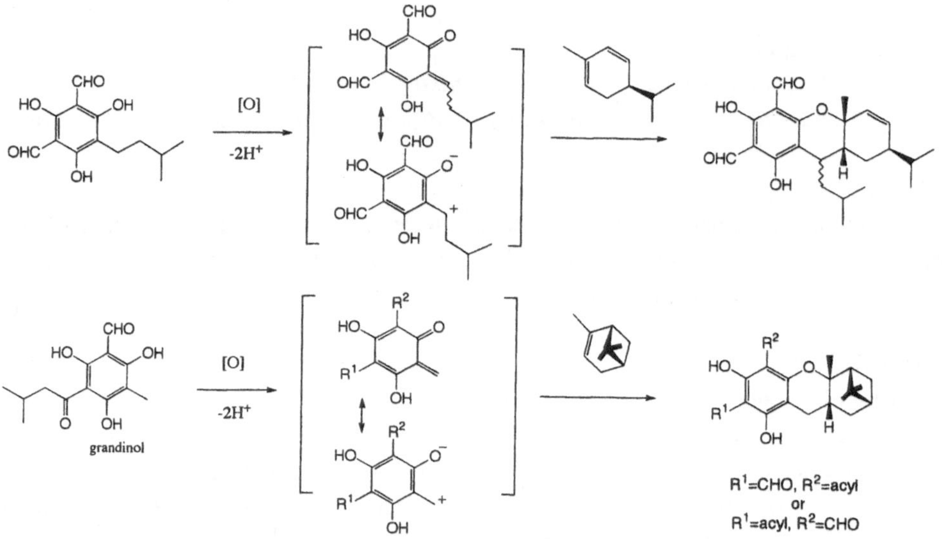

Fig.2   Proposed biogenetic pathway of euglobals and robustadials

It has been found that, however, to be difficult to accomplish the intermolecular cycloaddition of o-quinone methides with unactivated alkenes such as terpenes, since most o-quinone methides are liable to decompose just after their generation, and unactivated alkenes, including terpenes, generally show poor reactivity with the desired o-quinone methides in the usual reaction media. Moreover, terpenes are also apt to decompose under Lewis acid-catalysed conditions in which some o-quinone methides can be generated.   Previously, we accomplished varied cycloaddition reactions by using electrochemical system in lithium perchlorate/nitromethane,[4-8] and   envisioned that the synthesis of natural euglobals could be readily accessible by the cycloaddition of terpenes and *in situ* generated *o*-quinone methides which has been regarded as a biogenetic pathway.

METHOD

Presently, some electrochemical methods were introduced for the generation of quinone methides and their interemolecular cycloaddition with terpenes. First, o-quione methides were generated by the electrochemical oxidation of 2-[1-(phenylsulfanyl) alkyl]phenols to give corresponding cycloadducts with terpenes. In lithium perchlorate/nitromethane system, o-quinone methides were stabilized and promoted to give corresponding cycloadducts with unactivated alkenes via hetero-Diels-Alder reaction. By using these electrooxidation system, some euglobal skeletons were simply synthesized in moderate yields. In this electrooxidation system, however, coexisted terpenes were decomposed. On the other hand, it was found that catalytic amount of methylene blue played a role as a mediator of photo-electrochemical oxidation of sulfides. Important synthetic intermediates for *e.g.* euglobal Ia1 and Ia2 were synthesized to give *o*-quinone methide under lower potential condition.

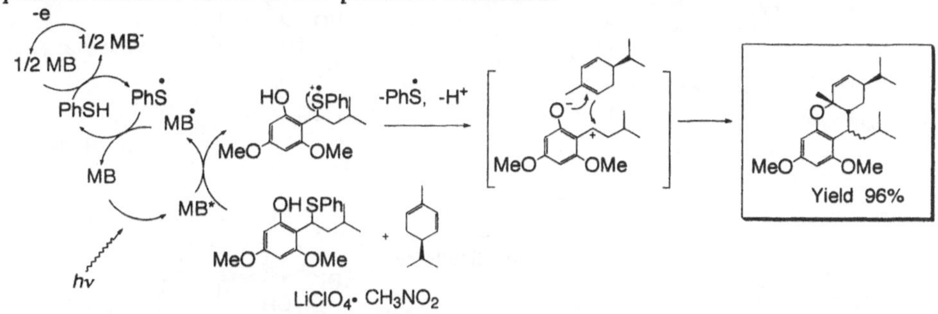

Fig.3   Photo-electrochemical synthesis of euglobal skeletons via hetero Diels-Alder reaction of *in situ* generated *o*-quinone methides

Secondly, euglobal G1, G2, G3, T1 and IIc were synthesized by the electrochemical oxidation of grandinol by using DDQ as a redox mediator on a PTFE (poly-tetrafluoroethylene) fiber coated Pt electrode. In this electrochemical system, hetero Diels-Alder reaction of quinone methides and terpenes was promoted, and DDQ was effectively regenerated. Spectral data of synthesized euglobals were completely identical with those of natural products, and their absolute configurations were also established.

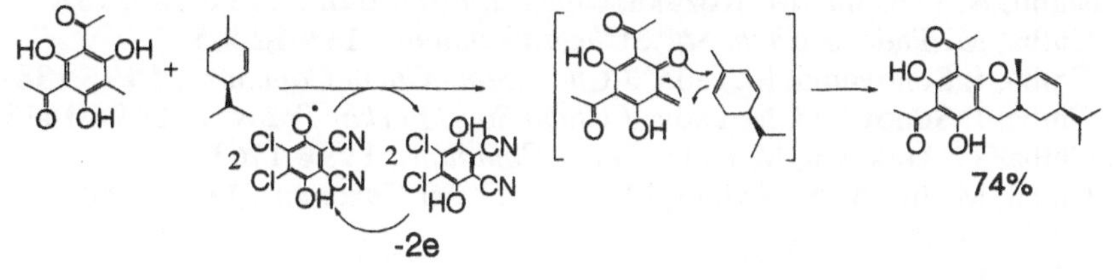

Fig. 4   Proposed reaction mechanism of the cycloaddition reaction of quinone methides and terpenes by electrochemical oxidation using DDQ as a redox mediator

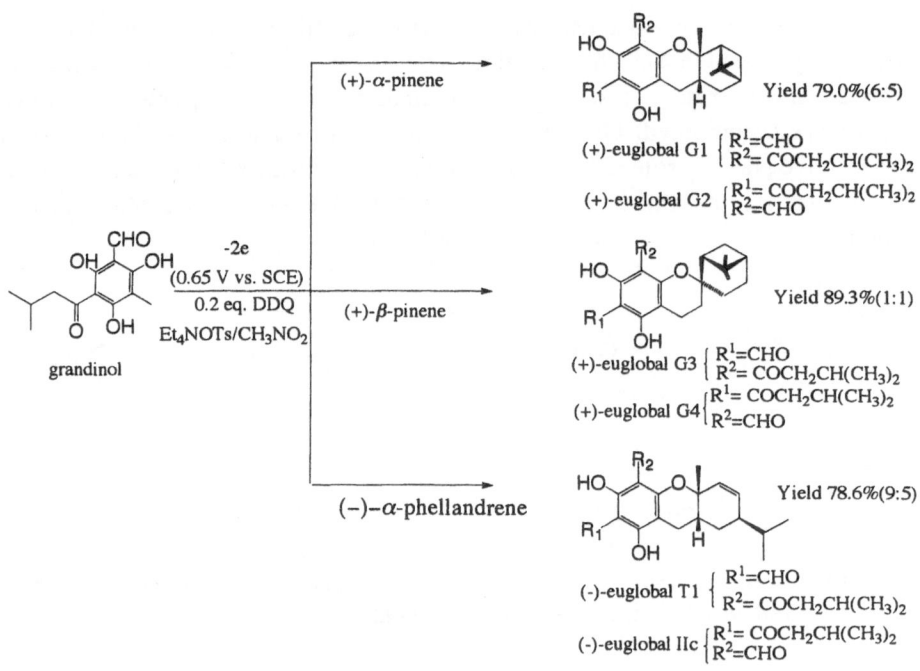

Fig. 5 Electrochemical synthesis of natural euglobals

References

1) M. Kozuka, T. Sawada, E. Mizuta, F. Kasahara, T. Amano, T. Komiyama and M. Goto, *Chem. Pharm. Bull.*, **1982**,30, 1964
2) R. Xu, J. K. Snyder, K. Nakanishi, *J. Am. Chem. Soc.*, **1984**,106, 734
3) M. Takahashi, T. Konoshima, K. Fujitani, S. Yoshida, H. Nishimura, H. Tokuda, H. Nishino, A. Iwashita, M. Kozuka, *Chem. Pharm. Bull.*, **1990**,38, 2737
4) K. Chiba, M. Tada, *J.Chem.Soc., Chem.Commun.*, **1994**,2485
5) K. Chiba, J. Sonoyama, M.Tada, *J.Chem.Soc., Chem.Commun.*, **1995**,1381
6) K. Chiba, J. Sonoyama, M.Tada, *J.Chem.Soc., Perkin Trans. 1*, **1996**,1435
7) K. Chiba, T. Arakawa, M.Tada, *Chem.Commun.*, **1996**,1763
8) K. Chiba, M. Jinno, A. Nozaki, M. Tada, *Chem.Commun.*, **1997**, 1403

ELECTROCHEMICAL CLEAVAGE OF BENZYLIC SINGLE AND DOUBLE CARBON-CARBON BONDS

Yury OGIBIN, Alexey ILOVAISKY, and Gennady NIKISHIN

N.D. Zelinsky Institute of Organic Chemistry, Russian Academy of Sciences, Moscow 117913, Russia

Electrochemical transformations of aromatic hydrocarbons, which are based on the cleavage of benzylic carbon-carbon π- and σ-bonds, were investigated. The study pursued the main goals: (1) to establish characteristics of the structural and electrochemical factors affecting the course of these type of reactions, (2) to estimate their scope and limitations; and (3) to carry out the methodological development of the reactions to make them a convenient and efficient tool of organic synthesis.

Aromatics with a benzylic $C=C$ bond, unsubstituted and alkoxy substituted arylalkanes, aryl- and arenocycloalkanes were the objects of the study. Electrolysis experiments were performed in undivided cells with a graphite or platinum anode at a constant current and 60°C using methanolic or ethanolic solutions of KF, Bu_4NBF_4, CF_3COONa or TsONa as a supporting electrolyte. Most of the tested compounds under these conditions undergo transformations induced by the electrochemical cleavage of benzylic π- and/or σ-bonds, dialkyl ethers of aromatic 1,2-diols or aromatic and aliphatic dialkyl acetals being formed as products.

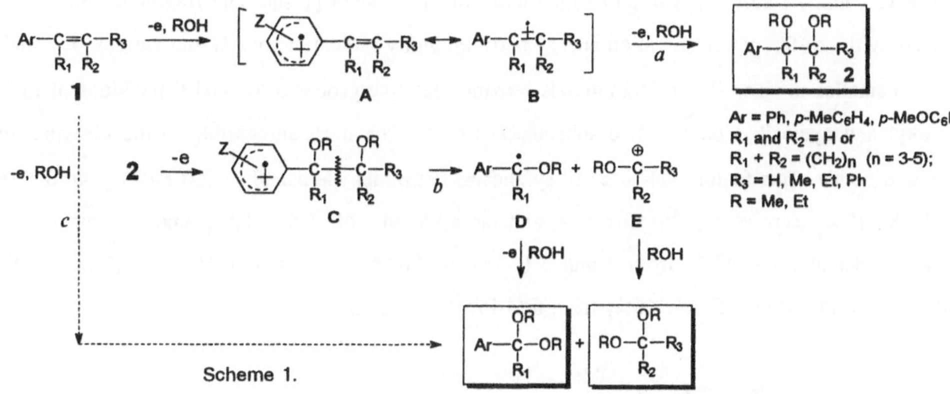

Scheme 1.

The mechanisms of their formation involve as key steps the electrogeneration of arenonium radical cations **A** and **C**, the isomerization of **A** into aliphatic radical cations **B** with the rupture of the benzylic π-bond, and the fragmentation of **C** with the scission of the σ-bond.

The electrochemical cleavage of the benzylic carbon-carbon π-bond. This reaction produces ethers of the type **2** (scheme 1, path *a*), the optimum values for the conversion of **1** (75-95%) and the yield of **2** (70-90%) being attained with passage of about 2 F/mol. On the basis of the reaction, a convenient and efficient one-step method was developed for the preparation of (1,2-dimethoxyalkyl)arenes **3** (yield 70-85%) [1], dimethyl ethers of 1-phenylcycloalkane-1,2-diols **4** (75-80%) [2], and of cycloalkane-1,2-diols fused with arene groups **5** and **6** (87-90%) [2,3].

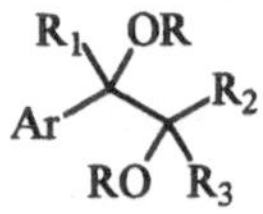

3

Ar = Ph, *p*-MeC_6H_4, *p*-$MeOC_6H_4$
R_1,R_2,R_3 = H, Me, Et, Ph

(n=1-3)
4

(n=1,2)
5

6

The electrochemical cleavage of the benzylic carbon-carbon σ-bond. The cleavage of this type results from the fragmentation of the radical cations **C** (scheme 1, path *b*). The driving force for this reaction is a considerable gain in thermodynamic energy. It is provided at the expense of forming stabilized benzylic radicals **D** and extremely stable carbocations **E**, the positive charge of which is efficiently stabilized with an oxygen atom of the α-MeO group. The fragmentation is also favoured by the difference in solvation energy between bulky radical cations **C** and cations **E** which are much smaller in size. Hence, compounds bearing methoxy groups at α- and β-positions of their side alkyl and cycloalkyl chains, like substances **3-6**, are the most susceptible to the cleavage of benzylic σ-bonds. This feature was used to synthesize benzaldehyde acetals **7** (60-80%) from **3** (2-6 F/mol, 80-95% conversion) [1], diacetals and acetal-ketals of 1,5-, 1,6-, and 1,7-dicarbonyl compounds **8** and **9** (60-85%) from **4** and **5** (4 F/mol, 80-95% conversion) [2], and 1,8-naphthalic dialdehyde acetals **10-11** (5.5:1, 46%) from **6** (4 F/mol, 92% conversion) [3].

7
Ar = Ph, *p*-MeC_6H_4,
p-$MeOC_6H_4$

8

9

10

11

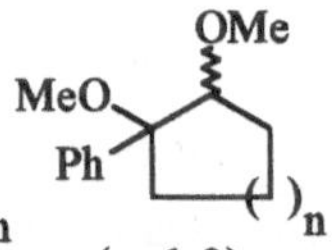

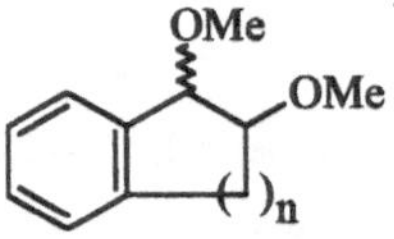

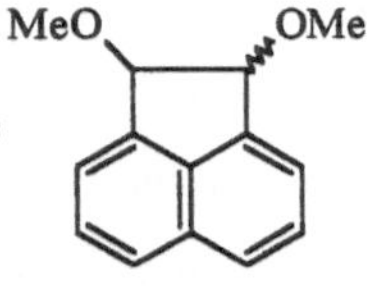

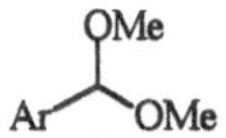

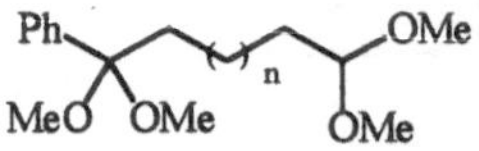

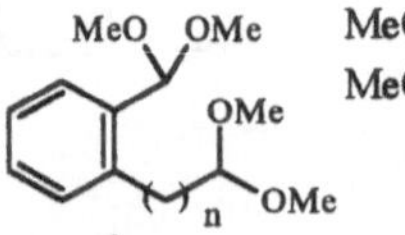

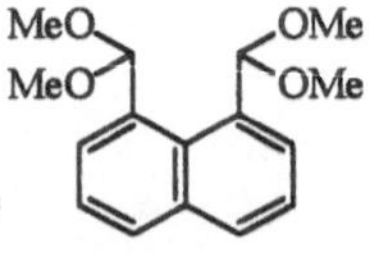

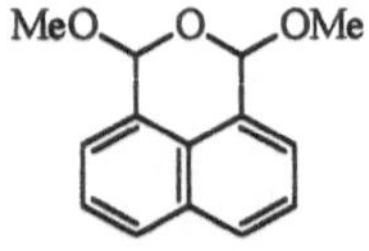

It is important to note that alternative chemical approaches to acetals **8-10** are unknown. These derivatives fail to synthesize by means of the acetalization reaction used usually for this purpose, because of the almost complete conversion of the appropriate dicarbonyl compounds into cyclic acetals of the type **11**, as it takes place in the case of *o*-phthalic, homophthalic, and 1,8-naphthalic dialdehydes.

α- And β-monomethoxyalkylbenzenes and 1,2-diphenylethane demonstrate moderate susceptibility to this cleavage, while α,β-dimethoxyalkyl substituted 3,4-dimethoxy- and 3,4-(methylenedioxy)benzenes are cleaved to a much lesser extent [4].

The electrochemical cleavage of the benzylic C=C bond. The cleavage of this type results in the subsequent scission of benzylic π- and σ-bonds in one process as an one-pot reaction (scheme 1, path c). The data of the prior study of these reactions have enabled a methodology to be worked out for the efficient direct transformation of the aromatics with a benzylic C=C bond into acetal derivatives of aromatic mono- and dicarbonyl compounds. Using this methodology, the electrosynthesis of benzaldehyde acetals [1-3] was performed and original approaches to aromatic 1,5-, 1,6-, and 1,7-dicarbonyl compounds[5] and to their diacetal, acetal-ketal derivatives [2,3] were elaborated.

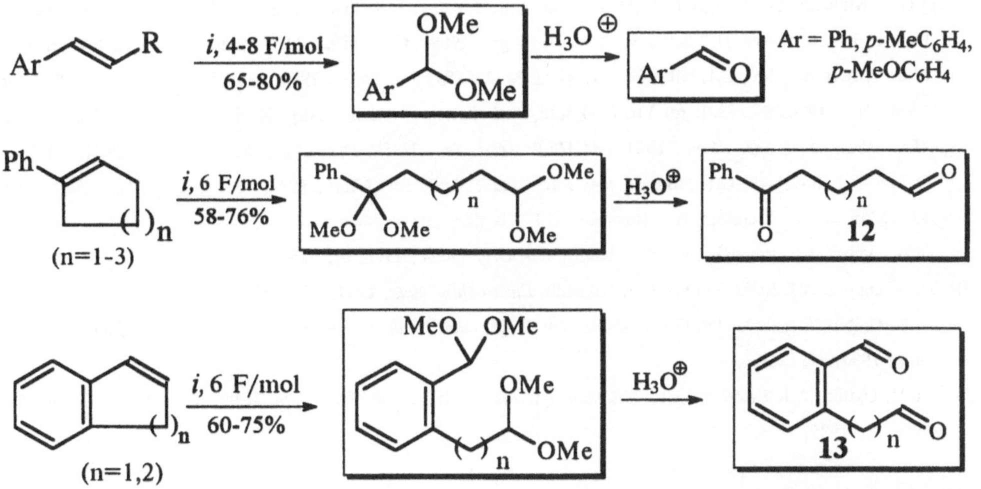

Conclusions. This presentation summarizes our study of the electrochemical cleavage of benzylic single and double carbon-carbon bonds and of the application of these reactions in organic synthesis [1-5]. Its results allow the following conclusions:

(1) Benzylic carbon-carbon bonds in aromatics with a benzylic C=C bond, arylalkanes, aryl- and arenocycloalkanes bearing methoxy groups at α- and β-positions of their side alkyl and cycloalkyl fragments may be simply and efficiently cleaved by direct anodic oxidation of these compounds in lower alcohols using undivided cells and KF or Bu_4NBF_4 as a supporting electrolyte.

(2) Products of the cleavage are dialkyl ethers of aromatic 1,2-diols or aromatic acetals, diacetals and acetal-ketals. The ethers are formed by the scission of benzylic carbon-carbon π-bond and the acetal derivatives result from the cleavage of benzylic σ-bond and from the subsequent scission of these bonds in one process.

(3) The methodology developed for the electrochemical cleavage of benzylic C=C bond has made the reaction a convenient and versatile tool for the prepration of oxygen functional aromatics, such as aromatic mono- and dicarbonyl compounds and their acetal derivatives. These syntheses essentially expand the range of known approaches to the compounds, and are, as a rule, more convenient and efficient than alternative chemical routes, and in a number of cases are the only methods for their synthesis.

References

1) (a) G. I. Nikishin, Yu.N. Ogibin, M.N. Elinson, A.B. Sokolov, T.B. Pribytkova, I.V. Makhova, L.F.Belova, A.I. Platova, L.A. Kheifets. U.S.S.R., SU 1692978 (Appl. 1989); C.A. **1992**, *117*, 26084; (b) Yu.N. Ogibin, M.N. Elinson, A.B. Sokolov, G.I. Nikishin. *Izv. Akad. Nauk SSSR, Ser. Khim.*, **1990**, 494 [*Bull. Acad. Sci., USSR, Div. Chem. Sci.*, **1990**, *39*, 432]; (c) Yu.N. Ogibin, A.B. Sokolov, A.I.Ilovaisky, M.N. Elinson, G.I. Nikishin. *Izv. Akad. Nauk SSSR, Ser. Khim.*, **1991**, 644 [*Bull. Acad. Sci., USSR, Div. Chem. Sci.* **1991**, *40*, 561]; (d) Yu.N. Ogibin, A.I. Ilovaisky, G.I. Nikishin. *Izv. Russ. Akad. Nauk, Ser. Khim.*, **1994**, 1624 [*Russ. Chem. Bull.*, **1994**, *43*, 1536]; (e) Yu.N. Ogibin, A.I. Ilovaisky, G.I. Nikishin. Unpublished data.

2) Yu.N. Ogibin, A.I. Ilovaisky, and G.I. Nikishin. *J. Org. Chem.*, **1996**, *61*, 3256.

3) Yu.N. Ogibin, A.I. Ilovaisky, and G.I. Nikishin. *Electrochim. Acta*, **1997**, *42*, 1933.

4) Yu.N. Ogibin, A.I. Ilovaisky, G.I. Nikishin. *Izv. Russ. Akad. Nauk, Ser. Khim.*, **1993**, 140 [*Russ. Chem. Bull.*, **1993**, *42*, 126].

5) Yu.N. Ogibin, A.I. Ilovaisky, G.I. Nikishin. *Izv. Russ. Akad. Nauk, Ser. Khim.*, **1996**, 2044 [*Russ. Chem. Bull.*, **1996**, *45*, 1939].

SOME OBSERVATIONS ON INTERMEDIATES OF c-HEXENE BIFUNCTIONALIZATION

Mihály NOVÁK, Csaba VISY

Institute of Physical Chemistry, József Attila University
6722 Szeged, P.O.Box 105, Hungary

ABSTRACT

Bifunctionalisation leading to oxygenated chloro-derivatives of c-hexene was studied in nitromethane. It was observed that surface conditions and solvation strongly affect the product distribution of the reaction.

INTRODUCTION

Much of the electroorganic work is carried out in so-called "non-aqueous" media, without proper indication of actual water content and the possible effects of its changing. Furthermore, it is well known that, the results obtained in various laboratories spread considerably for apparently no reason. It is believed, that in many cases the variance might be caused by trace contaminations, primarily by different water content of the anolytes. As an example, the single step transformation of olefins to bifunctional derivatives is presented. The anodic transformation of c-hexene is a long standing object of previous and recent studies [1].

In order to see the effect of water more closely and to understand the role of intermediates in formation of various products, a detailed study of anodic bifunctionalization of c-hexene was carried out in nitromethane (NM) solution.

EXPERIMENTAL

The experiments [2] were carried out in a divided electrochemical cell under controlled potential mode at 23 °C. Pt sheet was the working electrode. As reference Ag/AgClO$_4$ electrode was used.

Nitromethane was distilled and stored over 4A molecular sieve in dry N$_2$ atmosphere. Tetraalkylammonium salts (N(Et)$_4$Cl, N(Et)$_4$ClO$_4$) recrystallized from NM were the sources of Cl$^-$ ions and the conducting salts. Cyclohexene (CH) was freshly distilled before use. The solutions in the cell were deoxygenated with nitrogen. Carl-Fisher (CF) titration was used to determine the water content. The water content of purified NM was 40 ppm and it was increased to 150 ppm as the initial value of the electrolyte solution.

The starting concentration ratio of reactants was [Cl$^-$]/[CH] = 2/1 to ensure the Cl$^-$ excess.

The product analysis was carried out by direct injection of anolyte samples into GC. The electrolysis was carried out only till about 15% of starting CH was converted.

DISCUSSION

Product distribution. Analysis of reaction products under various conditions indicated that substitution reaction did not occur, only dichloro-c-hexene (Cl$_2$CH), and two oxygenated derivatives,

2-chloro-c-hexanol (ClCHOH), and 2-chloro-c-hexanon (ClCHO) were formed. No other derivatives were observed in the electrolyses carried out in the 200-1000 mV range. The current efficiency was around 100% at the beginning and it decreased to 80% with the increasing time of electrolysis.

Effect of potential. As it can be seen in Fig.1. at the start of electrolysis the potential influences the product distribution. Later with the increase of charge passed the spread of distribution decreases as it can be seen in Fig.2. in the case of ClCHOH.

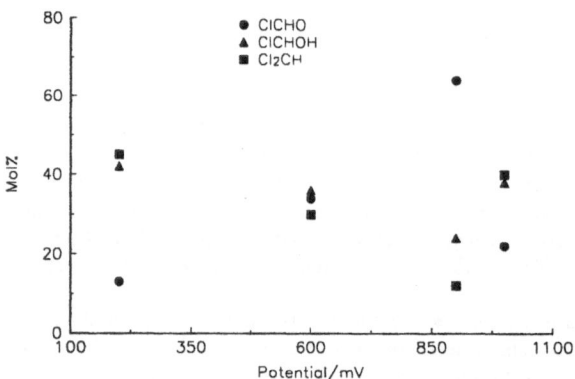

Fig.1.Relative amount of ClCHO, ClCHOH and Cl$_2$CH formed at: various potentials on passing 1 As charge

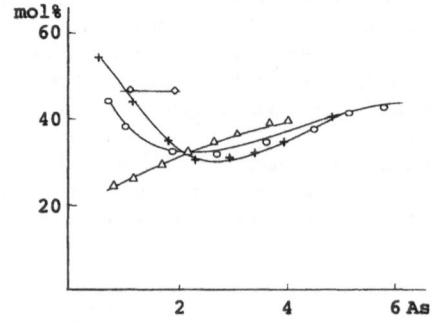

Fig.2. Relative amount of ClCHOH at +: 200 mV, ◊: 600 mV, △: 900 mV, ○: 1000 mV.

The concentration of Cl_2CH linearly increases with the charge but formation of other two products occures as it is given in Figs. 3.-4.

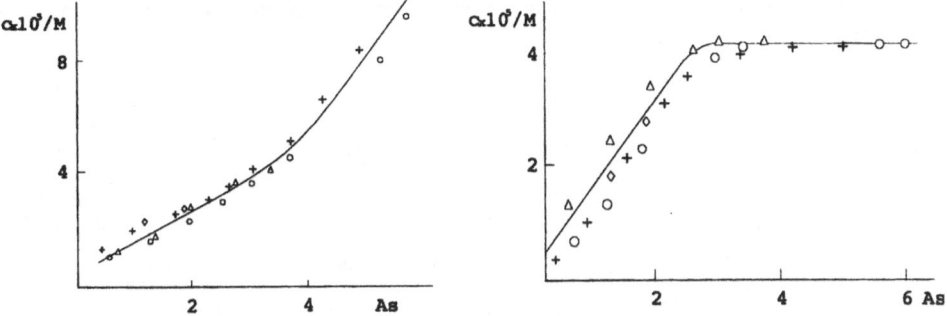

Fig.3. ClCHOH formation with charge passed,
+: 200 mV, ◊: 600 mV, ∆: 900 mV,
○: 1000 mV

Fig.4. ClCHO formation with charge passed
+: 200 mV, ◊: 600 mV, ∆: 900 mV,
○: 1000 mV

As the potential increases from 200 mV to 900 mV the concentration of Cl_2CH slightly decreases and at 1000 mV it increases again. ClCHOH formation is not sensitive to the potential, and its amount increases with the charge, Fig.3., but after a certain charge transferred the increase in concentration becomes faster. On the contrary, ClCHO formation takes place only till a certain value, then it stops, although the potential does not seem to affect the rate of formation.

Effect of water. The water content of the solution did not influence the Cl_2CH production, but formation of oxygenated species changed strongly. Thus, smaller amount of ClCHO formed in the presence of more water but the increase in sum of oxygenated species remained the same with the charge passed.

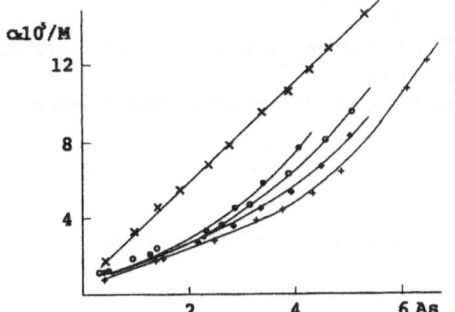

Fig. 5. Formation of ClCHOH at
+: 150 ppm, ◆: 300 ppm,
○: 1000 ppm, •: 3000 ppm,
×: 3000 ppm

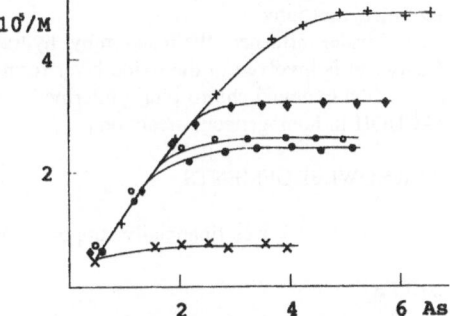

Fig..6. Formation of ClCHO at various
water content, marks as in Fig.5.

Surface coulometry reveiled that ceise in ClCHO formation is a consequence of oxide formation, inhibiting the oxidation of intermedite on free metall surface.

The voltammetric curves indicate that on addition of CH, the intermediates oxidize at nearly the same potential, (Fig.7), while an increase in water concentration influences the potential of the oxidation (Fig.8).

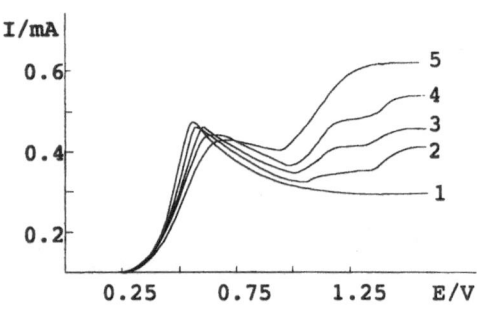

Fig.7. Change in voltammetric curves on increase in CH conc. ([Cl⁻] = 10^{-3}M): 1: 0.0 M, 2: 3x10^{-4}, 4: 9x10^{-4}, 3: 6x10^{-4}, 4: 9x10^{-4}, 5: 1.5x10^{-3} M.

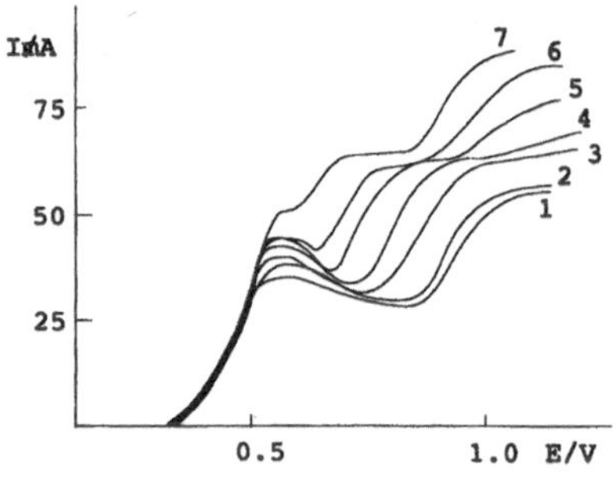

Fig.8. Effect of water, (1mM Cl⁻, 1.5 mM CH); 1: 80, 2: 300, 3: 800, 4: 1500, 5: 3000, 6: 6000, 7: 8000 ppm.

Considering the effect of potential and of water on product distribution and the voltammetric properties [3] of the reacting system it was concluded that
ClCHO formation occurs only on oxide free electrode surface,
ClCHOH and Cl$_2$CH form in homogeneous reactions,
ClCHOH is not a precursor of ClCHO formation, but both species have a common intermediate source,
water influences the reaction by hydrating the Cl⁻ ions and the chloro-olefin intermediates, but free water is involved in the oxide layer formation on the anode,
the hydrated chloro-olefin intermediates are converted to ClCHO by charge transfer and to ClCHOH in homogeneous reaction.

ACKNOWLEDGEMENTS

This work was financially supported by OTKA T021268 and F.K.F.P. 0733 Grants.

REFERENCES

1) K.-C. Möller and H. J. Schäfer, Electrochim. Acta, **1997**, *42*, 1971
2) M. Novák and Cs. Visy, Electrochim. Acta, **1983**, *28*, 507
3) Cs. Visy and M. Novák, J. Electroanal. Chem., **1990**, *296*, 571

FORMATION AND REACTIVITY OF RADICAL CATIONS OF CYCLOPROPANE DERIVATIVES

Kazuhiko MIZUNO, Shogo NISHIOKA, Hajime MAEDA, and Nobuyuki ICHINOSE

Department of Applied Chemistry, College of Engineering,
Osaka Prefecture University, Sakai, Osaka 599, Japan

ABSTRACT

Anodic oxidation of 1,2-bis(4-methoxyphenyl)cyclopropane (**1a**) and 1-{2',2'-(4-methoxyphenyl)-vinylidene}-2,3-dimethylcyclopropane (**2**) using Pt electrode in the presence of $Mg(ClO_4)_2$ in acetonitrile generate the radical cation species, respectivity, which colored purple. The former caused the *cis-trans* isomerization, but the latter did not. The anodic oxygenation of **1a** gave 4-methoxybenzaldehyde (**4a**) and 4-methoxyacetophenone (**5a**).

INTRODUCTION

Recently, much attention has been focused on the reactivities of the radical cations of small ring compounds, which are generated by one-electron oxidation using the electrochemical, photochemical, and other chemical methods. However, the electrochemical anodic oxidation often affords two-electron oxidation products. For example, Shono et al. previously reported the anodic oxidation of phenylcyclopropane in methanol to give 1,3-dimethoxy-1-phenylpropane via two electron-oxidation.[1] We now report the reactivities of the radical cation species generated by the electrochemical and one-electron oxidation of 1,2-diarylcyclopropanes and (2',2'-diarylvinylidene)cyclopropanes.

Generation of the Radical Cations of 1,2-Diarylcyclopropanes and (2',2'-Diarylvinylidene)-cyclopropanes: Anodic oxidation of *trans*-1,2-bis(4-methoxyphenyl)cyclopropane (t-**1a**) using Pt electrode in acetonitrile containing $Mg(ClO_4)_2$ under argon atmosphere immediately caused the coloration of red purple. The absorption maximum of this species at 561 nm was assigned to the radical cation of t-**1a**. Similar coloration was observed by the oxidation of t-**1a** using 2,3-dichloro-5,6-dicyanobenzoquinone (DDQ) or $Cu(BF_4)_2$. These results support the generation of the radical cation of t-**1a**. Similar anodic oxidation of c-**1a** occurred to generate the same radical cation species as that of

t-**1a-d**   c-**1a-d**   c-**2**   t-**2**

a ; Ar = 4-$MeOC_6H_4$, **b** ; Ar = 4-MeC_6H_4, **c** ; Ar = 4-ClC_6H_4, **d** ; Ar = C_6H_5

t-**1a**. However, 1,2-bis(4-methylphenyl)-, 1,2-bis(4-chlorophenyl)-, and 1,2-diphenylcyclopropanes (c-**1b-d** and t-**1b-d**) did not show any coloration by the anodic or chemical oxidations. Interestingly, the coloration of the radical cation species generated by the photoinduced electron transfer from **1a** to the excited singlet state of 9,10-dicyanoanthracene (DCA) was not observed at all.[2-4]

Anodic oxidation of *cis*- and *trans*-1-{2',2'-(4-methoxyphenyl)vinylidene}-2,3-dimethylcyclopropanes c-**2** and t-**2** also showed the coloration of purple, whose absorption maximum was at 511 nm. These species were also assigned to the radical cations of c-**2** and t-**2**.

Cis-Trans Isomerization of 1,2-Diarylcyclopropanes and (2',2'-Diarylvinylidene)cyclopropanes via Anodic Oxidation: *Cis-trans* isomerization efficiently occurred by the anodic oxidation of c-**1a** and t-**1a** using Pt electrode in acetonitrile containing $Mg(ClO_4)_2$ under argon atmosphere. The current efficiency was 1.5 and the chain process was involved. The *cis-trans* ratio was 5 : 95 from both isomers. Prolonged oxidation afforded polymeric materials. Similar anodic oxidation of **1b-d** also caused the *cis-trans* isomerization. However, the rates were slow (current efficiency = 0.1-0.2) and the starting cyclopropanes were rapidly consumed to give polymeric materials on the anodic Pt electrode. The anodic oxidation of c-**2** and t-**2** did not cause any isomerization, although the DCA-sensitized photoisomerization of c-**2** and t-**2** in acetonitrile occurred to give a 1 : 3-photostationary mixture.

Oxygenation of 1,2-Diarylcyclopropanes via Anodic Oxidation: Anodic oxidation of t-**1a** in the presence of molecular dioxygen in acetonitrile afforded 4-methoxybenzaldehyde (**4a**) and 4-methoxyacetophenone (**5a**) as major products, accompanied by a small amount of *trans*-3,5-bis-(4-methoxyphenyl)-1,2-dioxolane (t-**3a**). On the other hand, DCA-sensitized photooxygenation of t-**1a** afforded c-**3a** and t-**3a** in nearly quantitative yield.[3,4] Prolonged irradiation gave **4a** and **5a**. From these results, **4a** and **5a** were the secondary products generated by the decomposition of c-**3a** and/or t-**3a**.

References
1) T. Shono and Y. Matsumura, *J. Org. Chem.*, **35**, 4157 (1970).
2) K. Mizuno, N. Ichinose, and Y. Otsuji, *Chem. Lett.*, **1985**, 455.
3) K. Mizuno, N. Ichinose, and Y. Otsuji, *J. Org. Chem.*, **57**, 1855 (1992).
4) K. Mizuno, N. Kamiyama, N. Ichinose, and Y. Otsuji, *Tetrahedron*, **41**, 2207 (1985).

EPOXIDATION OF 1,4-NAPHTHOQUINONES BY ELECTROCHEMICALLY GENERATED HYPOHALOGEN ACIDS

Nobuhiro TAKANO and Manabu OGATA
Department of Applied Chemistry, Muroran Institute of Technology, Muroran 050, Japan

ABSTRACT

Epoxidation using electrochemically generated hypohalogen acids in organic solvent-water-halide salt systems was investigated. In the CH_3CN-H_2O-NaI system, the epoxidation of 1,4-naphthoquinones provided the corresponding epoxides in good yields.

INTRODUCTION

Recently, the epoxidation of cyclic unsaturated ketones such as naphthoquinones and flavones with chemical oxidants has been described[1]. On the other hand, the synthesis of organic compounds using electrochemically generated species is a current topic in synthetic organic chemistry[2]. However, there have been no reports on the electrochemical epoxidation of quinones. In this paper we report the epoxidation of 1,4-naphthoquinones **1a, 1b** using electrochemically generated hypohalogen acids HOX in organic solvent-water-halide salt systems.

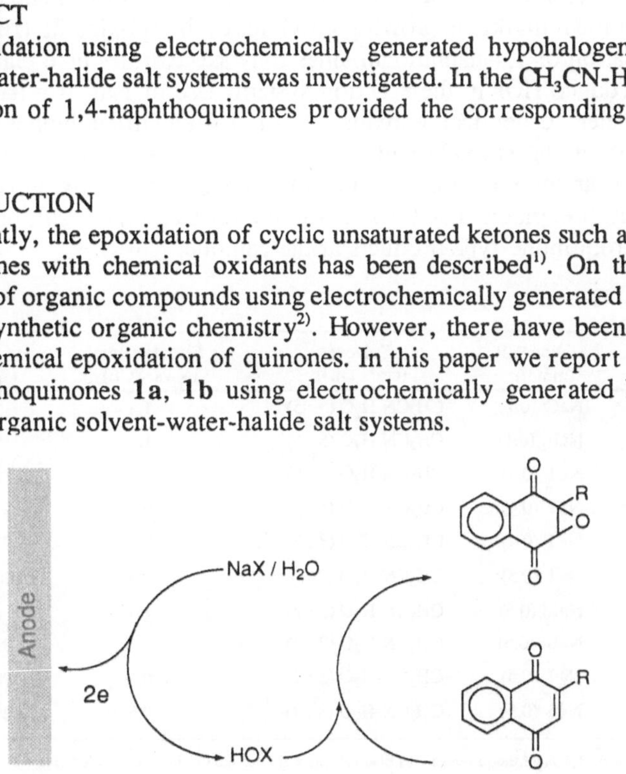

RESULTS AND DISCUSSION

When the oxidation potentials of the halide salts in an CH_3CN-H_2O solution were measured, the anodic current based on two-electron oxidation of halide ion X^- to halogen X_2 was observed. The halide ions were oxidized at ca. 1.2 V (Cl^-), 0.7 V (Br^-), and 0.3 V (I^-). In the presence of water the halogen X_2 formed produces hypohalogen acid HOX immediately.

We attempted the epoxidation of 1,4-naphthoquinones using electrochemical technique. Most of the controlled-potential electrolyses of 0.1 mmol 1,4-naphthoquinones in organic solvent-water containing halide salts were carried out in a divided cell to avoid the transfer of the substrate and product to a cathode. The results are summarized in Table 1. Using NaBr or NaI as a halide salt, the epoxides of 1,4-naphthoquinone **2a** and **2b** were obtained in good yields (Entries 2, 3, 8, and 9). However, using NaCl the electrolysis gave appreciably low yield and the substrates were recovered (Entries 1 and 7). Increase of the concentration of NaI and the content of water resulted in the remarkable growing on yields of the epoxide **2a** (Entries 4, 5, and 6). The epoxidation of 1,4-naphthoquinones may successfully take place when I_2 exists enough to produce HOI in the reaction system. Use of NaI as a halide salt is desirable for the electrochemical epoxidation of 1,4-naphthoquinones from the standpoint of yields of the epoxide and an oxidation potential of iodide ion. The electrolysis of **1b** in an undivided cell reduced the yield of the epoxide remarkably (Entry 10). This result explains that 1,4-naphthoquinones and the resulting epoxides are reduced to hydroquinone-type products at the cathode.

Table 1.

Entry	Compd	Halide Salt (mmol)	Solvent ($cm^3 : cm^3$)	Elec. Potential (V vs. SCE)	Yield of 2[a] (%)
1	1a	NaCl (0.1)	CH_3CN-H_2O (5 : 5)	1.5	2
2	1a	NaBr (0.1)	CH_3CN-H_2O (5 : 5)	1.0	15
3	1a	Na I (0.1)	CH_3CN-H_2O (5 : 5)	0.5	38
4	1a	Na I (0.4)	CH_3CN-H_2O (5 : 5)	0.5	91
5	1a	Na I (0.4)	CH_3CN-H_2O (3 : 7)	0.5	97
6	1a	Na I (0.5)	CH_3CN-H_2O (3: 7)	0.5	100
7	1b	NaCl (0.5)	CH_3CN-H_2O (3 : 7)	1.5	6
8	1b	NaBr (0.5)	CH_3CN-H_2O (3 : 7)	1.0	89
9	1b	Na I (0.5)	CH_3CN-H_2O (3 : 7)	0.5	98
10	1b	Na I (0.5)	CH_3CN-H_2O (3 : 7)	0.5	29[b]

a) Yield was based on 1,4-Naphthoquinones.   b) Electrolyses were carried out in an undivided cell.

References
1) (a) S. Colonna, A. Manfredi, R. Annunziata, and N. Gaggero, *J. Org. Chem.*, **55**, 5862 (1990); (b) J. A. Donnelly and D. E. Maloney, *Tetrahedron*, **35**, 2875 (1979); (c) W. Adam, D. Golsch and L. Hadjiarapoglou, *J. Org. Chem.*, **56**, 7292 (1991).
2) J. Simonet, "Electrogenerated Reagents" and M. M. Baizer, "Electrogenerated Bases and Acids" in "Organic Electrochemistry", H. Lund and M. M. Baizer, ed by Marcel Dekker, Inc., New York (1991), p1217, p1265.

EXPERIMENTAL AND THEORETICAL STUDIES ON PREPARATION OF HYDROQUINONE FROM BENZENE IN A PAIRED PACKED-BED ELECTRODE REACTOR

Xinsheng Zhang, Ping Ding, Wei-Kang Yuan

UNILAB Research Centre of Chemical Reaction Engineering, East China University of Science and Technology, Shanghai, 200237, China

ABSTRACT

A generalised reaction-diffusion model for describing preparation hydroquinone from benzene in a paired packed-bed electrode reactor is developed. Finite element method is applied for solving the model equations. Experiments were carried out in a rectangular electrode reactor packed lead shots as cathode and porous lead plats as anode. Optimal current efficiency of benzene oxidation and benzoquinone reduction (62.5 % and 95.9 %) was obtained at operating conditions: electrolytes rate, 0.19 m/s; total current, 15 A; packed-bed thickness of anode, 10 mm; cathode , 30 mm, and benzene and benzoquinone concentrations, 24 % and 4 % respectively.

INTRODUCTION

There exists a lot of papers relating to experiments of paired electrolysis of benzene to hydroquinone[1]. However, in contrary to a large number of papers on packed-bed electrode reactor modelling, reports dealing with a model of a paired packed-bed electrode reactor are extremely scarce. Oloman et al [2] and Scott[3] have set up a model to describe the process of anodic oxidation of benzene. The present paper aims at exploring the preparation of hydroquinone from benzene in a paired packed-bed electrode reactor from both computer simulation and experimental validation.

EXPERIMENTAL

Experiments were carried out in a flow-by paired packed-bed electrode reactor made of polypropylene. Each chamber (cathode and anode) has a size of 0.01-0.03m × 0.45m × 0.05m (thickness × length × width), with a ion exchange membrane in between. The cathodic electrode consisted of lead shots with 3-5 mm in diameter and anode was packed with porous lead alloy plates (Ag, 1 %).

Electrolytes were circulated by pumps. Benzene, dispersed in aqueous sulphuric acid, was passed through the anode compartment to produce benzoquinone was Anolysis contained benzoquinone was passed through the cathode chamber where hydroquinone was obtained. Sulphuric acid concentration of electrolyte was 10 % and reaction temperature 35°C. Concentrations of desired products benzoquinone (BQ) and hydroquinone (HQ) were determined with spectrophotometer. Optimal current efficiency of benzene oxidation and benzoquinone reduction (62.5 % and 95.9 %) was obtained at following operating conditions, electrolytes rate, 0.19 m/s; total current, 15 A; packed-bed thickness of anode, 10 mm; cathode , 30 mm, and benzene and benzoquinone concentrations, 24 % and 4 % respectively.

MATHEMATICS MODELLING

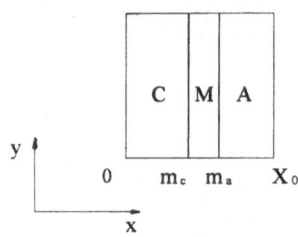

The sketch of the paired packed-bed electrode reactor under consideration is shown in Figure 1. In simplified anodic oxidation mechanism, benzene (B) first transfers into aqueous sulphuric acid phase and then diffuses to anodic surface where it is oxidised to benzoquinone (i_{1a}). Most of benzoquinone diffuses into aqueous phase from anodic surface and then into benzene phase, whereas a minority of benzoquinone is oxidised into maleic acid (M), (i_{2a}), which diffuses into aqueous phase. Maleic acid oxidisation to carbon dioxide is neglected. The side reaction is oxygen evolution (i_{3a}). In simplified cathodic reduction mechanism, benzoquinone transfers from benzene phase to aqueous phase and then diffuses to cathodic surface on which it is reduced to hydroquinone (i_{1c}) which diffuses into the aqueous phase. The side reaction is hydrogen evolution (i_{2c}). Benzoquinone concentration in the aqueous phase is in equilibrium with its concentration in the benzene phase.

Figure 1. The sketch of the paired packed-bed electrode reactor

Based on the above consideration, the following two-dimensional model can be applied for describing the process.

Anodic oxidation,

$$\frac{\partial^2 \phi}{\partial X^2} + \left(\frac{X_0}{Y_0}\right)^2 \frac{\partial^2 \phi}{\partial Y^2} = -\frac{F}{RT} X_0^2 \frac{a_a}{\gamma_a} i_{Ta} \tag{1}$$

$$\frac{\partial^2 C_{Qa}}{\partial X^2} - \frac{u_a X_0}{D_Q} \frac{X_0}{Y_0} \frac{\partial C_{Qa}}{\partial Y} = -\frac{X_0^2 K_{Qa} a_a (1-\varepsilon_{da})}{(1-\varepsilon_{da}+m\varepsilon_{da})D_Q}(C_{Qsa} - C_{Qa}) \tag{2}$$

$$\frac{\partial^2 C_M}{\partial X^2} - \frac{u_a X_0}{D_M} \frac{X_0}{Y_0} \frac{\partial C_M}{\partial Y} = -\frac{X_0^2}{D_M} k_{2a} a_a C_{Qsa} \tag{3}$$

$$C_{Qsa} = \frac{K_{Qs} C_{Qa} + K_{Bs}(1-C_{Bs})}{K_{Qs} + k_{2a}} \tag{4}$$

$$C_{Bs} = \frac{-K_{Bs} + \sqrt{K_{Bs}^2 + 4k_{1a}K_{Bs}C_{B0}}}{2k_{1a}C_{B0}} \tag{5}$$

$$I_{ja} = \int_0^{Y_0} \int_{ma}^{X_0} a_a Z_0 i_{ja} dx dy \qquad j=1,2,3 \tag{6}$$

Total current: $I_{Ta} = \sum_{j=1}^{3} I_{ja}$ \hfill (7)

Current efficiency: $E_a = \dfrac{I_{1a} - I_{2a}}{I_{Ta}}$ \hfill (8)

Benzoquione selectivity: $S_a = \dfrac{I_{1a} - I_{2a}}{I_{1a}}$ \hfill (9)

Cathodic reduction,

$$\frac{\partial^2 \phi}{\partial X^2} + \left(\frac{X_0}{Y_0}\right)^2 \frac{\partial^2 \phi}{\partial Y^2} = \frac{F}{RT} X_0^2 \frac{a_c}{\gamma_c} i_{Tc} \tag{10}$$

$$\frac{\partial^2 C_{Qc}}{\partial X^2} - \frac{u_c X_0}{D_Q} \frac{X_0}{Y_0} \frac{\partial C_{Qc}}{\partial Y} = \frac{K_{Qc} a_c (1 - \varepsilon_{dc})}{(1 - \varepsilon_{dc} + m \varepsilon_{dc}) D_Q} \left(C_{Qc} - C_{Qsc}\right) \tag{11}$$

$$\frac{\partial^2 C_{HQ}}{\partial X^2} - \frac{u_c X_0}{D_{HQ}} \frac{X_0}{Y_0} \frac{\partial C_{HQ}}{\partial Y} = -K_{Qc} a_c \left(C_{Qc} - C_{Qsc}\right) \frac{X_0^2}{D_{HQ}} \tag{12}$$

$$C_{Qsc} = \frac{K_{Qc}}{K_{Qc} - k_{1c}} C_{Qc} \tag{13}$$

$$I_{jc} = \int_0^{Y_0} \int_0^{m_c} a_c Z_0 i_{jc} dx dy \qquad j = 1,2 \tag{14}$$

Total current: $I_{Tc} = \sum_{j=1}^{2} I_{jc}$ \hfill (15)

Current efficiency: $E_c = \dfrac{I_{1c}}{I_{Tc}}$ \hfill (16)

Boundary conditions,

$$\left(\frac{\partial \phi}{\partial X}\right)_{X=0} = \left(\frac{\partial \phi}{\partial X}\right)_{X=1} = 0 \qquad (0 \le Y \le Y_0) \tag{17}$$

$$\left(\frac{\partial \phi}{\partial Y}\right)_{Y=0} = \left(\frac{\partial \phi}{\partial Y}\right)_{Y=1} = 0 \qquad (0 \le X \le X_0) \tag{18}$$

$$Y = 0 \qquad 0 \le X \le M_c \qquad C_{Qc} = 1 \qquad C_{HQ} = 0 \tag{19}$$

$$Y = 0 \qquad M_a \le X \le 1 \qquad C_{Qa} = 0 \qquad C_M = 0 \qquad C_B = C_{B0} \tag{20}$$

At the interfaces between membrane and electrodes, continuity boundary conditions of potential and current are assumed.

The model was solved numerically by the finite element method. Both anode and cathode regions were divided into 2000 elements (20×100), 20 in flow direction and 100 in current direction. The increment in the electrolyte flow direction was constant, but it decreased step by step towards ionic membrane in current direction. The parameters and kinetic equations in the model are shown in [4].

RESULTS AND DISCUSSION

Some of the computer simulation results and experimental data are illustrated in Figures 2, 3, 4, 5. As can been seen that, in Figures 2 and 3, although experimental results and predicted values are slightly different in magnitudes, their trends are consistent with each other. Current efficiency almost increases linearly with the increase of concentration of benzene or benzoquione. Effects of total current on current efficiency (Figure 4) shows that for benzene oxidation, predicted value is significantly different from experimental data. It is probably due to the assumption of constant potential and constant over potential of oxygen evolution. The former assumption and assumption of over potential of hydrogen evolution probably lead to the difference between calculated data and experimental results in Figure 5.

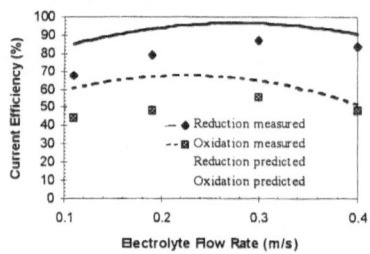

Figure 2 Effects of flow rate on current efficiency. B, 24%; BQ, 39%, t = 20 min; T = 35°C; I=10 A

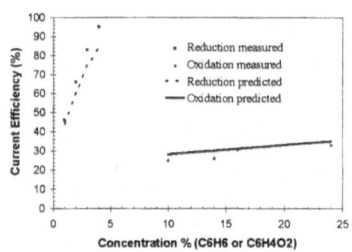

Figure 3 Effects of concentration on current efficiency. L=0.19 m/s; t = 20 min; T = 35°C; I=10 A; bed thickness = 30 mm.

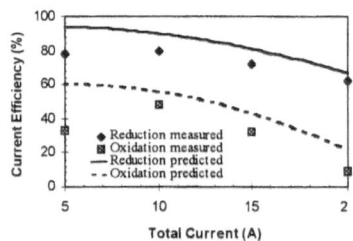

Figure 4 Effects of Total current on current efficiency. B, 24%; BQ, 39%, t = 20 min; T = 35°C; L=0.19 m/s

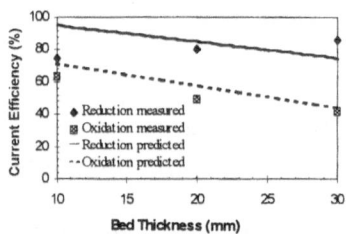

Figure 5 Effects of bed thickness on current efficiency. B, 24%; BQ, 39%; L=0.19 m/s; T = 35°C; I=10 A

ACKNOWLEDGEMENT

This work is supported by the National Nature Science Foundation, China, and Chinese Petroleum and Chemical Technology Company.

NOTATION

a	interfacial area, m2m-3	Ci	dimensionless concentration of
D	diffusion coefficient, m2s-1	E	potential, V
f	F / RT	i	current density, Am-2
F	Faraday constant, 96484 C mol-1	ki	ith reaction rate coefficient,
K	mass transfer coefficient, ms-1	I	total current, A
m	distribution coefficient	X_0	dimensionless electrode thickness
Y_0	dimensionless electrode length	Z_0	dimensionless electrode width

Subscript

a	anode	c	cathode	s	electrode surface

Greek symbols

ϕ	potential, V	ϵ	porosity	γ	conductivity of electrolyte solution, mho m-1

References

1) Millington J. P., Tortman J., J. Electrochem. Soc., 149th Meeting, Abstract, **1976,** 279.
2) Oloman C., Reilly P., J. Electrochem. Soc., **1987**, 134(2), 859.
3) Scott K., Chem. Eng. Proc., **1992,** 31, 21.
4) Zhang X., Ph.D. Thesis, East China University of Science and Technology, China, **1996**

USEFUL ELECTRO-OXIDATIVE CONVERSIONS OF AROMATIC COMPOUNDS

James H.P. UTLEY and Gregor G. ROZENBERG

Department of Chemistry, Queen Mary and Westfield College (University of London), Mile End Road, London E1 4NS, UK

ABSTRACT: Conditions for the direct and indirect side-chain oxidation of methylnaphthalenes were explored, together with substituent effects likely to influence the acidity of their radical-cations. That substituent effects alter the electron demand at side-chain positions was shown using chemical shifts (^{13}C) at the benzylic carbons of 2-benzylnaphthalenes substituted at the *para* position of the benzyl group. However, the reactivity of 2-methylnaphthalenes in direct anodic oxidation ranges from exclusive dimerisation to exclusive nuclear substitution with negligible attack at the methyl group.

In contrast DDQ is an effective side-chain oxidant for electron-rich 2-methylnaphthalenes. An indirect method was devised in which non-stoichiometric amounts of DDQ are used with continuous anodic regeneration in an undivided cell and at constant current.

INTRODUCTION: Side-chain oxidation of alkylaromatics to aromatic aldehydes is a highly desirable electrosynthetic conversion and is believed to be the basis for commercial production by BASF of fine chemicals for flavourings and perfumes. The BASF conditions for anodic oxidation are relatively simple[1], using constant current, graphite anodes, methanol as solvent/nucleophile, and potassium fluoride as electrolyte.

However, such side-chain oxidations appear to be very substrate-specific and, for toluenes, best results require activation by electron-donating groups at the *para*-position. Methylnaphthalenes tend to undergo nuclear substitution, oxidation to quinones or, in the absence of good nucleophiles, dimerisation. It is widely accepted that the key step in side-chain oxidation is proton loss from the first-formed radical-cations. And yet there is no clear relationship between the expected relative thermodynamic acidities (pK$_a$) of radical-cations of alkylaromatics and their propensity to undergo side-chain oxidation. This paper reports on conditions under which anodic oxidation might be used to convert methylnaphthalenes into the corresponding aldehydes.

RESULTS AND DISCUSSION: The dichotomy referred to above is illustrated by our first experiments. Using, except for the substitution of graphite by platinum, the BASF conditions [Pt anode, MeOH-KF(0.15M) constant current], we confirm that 4-methoxytoluene gives a high yield of 4-methoxybenzaldehyde (81%). However, toluene is converted into a complex mixture of nuclear substitution products with no indication of side-chain oxidation despite the likelihood that the acidity of the radical-cation of toluene is higher than that of the radical-cation of 4-methoxytoluene which is stabilised by the electron-donating methoxy group.

Direct oxidation of 6-substitued-2-methylnaphthalenes: In the expectation that electron demand at the side-chain of 2-methylnaphthalenes would be greatly affected by substitution, the direct anodic oxidation of compounds (**1a**) - (**1d**) and (**4**)was explored. Cyclic voltammetry in CH_3CN-Bu_4NBF_4 (0.1M) confirmed that these compounds are oxidised irreversibly and that the potentials are in the expected order, i.e. (**1d**), R = SO_2Et, was the most difficult to oxidise (E_p = 2.17 V *vs*. SCE) and (**1b**), R = MeO, the easiest (E_p = 1.41 V). The point is made further by the plot in Figure 1 (no reliable σ_p value for SO_3Et was available).

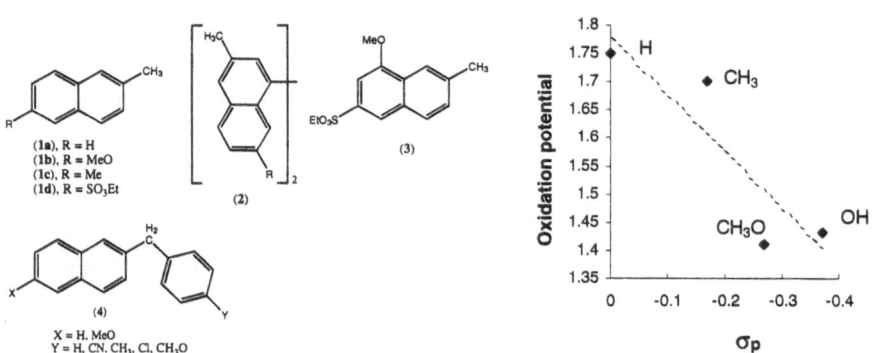

Figure 1. Substituent effects on oxidation potentials of 6-substituted-2-methylnaphthalenes

Preparative-scale oxidation of the compounds (**1a**), (**1b**) and (**1d**) under comparable conditions showed that (**1a**) and (**1b**) gave the dimers (**2**), whereas (**1d**) gave (**3**) in 28% yield. The yields of dimers are a function of oxidation potential (Figure 2); the trend is in line with other observations[2] in which stabilised radical-ions are more reactive towards dimerisation than those related but less stabilised.

A determined attempt at influencing the probability of side-chain proton loss from 2-methylene groups in naphthalene radical-cations was made in a study of the electrochemistry of the 2-benzylnapthalenes (**4**). Electron demand at the 2-benzyl position could be influenced by substitution at the 6-position (in the napthalene unit) and in the 4-position (in the attached benzyl group). Confirmation that such an influence is real is the correlation found between polar substituent constants and ^{13}C chemical shifts (Figure 3). However, even under the best conditions established for direct anodic oxidation of compounds (**4**) [Pt anode, divided cell, MeOH-KF(0.2M)-CH_2Cl_2 (10% v/v) at 6.5 mA cm^{-2}] the only products were the 1,4-naphthaquinones or corresponding quinone ketals (all in modest yield). In one case (**4**, Y = MeO) a small amount (12%) of side-chain methoxylated product was found.

Success was achieved using another type of side-chain oxidant, the high potential quinone, 2,3-dichloro-5,6-dicyanobenzoquinone (DDQ). However, this reagent must normally used in considerably more than stoichiometric amounts (Table 1). And it is noteworthy that, even though a hydride-abstraction mechanism is widely believed to operate, the effectiveness of the method is related to ease of electrochemical oxidation of the substrate.

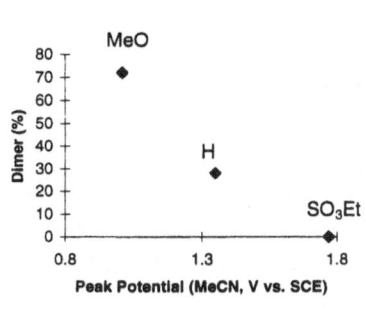

Figure 2. Dimerisation as a function of potential

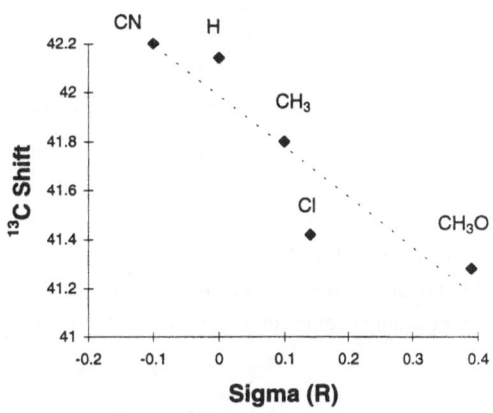

Figure 3. Polar effects on ^{13}C shifts in 2-benzylnaphthalenes (**4**, X=H, Y indicated on line)

Table 1. Side-chain oxidation with DDQ

X	DDQ (equivs)	%(**5**)	%(**6**)	Conversion (%)
H	3	0	0	0
CH$_3$	2	50	30	>95
CH$_3$O	3	90	0	>95
EtOSO$_2$-	3	0	0	0

The results from the batchwise oxidations (Table 1) prompted a study of the anodic regeneration of DDQ from the spent reagent (DDQH$_2$). This has previously been reported[3] in non-aqueous conditions. Under the conditions described in Table 1 we observe by cyclic voltammetry reversible oxidation of DDQH$_2$ at a potential sufficiently less anodic than that of 6-methoxy-2-methylnaphthalene (**1b**) to permit in situ regeneration of the spent reagent (details in Scheme 1).

Systematic exploration of reaction conditions shows that in aqueous acetic acid excellent yields of the side-chain products (**5**) and (**6**) are obtained in continuous electrolysis with much less than stoichiometric proportions of DDQ (Table 2). Furthermore the process has been developed to the

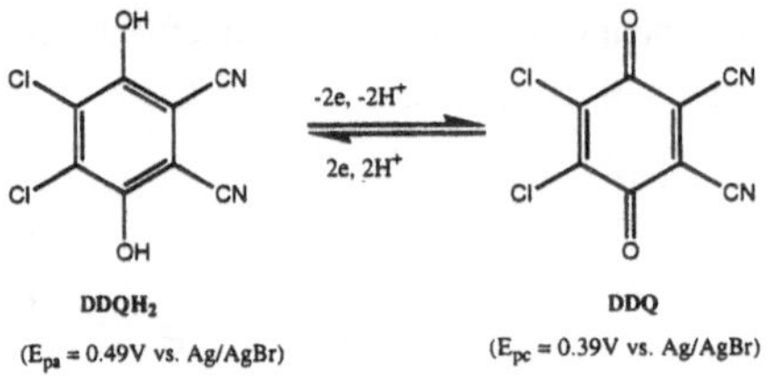

48

DDQH₂
(E_{pa} = 0.49V vs. Ag/AgBr)

DDQ
(E_{pc} = 0.39V vs. Ag/AgBr)

Reversible at Pt anode at 0.3V s⁻¹; H₂O-HOAc-Et₄NOTs (0.5M)

N.B. Oxidation of 6-methoxy-2-methylnaphthalene (**1b**), at high concentration, becomes significant at *ca.* 0.9V

Scheme 1. Anodic regeneration of DDQ

point where dimer formation has been suppressed and the desired product, the carboxaldehyde (**5**) is obtainable in high yield at constant current in an undivided cell and on a reasonable scale. Interesting other features of these results, perhaps of some mechanistic significance, are the formation of the dimer (**2**) under some conditions and of the adduct (**7**).

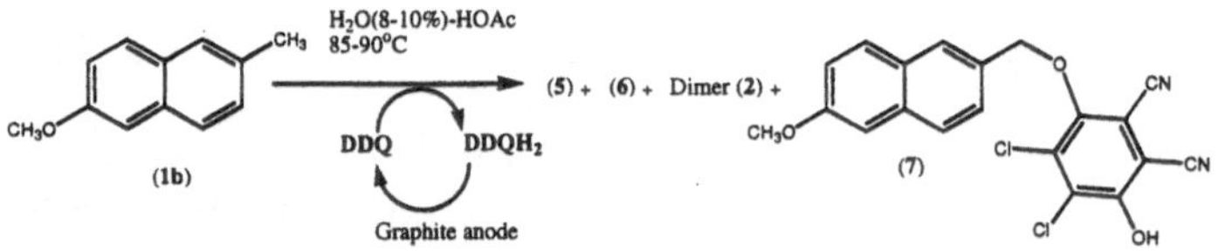

Table 2. Mediated side-chain oxidation of 6-methoxy-2-methylnaphthalene [graphite anode, constant potential (0.8 V Ag/AgBr)]

[(**1b**)]ᵃ	[TEATs]	F	(**5**)	(**6**)	(**2**)	Conversion
0.1M	0.1M	3.5	30	-	44	74
0.1M	0.5M	2.2	37	-	14	65
0.1M	0.8M	9	70	-	[*ca.* 15% (**7**)]	92
0.4M	0.8M	4	72	-	[*ca.* 15% (**7**)]	90
0.4M	0.7M	6ᵇ	71	-	[*ca.* 15% (**7**)]	87

(a) 0.25 equivalents DDQ; (b) constant current (7 mA cm⁻²), cell vol. 500 cm³

REFERENCES: *(1). BASF AG-GE patent* 2848397 (1980); *BASF AG-GE patent* 2855508 (1980); *BASF AG-GE patent* 2935398 (1981); *(2).* I. Fussing, M. Güllü, O. Hammerich, A. Hussain, M.F. Nielsen and J.H.P. Utley, *J. Chem. Soc., Perkin Trans. 2,* 1996, 649; O. Hammerich and M.F. Nielsen, *Acta Chem. Scand.,* in press; J. Heinze, A. Smie, R. Müller, P. Hübler, P. Tschunky and K. Meerholz, Abstract 1272, 192ⁿᵈ ECS Meeting, Paris 1997; *(3).* U.H. Brinker, M. Tyner III and W.M. Jones, *Synthesis,* 1975, 671

ACKNOWLEDGEMENT: One of us (GGR) is grateful to SmithKline Beecham Pharmaceuticals for the provision of a studentship.

THE ELECTROCHEMICAL TRIFLUOROMETHYLATION AND PERFLUOROALKOXYLATION OF AROMATIC COMPOUNDS

Vitali A. GRINBERG [a], Cynthia A. LUNDGREN [b], Sergey R. STERLIN [a]

[a] *A.N.Frumkin Institute of Electrochemistry Russian Academy of Sciences, 31 Leninsky prospekt, 117071 Moscow, Russian Federation*
[b] *Central Research & Development E.I. Du Pont de Nemours & Company, Inc., Experimental station P.O. Box 80262, Wilmington, DE 19880-0262, USA*

ABSTRACT

The anodic radical trifluoromethylation on platinum electrode in aqueous acetonitrile of a number of aromatic compounds ($CF_3C_6H_4NO_2$, $1,3-(CF_3)_2C_6H_4$, $CF_3C_6H_4COCH_3$) containing two electron-withdrawing groups was studied and a new approaches of anodic perfluoroalkoxylation of arenes were proposed.

INTRODUCTION

Previously we have established that coelectrolysis of CF_3COOH and substituted benzenes containing one electron-withdrawing group in aqueous acetonitrile on a platinum electrode results in the products of trifluoromethylation, along with the normal products of trifluoroacetoxylation of the benzene ring [1], especially in the case of aromatic compounds with strong electron-acceptor groups [2]. The electrochemical methods of the synthesis of arylperfluoroalkyl ethers is a poorly investigated field of chemistry.

RESULTS AND DISCUSSION

It was shown that the decrease of adsorbability of arenes containing two electron-withdrawing groups in spite of the simultaneous decrease of their oxidation rate in the range of the potential trifluromethyl radical generation led to the decrease of the yield of the trifluoromethylation products in comparison with arenes containing one electron-withdrawing substituent ($C_6H_5CF_3$, C_6H_5CN, $C_6H_5NO_2$, $C_6H_5COCH_3$). It is possible that the additional reason for the decrease of trifluoromethylation products yield can be connected with steric hindrance preventing the attack by trifluoromethyl radicals. At the same time as it can be expected the yield of the trifluoroacetoxylation products of arenes with two electron-withdrawing substituents decreases sharply. The mechanism of electrochemical radical trifluoromethylation of arenes, containing two electron-withdrawing groups apparently doesn't differ from described in [2]. The introduction of the fourth substituent into the molecule of aromatic compound by the anodic radical trifluoromethylation leads to the violation of aromatic structure and the formation of cyclohexadiene derivatives; similar results have been obtained in [3,4] by the interaction of thermally generated C_6F_{13} radicals with benzene or by the reaction on cathodically generated perfluoroalkyl radicals under the electroreduction of the higher perfluoroalkylhalides in the presence of benzonitrile as a solvent.

It was determined that preparative trifluoromethylation on glassy carbon anode, which had a small adsorbing ability [5], leads to the sharp decrease of the trifluoromethylation products yield, even in the case of aromatic compounds, containing one electron-withdrawing group. The last circumstance indicates the significant role of aromatic compounds adsorption in the arenes anodic trifluoromethylation.

The possibility of electrochemical synthesis of arylperfluoroalkyl ethers was investigated on the following examples: a) the electrooxidation of perfluorocarboxylic acid on the consumable oxide anode from PbO_2 in the presence of aromatic compounds and b) anodic oxidation of the system aromatic compounds/alkali metal perfluoroalkoxylate on platinum electrode. The electrooxidation of the system CF_3COOK/CF_3COOH on PbO_2 anode in the range of the potentials where Kolbe reaction takes place (anodic current density 200 mA/cm^2) in aqueous acetonitrile in the presence of benzonitrile led to the formation of trifluoromethoxylic derivatives of benzonitrile. This result showed only the principal possibility of the idea to use consumable electrode enriched by oxygen for the generation of $CF_3O\cdot$ radical (the main process which occurred on PbO_2 anode was oxidation of aromatic compound). The absence of electron-withdrawing groups in the molecule of aromatic compounds led to the oxidation of arenes at less positive potentials. That prevented the potential to reach the values where Kolbe reaction and generation of $CF_3O\cdot$ -radical could occur. The electrooxidation of the system CF_3COOK/CF_3COOH on PbO_2 anode in aqueous acetonitrile in the presence of benzene led to the formation of phenol and polymeric products.

The anodic oxidation of benzotrifluoride in the preparative electrolysis with supporting electrolyte $(CF_3)_2CFOCs$ led to the formation of the mixture of three isomers of (heptafluoro-isopropoxy)benzotrifluoride and two isomers of bis(heptafluoroisopropoxy)benzotrifluoride. The process apparently occurs in accordance with ECE- mechanism:

$$ArH + R_fO^- - 2e \longrightarrow Ar\,OR_f + H^+; \qquad Ar\,OR_f + R_f\,O - 2e \longrightarrow Ar\,(OR_f)_2 + H^+$$

It must be underlined that in contrast to anodic trifluoromethylation of the aromatic compounds[2], in this case the introduction of two R_fO groups into aromatic ring takes place. These results apparently indicate at more easy electrooxidation of perfluoroalkoxy derivatives, produced at the first stage in comparison with trifluoromethyl derivatives.

Similarly we could realize trifluoromethylation of $PhCF_3$ under it electrooxidation in the solution of CF_3OCs in unaquaous MeCN. The developed reaction is obviously a first example of the electrochemical perfluoroalkoxylation of aromatic compounds.

References
1) Z.Blum, L.Cedhein, K.Nyberg , *Acta chem.scand.Ser.* B 1975, **92**, 715.
2) V.A.Grinberg, V.A.Polischshyk , L.S.German, et al, *Izv.AN SSSR ser. khim.* **1978**, 3, 677.
3). C.G.Krespan , *J. Fluorine Chem.* **1988**, 40, 129.
4). C.P.Andrieux, L.Gells, M. Medebielle, J. Pinson, J.M. Saveant, *J. Am. Chem. Soc.* **1990**, 112, 3509.
5) V.F.Cherstkov , V.A.Grinberg , S.R.Sterlin et al. , *Izv.AN SSSR ser. khim* **1991**, 5, 1134.

ANODIC ELECTROCHEMISTRY: RECENT ADVANCES IN THE TOTAL SYNTHESIS OF COMPLEX ORGANIC MOLECULES

Kevin D. MOELLER, Dean FREY, Laura MATSON-BEAL, Santhaparam H. K. REDDY, and Yunsong TONG

Department of Chemistry, Washington University, St. Louis, MO 63130

ABSTRACT

Anodic electrochemistry holds great promise as a tool for organic synthesis. Recently, four new observations have opened the door for a variety of total synthesis efforts in areas ranging from the development of conformationally constrained peptidomimetics to the synthesis of complex organic molecules with central quaternary carbons. These four new observations are highlighted here.

INTRODUCTION

Oxidative electrochemistry provides an attractive method for making complex organic molecules because it offers the opportunity to selectively functionalize organic molecules and generate reactive intermediates under neutral conditions. In addition, the ability of electrochemistry to conduct oxidation reactions at a wide range of potentials suggests that electrochemical techniques should be generally useful for meeting a variety of synthetic challenges. In order to illustrate these ideas, a number of syntheses utilizing electrochemical synthetic methods to effect key transformations have been undertaken. Several of the targets being synthesized are outlined in Figure 1. Each of these molecules present a unique challenge for the use of organic electrochemistry as a synthetic tool. Reported herein are preliminary results that "open the door" for the construction of each of these compounds.

Figure 1

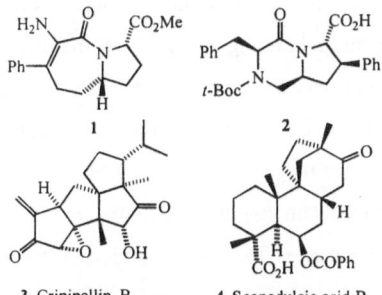

1

2

3. Crinipellin B

4. Scopadulcic acid B

RESULTS AND DISCUSSION

<u>Tandem anodic oxidation-olefin metathesis reaction sequences: An approach to the constrained Phe-Pro building block **1**</u>: The anodic methoxylation of amides is one of the most useful electrochemically based synthetic methods.[1] It has been used in the synthesis of a variety of alkaloid

natural products[1] and constrained peptidomimetics.[2] However, to date a convenient method for using electrochemistry to convert proline into bicyclic lactam peptidomimetics containing seven-membered ring lactams has not been found.[3] The seven-member ring lactam based peptidomimetics are important because they provide a means for adding conformational flexibility back into six-member ring lactam based peptidomimetics that have proven to be constrained to tightly. Outlined in Scheme 1 is a convenient approach to this problem.

Scheme 1

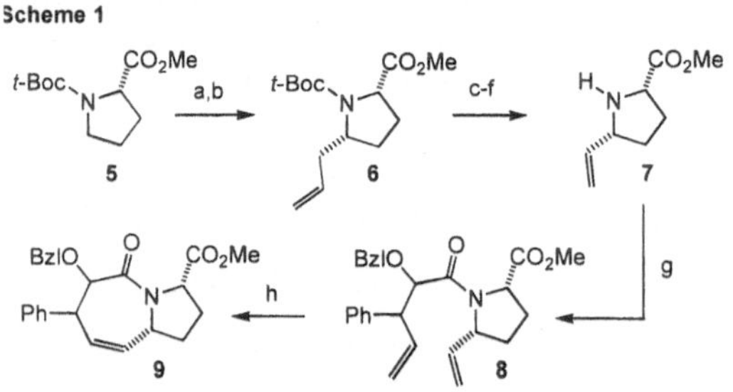

Reagents: (a) carbon anode, Et$_4$NOTS, MeOH, 3.0 F/mole, 26.8 mA, 99%; (b) BF$_3$·Et$_2$O, allylTMS, Et$_2$O, 77% (2:1 ratio of cis:trans isomers); (c) i. O$_3$, 1:1 MeOH:CH$_2$Cl$_2$, ii. NaBH$_4$, -78°C to RT, 96%; (d) o-NO$_2$(C$_6$H$_4$)SeCN, (n-Bu)$_3$P, THF, 85%; (e) i. MCPBA, CH$_2$Cl$_2$, -70°C, ii. Me$_2$S, Et$_3$N, -70°C to RT, 96%; (f) TFA, CH$_2$Cl$_2$, 0°C to RT, 65%; (g) RCO$_2$H, EDC, HOBt, Et$_3$N, CH$_2$Cl$_2$, 0°C to RT; (h) Cl$_2$Ru(PCy$_3$)$_2$=CHPh, CH$_2$Cl$_2$, 40°C, 80%.

In this scheme, the protected proline **5** was oxidized and then the subsequent methoxylated product treated with BF$_3$·Et$_2$O and allyltrimethylsilane in order to form the allyl substituted product **6** as a 2:1 mixture of cis to trans isomers. The isomers were separated and then the allyl moiety of the cis product converted into the desired vinyl substituent in building block **7**. If the trans stereochemistry is needed for the vinyl substituted proline, then the vinyl group can be added directly to the methoxylated amide with the use of a cuprate reagent.[4] The substituted proline was converted to diene **8** and then the desired seven-member ring lactam completed with the use of an olefin metathesis reaction.[5] Compound **9** can be converted into the desired TRH building block **1** using chemistry that has already been reported.[2,3] The olefin metathesis reaction ran equally well for the trans isomer allowing for the synthesis of building blocks with either stereochemistry.

Scheme 2

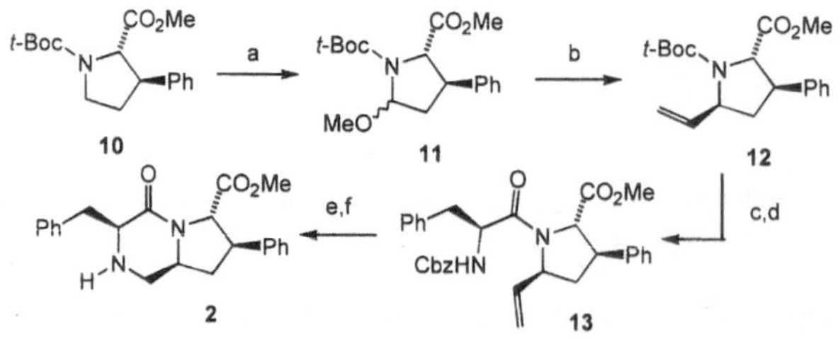

Reagents: (a) Carbon anode, Pt cathode, 0.03 M Et$_4$NOTs, MeOH, 3 F/mole 87%; (b) CH$_2$=CHLi, CuBrMe$_2$S, BF$_3$·Et$_2$O, 34%; (c) BF$_3$·Et$_2$O; (d) CbzPheF, N-ethylmorpholine, CH$_2$Cl$_2$, 50% over two steps; (e) i. O$_3$, MeOH; ii. Me$_2$S; (f) H$_2$, Pd, MeOH, 75% over two steps.

Anodic Oxidation of Substituted Prolines: An Approach to the Synthesis of a Constrained Phe-Phe Building Block: One of the most important reasons to develop electrochemistry as a synthetic method for making peptidomimetics is that it allows one to capitalize on existing amino acid based starting materials. For example, consider the synthesis of **2** outlined in Scheme 2. In this case, the peptidomimetic targets a Phe-Phe region in the hormone being studied. Fortunately, a conformationally constrained analog of phenylalanine **10** has already been synthesized.[6] The oxidation of **10** led to a high yield of the methoxylated amide **11**. Treatment of **11** with a vinylcuprate reagent in the presence of $BF_3 \cdot Et_2O$ led to the formation of the vinyl substituted derivative **12**. While the yield of this reaction has not yet been optimized, the speed with which **12** can be synthesized from the known starting material already makes the route very attractive. The synthesis of **2** was finished using chemistry directly analogous to earlier related examples.[7] In this way, the desired bicyclic peptidomimetic **2** was made in only six steps from the existing amino acid derived starting material.

Anodic Olefin Coupling Reactions: The Synthesis of Quaternary Carbons from Elimination Sensitive Substrates: We have demonstrated that intramolecular anodic olefin coupling reactions can lead to the formation of five-member ring products even when the substrate contains an allylsilane group having an allylic oxygen substituent.[8] The reactions did not lead to elimination of the oxygen group and formation of a diene product. The success of these cyclization reactions suggested the use of an intramolecular anodic olefin coupling reaction for the synthesis of crinipellin B. However, the synthesis of crinipellin would require that the anodic olefin coupling reaction form a sterically difficult to synthesize quaternary carbon while still not eliminating the alkoxy group from the allylic carbon of an allylsilane. Is this possible? In order to address this question, substrate **14** was synthesized and

oxidized (Scheme 3). This reaction smoothly led to the formation of the desired bicyclic product **15** without any elimination of the allylic alkoxy group. Two stereoisomers were formed. Both isomers had the -OTBDMS group and the vinyl substituent trans to each other and a cis-ring fusion.

Scheme 3

the vinyl substituent trans to each other and a cis-ring fusion. Clearly, the electrochemical anodic olefin coupling reaction leading to the formation of a five-member ring was gentle enough to preserve an allylic alkoxy group even when the reaction was forced to overcome the steric barriers associated with quaternary carbon formation. It would appear that the anodic olefin coupling reaction will be able to overcome the synthetic challenges associated with the crinipellin ring skeleton.

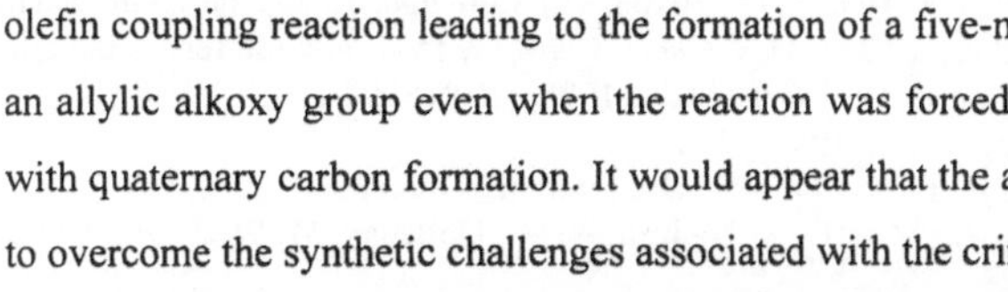

Intramolecular Anodic Olefin Coupling Reactions and the Formation of Bridged Bicyclic Ring Skeletons: The success of intramolecular anodic olefin coupling reactions for the formation of quaternary carbons led to suggestions that the reaction would be useful for constructing natural products containing bridged bicyclic ring skeletons. Analogous radical anion intermediates have proven useful for synthesizing these ring systems.[9] Do radical cation intermediates undergo the same types of transformations? If so, then cyclizations to form bridged bicyclic ring systems could be initiated from either electron-rich or electron-poor olefins. In order to investigate this possibility, a synthesis of scopadulcic acid B has been undertaken. Preliminary results were obtained for this synthesis by studying the oxidation of substrates **16a** and **16b** (Scheme 4). In both cases, the anodic oxidation led to the formation of the desired bicyclic ring skeleton. The

Scheme 4

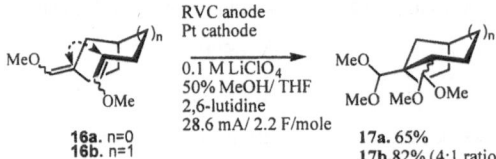

16a. n=0
16b. n=1

RVC anode
Pt cathode
0.1 M LiClO$_4$
50% MeOH/ THF
2,6-lutidine
28.6 mA/ 2.2 F/mole

17a. 65%
17b. 82% (4:1 ratio of isomers)

oxidation of **16a** led to a lower yield than the oxidation of **16b** and was not as clean by ^{1}H NMR analysis of the crude reaction product. Presumably, this was a more difficult cyclization because of the added strain associated with the bicyclo[2.2.1]heptane ring skeleton. The cleanliness of the cyclization resulting from the oxidation of **16b** bodes well for the development of a general route to bicyclo[3.2.1]octane ring skeletons. In addition, it was found that the oxidation of **16b** proceeded smoothly (70% isolated yield of **17b**) with the use of a 6V lantern battery as the power source.[10]

Conclusions and Future Plans: Recent reactions have demonstrated the potential utility for using anodic oxidation chemistry as a key tool for constructing a variety of complex organic molecules. Work aimed at synthesizing the target molecules illustrated in Figure 1 is underway.

Acknowledgements: We thank the National Institutes of Health (R01 GM53240-01) and the National Science Foundation (CHE-9628881) for their generous support of this work. We also gratefully acknowledge the Washington University High Resolution NMR facility, partially supported by NIH grants RR02004, RR05018, and RR07155, and the Washington University Mass Spectrometry Resource Center, partially supported by NIH RR00954, for their assistance.

References:
1. Shono, T. *Topic in Current Chemistry* **1988**, *148*, 131 and references therein.
2. (a) Li, W.; Moeller, K. D. *J. Am. Chem. Soc.* **1996**, *118*, 10106, (b) Rutledge, L. D.; Perlman, J. H.; Gershengorn, M. C.; Marshall, G. R.; Moeller, K. D. *J. Med. Chem.* **1996**, *39*, 1571, and references therein.
3. Li, W.; Hanau, C. E.; d'Avignon, A.; Moeller, K. D. *J. Org. Chem.* **1995**, *60*, 8155.
4. Collado, I.; Ezquerra, J.; Pedregal, C. *J. Org. Chem.* **1995**, *60*, 5011.
5. Miller, S. J.; Grubbs, R. H. *J. Am. Chem. Soc.* **1995**, *117*, 5855 and references therein.
6. Chung, J. Y. L.; Wasicak, J. T.; Arnold, W. A.; May, C. S.; Nadzan, A. M.; Holladay, M. W. *J. Org. Chem.* **1990**, *55*, 270.
7. Fobian, Y. M.; d'Avignon, D. A.; Moeller, K. D. *Bioorg. Med. Chem. Lett.* **1996**, *6*, 315.
8. Frey, D. A.; Marx, J. A.; Moeller, K. D. *Electrochimica Acta* **1997**, *42*, 1967.
9. Sowell, C. G.; Wolin, R. L.; Little, R. D. *Tetrahedron Lett.* **1990**, *31*, 485.
10. Frey, D. A.; Wu, N.; Moeller, K. D. *Tetrahedron Lett.* **1996**, *37*, 8317.

ELECTROGENERATED *N,O*-ACETALS IN CYCLIZATION AND CYCLO-ADDITION REACTIONS

Karsten Danielmeier, Doris Kolter, Masahiro Sadakane, Kerstin Schierle, Annette Stahl, Ruth Vahle, Andrea Zietlow and Eberhard Steckhan*

Kekulé-Institut für Organische Chemie und Biochemie der Universität Bonn, Gerhard-Domagk-Str. 1, D-53121 Bonn, Germany

Abstract. Electrogenerated chiral *N,O*-acetals proved to be excellent precursors for the synthesis of chiral enantiomerically pure polycyclic compounds of potential biological activity.

Introduction. Cyclization and cycloaddition reactions play an important role in the construction of bioactive *N*-heterocyclic compounds such as enzyme inhibitors, constrained peptide mimetics and alkaloids. As reactive intermediates, imines, *N*-acyl imines, iminium ions and *N*-acyl iminium ions are commonly employed. Among the many methods for the formation of such amidoalkylation intermediates[1], the electrochemical pathway *via* the stable *N,O*-acetals as *N*-acyliminium ion precursors is one of the most efficient and flexible ones. This has been very nicely developed in the case of intramolecular cyclizations for the construction of constrained peptide mimetics by *Moeller.*[2]

The following principles for cyclization and cycloaddition reactions using electrogenerated *N,O*-acetals can be envisioned (Scheme 1):

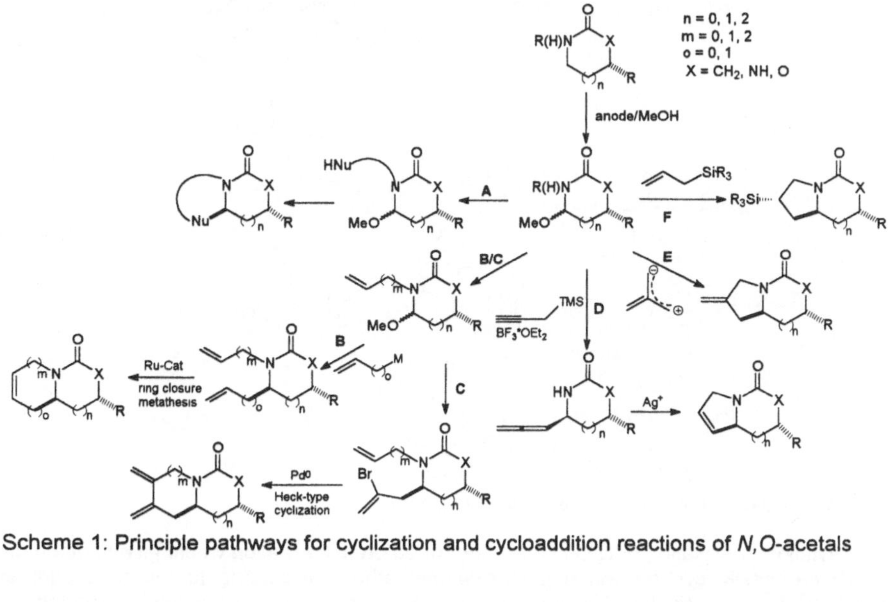

Scheme 1: Principle pathways for cyclization and cycloaddition reactions of *N,O*-acetals

55

Results and Discussion.

A] The *N,O*-acetal which is obtained *via* anodic oxidation is *N*-substituted either before or after the electrolysis by a group containing a nucleophilic function. Under *Lewis*- or *Broenstedt* acid catalysis the intermediate *N*-acyl iminium ion cyclizes to the nucleophile. Such reactions have been very efficiently applied by *Moeller*[2] for the synthesis of peptide mimetics. We applied similar reactions starting from L- or D-threonine via a *Hoefer-Moest* reaction (Scheme 2). The thus obtained *N,O*-acetal has been *N*-alkenylated either by homoallyl bromide or phenethyl bromide. Cyclization of the homoallyl substituted compound was initiated by formic acid according to similar procedures by *Speckamp*[3] and *King*[4] giving mainly the enantiomerically pure (*1R,7S,8aR*)-1-methyl-3-oxo-1,5,6,7,8,8a-hexahydro-oxazolo-[3,4-a]pyrid-7-yl formate together with minor amounts of the three other diastereomers.[5] The phenylethyl substituted compound was cyclized with titanium tetrachloride similar to a formic acid catalyzed cyclization by *Kano*[6] to give only one enantiomerically pure diastereomer of the substituted tetrahydro isoquinoline derivative.[7]

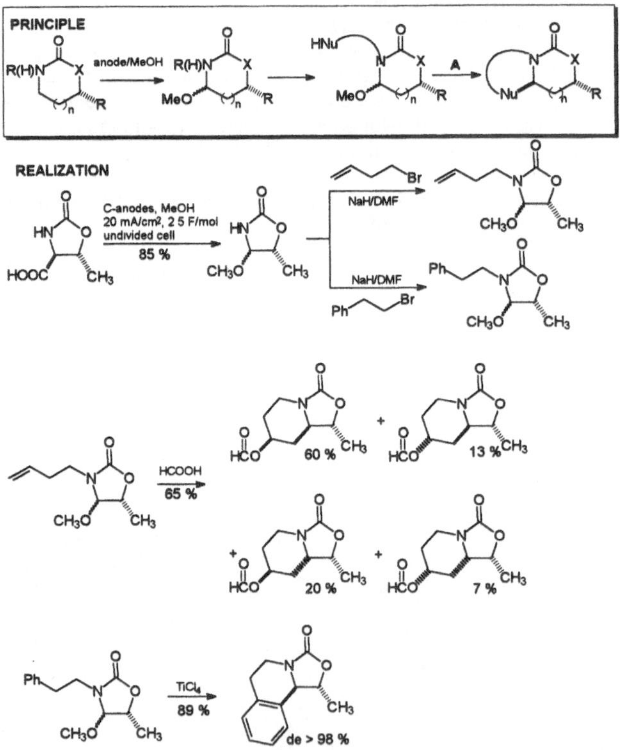

Scheme 2: Cyclization reactions according to path **A**.

B] Vinylation or allylation of *N,O*-acetals after or before alkenylation would give compounds which are easily cyclized *via* ring closure methathesis according to the procedure of *Blechert*[8] for very similar structures. In the proline case, *Moeller* recently reported an example for the formation of a pyrrolo[1,2-a]azepine structure by ring closure metathesis reaction starting from an electrochemically generated building block.[9] One fully developed example and some more examples for the first step are given in Scheme 3. Further development of the principle is currently under way.

C] If the allylation of the *N,O*-acetals is performed with an allyl silane containing a vinyl bromide function, a compound is formed which might undergo a Heck-type cyclization after alkenylation of the nitrogen according to path C. A realization of the first step is shown in Scheme 4. Follow-up alkenylation and ring closure are underway.

It was shown that pyrrolidine sugar analogs are very good mimics of the transition state for glycoside hydrolysis, both in respect to the flattened half chair conformation and the charge distribution while the position of the hydroxyl groups is less important. Therefore, they can bind tightly to the enzyme and inhibit it competitively. Because of their pharmacological importance, a large number of synthetic routes to such compounds have recently been developed. Efficient precursors for such compounds are pyrrolo-[1,2-c]-oxazolidinones which can be easily converted to 2-hydroxymethyl pyrrolidines. Here we present two economical ways to differently substituted chiral pyrrolo oxazolidinones derived from low-priced chiral precursors via stereocontrolled routes. The *N,O*-acetals were prepared electrochemically.[10]

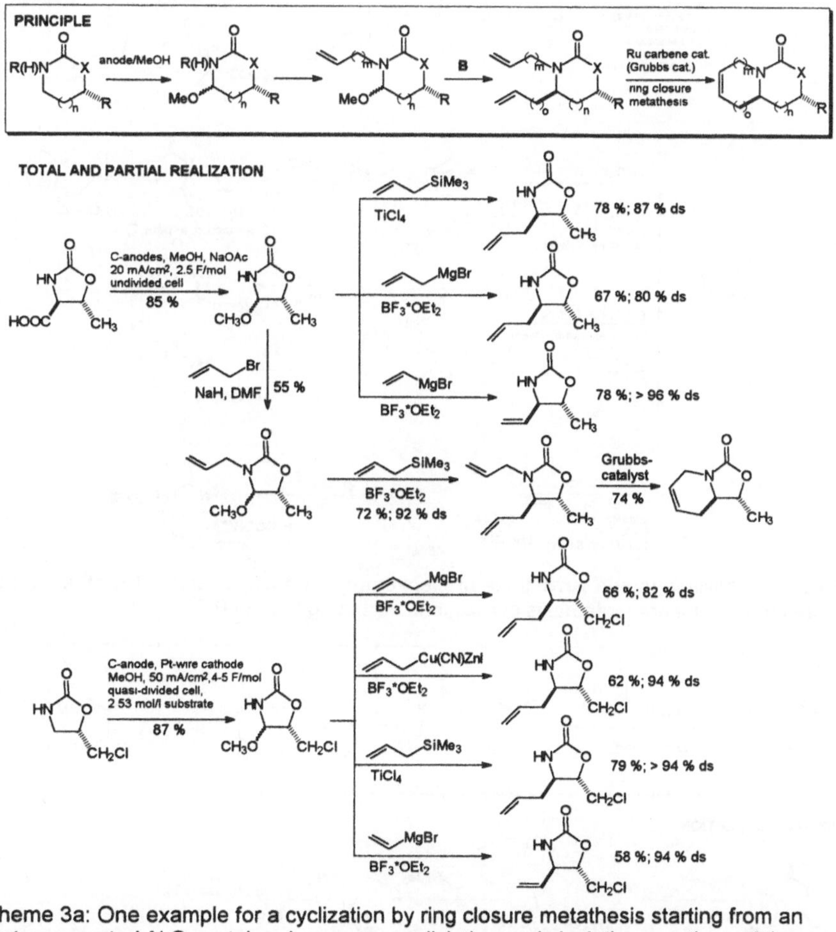

Scheme 3a: One example for a cyclization by ring closure metathesis starting from an electrogenerated *N,O*-acetal and some more allylation and vinylation reactions of those starting materials as a first step to ring closure metathesis cyclizations according to path B.

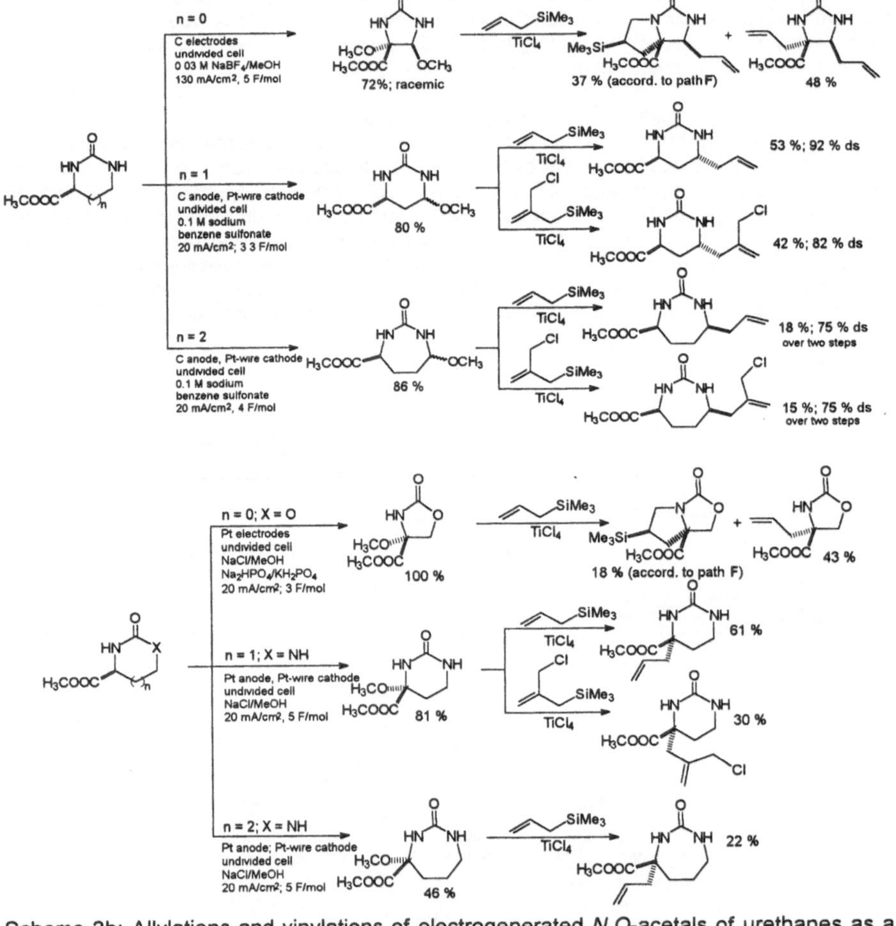

Scheme 3b: Allylations and vinylations of electrogenerated N,O-acetals of urethanes as a first step to ring closure metathesis cyclizations according to path B.

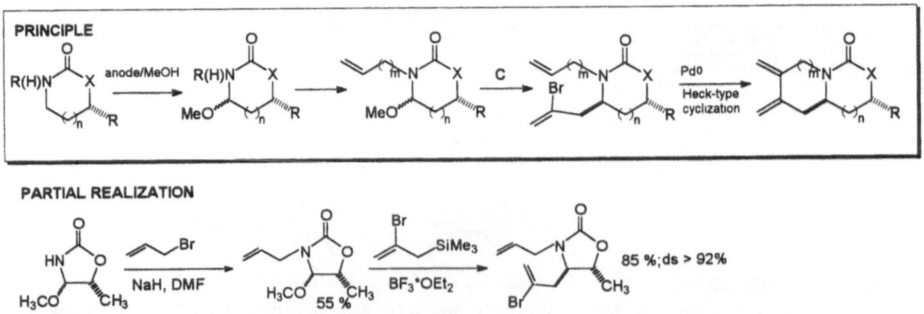

Scheme 4: First steps in the synthesis of a precursor for a Heck-type cyclization according to path C.

D] The first approach according to path **D** starts from for example (4*RS*,5*S*)-5-chloromethyl-4-methoxy-2-oxazolidinone. The methoxy group of the oxazolidinone is substituted by an allene-function through the use of propargyl trimethylsilane as nucleophile and BF$_3$*OEt$_2$ as *Lewis*-acid via the intermediate *N*-acyliminiumion. Cyclization with 0.5 eq silver tetrafluoroborate leads to the enantiomerically pure pyrrolo-[1,2-c]oxazolidinon in 81 % yield. Further transformations of the double bond by osmium tetroxide result in the diastereoselective formation of a *cis*-diol. The relative configuration has yet to be determined (Scheme 5). The product is an interesting azasugar precursor.

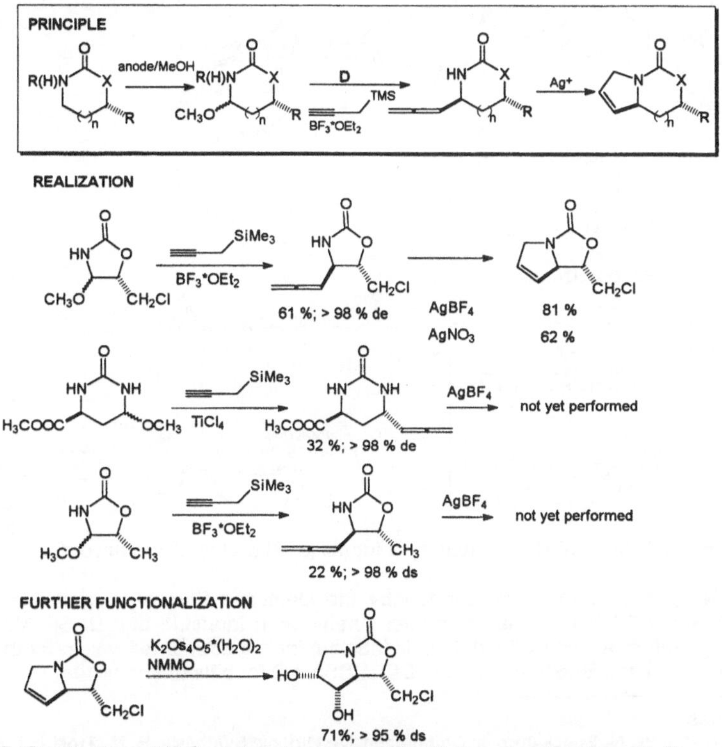

Scheme 5: Cyclization of *N,O*-acetals by propargylic silane according to path **D**.

E] In the second approach a formal [3+2] cycloaddition to *N,O*-acetals is performed in a two-step process using 2-chloromethyl-3-trimethylsilyl-1-propene as reagent according to path **E**. *Lewis*-acid catalyzed alkoxy substitution by the allylsilane is followed by base-induced nucleophilic substitution of the chloro substituent. Thus, bicyclic methylenepyrrolidines, which are building blocks for pharmacologically interesting compounds, are obtained in good yields and high stereoselectivities.[11] The *N,O*-acetals are easily prepared by anodic oxidation of the cyclic carbamates, ureas or lactams. (Scheme 6). Some transformations are shown below (Scheme 6).

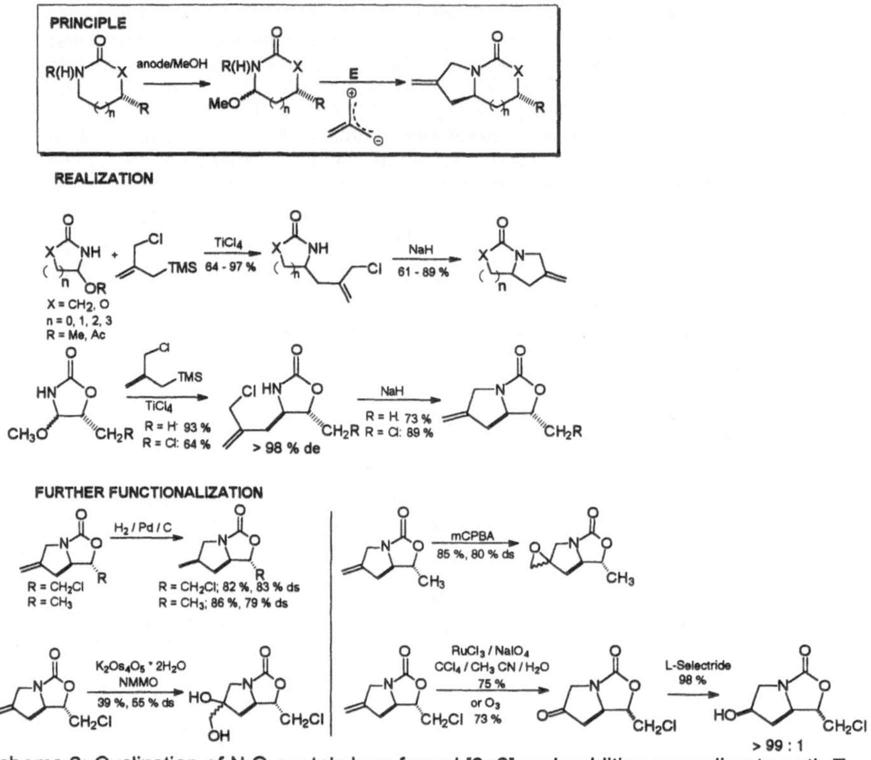

60

Scheme 6: Cyclization of N,O-acetals by a formal [3+2] cycloaddition according to path E.

Acknowledgements: Financial support by the Deutsche Forschungsgemeinschaft (Ste 227/15-3 and 227/19-1-3), the Fonds der Chemischen Industrie and BASF Aktiengesellschaft is gratefully acknowledged. K.S. is thankful for a graduate fellowship by the state of NRW. We thank the BASF AG and the DEGUSSA AG for gifts of chemicals.

References
[1] H. Hiemstra, W. N. Speckamp, in: Comprehensive Organic Synthesis, B. M. Trost, I. Fleming, Eds., Vol. 2, p. 1047 ff. Pergamon Press, Oxford 1991.
[2] K. Moeller, *Topics Curr. Chem.* **1997**, *185*, 49 (Ref. 13-15).
[3] W. N. Speckamp, H. Hiemstra, *Tetrahedron* **1984**, *41*, 4367; W. N. Speckamp, *Recl. Trav. Chim. Pays-Bas* **1981**, *100*, 345.
[4] M. S. Hadley, F. D. King, R. T. Martin, *Tetrahedron Lett.* **1983**, *24*, 91.
[5] C. Herborn, A. Zietlow, E. Steckhan, *Angew. Chem.* **1989**, *101*, 1392; *Angew. Chem. Int. Ed. Engl.* **1989**, *28*, 1399.
[6] S. Kano. Y. Yuasa, S. Shibuya, *Heterocycles* **1985**, *23*, 395.
[7] R. Vahle, E. Steckhan, unpublished results.
[8] M. Huwe, S. Blechert, *Tetrahedron Lett.* **1995**, *36*, 1621.
[9] K. Moeller, communication at the 191st Meeting of The Electrochemical Society, Motreal, May 4 - 9, 1997.
[10] a) K. Danielmeier, E. Steckhan, *Tetrahedron: Asymmetry* **1995**, *6*1181; b) K. Danielmeier, K. Schierle, E. Steckhan, *Tetrahedron* **1996**, *52*, 9743; c) A. Zietlow, E. Steckhan, *J. Org. Chem.* **1994**, *59*, 5658; d) K. Danielmeier, K. Schierle, E. Steckhan, *Angew. Chem. Int. Ed.* **1996**, *35*, 2247.
[11] M. Sadakane, R. Vahle, K. Schierle, D. Kolter, E. Steckhan, *Synlett* **1997**, 95.

ANODIC OXYDATION OF HETEROCYCLES : ALKALOIDS SYNTHESIS

Corinne BACQUE-VANUCCI, Catherine CELIMENE, Marc DAVID,
Hamid DHIMANE and Gérard LHOMMET

Université P.& M. Curie, Laboratoire de Chimie des Hétérocycles,
4 Place Jussieu, F-75252 Paris, France

Abstract: The reductive amination of appropriate ketopyrrolidines leading to 3,5-dialkylindolizidines usually gives stereoselectively the indolizidine with a cis relative stereochemistry on the piperidine moiety. We report herein unexpected results concerning the effect of the nature of the C3 substituent on this stereoselectivity.

The indolizidine alkaloids have been detected in animals[1] as well as in vegetables[2]. We have been interested in preparing natural indolizidines extracted from the skin of neotropical frogs[3]. One way to build the indolizidine framework is an intramolecular reductive amination under catalytic hydrogenation conditions. This method has often been used in the synthesis of natural 3,5-disubstituted indolizidines[4] from appropriate ketopyrrolidines (Scheme 1). The main stereomer (de > 70%) has a cis arrangement of C-5 and C-9 hydrogens (indolizidine numbering). Only the center C-9 seems to direct the configuration of C-5 since this cis relationship between C-5 and C-9 hydrogens takes place selectively whatever the C-3 absolute configuration is[4]. This stereoselective cyclisation is presumed to proceed *via* a catalytic reduction of the iminium intermediate formed *in situ* from the free ketoamine (Scheme 1).

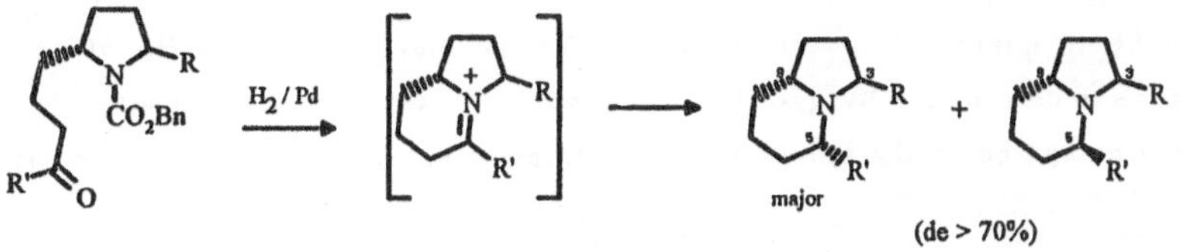

Scheme 1

61

Within the context of our search to use the (S)-proline as a chiral building block in the synthesis of (-) indolizidine 195B[5] we considered two slightly different routes both starting from the common pyrrolidine 1. In route A, we first decided to construct the indolizidine skeleton by intramolecular reductive amination from substrates 4 (R = CH₂OAc) or 5 (R = CO₂Me) before the C-3 appendage elongation. In route B, the C-3 n-butyl substituent elaboration from the carbomethoxy group should be performed before the final reductive amination step (Scheme 2).

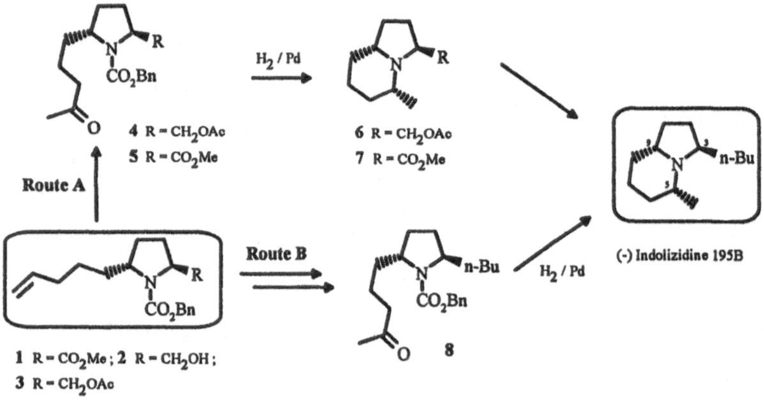

Scheme 2

The pyrrolidine 1 was synthesized in a stereoselective manner (de > 90%), in four steps from *L*-proline using the acyliminium methodology[5].

Route A was first examined starting from the trans ketopyrrolidine 4 (R = CH₂OAc). Chemoselective reduction[6] of the ester moiety of 1 and subsequent acylation of the resulting alcohol 2 gave acetate 3. This latter was submitted to the Wacker oxidation process providing the desired pyrrolidine 4[6]. Exposure of this compound to an atmosphere of hydrogen in the presence of palladium on charcoal (5%) as catalysts in methanol, caused the amine deprotection, annulation and finally the reduction of the resulting iminium, leading to a mixture of 6a[7] along with its C5 epimer 6b in quantitative yield (Scheme 3). The diastereomeric ratio of 6a and 6b (64:36) was determined by gas chromatography. This diastereoselectivity (de = 28%) was low compared to the known results with alkyl substituent at the C-3 position (de

> 70%)[4]. Attempts to improve this stereoselectivity by using various solvents and palladium catalysts were unsuccessful.

This drop in selectivity seemed to be due to the presence of the methyleneacetate group. Before we undertook the synthesis of (-) indolizidine 195B according to route B, we decided to carry out the reductive amination on the substrate 5 bearing a carbomethoxy group on C3 carbon in order to check the effect of such a substituent.

Oxidation of the key intermediate 1[8] under the Wacker type procedure [PdCl$_2$(PhCN)$_2$/CuCl/O$_2$] furnished the ketopyrrolidine 5 in 88% isolated yield[8]. The reductive amination of 5 under catalytic hydrogenation conditions was carried out in ethyl acetate as solvent. The crude analysis (GC / MS) showed the presence of four diastereomers of 7 among which the predominent represented more than 90%.

The presence of four isomers was not surprising since the starting material 5 was contaminated with its cis isomer[8]. More surprising was the relative stereochemistry of the major indolizidine isolated from the mixture of 7. Indeed the NOE experiments of this stereomer[9] indicated a cis relative arrangement of C-5 and C-3 hydrogens. No signal enhancement was observed neither between C-9 and C-3, nor between C-9 and C-5 hydrogens. These NOE results were in favor of the epimer 7b for the major indolizidine obtained (Scheme 3). Moreover a second isolated diastereomer[10] was shown to have a cis arrangement between the C-3, C-5 and C-9 hydrogens on the basis of NOE experiments. The formation of this stereomer could result from the reductive amination of the cis isomer of 5[8].

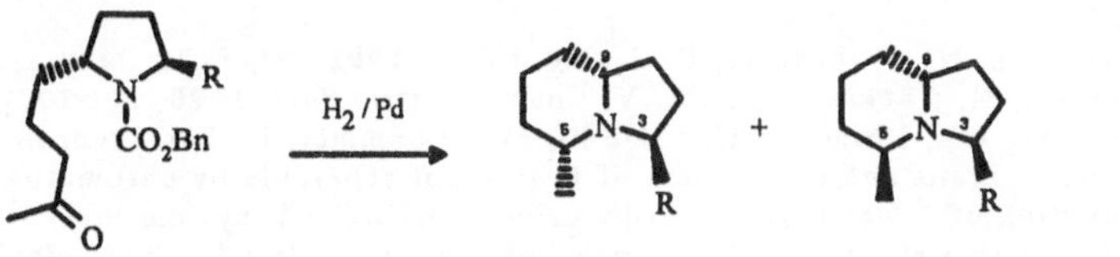

		6a (64%)	6b (36%)
4	R = CH$_2$OAc	6a (64%)	6b (36%)
5	R = CO$_2$Me	7a (see text)	7b (> 90%)
8	R = n-Bu	(-) 195B (86%)	5-epi (-)195B (14%)

Scheme 3

To check this unexpected stereochemistry of the major stereomer 7b, we decided to transform its ester moiety into a methyleneacetate group and to compare the obtained product with **6a** and **6b**. The major ester isolated from the mixture of 7 was quantitatively reduced[11] into its primary alcohol **9** which was then protected as an acetate[12] in 93% isolated yield. The acetate thus obtained was found to be identical to **6b** in all respects (Scheme 4).

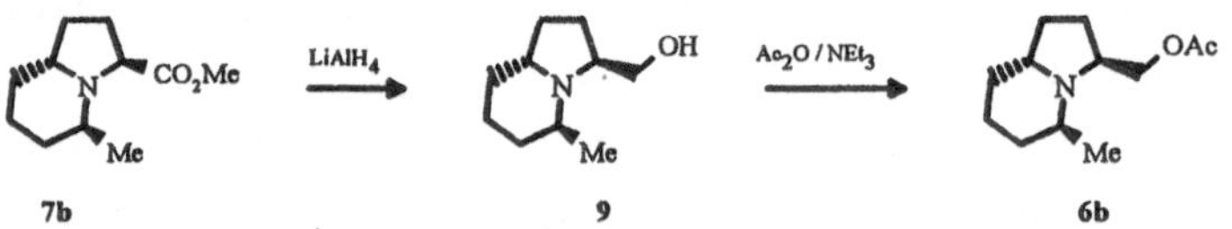

7b   **9**   **6b**

Scheme 4

This result confirmed the absolute stereochemistry of the major stereomer **7b** and hence the surprising stereoselectivity in the reductive amination of **5**. At this stage we have no serious evidences to explain this unexpected stereoselectivity.

As we see, route A gave low or reverse stereoselectivity depending on the nature of the C-3 substituent. Finally the synthesis of (-) indolizidine 195B was achieved[5] according to route B via the trans ketopyrrolidine **8** with 72% diastereoselectivity (Scheme 3).

References and notes.

1. Tokuyama, T. ; Nishimori, N. ; Karle, I. L. ; Edwards, M. W. ; Daly J. W. *Tetrahedron* **1986**, *42*, 3453.
2. Michael, J. P. *Natural Products Reports* **1990**, 485.
3. Fleurant, A. ; Célérier, J. P. ; Lhommet, G. *Tetrahedron : Asymmetry* **1993**, *4*, 1429.
4. Machinaga, N. ; Kibayashi, C. *J. Org. Chem.* **1992**, *57*, 5178. Momose, T ; Toyooka, N. ; Seki, S. ; Hirai, Y. *Chem. Pharm. Bull.* **1990**, *38*, 2072.
5. Célimène, C. ; Dhimane, H. ; LeBail, M. ; Lhommet, G. *Tetrahedron Lett.* **1994**, *35*, 6105.Trans and cis isomers of **1** were not separable by chromatography.
6. Reduction of **1** was carried out by using $NaBH_4/CaCl_2$ system in THF/EtOH (1/1). At this stage the pure trans alcohol **2** was isolated in 67% yield. Acetate **3** was prepared from **2** using acetyl chloride in pyridine (92% yield).
7. Epimers **6** were isolated by column chromatography (SiO_2, Et_2O/Pentane, 3/1).
8. The key intermediate **1** could not be separated from its cis isomer (4%) and therefore **5** was used as a trans/cis (96/4) mixture.
9. Product **7b** was purified by column chromatography (SiO_2, Et_2O/Pentane, 3/1)
10. The two others stereomers were present in too low concentration to be isolated.
11. The ester moiety reduction was performed at -78°C using $LiAlH_4$ in THF.
12. This acetate was obtained by refluxing **9** in THF in the presence of Ac_2O (1.1 equiv.) and Et_3N (4 equiv.). Otherwise Ac_2O/Pyridine or Ac_2O/DCC/DMAP left **9** unchanged. The use of AcCl instead of Ac_2O gave a complex mixture.

IMINOQUINONES AS ELECTROGENERATED ELECTROPHILIC SPECIES

Ian MARCOTTE[1], Jean Marc CHAPUZET[1], Yves DORY[2] and Jean LESSARD[1]

[1]Centre de Recherche en Électrochimie et Électrocatalyse, [2]Laboratoire de Modélisation Moléculaire

Département de Chimie, Université de Sherbrooke, Sherbrooke, Québec, Canada J1K 2R1

ABSTRACT

The electrochemical reduction of 4-, 5-, 6- and 7-nitroindoles, at Hg, in basic (KOH 0.15 M, pH > 13) and acidic (HX 0.15 M, pH = 0.3) aqueous methanol (MeOH-H$_2$O 95:5, v/v) gave substituted aminoindoles (mono and disubstituted) and/or the corresponding aminoindole in a ratio depending on the nitroindole, the pH and the nucleophile present in the medium. A mechanistic hypothesis supported by theoretical calculations is proposed.

INTRODUCTION

Studies on the electrochemical behavior of nitroindoles **1**, performed at Hg in basic (KOH 0.15 M pH > 13) and acidic (HX 0.15 M pH = 0.3) aqueous methanol (MeOH-H$_2$O 95:5, v/v), have shown that mono substituted aminoindoles **2** are the main products in the presence of good nucleophiles (RS$^-$, Br$^-$), the other compounds formed being disubstituted aminoindoles **3** and aminoindoles **4** (Scheme 1) (1).

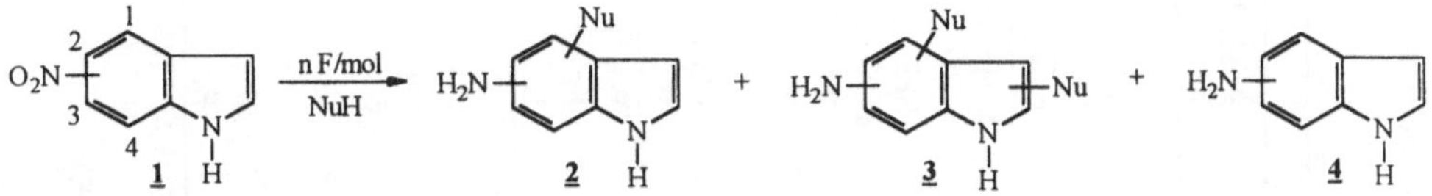

SCHEME 1

RESULTS AND DISCUSSION

Some results and the experimental conditions are presented in Table 1. They show that the ratio of monosubstituted aminoindoles **2** to aminoindole **4** varies with the nitroindole **1**, the pH and the strength of the nucleophile (1). The formation of monosubstituted (**2**) and disubstituted (**3**) aminoindoles can be explained by the mechanism presented in Scheme 2 in the particular case of the reduction of 6-nitroindole (**1**, NO$_2$ at C-6) in basic medium. The key intermediates reacting with the nucleophile are **5**, **6** and **7**. In acidic medium, a similar mechanism with protonated intermediates would be involved.

Table 1. Preparative electrolyses of 4-, 5-, 6- and 7-nitroindoles, at Hg, in acidic (pH = 0.3) and basic (pH > 13) media: the influence of nucleophiles

Substrate	Electrolyte[a]	E^b (V)	Charge (F/mol)	Monosubstituted aminoindole Nu	Position	Yield (%)	Disubstituted aminoindole Nu	Position	Yield(%)	Amine Yield (%)
6-Nitroindole	KOH	-1.24	6							53-65
	KOH/EtSH	-1.48	4.4	SEt	C-2	31	SEt	C-2/C-7	2	
				SEt	C-7	24				
	KOH/PhsH	-1.48	4.8	SPh	C-7	34				
5-Nitroindole	KOH	-1.31	5.6	OMe	C-4	43-47				30-34
	KOH/EtSH	-1.31	4.7	SEt	C-4	70				15
	KOH/PhSH	-1.31	4.7	SPh	C-4	59				
	H_2SO_4	-0.77	5	OMe	C-4	14				12
	H_2SO_4/EtSH	-0.77	5.1	OMe	C-4	15				18
				SEt	C-4	5				
	H_2SO_4/PhSH	-0.77	4.3	SPh	C-4	20				
				SPh	C-3	15				
7-Nitroindole	HBr	-0.61	4.3	Br	C-4	71-86				traces
	H_2SO_4	-0.56	4.2	OMe	C-6	71				traces
	HBr	-0.56	4.4	Br	C-6(C-4)	44-49	Br	?	traces	traces
				Br	C-4(C-6)	traces				
4-Nitroindole	KOH[c]	-1.40	5.1	SEt	C-7	59				37-44
	KOH/EtSH	-1.40	4.0	SPh	C-7	56-69				
	KOH/PhSH	-1.40	4.3	SPh	C-3	2-12	SPh	?	3-5	
				SEt	C-7	5-7	SPh	?	3-6	
	H_2SO_4/EtSH	-0.62	4.3	SEt	C-7	35-40				0-6
	H_2SO_4/PhSH	-0.62	4.0	SPh	C-7	15				
				SPh	C-3	15				
	HBr	-0.62	4.3	Br	C-7	37-47				traces

a: Solvent: MeOH-H_2O (95:5, v/v): electrolytes: KOH 0.15 M, H2SO4 0.15 M, HBr 0.15 M; added nucleophile: EtSH 48 mM, PhSH 12 mM; substrate concentration: 6 mM.

b: Reference electrode: saturated calomel electrode (SCE).

c: An azo dimer has been isolated in a 10-18 % yield.

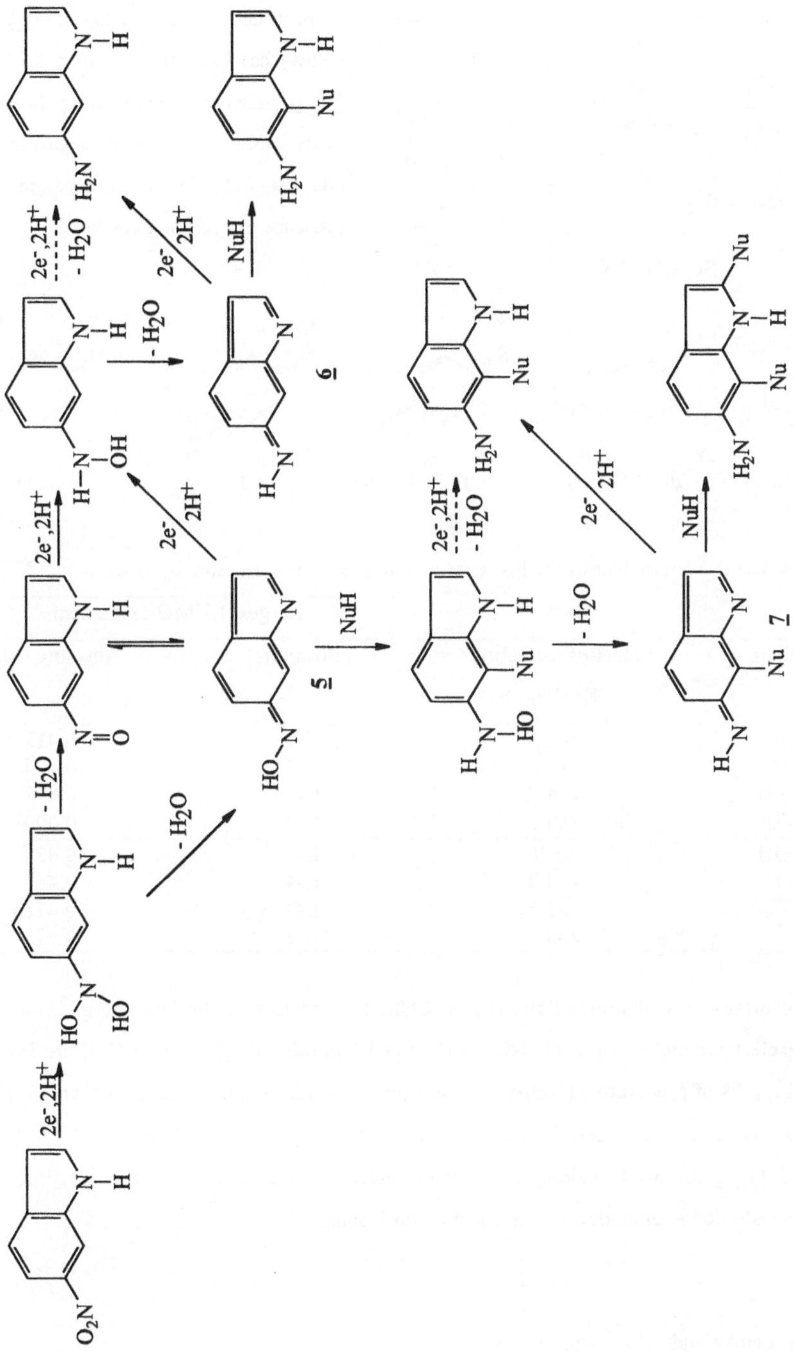

SCHEME 2

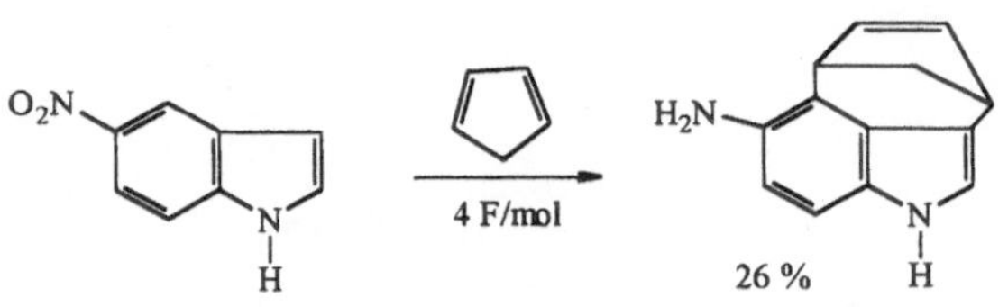

A protonated diiminoquinone (**12** or **13** below) has been trapped in a 26 % yield (see **8**, Scheme 3), via a pericyclic reaction, in the electroreduction of 5-nitroindole (**1**, NO$_2$ at C-5) in acidic medium, in the presence of cyclopentadiene.

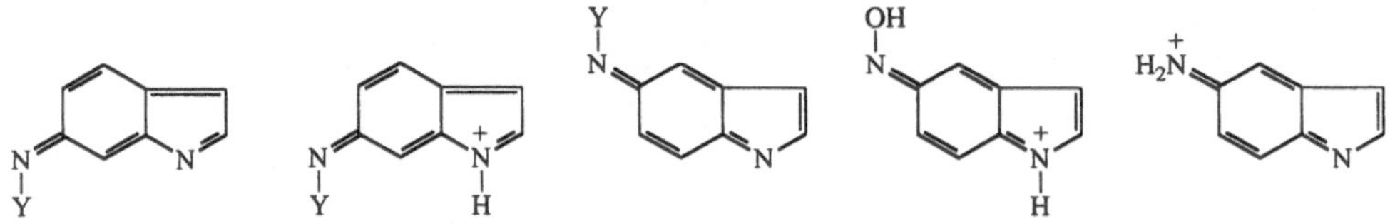

1 (NO$_2$ at C-5)   **8**

SCHEME 3

9 (Y=OH, H)   **10** (Y=OH, H)   **11** (Y=OH, H)   **12**   **13**

Table 2. Largest LUMO coefficient of the less energic intermediates calculated by AM1 within MOPAC

Intermediate	Heat of formation (Kcal/mol)	Largest LUMO coefficient	
		Position	Absolute value
9, Y=OH	94.4	C-7	0.413
9, Y=H	113.2	C-7	0.424
10, Y=OH	234.7	C-7	0.478
10, Y=H	258.7	C-7	0.500
11, Y=OH	95.9	C-4	0.426
11, Y=H	114.7	C-4	0.441
12	232.6	C-4	0.472
13	255.4	C-3	0.473

Results of theoretical calculations of the largest LUMO coefficient of the less energic intermediates involved in the electroreduction of 6- (**1**, NO$_2$ at C-6) and 5-nitroindole (**1**, NO$_2$ at C-5) are gathered in Table 2. In all results of preparative electrolyses with these two nitroindoles (Table 1) except for one (6-nitroindole, KOH/EtSH), the largest LUMO coefficient of the presumed intermediates (**9**, **10** for 6-nitroindole and **11**-**13** for 5-nitroindole) corresponds indeed to the position of the substituent in the isolated monosubstituted aminoindoles (major and minor isomers).

REFERENCE

I. Marcotte, A. Lemire and J. Lessard, *This issue*.

THE MECHANISM OF THE C-S BOND CLEAVAGE IN ELECTRON TRANSFER REACTIONS OF α-ORGANOTHIO-SUBSTITUTED HETEROATOM COMPOUNDS

Jun-ichi YOSHIDA, Masanobu SUGAWARA, and Masayuki IMAGAWA

Department of Synthetic Chemistry & Biological Chemistry, Graduate School of Engineering, Kyoto University, Kyoto 606-01, Japan

Electron transfer reactions of α-organothio-substituted heteroatom compounds lead to the cleavage of the C-S bond and the introduction of nucleophiles to the carbon. For example, It has been reported that the oxidation of α-organothio ethers in acetic acid results in the cleavage of the carbon-sulfur bond and the introduction of the acetoxyl group onto the carbon.[1] The anodic oxidation of α-organothio ethers in methanol also gives the dimethyl acetal.[2] In the anodic oxidation of benzyl sulfides, the C-S bond is cleaved and a nucleophile is introduced onto the carbon.[3] Photoelectron transfer is also effective to cleave the carbon-sulfur bond.[4]

Recently we have developed the effective inter- and intramoleclar C-C bond formation by the anodic oxidation of α-organothio ethers[5a] and carbamates.[5b] The C-S bond is cleaved and the carbon nucleophiles such as allyltrimethylsilanes and carbon-carbon double bonds attack the carbon to give the substitution product.

Although the synthetic utility of such reactions has been studied rather extensively, the mechanisms of action of organothio groups in the anodic oxidtion of heteroatom compounds have not been fully clarified as yet. This paper describes how the C-S bond is cleaved in the electron transfer reactions of organothio-substituted heteroatom compounds.

Structure of the Cation Radicals

The molecular orbital calculations indicate that the HOMO of the neutral α-organothio ethers is the nonbonding p orbital of the sulfur. This is consistent with the fact that the oxidation potentials of α-organothio ethers are similar to those of the corresponding sulfides. Therefore, there is no orbital interaction between the sulfur and the oxygen. As a matter of fact, the HOMO energy of the neutral $HSCH_2OH$ obtained by the *ab initio* calculations does not vary with the torsion angle of S-C-O-H.

We also carried out the molecular orbital calculations of the cation radical of $HSCH_2OH$. In this case the total energy of the cation radical varies with the torsion angle of S-C-O-H, and becomes the minimum when the torsion angle is about 90 degrees. This result suggests that there is some interaction between the C-S σ orbital and the p orbital of the oxygen. As a matter of fact, the C-S bond is significantly longer than that in the neutral molecule and the C-O bond in the cation radical is shorter than that in the neutral molecule. This means that the C-S bond in the cation radical is weakened and is ready to be cleaved, and the partial double bond character is developing between the carbon and the oxygen.

The Mode of the Bond Cleavage

The next question is how the C-S bond in the cation radical is cleaved. In principle, there are two possibilities for the modes of the cleavage. The first one produces the carbon radical and sulfur cation. The second one produces the carbocation and sulfur radical.

In order to determine the mode of the C-S bond cleavage, the product analysis of the anodic oxidation of α-phenylthio substituted heteroatom compounds was carried out. Anodic oxidation reactions of phenylthio substituted heteroatom compounds (1) were carried out in Bu_4NClO_4 / CH_2Cl_2 in the presence of alcohols (2) such as menthol and methanol. The reaction gave the acetal (3) and diphenyl disulfide (4). The formation of 4 indicates that the cleavage of the C-S bond in the cation radical intermediate leads to the carbocation and the organothio radical that dimerizes to produce the diorgano disulfide.

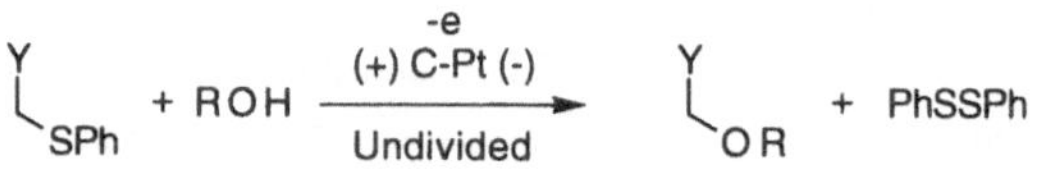

Y	R	electricity (F/mol)	product(%) 3	product(%) 4	unchanged 1 (%)
MeO	Menthyl	1.0	35	41	50
MeO	Menthyl	2.0	91	87	13
Me(MeO$_2$C)N	Me	2.9	90	54	0

According to the molecular orbital calculations, the formation of the organothio radical and the carbocation stabilized by the adjacent oxygen atom is more favorable than the formation of the organothio cation and the carbon radical.

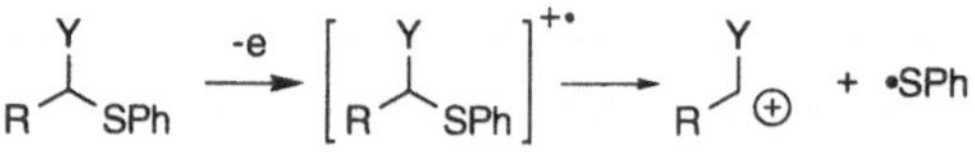

References

1) T. Mandai, H. Irie, M. Kawada, J. Otera, *Tetrahedron Lett.*, **25**, 2371 (1984).

2) J. Yoshida, S. Isoe, *Chem. Lett.*, 631 (1987).

3) M. Kimura, Y. Yamamoto, Y. Sawaki, *Denki Kagaku*, **61**, 862 (1993).

4) I. Saito, M. Takayama, T. Sakurai, *J. Am. Chem. Soc.*, **116**, 2653 (1994).

5) (a) J. Yoshida, M. Sugawara, N. Kise, *Tetrahedron Lett.*, **37**, 3157 (1996). (b) M. Sugawara, K. Mori, J. Yoshida, *Electrochimica Acta*, **42**, 1995 (1997).

MODIFICATION OF CEPHALOSPORINS BEARING SUBSTITUENTS AT C(3')-POSITION VIA ELECTROGENERATED RADICAL-CATION OF SULFIDE

Hideo TANAKA, Yoshihisa TOKUMARU, Ken-ichi FUKUI, Sigeru TORII*, Christian AMATORE [†], and Anny JUTAND [†]

Department of Applied Chemistry, Faculty of Engineering, Okayama University, Okayama 700, Japan

[†] Ecole Normal Superieure, Labotatorire de Chimie, URA CNRS 1679
24 rue Lhomond, 75231 Paris Cedex 05, France

ABSTRACT

Chemoselective electrooxidations of 3-(3,5-Di *tert*-butyl-4-methoxyphenylsulfenylmethyl)-Δ^3-cephem **1a** was successfully achieved to afford 3-methoxymethyl-Δ^3-cephem **2**. On the other hand, electrooxidations of 3-(3,5-dimethyl-4-methoxyphenylsulfenylmethyl)-Δ^3-cephem **1b** was afforded 3-dimethoxymethyl-Δ^3-cephem **3**.

Electrogeneration and reaction of radical-cation of sulfide have been intensively investigated, and the reactions offer various kind of applications in organic synthesis. On the other hand, substitution at C(3)-methyl group of cephalosporins is an essential step for the synthesis of useful cephalosporin antibiotics. A number of cephalosporins have been prepared by modification of the 3'-substituents of 3-acetoxymethyl-Δ^3-cephem (7-ACA) and 3-halomethyl-Δ^3-cephems with nucleophiles. Introduction of oxygen substituents at the C(3')-position, however, has been hardly done owing to their low nucleophilic nature. As a new device for introduction of the oxygen substituents at the C(3')-position of the cephalosporin, we investigated a two-step transformation of 3-halomethyl-Δ^3-cephem, consisting of replacement of the halogen atom with arylsulfenyl groups and subsequent electrooxidation of the sulfide moiety resulting in C-S bond fission followed by nucleophilic substitution. Herein, we describe chemoselective electrooxidations of 3-arylsulfenylmethy-Δ^3-cephems leading to either 3-methoxymethylcephem **2** or 3-dimethoxymethyl-Δ^3-cephem **3** (Scheme 1).

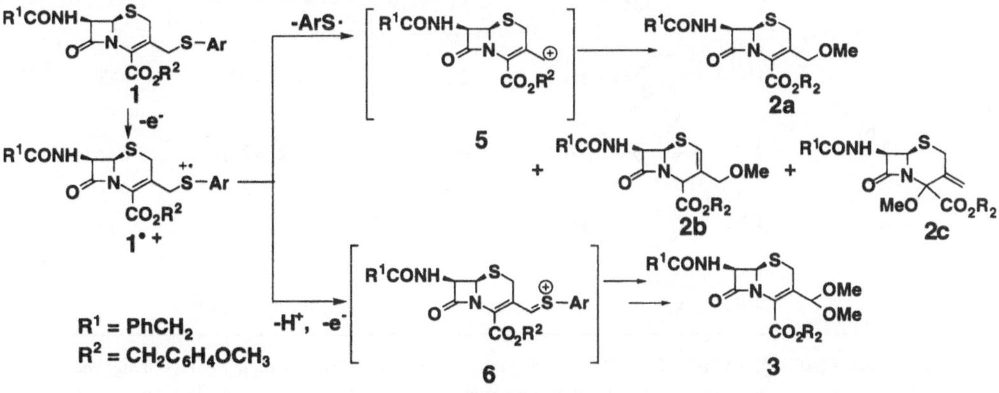

Scheme 1

3-(3,5-Di-*tert*-butyl-4-methoxyphenylsulfenylmethyl)-Δ^3-cephem **1a**, 3-(3,5-dimethyl-4-methoxyphenylsulfenylmethyl)-Δ^3-cephem **1b**, 3-(2,6-dimethyl-4-methoxyphenylsulfenylmethyl)-Δ^3-cephem **1c**, 3-(4-methoxyphenylsulfenylmethyl)-Δ^3-cephem **1d**, and 3-phenylsulfenylmethyl-Δ^3-cephem **1e** were prepared successfully by the reaction of 3-chloromethyl-Δ^3-cephem with the corresponding thiophenols / $KHCO_3$ / aq acetone. Electrolysis of the cephem **1a** in a CH_2Cl_2 / MeOH mixed solvent containing Bu_4NBF_4, $MgSO_4$ and 2,2,2-trifluoroethanol at -5 °C afforded methoxylated cephems **2** (82%; **2a/2b/2c**: 72/3/25) together with a small amount of dimethylacetal **3** (5%) (Table 1, entry 1). The product selectivity of the electrolysis was highly dependent on the substituents on the arylsulfenyl moiety. Electrolysis of the cephems **1b** and **1d** afforded 3-(dimethoxymethyl)-Δ^3-cephem **3** (48%, 32%) as a major product (entries 2, 4), while electrolysis of cephem **1c** afforded **2** (62%), predominantly (entry 3). The 4-methoxy substituent on the arylsulfenyl moieties is essential for chemoselective electrooxidation of the sulfenyl moiety. Indeed, electrolysis of phenylsulfenyl-substituted cephem **1e** afforded only a complex mixture of decomposition products (entry 5).

The results, discribed above, suggest that both steric and electronic effects of the substituents on the arylsulfenyl moiety play an important role in the chemoselective and product-selective electrooxidation of **1**. Although the mechanism is not clear at present, it is likely that chemoselective electrooxidation occurs at the C(3')-sulfenyl moieties leading to radical cations **1**$^{\cdot+}$ which are subsequently converted to **2** or **3** through **5** or **6**.

Table 1. Electrooxidation of 3-Arylsulthenylmethyl-Δ3-Cephems **1**[a]

Entry	Substrate Ar		Yield, %		
			Methoxycephems	(2a/2b/2c)	Acetal 3
1	**1a:**		82	(72/3/25)	5
2	**1b:**		22	(95/4/1)	48
3[b]	**1c:**		62	(72/3/25)	trace
4	**1d:**		22	(59/23/18)	32
5	**1e:**			(a complex mixture)	

[a] Carried out with **1** (200 mg) in CH_2Cl_2/CH_3OH (3.7mL/0.3 mL) contaning Bu_4NBF_4 (50 mg), $MgSO_4$ (50 mg) and CF_3CH_2OH (0.05 mL) in an undivided cell fitted with two platinum plate electrodes at a constant current density (10 mA/cm$_2$, 5 F/mol) under Ar atomos.

[b] Recovered **1c** (18%)

ELECTROLYTIC C-S CLEAVAGES GIVING RISE TO KETONES AND CARBOXYLIC ACIDS

Makoto KIMURA, Naotaka SAITOH, Hirotaka KAWAI and Yasuhiko SAWAKI

Department of Applied Chemistry, Graduate School of Engineering, Nagoya University, Chikusa, Nagoya 464-01, Japan

ABSTRACT

The electrooxidation of sulfides has been utilized to selective and mild reactions relevant to ketones and carboxylic acids through the reaction control of C-S bond cleavages based on the advantages of the electrochemical method.

INTRODUCTION

Sulfides are synthetically useful. The electrooxidation of sulfides has been developed to useful transformations in electroorganic synthesis.[1,2] The electrochemical method is advantageous because of ease in the reaction control and mildness of reaction conditions, [3] especially suitable for the carbon-sulfur bond cleavage. Here we wish to illustrate these aspects of electrooxidative C-S cleavages in a selective deprotection of thioacetals to ketones, a novel deprotection for carboxylic acids and the preparation of ketones from β-hydroxy sulfides.

ELECTROOXIDATION OF THIOACETALS[4]

Thioacetals are useful for the synthesis of aldehydes and ketones. The deprotection of S,S-acetals to carbonyl compounds has been carried out oxidatively, *e.g.*, with Hg(II) or Cu(II) salts and also electrooxidatively. However, little has been developed for the selective deprotection of various thioacetals. Examination on the electrooxidation of thioacetals led us to find several selective deprotections (1) in the presence of an acid-sensitive compound such as an O,O-acetal, (2) between two different S,S-acetals and (3) between an S,S-acetal and an S,S-ketal.

The first electrooxidative deprotection of thioacetals was easily achieved by the addition of a suitable base like 2,6-lutidine; an O,O-acetal remained untouched. The second was effected by use of the five-membered S,S-acetal, which was found relatively unreactive under conditions of a decreased water content (Eq. 1).

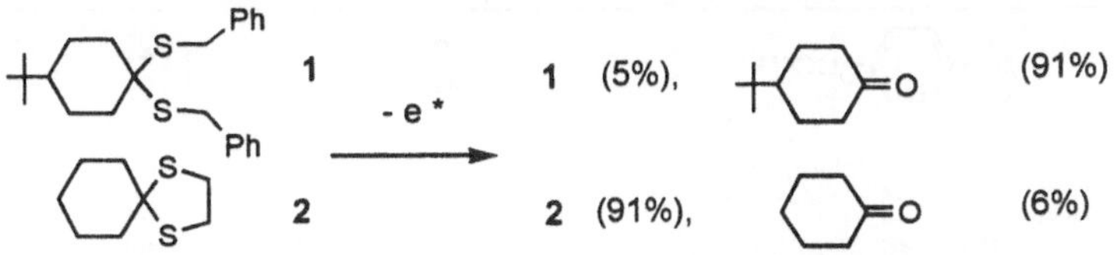

* MeCN-H₂O (0.3%), Bu₄NClO₄, 20 mA, 2.0 F/mol

The present electrolysis was also effective for the selective deprotection of a ketone S,S-acetal in the presence of an aldehyde S,S-acetal; the latter being left intact (Eq. 2).

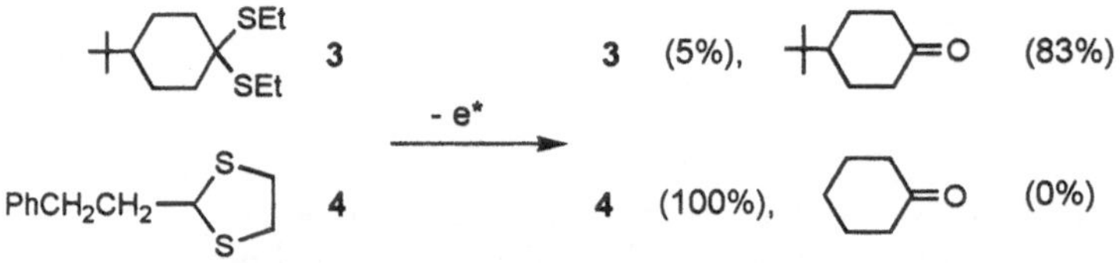

* MeCN-H$_2$O (1%), Bu$_4$NClO$_4$, 20 mA, 2.0 F/mol

ELECTROLYTIC DEPROTECTION FOR CARBOXYLIC ACIDS

A large number of esters such as methyl, ethyl and their substituted derivatives have been explored for the protective group for carboxylic acids.[5] Methylthio-ethyl esters are known to undergo the oxidative deprotection with H$_2$O$_2$ and molybdenum oxide catalyst, followed by treatment with dilute alkali. Previously we reported the electrooxidative deprotection of the arylthiomethyl esters to carboxylic acids.[6] Since t-butyl esters are commonly used as an alkaline-resistent protective group, we became interested in alkylthio substituted t-butyl esters and their electrooxidative deprotection.

Electrolyses of alkylthio substituted esters **5** in aqueous acetonitrile afforded the expected p-toluic acid; good yields of the carboxylic acid were obtained when an electrolyte involving chloride ion was used (Eq. 3).

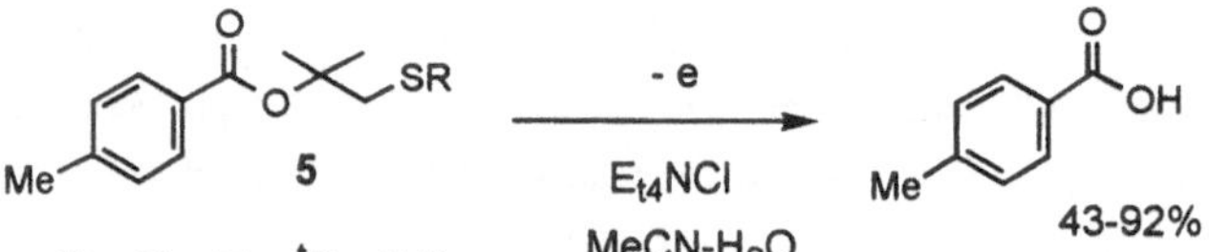

Table 1. Electrooxidative deprotection of t-butylthio substituted t-butyl esters to carboxylic acids [a]

Carboxylic acid	F/mol	Yield, %
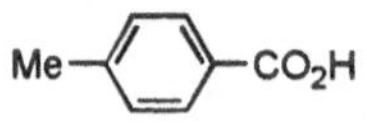 Me—⬡—CO$_2$H	8 10 [b]	92 72
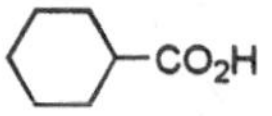 ⬡—CO$_2$H	8	80
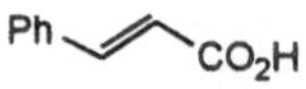 Ph~CO$_2$H	8	73
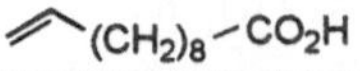 ~(CH$_2$)$_8$—CO$_2$H	6	20

a) Esters 0.5 mmol, MeCN-H$_2$O (9 : 1) 25 mL, Et$_4$NCl 2 mmol, Pt-Pt electrodes, undivided cell, 50 mA. b) Added 5 equiv. of 2,6-lutidine.

A simple *t*-butyl ester remained untouched in the mixed electrolysis with **5** (R = ^{t}Bu). Several esters prepared from acid chlorides and the *tert*-alcohol were electrolyzed under the conditions (Table 1). The deprotection of Eq. 3 using an added base like 2,6-lutidine proceeded with only slight decrease in the yield. Yields of carboxylic acid are good with the cases for aliphatic and aromatic carboxylic acids except for an acid having a terminal olefin. Thus, the present electrooxidation can provide a selective and mild deprotection for carboxylic acids.

ELECTROOXIDATION OF β-HYDROXY SULFIDES TO KETONES

The electrooxidation of alcohols carring a β-methoxy or β-diethylamino functional group has been known to give ketones via an efficient C-C cleavage in methanol.[7] 2-Phenylthio-cyclohexanols are also reported to undergo the ring opening to ketones or their acetals upon electrolysis in methanol.[8] Our interest in these cation radical induced C-C cleavages prompted us to extend the electrooxidation of β-hydroxy sulfides (**6**) to the preparation of ketones (Eq. 4).

$$
\begin{array}{c}
\underset{\substack{|\ \ \ | \\ \text{HO} \quad \text{SR}}}{\overset{\substack{R^2 \ \ R^3 \\ |\ \ \ |}}{R^1\text{-C}\text{-C}\text{-R}^4}} \quad \xrightarrow[\substack{\text{MeOH} \\ \text{or MeCN-H}_2\text{O}}]{\ -\text{e}\ } \quad \underset{\overset{\|}{\text{O}}}{\overset{R^2}{\underset{|}{R^1\text{-C}}}} \ + \ \underset{\overset{\|}{\text{O}}}{\overset{R^3}{\underset{|}{\text{C}\text{-R}^4}}} \qquad (4)
\end{array}
$$

6

Table 2. Electrooxidative preparation of ketones from β-hydroxy sulfides (**6**)

6	Conditions	F/mol	Product	Yield, %
Ph–C(Bu)(HO)–C(Me)(SEt)–Me	MeCN-H$_2$O (9 : 1)	4.5	Ph–C(=O)–Bu	81
^{i}Bu–C(^{i}Bu)(HO)–CH$_2$(SEt)	MeOH	9	^{i}Bu–C(=O)–^{i}Bu	64
Ph–CH(HO)–C(R^1)(SEt)–R^2 (R^1, R^2 = Me)	MeOH	7	Ph–C(OMe)(OMe)–H	76
H$_2$C–C(Me)(HO)(SEt)–C$_6$H$_4$–OMe	MeOH	2	MeO–C$_6$H$_4$–C(OMe)(OMe)–H	69
Me–CH(HO)–C(Bu)(SEt)–Bu	MeOH *)	8	Bu–C(OMe)(OMe)–Bu	82

6 1 mmol, Et$_4$NOTs 2 mmol, solvent 25 mL, Pt-Pt electrodes, 50 mA. *) 2,6-lutidine added.

Though the electrolysis of **6** can give rise to two types of ketones, it is possible to obtain an either ketone if the substituents R^1-R^4 are properly chosen to give a volatile ketone like acetone. In fact, electrolyses of several **6** in methanol or aqueous acetonitrile allowed us to isolate single ketones or their dimethyl acetals (Table 2). The moiety of *tert*-alcohol in **6** tends to give ketones, whereas the sulfide part usually do dimethyl acetals in methanol solvent. The results on the substituents R^1-R^4 suggests that at least two alkyl groups with R^1 and R^2 or R^3 and R^4 should render **6** to undergo an efficient C-C cleavage including C-S cleavage, otherwise would cause a trivial sulfoxide formation.

References
1) S. Torii, "*Electroorganic Syntheses*", Kodansha, Tokyo, 1985, Chapter 6; B. Svensmark and O. Hammerich, "*Organic Electrochemistry*," Ed H. Lund and M. M. Baizer, Marcel Dekker, New York, **1991**. Chapter 17.
2) Recent examples, A. W. Erian, A. Konno, and T. Fuchigami, *J. Org. Chem.*, **1995**, *60*, 7654; M. Kimura, K. Kobayashi, Y. Yamamoto, and Y. Sawaki, *Tetrahedron*, **1996**, *52*, 4303.
3) Revs. for electrochemical deprotection, V. G. Mairanovsky, *Angew. Chem. Int. Ed. Engl.*, **1976**, *15*, 281; M. I. Montenegro, *Electrochim. Acta*, **1986**, *31*, 607.
4) M. Kimura, H. Kawai, and Y. Sawaki, *Electrochim. Acta*, **1997**, *42*, 497.
5) T. W. Green and P. G. M. Wuts, "*Protective Groups in Organic Synthesis*." 2nd Ed., John-Wiley, New York, 1991.
6) M. Kimura, S. Matsubara, Y. Yamamoto, and Y. Sawaki, *Tetrahedron*, **1991**, *47*, 867.
7) T. Shono, H. Hamaguchi, Y. Matsumura, and K. Yoshida, *Tetrahedron, Lett.*, **1977**, 3625.
8) S. Torii, T. Inokuchi, and Araki, *Synth. Commun.*, **1987**, *17*, 1797.

RING OPENING REACTION OF PHENYLTHIO-CYCLOPROPANES BY ANODIC OR PHOTOCHEMICAL ONE-ELECTRON OXIDATION

Makoto KIMURA, Hirotaka KAWAI and Yasuhiko SAWAKI
Department of Applied Chemistry, Graduate School of Engineering,
Nagoya University, Chikusa, Nagoya 464-01, Japan

ABSTRACT

The ring opening of several phenylthiocyclopropane cation radicals generated electrochemically or photochemically has been studied with the product analysis and laser flash photolysis.

INTRODUCTION

The ring opening reaction of cyclopropane derivatives has recently been studied on the cation radical induced reactions.[1-3] Sulfieds can readily give their cation radicals upon one-electron oxidation.[4] We became interested in mono- and bis-phenylthio substituted cyclopropanes (**1** and **2**, respectively) and their cation radicals, and studied the electrooxidation of **1** and **2** in methanol solvent. Furthermore, photochemical reactions and laser flash photolyses of **2** were carried out in order to detect their cation radical species.

RESULTS AND DISCUSSION

Constant-current electrolysis of monothio **1** in MeCN-MeOH (1:1) generally resulted in the formation of the corresponding sulfoxides. In contrast, 1,1-bis(phenylthio)cyclopropanes **2** underwent the electrtrooxidative ring opening reaction to give methanol addition products (Table). Though the electrooxidation of thioacetals effects a selective deprotection to carbonyl compounds,[5] no product for cyclopropanones or their acetals was detected. Products in Table show that the electrooxidative ring opeining of **2** takes place at the position of C(1); the acid-catalyzed reaction of 1,1-bis(methylthio)cyclopropanes is reported to give products ring-opened at C(2) and C(3).[6]

Irradiation of **2** in MeCN-MeOH containing 1,4-dicyanonaphthalene (DCN) as a one-electron transfer sensitizer gave methanol addition products ring-opened similarly at C(1) position; *e.g.*,

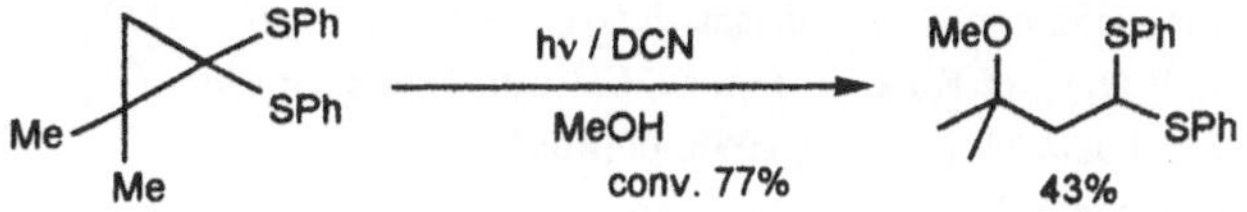

Laser flash photolyses with nano-second pulse of **2** in acetonitrile allowed us to detect the incipient intermediates indicative of two-center three-electron bonded sulfide cation radicals.[7] For examle, 2-methyl

Table   Electrolysis of bis(phenylthio)cyclopropanes **2** in methanol[a]

2	F/mol	Product (yield/%)	
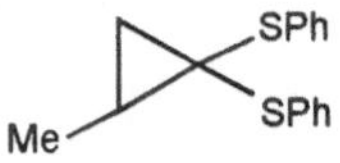	4	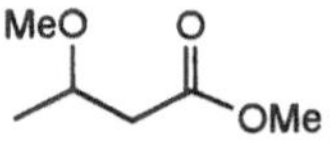 (36)	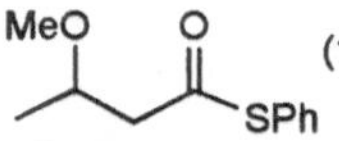 (15)
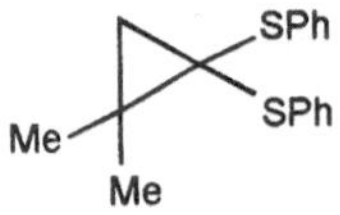	3	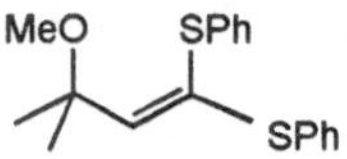 (44)	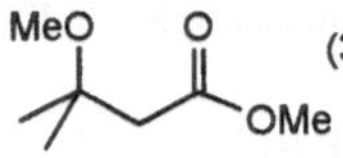 (33)
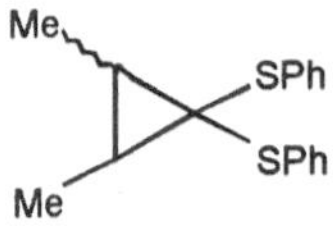 b)	6	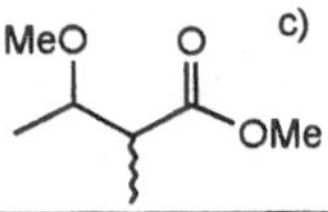 c) (72)	

a) Substrate **2** 1 mmol, MeCN-MeOH (1 : 1) 25 mL, Bu$_4$NClO$_4$ 2 mmol, undivided cell, 20 mA.   b) A mixture of *cis* and *trans* isomers (3 : 1).   c) An *erythro* and *threo* (2 : 3) mixture.

substituted **2** showed absorption maxima at 520 and 840 nm.   When varying amounts of methanol were added to the acetonitrile solution, the transient absorption gradually reached a maximum at 440 nm, which was assigned as a cyclopropane cation radical as depicted in the following equation.   Kinetic analyses of these absorptions suggested the occurrence of an equilibrium between two distinct cation radicals. The rate constants for the ring opening of **2** (k_1) can be estimated.   It is shown that the present ring opening should be greatly accelerated by the substitution of the methyl group in **2**.

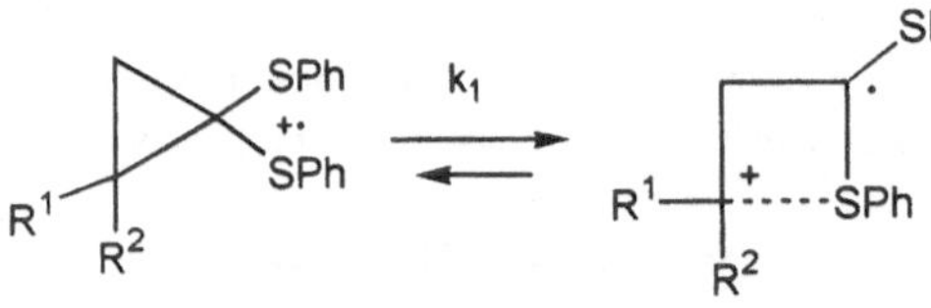

R^1, R^2	H,H	H,Me	Me,Me
k_1/sec	$<10^4$	2.5×10^5	$>10^7$

References
1) S. Torii, T. Inokuchi, and N. Takahashi, *J. Org. Chem.*, **1978**, *43*, 5020.
2) Y. Takemoto, T. Ohra, H. Koike, S. Furuse, C. Iwata, and H. Ohishi, *J.Org.Chem.*, **1994**, *59*, 4727.
3) A. Abe and A. Oku, *J. Org. Chem.*, **1995**, *60*, 3065.
4) M. Kimura, K. Kobayashi, Y. Yamamoto, and Y. Sawaki, *Tetrahedron*, **1996**, *52*, 4303.
5) M. Kimura, H. Kawai, and Y. Sawaki, *Electrochim. Acta*, **1997**, *42*, 497.
6) M. Braun and D. Seebach, *Chem. Ber.*, **1976**, *109*, 669.
7) S. A. Chaudhri, H. Mohan, E. Anklam, and K.-D. Asmus, *J. Chem. Soc., Perkin 2*, **1996**, 383.

ADDITION OF ELECTROCHEMICALLY GENERATED CHALCOGENOLATE ANIONS AND/OR RADICALS ON THE ACTIVATED DOUBLE BOND

JIŘÍ LUDVÍK [a], DAVID DVOŘÁK [a,b], EVA NEUGEBAUEROVÁ [b] and FRANTIŠEK LIŠKA [b]

[a] J. Heyrovský Institute of Physical Chemistry, Academy of Sciences of the Czech Republic, Dolejškova 3, 182 23 Prague 8, Czech Republic
[b] Department of Organic Chemistry, Prague Institute of Chemical Technology, Technická 5, 166 28 Prague 6, Czech Republic

INTRODUCTION

The electrochemical reduction of aromatic dichalcogenides (ArEEAr, where E=S,Se,Te) in aprotic media proceeds bielectronically yielding corresponding chalcogenolates ArE^- in the overall reaction

$$Ar\text{-}E\text{-}E\text{-}Ar \xrightarrow{+2e^-} 2\ Ar\text{-}E^-$$

The more detailed comparison of analogous S-, Se- and Te- derivatives[1] shows differences in the reduction mechanism and in the reactivity in electrosynthesis, especially when mercury electrodes are used. Disulfides do not react spontaneously with mercury metal and their reduction proceeds irreversibly following an ECE mechanism where the existence of a radical intermediate is assumed[2].

$$Ar\text{-}S\text{-}S\text{-}Ar \xrightarrow{+e^-} Ar\text{-}S\text{-}S\text{-}Ar^{-\cdot} \longrightarrow Ar\text{-}S^- + Ar\text{-}\dot{S} \xrightarrow{+e^-} 2\ Ar\text{-}S^-$$

On the other hand, diselenides (and ditellurides) in contact with mercury form spontaneously complexes of the type Ar-Se-Hg-Se-Ar[3] which then get reduced in two reversible steps:

Anteceding reaction: $Ar\text{-}Se\text{-}Se\text{-}Ar + Hg \longrightarrow Ar\text{-}Se\text{-}Hg\text{-}Se\text{-}Ar$

1st wave: $Ar\text{-}Se\text{-}Hg\text{-}Se\text{-}Ar + 2e^- \longrightarrow 2\ Ar\text{-}Se^- + Hg$
$2\ Ar\text{-}Se^- + 2\ Ar\text{-}Se\text{-}Hg\text{-}Se\text{-}Ar \longrightarrow 2\ [Hg\,(Ar\text{-}Se)_3]^-$

2nd wave: $2\ [Hg\,(Ar\text{-}Se)_3]^- + 4e^- \longrightarrow 6\ Ar\text{-}Se^- + 2\ Hg$

The above mentioned cathodically generated S- and Se- precursors (Ar = phenyl) were used for addition reactions on the double bonds of unsaturated organic molecules. As substrates we used chlortrifluoroethene, 1,2-dichlorodifluoroethene and methyl-2,5-dihydro-2,5-dimethoxy-3-furancarboxylate. Besides diphenyldichalcogenides, a cyclic diselenide was also used. Reactions proceeded in non-aqueous acetonitrile on the mercury pool electrode. Products were characterized by IR and ^1H, ^{19}F and ^{77}Se NMR spectra.

ADDITION REACTIONS

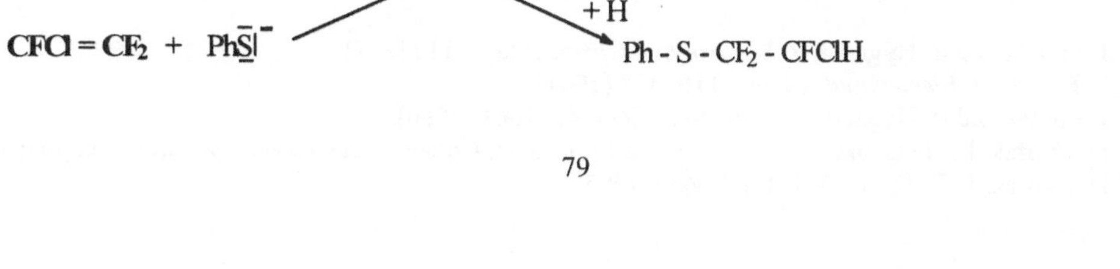

$$CFCl = CF_2 + PhSe^- \longrightarrow Ph-Se-CF_2-\bar{C}FCl^-$$

with branches:
- $+ PhSeSePh \longrightarrow Ph\,Se-CF_2-CFCl-Se\,Ph$
- $- PhSe^- $
- $+ H^+ \longrightarrow Ph-Se-CF_2-CFClH$

$$CFCl = CFCl + PhS^\cdot \xrightarrow{+e^-} Ph-S-CFCl-\bar{C}FCl^-$$

with branches to:
- $Ph-S\diagup C=C \diagdown$ with F, F, Cl ($- Cl^-$)
- $Ph-S \diagup C=C \diagdown$ with Cl, F, F

$$CFCl = CFCl + PhS^- (PhSe^-) \nrightarrow$$

$$2\,CFCl = CF_2 + \begin{array}{c} {}^-Se-CH_2 \\ {}^-Se-CH_2 \end{array} C \begin{array}{c} CH_2OH \\ CH_2OH \end{array} \xrightarrow{+2H^+} \begin{array}{c} HClFC\cdot CF_2-Se-CH_2 \\ HClFC\cdot CF_2-Se-CH_2 \end{array} C \begin{array}{c} CH_2OH \\ CH_2OH \end{array}$$

Furancarboxylate scheme:

$$\text{(CH}_3\text{O, O, OCH}_3\text{ furan with COOCH}_3) + Ph-Se^- \longrightarrow \text{(Ph-Se, COOCH}_3, \ominus, CH_3O, O, OCH_3) \xrightarrow{+H^+} \text{(Ph-Se, COOCH}_3, CH_3O, O, OCH_3)$$

CONCLUSIONS

From the mechanistic point of view, the presented electrosynthetic study confirmed, that during electroreduction disulfides form radical (ArS$^\cdot$) as well as anionic (ArS$^-$) electrochemically generated intermediates, whereas diselenides form only anions ArSe$^-$. This evidence is based on different reactivity of CFCl=CF$_2$ and CFCl=CFCl towards nucleophilic and radical intermediates: disulfides being reduced react both with 1,2-dichlordifluorethene and chlortrifluorethene, reduced diselenides are added only as nucleophiles on the asymmetric substrate.

The latter reactions represent a new synthetic route to aryl(fluorethyl)selenides and -sulfides [4]. The addition proceeds even with reduced cyclic diselenides. Another unsaturated substrate used for addition reactions was methyl-2,5-dihydro-2,5-dimethoxy-3-furancarboxylate. By addition of ArSe$^-$ it is possible to introduce the selenium atom in position 4 of the β-substituted tetrahydrofurane ring, which is in a classical way a very difficult problem[5]. The addition reactions of cathodically generated chalcogen containing precursors on compounds bearing activated double bonds have been thus found as a suitable electrosynthetic procedure due to its selectivity and mild reaction conditions.

REFERENCES

1. J. Ludvík and B. Nygaard, *J. Electroanal. Chem.*, **423**, 1-11 (1997).
2. F. Pragst - *J. Electroanal. Chem.*, **119**, 315 (1981).
3. J. Ludvík and B. Nygaard, *Electrochim. Acta*, **41**, 1661 (1996).
4. D. Dvořák, E. Neugebauerová, F. Liška and J. Ludvík, *Collect.Czech Chem.Commun.*, - in print
5. D. Dvořák, PhD Thesis, VSCHT Prague 1995.

ANODIC OXIDATION OF BIS(2,6-DIMETHOXYPHENYL)-CHALCOGENIDES

Tatsuo ERABI, Eiichiro FUCHIOKA, Shuichi HAYASE, and Masanori WADA
Department of Materials Science, Faculty of Engineering, Tottori University, 4-101, Koyama Minami, Tottori 680, Japan

ABSTRACT

Cyclic voltammetric measurements were made for bis(2,6-dimethoxyphenyl)chalcogenides (Φ_2X, Φ = 2,6-dimethoxyphenyl, X = S, Se, Te). It is expected that the active species generated at the first oxidation wave are the corresponding cation radicals for Φ_2Te and Φ_2S, whereas it is the cation for Φ_2Se.

INTRODUCTION

Organochalcogeno compounds such as sulfides, selenides and tellurides have been used for organic synthesis. Especially in the field of electrochemistry, the anodic oxidation of these compounds has been attempted to generate corresponding cations or radicals and the possibility of regenerative use of the cations or radicals has been suggested for the hydroxylation of olefines.[1] However, it becomes clear that they are, unfortunately, highly ill-smelling and great care must be taken in handling them. During the course of our study on 2,6-dimethoxyphenyl derivatives, we have attempted the preparation of the chalcogenides, and found[2] that these compounds are air-stable crystalline with no appreciable odor. So, we have studied anodic oxidation of bis(2,6-dimethoxyphenyl)dichalcogenides (Φ_2X_2, Φ = 2,6-dimethoxyphenyl, X = S, Se, Te) at a Pt electrode in acetonitrile in order to generate an active species being useful in organic synthesis, and found[3] that the primary step of the oxidation proceeds with one-electron transfer from chalcogen atom to generate Φ_2X_2 cation radicals and the active species generated at the step is 2,6-dimethoxyphenylchalcogeno cation, ΦX^+. But the regenerative use of the cation has been failed in acetonitrile solution because of the dechalcogenation to generate bis(2,6-dimethoxyphenyl)chalcogenides (Φ_2X) and chalcogen atom. The present paper deals with cyclic voltammetric behavior of Φ_2X in nitromethane, prior to the regenerative use of corresponding cation radicals.

RESULTS AND DISCUSSION

Cyclic voltammetric measurements were carried out in a conventional three electrodes cell with platinum wire working electrode, platinum plate counter electrode and aqueous Ag/AgCl reference electrode separated from the electrolytic solution with Luggin capillary and salt bridge. Figure 1 shows cyclic voltammograms of Φ_2X in nitromethane containing 0.1 M tetrabutylammonium tetrafluoroborate at a sweep rate of 5 mV s^{-1}. When the sweep started from 0 V towards the positive direction, the first anodic peaks for Φ_2Te, Φ_2Se and Φ_2S oxidations appeared at 0.95, 1.15 and 1.25 V vs. aq. Ag/AgCl, respectively. No cathodic counterpart corresponding to these peaks was found even at higher sweep rates up to 0.1 V s^{-1}. The first anodic peak currents were proportional to the concentration of Φ_2X's and to the square root of the potential sweep rate, indicating that the waves were diffusion-controlled. The number of electrons involved in the first waves was estimated to be about 1 for oxidation of Φ_2Te and Φ_2S, and about 2 for Φ_2Se, by comparing the anodic peak heights with that of ferrocene. These results were somewhat different from those for oxidation of Φ_2X_2's.[3] A linear relationship was observed between the anodic peak potential of Φ_2X_2's and the first ionization potential of these chalcogen atoms, and the anodic reaction was one-electron transfer from the chalcogen atoms to generate the corresponding cation radicals, $\Phi_2X_2^+$. On the contrary, such a linear relation could not observed for oxidation of monochalcogenides. That is, it is presumed that the active species generated at the first wave are Φ_2Te and Φ_2S cation radicals for oxidation of Φ_2Te and Φ_2S, whereas it is Φ_2Se cation for Φ_2Se oxidation.

A controlled potential electrolysis of Φ_2X's was carried out at the first oxidation potentials using a divided cell in order to generate an active species being useful in organic synthesis. The products were identified by HPLC and NMR. But 70—90% of the starting materials, monochalcogenides, were recovered in all cases, despite the passage of quantity of electricity of 2 F/mol. For the oxidation of Φ_2Te, the anolyte has gradually become orange-colored during the electrolysis. Since the absorbance at 430 nm increased with increasing quantity of electricity up to 2 F/mol for the oxidation of Φ_2Te, an oxidation product seems to be fairly stable in nitromethane. In addition, since the electrolytic solution was subjected to the post-treatment with water or methanol, followed by analysis by NMR or HPLC, it seems possible that the oxidation product reacts with water or methanol to form the starting materials. Although the other direct analyses of the electrolytic products such as GC-MS are necessary, it seems reasonable that the possibility of using these materials as an oxidation catalyst is present in the present electrolytic system.

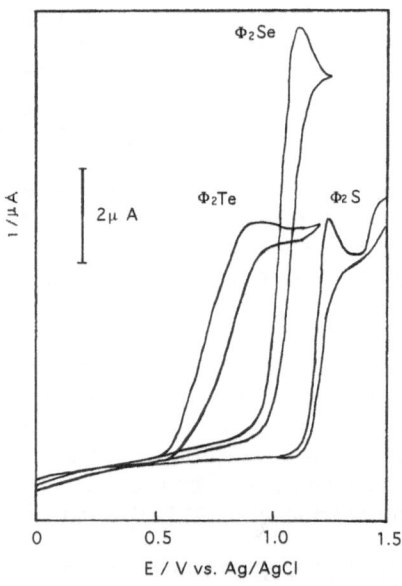

Fig. 1   Cyclic voltammograms of 1 mM Φ_2X in nitromethane.

References
1) S.Trii,Electroorganic Syntheses, Part 1, Oxidations Methods and applications, **1985**, Kodansha, Tokyo, Chaps.6 and 10.
2) (a)M.Wada, K.Tenma, K.Kajihara, K.Hirata, and T.Erabi, *Chem. Express*, **1989**, *4*, 109; (b) M.Wada, K.Kajihara, T.Morikawa, and T.Erabi, *Chem. Express*, **1991**, *6*, 875.
3) T.Erabi, K.Shimizu, S.Hayase, and M.Wada, *Denki Kagaku*, **1995**, *63*, 960.

REDOX PROPERTIES OF HYDROXYPHENYL NITRONYL NITROXIDE

Katsuya ISHIGURO, Mitsutoshi OZAKI, Nobuyuki SEKINE, and Yasuhiko SAWAKI

Department of Applied Chemistry, Graduate School of Engineering, Nagoya University, Chikusa-ku, Nagoya 464-01, Japan

ABSTRACT

Hydroxyphenyl nitronyl nitroxide, a cross-conjugated biradical, was prepared by the lead dioxide oxidation of p-hydroxyphenyl nitronyl nitroxide, and was characterized by ESR spectroscopy. The anodic or chemical oxidation did not generate the biraidcal but lead to the formation of the hydroxyphenyl substituted nitronyl nitroxide cation. Thus, a stable acid/base pair with different spin multiplicities could be prepared separately.

INTRODUCTION

Spin-crossover is one of attracting targets in organic chemistry and may be useful as a novel functional organic ferromagnet. A simple model is a cross-conjugated unsymmetrical biradical with donor and acceptor characters, and spin-crossover may be induced when the triplet diraidcal state is energetically close to the singlet zwitterionic state. Since a large difference is expected in dipole moments or an acid-base properties between the biradical and zwitter ions, the transition may bring about specific intermolecular interactions which affect the relative stabilities of low- and high-spin states.

RESULTS AND DISCUSSION

Previously, we reported that one-electron oxidation of a p-hydroxyphenyl nitronyl nitroxide lead to the generation of a closed-shell cations rather than the corresponding biradical.[1,2] The two species were not stable enough to examine such a cross-over induced by pH change. Here we report the stable biradical and cations of nitronyl nitroxide with bulky t-butyl substituents.

83

Synthesis of 3',5'-di-t-butyl-4-hydroxyphenyl nitronyl nitroxide. A phenol-
-substituted nitronyl nitroxide radical (**1**) was obtained by the m-chloro-
-peroxybenzoic acid (mCPBA) oxidation of the corresponding bishydroxylamine
prepared from 3,5-di-t-butyl-4-hydroxybenzaldehyde (Scheme 1). The ESR
spectrum of **1** showed a five-line splitting pattern with hyperfine coupling
constant (a_N) of 7.4 G (2N) as observed common 2-aryl nitronyl nitroxides.

Scheme 1.

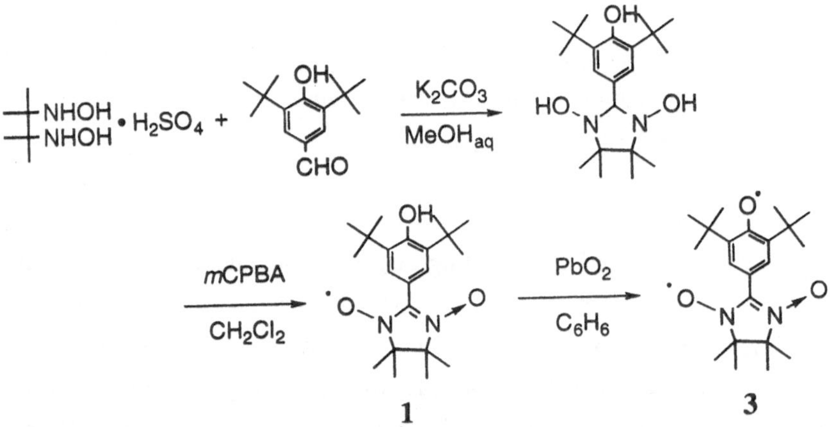

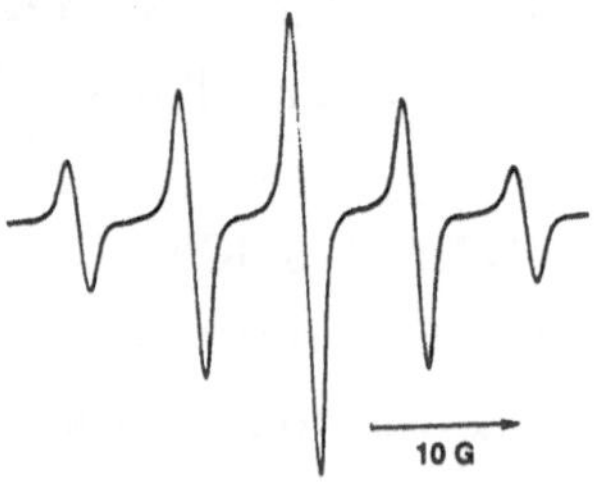

Figure 1. ESR spectrum of **1** in
toluene at room temperature

When **1** was oxidized with lead dioxide in toluene, the ESR spectrum changed
to a five-line spectrum. The resulting coupling constant $a_N = 3.7$ G (2N) is

consistent to that of biradical **3**. When the solution was frozen at 77 K, an ESR spectrum characteristic for triplet biradical species was observed.

The biradical **3** was stable over several days in solutions. By removing the solvent carefully under vacuum, **3** could be isolated as deep blue powders. the absorption spectrum of **3** in acetonitrile were observed at 318, 359, and 604 nm.

Generation of Closed-Shell Cation 2 by One-Electron Oxidation of 1. The oxidation of **1** by gamma-radiolysis in $CFCl_3$ matrix at 77 K resulted in the decrease of ESR intensity of **1** without generation of additional signals. In order to characterize the species generated, anodic oxidation of **1** was investigated by cyclic voltammetry. The voltammogram of 1 in acetonitrile showed a reversible couple at +0.71 V vs SCE (Figure 2a), which corresponds to the redox of **1** and **2**.

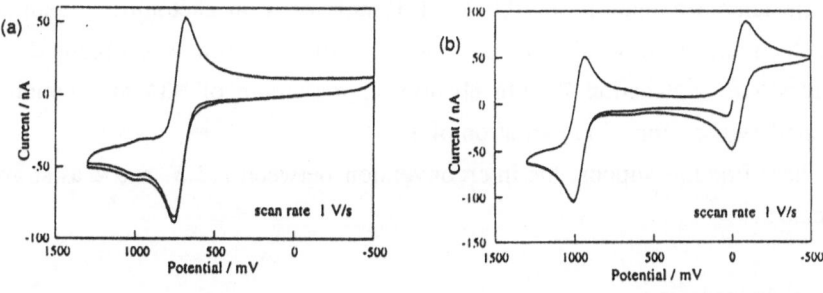

Figure 2. CV of (a) **1** and (b) **3** in Acetonitril-Benzene (4:1) in the Presence of 0.1 M n-Bu_4BF_4 (V vs SCE).

The reversible oxidation of **3** occurred at +1.00 V (Figure 2b). These results indicates that singlet cation **2**, biradical **3**, and anion radical **4** are stable during the CV analyses.

Scheme 2.

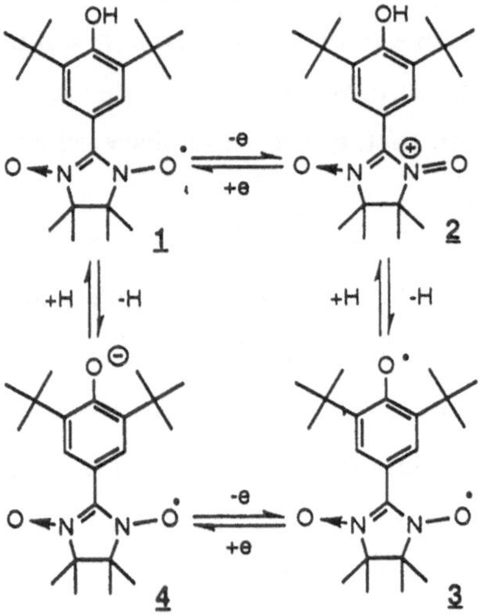

The anodic oxidation at +0.9 V (1 F/mol) of **1** in acetonitrile resulted in almost complete conversion of **1**, but no ESR spectrum was obtained. The product here was cation **2** with absorption maximum of 334 nm, which was obtained by the cupric ion oxidation of **1**.

These findings support the interconversion between **1,2,3**, and **4** as shown in Scheme 2.

Acknowledgement.

We would like to thanks Professor Tetsuo Miyazaki for ESR measurements.

References
1) K. Ishiguro, Y. Kamekura, and Y. Sawaki, *Mol. Crys. Liq. Cryst.*, **1993**, *232*, 113.
2) K. Ishiguro, M. Ozaki, N. Sekine, and Y. Sawaki, *J. Am. Chem. Soc.*, **1997**, *119*, 3625.

SYNTHESIS OF TERMINAL-BIRADICAL COMPOUNDS CONSISTING OF TWO *N*-OXYL GROUPS CONNECTED WITH CONJUGATED π-SYSTEMS

Sigeru Torii,* Tomoyuki Hase, Manabu Kuroboshi, Toshimasa Katagiri, Christian Amatore,[†] Anny Jutand,[†] and Hiroyuki Kawafuchi[†]
Department of Applied Chemistry, Faculty of Engineering, Okayama University
Tsushima-Naka 3-1-1, Okayama 700, Japan
[†]Ecole Normal Superieure, Departoment de Chimie, URA CNRS 1679
24 rue Lhomond, 75231 Paris Cedex 05, France

The design and synthesis of molecular magnets are one of the current interests. Essentials are how to increase the spin density and to regulate the direction of the spins. For latter purpose, the radicals are attempted to be joined by π-electron systems to allow to interact each other. Poly-radical compounds connecting the radicals by aromatic structure are reported. However, the overlap with aromatic ring seems not efficient because of a steric repulsion. We planed to connect radicals by other type of π-system, *e.g.*, acetylenic and vinylic moieties. Herein, we describe synthetic results on terminal-biradical compounds **1** containing two *N*-oxyl groups connected with conjugated π-systems (Scheme 1) by (1) addition of magnesium acetylide to 4-oxo-TEMPO derivatives **2** followed by dehydration (Route A), or (2) Pd-catalyzed cross-coupling between acetylene derivatives and vinyl triflate **4** derived from **2** (Route B).[1] The vinyl triflate **4** was found to be a potent intermediate to synthesize a variety of biradical compounds **1**. Electronic and magnetic properties of **1** estimated by cyclic voltammetry and ESR spectroscopy are as follows (Table 1):

Scheme 1.

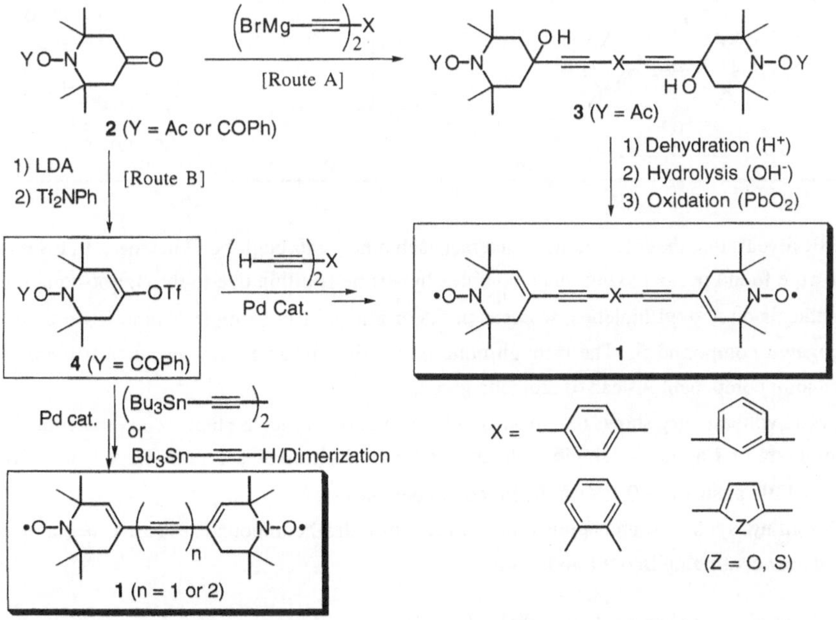

Table 1. Representative ESR spectra and Cyclic Voltammograms of Bis(N-Oxyl) Compounds

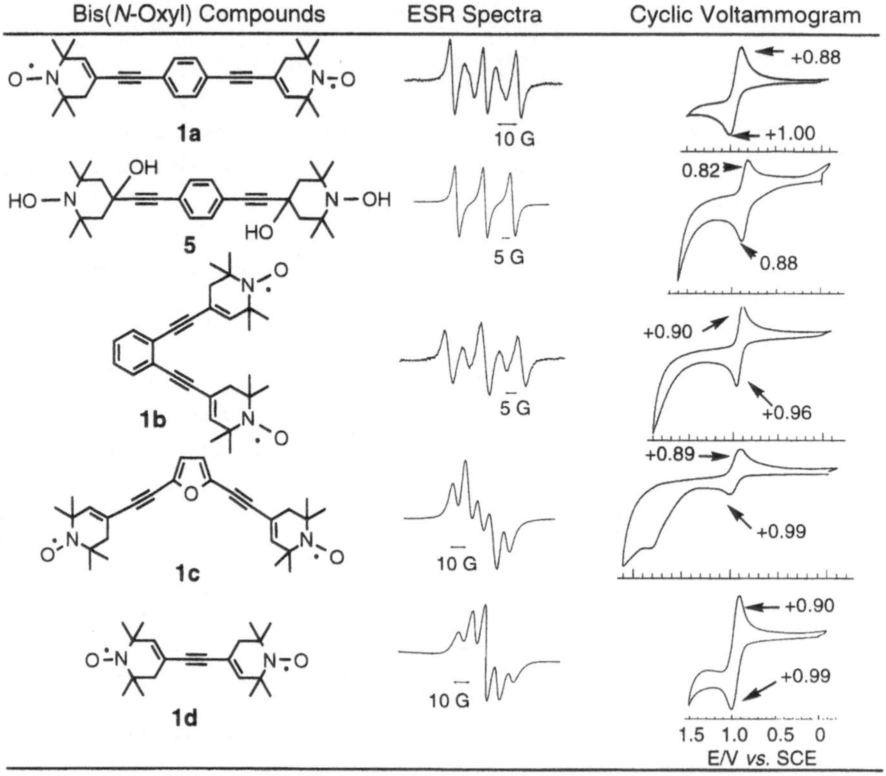

Bis(N-Oxyl) Compounds	ESR Spectra	Cyclic Voltammogram

ESR reveals that these two radicals interact each other in through-bond manner: In ESR spectra of **1a ~ 1d** were found peaks having much complex hyperfine splitting due to through-π-bond interaction of two radicals on N-oxyl moieties, whereas peaks originated from simple N-radical were observed in de-conjugated compound **5**. The through-bond interaction became stronger when triple bonds were used as a joint component instead of aromatic groups.

Cyclic voltammetry shows that these biradical compounds act as stable redox materials: In cyclic voltammogram of **1** appears a single 2-electron redox peaks around 1 V vs SCE. π-Conjugation (**5 →** **1a**) made a little positive (~ 0.1 V) shifts in redox potentials.

Designing, synthesis, and applications of these bi-radical compounds as redox mediators and spin labels are now continuing in our Laboratories.

1. S. Torii, et al., Tetrahedron Lett. **1997**, 38, 7391.

INTERFACIAL REDOX BEHAVIOR OF MONO- AND MULTILAYERS OF TETRATHIAFULVALENE TERMINATED ALKANE-TETRATHIOL ON GOLD ELECTRODE

Hidetaka NAKAI, Hisashi FUJIHARA, Masakuni YOSHIHARA, and
Toshihisa MAESHIMA
*Department of Applied Chemistry, Faculty of Science and Engineering,
Kinki University, 3-4-1, Kowakae, Higashi-Osaka 577, Japan*

ABSTRACT
    A new tetrathiol bearing tetrathiafulvalene (TTF) as a redox active group has
been synthesized.   The tetrathiol provided the monolayer by self-assembly method
and the multilayer by repeated electrochemical oxidation on gold electrode.   The
properties of mono- and multilayers with TTF-tetrathiol have been studied by cyclic
voltammetry.   It was found that the mono- and multilayers containing TTF molecule
induced by alkyl-tetrathiol were remarkably stable films.

INTRODUCTION
    There is considerable current interest in the preparation and properties of
organized mono- and multilayer thin films.   Molecular self-assembly methods are
commonly used to prepare organized molecular assemblies on surfaces, in which the
assemblies can serve as model systems for studying a variety of interfacial processes,
e.g., electron-transfer.   As typical example, self-assembled monolayers (SAMs) of
thiolates on gold surfaces are the forcus of intence investigation.   The majority of
studies has been carried out on *n*-alkane-monothiol possessing electroactive
headgroup such as ferrocene.   In contrast, much less is known about the surface
modification of gold electrode with poly-thiols.   We propose that the multiple
attachment sites of polythiols might indeed significantly enhance the stabilities of S-
tethered films.   We have now synthesized a new tetrathiol bearing tetrathiafulvalene
(TTF) as an electroactive group which provided the remarkably stable monolayer by
self-assembly method and the multilayer by repeated electrochemical oxidation.   This
paper presents the stable mono- and multilayers containing TTF molecule induced by
alkyl-tetrathiol.

1 : R = (CH₂)₃SH   2 : R = (CH₂)₃SH

RESULTS AND DISCUSSION
    The immersion of a clean gold electrode in a solution of tetrathiol **1** (1.0 mM in
CH_2Cl_2, 40 °C, 24 h) resulted in the formation of a monolayer of **1**, which
chemisorbed on the gold surface *via* thiolate-Au bonds.   After rinsing of the
electrode, the cyclic voltammogram in CH_2Cl_2 clearly showed the anodic and cathodic
peaks characteristic of the reversible two-electron oxidation of the immobilized

TTF-group (Figure 1). The oxidation potential was +0.40 V and +0.71 V (*vs.* Ag/0.1 M AgNO₃). The peak current is directly proportional to scan rate (over the range 10-500 mV/s), indicative of a surface-confined species.

Similar immersion of a gold electrode in a dichloromethane solution of monothiol **2** at 40 °C for 24 h led to the monolayer. The cyclic voltammogram of the monolayer showed the oxidation potential at +0.25 V and +0.65 V (vs. Ag/0.1 M AgNO₃). The surface coverage (Γ) of the monolayer of monothiol **2** is 1.1×10^{-10} mol/cm² which is smaller than that of the monolayer of tetrathiol **1** ($\Gamma = 3.2 \times 10^{-10}$ mol/cm²). The monolayer of tetrathiol **1** was remarkably stable compared to the monolayer of monothiol **2** as evidenced from the repeated electrochemical cycling of those monolayers. These findings are ascribed to the strong fixation between the gold and the TTF molecule by four-point attachment of the thiol.

Meanwhile, the cyclic voltammogram (CV) of tetrathiol **1** in CH₂Cl₂ + 0.1 M Bu₄NPF₆ at a glassy carbon electrode exhibited the reversible waves corresponding to the TTF/TTF⁺/TTF²⁺ couple at +0.40 V and +0.71 V, and the irreversible peak due to the thiol at +0.92 V. Interestingly, repeatedly scanning the potential over the range 0.00 V to +1.00 V resulted in a continuous increase in the size of cyclic voltammetric TTF/TTF⁺/TTF²⁺ waves, as shown in Figure 2, indicating the growth of a poly(thio-TTF) film on the electrode. In contrast to **1**, none of the polymer film from monothiol **2** was formed upon repeated cyclic scanning.

The electropolymerization of tetrathiol **1** on gold electrode was performed by repeated cyclic scanning of potential (0.00 V - +1.00 V). After twenty scans, the electrode was rinsed with solvent and dipped into fresh CH₂Cl₂ solution. The cyclic voltammogram thus obtained is shown in Figure 1. Many scans can be repeated without any change of the CV curve. This is a new method for the preparation of polymerized films. In general, the modified electrodes by electropolymerization were obtained from the monomers containing pyrrole and thiophene groups as electroactive group.

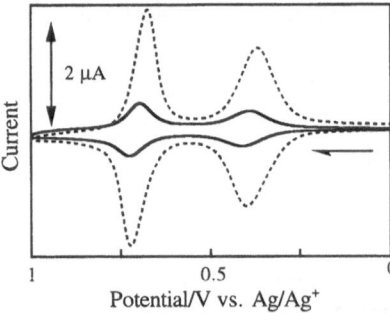

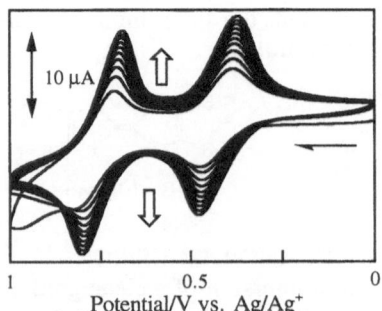

Fig. 1. Cyclic voltammograms of Au electrode modified with **1** by self-assembly method (solid line) and by electropolymerization (dashed line); solution, 0.1 M Bu₄NPF₆ in CH₂Cl₂; scan rate, 100 mV/s.

Fig. 2. Electropolymerization upon repeated cycling of **1** (1 mM) in CH₂Cl₂ + 0.1 M Bu₄NPF₆ on GC electrode. The potential was cycled continuously at 100 mV/s beween 0 and +1.0 V.

ELECTROCHEMICAL REACTIONS
OF PHOSPHOROUS ACID ESTERS.

A.P. Tomilov, B.I. Martynov, I.N. Chernyh, V.V. Turygin

State Research Institute of Organic Chemistry & Technology
23, Shosse Entuziastov, Moscow, 111024 Russia

ABSTRACT

The electrochemical synthesis of dialkylphosphites salts, pyrophosphates, trialkylphosphates and dialkylalkylphosphonates from the available dialkylphosphites has been studied. The procedures affording transition metals salts of phosphorous acid and trialkylphosphate has been elaborated.

INTRODUCTION

It is well known that dialkylphosphites are key compounds in the preparation of organophosphorus compounds of various structure. In this connection increasing attention on the roles and significance of phosphoryl transfer has resulted in the electrochemical methods as organophosphorus modern technology preparations.

According to the literature data, dialkylphosphites don't oxidize in the aprotic solvents over approachable potential range [1], nevertheless they allow to react readily with anodically generated halogens. In this case it may be supposed that the corresponding phosphoric chlorides are formed as primary products. Recently it has been noted that dialkylphosphites demonstrate electrochemical activity at cathodically polarised transitianal metal as iron, platinum, nickel and copper [2]. Generally the electrochemical reactions of dialkylphosphites in aprotic solvents may be presented by the following scheme:

$$(RO)_2P(O)Hal + HHal \underset{+2Hal^-}{\overset{-2e}{\longleftarrow}} (RO)_2P(O)H \overset{+e}{\longrightarrow} (RO)_2P(O)^- + 1/2\ H_2$$

91

RESULTS

It has been established that addition diethylphosphite to the solution of the supporting electrolyte in acetonitrile gives rise to a slight increase in the cathodic current at the potential of about -2.1 V as compared with the background solution. However the distinct reductive wave early reported [2] we could not reproduce.

We investigated electrolysis of the acetonitrilic solutions with diethylphosphite in the presence of various supporting electrolytes in undivided cell. Many metals were studied as anode materials, including platinum. It has been found that they dissolved at the presence of sodium iodide or hydrogen iodide as supporting electrolytes. This process leads to diethylphosphite salts.

It is interesting to note that the structure of the salts depends on the nature of metal. The salts of the transition metals are solid state compounds which are insoluble in organic solvents and their structure is likely to correspond with the formula $[(AlkO)_2P(O)]_nM$, were n is the metal valency. Their yields based on the losses of anode mass and are the following [3]:

Metals	Fe	Cd	Ni	Hg	Pt	Cu
Current efficiency, %	88.5	83.7	80.7	71.4	69.5	81.7

The salts formation with P-M bond may be represented as the combination of the following processes:

at the cathode: $(RO)_2P(O)H + e \longrightarrow (RO)_2P(O)^- + 1/2\ H_2$

at the anode: $M \longrightarrow 1/nM^{n+1} + ne$

in the solution: $n(RO)_2P(O)^- + M^{n+} \longrightarrow [(RO)_2P(O)]nM$.

In contrast with the salts of the transitional metals magnesium or aluminium obtained at the same conditions have the phosphorus - hydrogen bond in their structure and according to the analysis data they are monoalkylphosphorous acid derivatives:

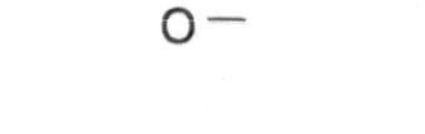

$$[(AlkO)\underset{\overset{|}{O}^-}{P}(O)H]_n M$$

It was very interesting to inspect the electrolysis of diethylphosphite at the same conditions in undivided cell equipped with the inert graphite electrodes. These experiments were carried out in the acetonitrile as solvent and at the presence of NaJ, HCl and Bu_4NBF_4 as the supporting electrolytes. Triethylphosphate and teraethylpyrophosphate were isolated as the main products of the electrolysis in halogen ions containing solutions [4]. At the presence of tetraethylammonium fluoroborate as the supporting electrolyte teraethylpyrophosphate was not observed, but there was many monoethylphosphite. The obtained results show that the anodically generated halogen directly participates in the electrode process. Generated halogen radicals may interact with diethylphosphite resulting the phosphoric halogenides:

$(RO)_2P(O)H + 2Hal^- \longrightarrow (RO)_2P(O)Hal + HHal + 2e$,

$(RO)_2P(O)Hal + (RO)_2POH \longrightarrow (RO)_2P(O)-O-P(OR)_2$

Oxidation converted the last substance to teraethylpyrophosphate.

The alkoxylation of dialkylphosphites described in the literature and leads to phosphoric acid esters [5]:

$(RO)_2P(O)H + R'OH \longrightarrow (RO)_2(R'O)PO$

The process was recommended to carry out in the anode compartment of the diaphragm cell in the medium of anhydrous alcohol and lithium chloride as supporting electrolyte. It may be noted that in this case the solution has a low conductivity and the isolation of target phosphate from the lithium chloride alcohol solutions is very difficult because of complexing. That is why we examined the influence of the halogen nature both diaphragm and undivided* cell:

Supporting electrolytes	HCl	Et₄NCl	Et₄NBr	NaJ
Triethylphosphate yields, %	80.6	67.2	66.8	51.2
	91.2*	-	84.5*	67.1*

The highest yield was achieved in the presence of hydrogen chloride and the lowest one was obtained with sodium iodide.

Thus, it has been established that the alkoxylation of diethylphosphite can be realised in a simple undivided cell with graphite electrodes and hydrogen chloride as the supporting electrolyte. The procedure of isolation of triethylphosphate from the reaction mixture in this case has no difficulties and the yield comes to 90%.

The possibility of the cathodic reduction of dialkylphosphites in the presence of alkyl halides to alkyl phosphonates with the yields 55-65% in diaphragm electrolyzers has been reported [6]. However, the diaphragm electrolyzers are hardly applicable for operation in non-aqueous solutions.

Therefore we studied this process in undivided cell. Addition of hexamethyl-phosphorus triamide to the other components of this reaction allows to obtain phos-phonates in small amounts The action of hexamethylphosphorus triamide can be probably explained by its complexing with the anode metal cations allowed generate phosphites salts in previous stage. Unfortunately in these experiments the yields of phosphonates were rather small, not higher than 10-12%. The better results were obtained in the experiments with the inertle anodes. Thus, in acetonitrile and lithium fluoroborate solution at graphite anode and zinc cathode yield of the phosphonate may be 25%. Further research in this field seems to be useful.

We wish to thank the International Science & Technology Center

(ISTC project 136-94) for financial support of this research.

References
1) A.P. Tomilov, Yu.M. Kargin, I.N. Chernyh, in book: *Electrochemistry of Organoelement Compounds*" (Rus.), Moscow, Nauka, **1986**, 118.

2) A.V. Bukhtiarov, V.V. Mikheev, A.V. Lebedev and Yu.G. Kudryavtsev, *Zh. Obshch. Khim.*, (Rus.), **1991**, *61,* 889.

3) V.F. Pavlichenko, A.E. Presnov, A.P. Tomilov, *Elektrokhimia.*, (Rus.),**1995**, *31*, 538.

4) V.F. Pavlichenko, A.E. Presnov, A.P. Tomilov, *Zh. Obshch. Khim.*, **1996,** *66,* 1347.

5) Ohmori M., Nakai S., Sekiguchi V., Masui M., *Chem. Pharm. Bull.*, **1973**, *27,* 1700.

6) V.A. Petrosyan, M.E. Niazimbetov, T.K. Baryshnikova, V.A. Dorokhov, *Izv. Akad. Nauk USSR., ser. khim.*, (Rus.), **1988**, *N 6,* 1985.

REACTIONS OF ELECTROCHEMICALLY GENERATED ALKOXY PHOSPHONIUM IONS

Hidenobu OHMORI, Hatsuo MAEDA, Sayaka MATSUMOTO, Takashi KOIDE,and Toshihide MAKI[#]

Faculty of Pharmaceutical Sciences, Osaka University, Osaka 565, Japan
([#]Facult of Pharmaceutical Sciences, Nagasaki University)

ABSTRACT
    Reactions of alkoxy triphenylphosphonium salts (**1**), prepared by a simple electrochemical procedure from primary and secondary alcohols, with nucleophiles indicated that a soft nucleophile is apt to attack at the α-carbon to give the corresponding S_N2 product, while a hard one at the phosphorus atom to regenerate the original alcohol. Thermal decomposition of **1** with BF_4^- as counter anion has been shown to afford alkyl fluorides. One-pot dehydroxy-substitution reactions at the anomeric carbons of protected sugars has been effected via *in situ* decomposition of the corresponding electrochemically generated alkoxy phosphonium ions.

INTRODUCTION
    Alkoxy phosphonium ion has been suggested as the crucial intermediate in the Mitsunobu reaction to afford the desired products, and numerous efforts have been paid to control its effective formation in the reaction system.[1]   Meanwhile we have found that alkoxy triphenylphosphonium salts (**1**) can be obtained by electrolysis of Ph_3P in the presence of alcohols.   Since most of the isolated phosphonium salts **1** are stable on storage, the property of an alkoxy phosphonium ion can be explored under various conditions beyond those of the Mitsunobu reactions, and hence wider synthetic application of **1** is expected.

RESULTS AND DISCUSSION

1. Preparation of alkoxy triphenylphosphonium salts (1)
    The phosphonium salts **1** had been prepared at first (a) by controlled potential electrolysis of Ph_3P in MeCN containing $NaClO_4$ and ROH in a divided cell,[2] then (b) by constant current electrolysis (CCE) in MeCN containing $NaClO_4$, ROH, and PhCOOH in an undivided cell.[3]   Recently, however, the process shown in Scheme 1 has been found highly effective, and **1** from various alcohols have been obtained (Table 1).[4]   In the previous two methods, successful isolation was limited only for **1** derived from simple primary alcohols such as MeOH (51%), EtOH (51%), and PrOH (45%).[3]

Scheme 1

$$ROH \xrightarrow[\substack{Ph_3P \ and \ Ph_3P^+H \cdot X^- \\ in \ CH_2Cl_2 \\ Graphite - Pt}]{\substack{Constant \ Current \ Electrolysis \\ in \ an \ undivided \ cell}} Ph_3P^+\text{-}OR \cdot X^- \quad (1)$$

$(X = ClO_4 \ or \ BF_4)$

Table 1 Yield/% of **1** prepared according to Scheme 1

−OR	X = ClO$_4$	X = BF$_4$	−OR	X = ClO$_4$	X = BF$_4$
CH$_3$CH$_2$O−(**1a**)	78 (68)	67	Ph–CH(CH$_3$)O− (**1g**)	97 (75)	90 (72)
CH$_3$(CH$_2$)$_2$O−(**1b**)	85 (81)		(menthyl)O− (**1h**)		(56)
PhCH$_2$CH$_2$O−(**1c**)	90 (84)	93			
t-Bu–C$_6$H$_{10}$–O−(**1d**)	87 (20)	84 (33)	(bornyl)O− (**1i**)		(80)
(cholestanyl, 3β) −O (**1e**)	100 (60)	86 (76)	(CH$_3$)$_3$CO−	Not formed	
(cholestanyl) −O (**1f**)		(49)	CH$_2$=CHCH$_2$O−	Not formed	
			p-Br-Ph-CH$_2$O−	Not formed	

[The number in ( ): isolated yield]

2. Reactions of **1** with nucleophiles

Three types of reaction courses can be envisaged in the reaction of **1** with a nucleophile: attack of a nucleophile at the carbon α to the oxygen (path A, which will include either S$_N$1 or S$_N$2 mode), at a β-hydrogen to cause an elimination (path B, which will be influenced by the conformation of **1**), and at the phosphorus atom to regenerate the original alcohol (path C).   Taking these points in to consideration, the phosphonium ion **1e** and **1f** (Table 1), whose absolute configurations and conformations are known, were chosen as model compounds to examine the reactivity of **1** toward nucleophiles, although it was already demonstrated that reactions of **1** (R = Me, Et, Pr; X = ClO$_4$) with PhSH, imidazole, and carboxylic acids proceed smoothly to give the corresponding PhSR, N-alkyl imidazoles, and esters.[2,3] The results obtained by using Bu$_4$NY (Y = Br, Cl, F, N$_3$, SCN), LiY(Y = Br, Cl, F), and PhZH (Z = O, S) have indicated that a soft nucleophile prefers path A giving the corresponding S$_N$2 product, while a hard one tends to take path C.   Elimination (path B) seems favored when the oxy phosphonium group occupies an axial position.   For details of the reactions, the original report should be referred to.[4]

3. Fluorodehydroxylation of alcohols [5]

As mentioned in the preceding section, the reaction of **1** with fluoride ion (a hard nucleophile) mainly proceeded via path C, and formation of alkyl fluoride (path A) was hardly observed.   However, thermal decomposition of **1** (X = BF$_4$) has been found effective for the replacement of hydroxyl groups in primary and secondary alcohols with a fluorine atom (Scheme 2).   The whole procedure from an alcohol is quite simple, involving: (1) CCE of a mixture of ROH, Ph$_3$P, and Ph$_3$PH · BF$_4$ in dichloromethane in an undivided cell; (2) refluxing a tetrahydrofuran or dioxane solution of the residue obtained by evaporation of the solvent in the electrolysis solution under reduced pressure. Typical examples of alkyl fluorides obtained by the procedure are included in Scheme 2.

Scheme 2

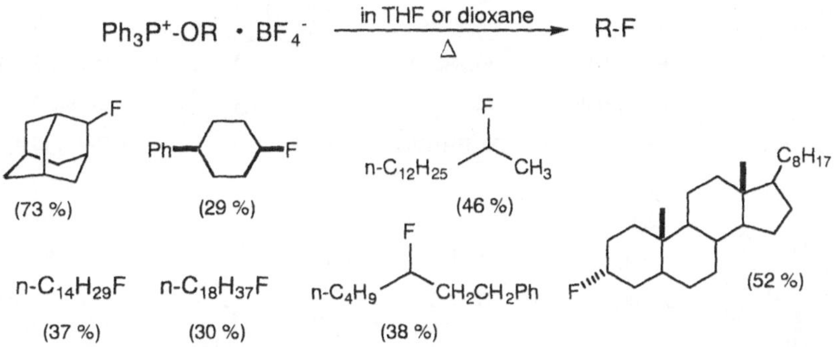

$$Ph_3P^+\text{-}OR \cdot BF_4^- \xrightarrow[\Delta]{\text{in THF or dioxane}} R\text{-}F$$

(73 %)   (29 %)   (46 %)

(52 %)

n-$C_{14}H_{29}F$ n-$C_{18}H_{37}F$

(37 %)   (30 %)   (38 %)

4. Dehydroxy-substitution reactions at the anomeric carbons of protected sugars

Isolation of the phosphonium salts **1** from a tertiary alcohol or a benzyl alcohol was not successful, probably because the phosphonium ion is decomposed in the electrolysis solution by an S_N1 process. However, the carbocation generated from such process might be trapped by a suitable nucleophile deliberately added to the electrolysis system, if the cation would have an appropriate lifetime. We have examined the possibility by focusing attention on the anomeric hydroxyl groups of protected sugars, since the carbocation generated at the anomeric position would be stabilized to some extent by the neighboring oxygen atom (Scheme 3). The procedure is again very simple: CCE of a mixture of Ph3P, a protected sugar, a supporting electrolyte, and a nucleophile in dichloromethane at room temperature in an undivided cell followed by work-up can afford the expected product . Some of the results obtained are given in Table 2.

Scheme 3

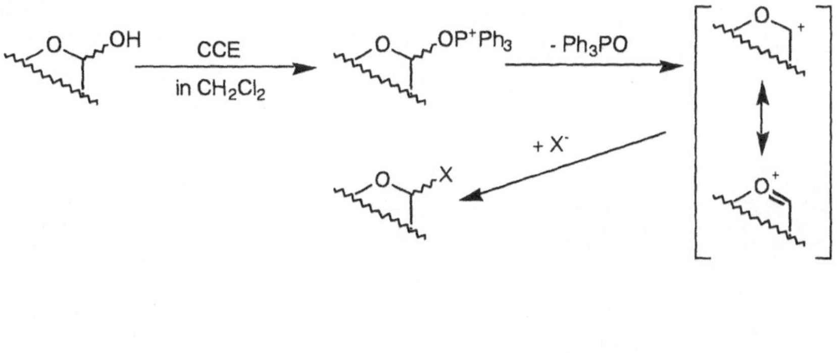

Table 2 Dehydroxy-substitution reaction of protected sugars

Substrate	Additive	Supporting electrolyte	Product (%) X =	
	DME	$Ph_3PH \cdot BF_4$		α-F (72)
	none	Et_4NCl		α-Cl (94)
	CF_3CH_2OH	Bu_4NClO_4		α-OCH$_2$CF$_3$ (92)
	$(CF_3)_2CHOH$	Bu_4NClO_4		α-OCH(CF$_3$)$_2$ (57)
	t-BuOH	$Ph_3PH \cdot ClO_4$		α-OBu-t (54)
	DME	$Ph_3PH \cdot BF_4$		α-F (25)
	none	Et_4NCl		α-Cl (88)
	CF_3CH_2OH	Bu_4NClO_4		-OCH$_2$CF$_3$ (82) (α: β = 67 : 33)
(α: β = 82 : 18)	t-BuOH	$Ph_3PH \cdot ClO_4$		α-OBu-t (62)

References

1) C. M. Afonso, M. T. Teresa, L. S. Godinho, and C. D. Maycock, *Tetrahedron*, **1994**, 50, 9671; A. R. Katritzky, D. C. Oniciu, and I. Ghiviriga, *Synth. Commun.*, **1997**, 27, 1613; D. L. Hughes, and R. A. Reamer, *J. Org. Chem.*, **1996**, 61, 2967; P. J. Harvey, M. von Itzstein, and I. D. Jenkins, *Tetrahedron,* **1997**, 53, 3933 and references therein.
2) H. Ohmori, S. Nakai, M. Sekiguchi, and M. Masui, *Chem. Pharm. Bull.*, **1980**, 28, 910.
3) H. Ohmori, S. Nakai, H. Miyasaka, and M. Masui, *Chem. Pharm. Bull.*, **1982**, 30, 4192.
4) H. Maeda, T. Koide, T. Maki, and H. Ohmori, *Chem. Pharm. Bull.*, **1995**, 43, 1076.
5) H. Maeda, T. Koide, S. Matsumoto, and H. Ohmori, *Chem. Pharm. Bull.*, **1996**, 44, 1480.

A NEW CONCEPT AND TECHNOLOGY IN ELECTROOXIDATIVE CARBON-CARBON BOND FORMATION

Jun-ichi YOSHIDA, Mitsuru WATANABE, Hideaki TOSHIOKA, Masayuki IMAGAWA, and Seiji SUGA

Department of Synthetic Chemistry and Biological Chemistry

Graduate School of Engineering, Kyoto University

Sakyo-ku, Kyoto, 606, Japan

ABSTRACT: The concept of electroauxiliary serves as a new methodology for the electroxidative carbon-carbon bond formation. The intramolecular and intermolecular carbon-carbon bond formation reactions have been achieved using single electroauxiliary such as tin and sulfur. The selective sequential transformation has achieved using two different electroauxiliaries.

INTRODUCTION

Carbon-carbon bond formation is an important process in organic synthesis, and various types of electroorganic reactions involving carbon-carbon bond formation have been developed so far. We have been interested in electrooxidative carbon-carbon bond formation of heteroatom compounds with carbon nucleophiles. In the oxidation of heteroatom compounds, the carbocation intermediates adjacent to the heteroatom is generated by the initial one-electron transfer followed by the elimination of a proton on the α-carbon. The reaction of this carbocation with suitable carbon nucleophiles would lead to the effective carbon-carbon bond formation (Scheme 1).

Scheme 1.

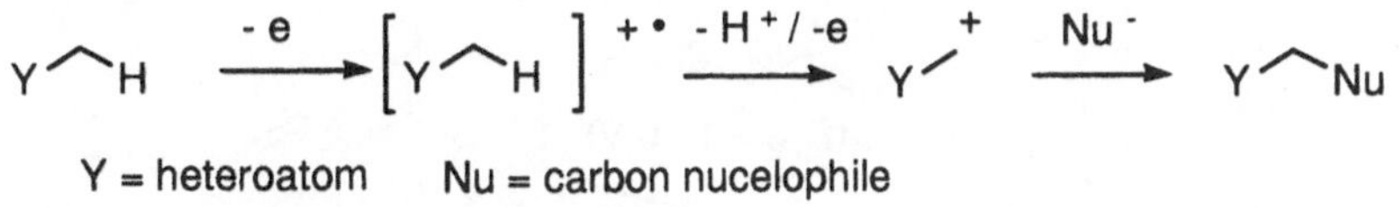

Y = heteroatom   Nu = carbon nucelophile

However, there are two major problems:

(1) It is rather difficult to oxidize the heteroatom compounds without affecting carbon nucleophiles, because the oxidation potentials of carbon nucleophiles are often lower than those of heteroatoms.

(2) The carbon-carbon bond formation products are also heteroatom compounds. Therefore, the oxidation potentials of the products are comparable to those of the starting material. So, it is quite

difficult to avoid the overoxidation.

In order to solve these problems the selective activation of the starting materials toward electron transfer by auxiliaries has proved to be quite effective (Scheme 2). We call such auxiliaries electroauxiliaries. In this paper we describe the principles of electroauxiliaries and their applications to electroorganic synthesis.

Scheme 2.

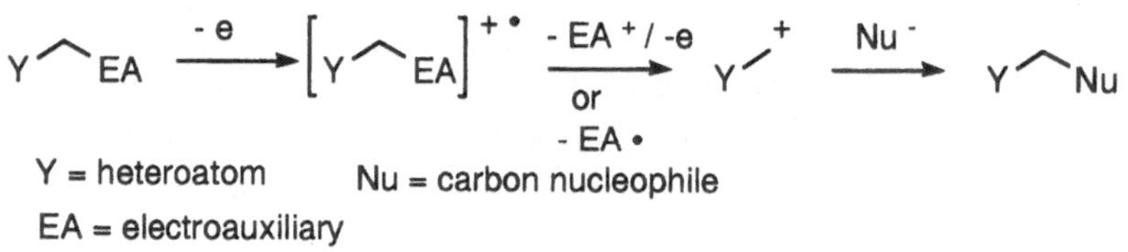

Y = heteroatom   Nu = carbon nucleophile
EA = electroauxiliary

CARBON-CARBON BOND FORMATION USING SINGLE ELECTROAUXILIARY

The most straightforward method for the activation of organic molecules toward oxidation is the raising the HOMO (highest occupied molecular orbital) level. We have demonstrated that group 14 elements such as silicon and tin serve as effective electroauxiliaries for the oxidation of heteroatom compounds. The interaction of the carbon-metal σ orbital with the nonbonding p orbital of the heteroatom causes the increase of the HOMO level which in turn favors the electron transfer. We call this type of electroauxiliaries Type I electroauxiliaries. Tin is especially effective for the electrooxidative carbon-carbon bond formation because the oxidation potentials of tin-substituted heteroatom compounds are usually less positive than those of carbon nucleophiles such as allylsilanes and enol silyl ethers (Scheme 3).[1]

Scheme 3.

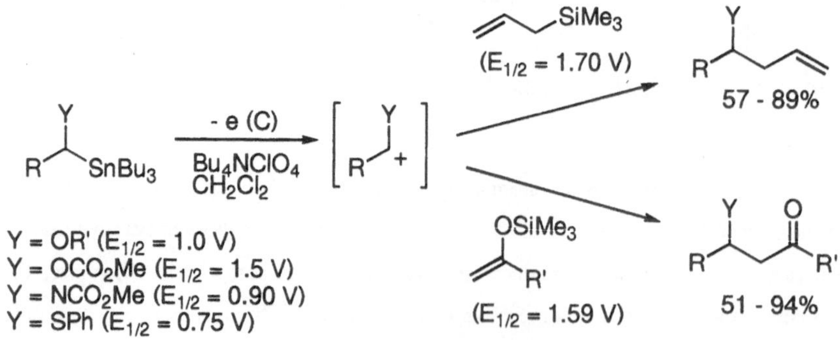

$Y = OR'$ ($E_{1/2} = 1.0$ V)
$Y = OCO_2Me$ ($E_{1/2} = 1.5$ V)
$Y = NCO_2Me$ ($E_{1/2} = 0.90$ V)
$Y = SPh$ ($E_{1/2} = 0.75$ V)

There is another type of electroauxiliary (Type II electroauxiliary). In this case, the HOMO of an auxiliary becomes the HOMO of the substrate molecule. The electron transfer takes place from the orbital of the auxiliary. We have shown that group 16 elements such as sulfur and tellurium serve as effective electroauxiliaries.[2]

SEQUENTIAL OXIDATION REACTIONS USING TWO ELECTROAUXILIARIES

The anodic oxidation of compounds having two different electroauxiliaries (EA_1 and EA_2) results in the selective cleavage of the C-EA_1 bond and the introduction of a nucleophile on the carbon (Scheme 4). For example, in the anodic oxidation of ethers having both silicon and tin as electroauxiliaries, the carbon-tin bond is expected to be cleaved selectively. The product thus obtained is subjected to the second anodic oxidation to cleave the C-EA_2 bond and the introduction of the second nucleophile. Therefore, the sequential introduction of two different nucleophiles can be achieved using two different electroauxiliaries.

Scheme 4.

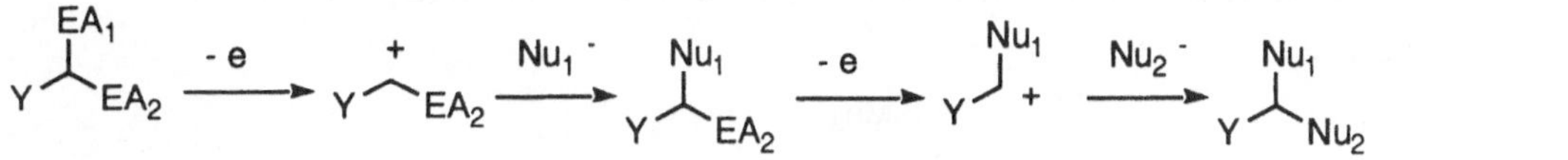

Y = heteroatom

EA_1, EA_2 = electroauxiliary Nu_1, Nu_2 = nucleophile

Thus, we examined the sequential anodic oxidation reactions of heteroatom compounds having both silicon and tin as electroauxiliaries. The oxidation potential of the ethers and sulfides having both silicon and tin at the α-position are slightly less positive than those having only tin as electroauxiliary. Therefore, organosilicon nucleophiles such as allylsilanes and enol silyl ethers can be utilized as carbon nucleophiles.

The anodic oxidation of methoxy(trimethylsilyl)(tributystannyl)methane in the presence of the enol silyl ether derived from cyclohexanone proceeds smoothly, and the carbon-tin bond is cleaved selectively without affecting the carbon-silicon bond (Scheme 5). The carbon-carbon bond formation product thus obtained is subjected to the second anodic oxidation. The carbon-silicon bond is cleaved and the methoxy group is introduced to the carbon. Therefore, two different types of nucleophiles can be introduced sequentially to the carbon to which two different electroauxiliaries have been attached.

Scheme 5.

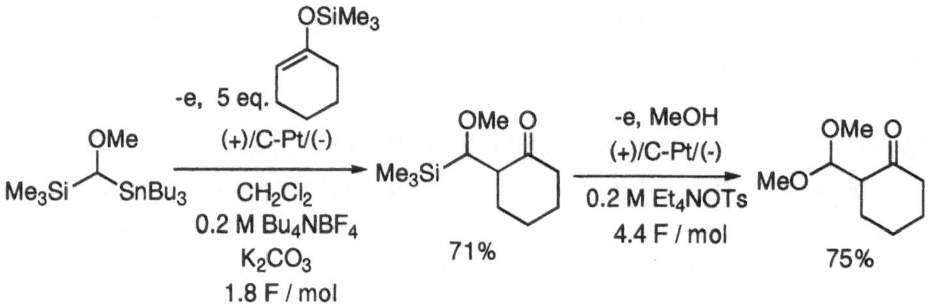

The sequential anodic oxidation can also be achieved with sulfides having silicon and tin as electroauxiliaries. Thus, the anodic oxidation of phenythio(trimethylsilyl)(tributylstannyl)methane in the presence of the enol silyl ether derived from cyclohexanone followed by the second anodic oxidation in methanol proceeds smoothly as shown in Scheme 6. It is noteworthy that in the second anodic oxidation the carbon-sulfur bond is also cleaved to give the corresponding acetal.

Scheme 6.

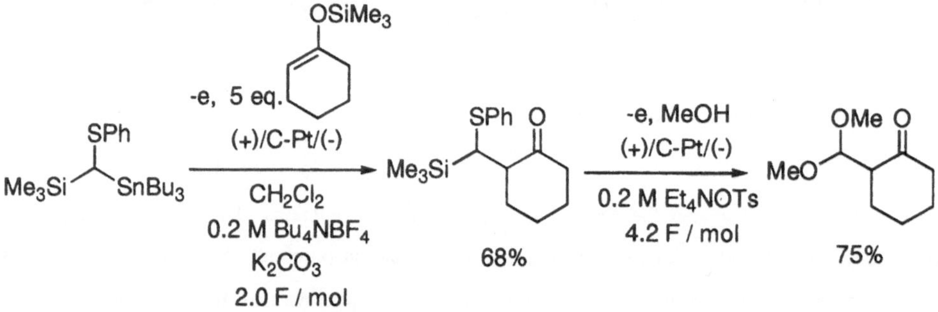

CONCLUSION

The present study demonstrates the potentiality of the concept of electroauxiliary as a powerful tool in electrooxidative carbon-carbon bond formation. It is hoped that various kinds of electroauxiliaries will be widely utilized in electroorganic synthesis.

1) J. Yoshida, *J. Syn. Org. Chem. Jpn.* **1995**, *53*, 53 and the references cited therein.

2) (a) J. Yoshida, M. Sugawara, and N. Kise, *Tetrahedron Lett.* **1996**, *37*, 3157; (b) M. Sugawara, K. Mori, and J. Yoshida, *Electrochim. Acta*, **1997**, *42*, 1995; (c) S. Yamago, K. Kokubo, and J. Yoshida, *Chem. Lett.* **1997**, 111.

ANODIC OXIDATION OF CYCLIC ORGANIC POLYSILANES

James Y. BECKER, Mequin SHEN, Zang-Rong ZHENG and Robert WEST

Chemistry Department , Ben-Gurion University of the Negev, Beer-Sheva 84105, Israel

Introduction

Cyclic polysilanes are of special interest because they exhibit unusual electronic, physical and chemical properties, resembling those of aromatic and poly-unsaturated compounds [1]. For example, cyclic polysilanes have electronic transitions in the UV-VIS region, form electron-delocalized anion radicals upon reduction and cation radicals upon oxidation, in which the odd electron is distributed over the ring, form charge-transfer complexes with π-acceptors, and exhibit substituent effects. It seems that delocalization of the Si-Si σ-electrons over the ring, as first suggested by Pitt [2], accounts for these unusual properties. The synthesis, structure, chemical reactions and photolysis of cyclic polysilanes are described in a recent review[3].

The present work describes the anodic oxidation of $(Mes_2Si)_3(I)$, $[t-Bu(Me)Si]_4(II)$, $[(n-Pr)_2Si]_5(III)$, $(Me_2Si)_6(IV)$ and $(Et_2Si)_7(V)$, under various electrolytic conditions. The effects of ring-size, electrolyte, solvent, atmosphere, applied potential and anode material are discussed.

Results and Discussion

A few organic compounds of silicon have previously been investigated by cyclic voltammetry. For instance, the oxidation potentials of linear permethylsilanes, $Me(SiMe_2)_nMe$ (n = 2-6) was found to decrease with increasing chain length [4]. The oxidation potentials of a series of cyclosilanes display at least two irreversible anodic waves in the range of 1.1 to 1.9V (*vs* SCE) [5]. The first oxidation potential depends both upon ring size and the nature of substituents on silicon. The present work describes the results obtained by preparative anodic oxidation of cyclic ppolysilanes I-V.

a) Electrolysis in the presence of BF_4^-

Previous results of preparative electrochemical oxidation of cyclic peralkylsilanes, e.g., dodecamethyl- cyclohexasilane ($[Me_2Si]_6$) and deca-n-propylcyclopentasilane ($[n-Pr_2Si]_5$) in CH_2Cl_2-TBABF$_4$ showed that they undergo ring opening, followed by further Si-Si bond cleavage and reaction with BF_4^- to form α,ω-difluorosilanes, $F-(SiR_2)_n-F$, as major products [6]. This work has been extended to other cyclic peralkylsilanes, and all the results are summarized in Table 1. In general three major types of linear silanes have been generated:

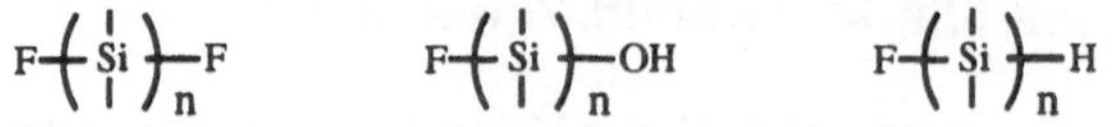

$$\text{F} \!-\!\!\Big(\!\!\underset{|}{\overset{|}{\text{Si}}}\!\!\Big)_{\!n}\!\!-\!\text{F} \qquad\qquad \text{F} \!-\!\!\Big(\!\!\underset{|}{\overset{|}{\text{Si}}}\!\!\Big)_{\!n}\!\!-\!\text{OH} \qquad\qquad \text{F} \!-\!\!\Big(\!\!\underset{|}{\overset{|}{\text{Si}}}\!\!\Big)_{\!n}\!\!-\!\text{H}$$

It is noteworthy that $(Mes_2Si)_3$ (**I**) afforded few other products which stem from aryl migration process, whereas $[t\text{-Bu(Me)Si}]_4$ (**II**) yielded a cyclic siloxane, $(t\text{-Bu(Me)Si})_4O$, additionally.

Table.[a] Product distribution from anodic oxidation[a] of **I - V**

$(R_2Si)_n$	F/mol	$n=1$	2	3	4	5	6	Other products (%)
				F- $(R_2Si)_n$ -F (%)[b]				
–								
I			15					Mes_2SiH_2 (5)
								Mes_3SiOH (7)
								Mes_3SiF (21)
								$H[Mes_2Si]F$ (13)
								$F[Mes_2Si][F_2Si]Mes$ (6)
II					60			$F[(t\text{-Bu(Me)Si})_4O]F$
					36			$(t\text{-Bu(Me)Si})_4O$
III	2.5	1	54	30	10	–		F- $(n\text{-}Pr_2Si)_3$-OH
								F- $(n\text{-}Pr_2Si)_3$-H
								H-$(n\text{-}Pr_2Si)_3$-OH
								F- $(n\text{-}Pr_2Si)_5$-OH
IV	4.0	5.5	27	12	49.5	1.5	0.5	F- $(Me_2Si)_2$-OH
								F- $(Me_2Si)_3$-OH
								F- $(Me_2Si)_5$-OH
								F- $(Me_2Si)_5$-H

[a] Controlled potential electrolysis, in an 'H'-type divided cell, on Pt anode; Conc. 12-14-mM in CH_2Cl_2-CH_3CN(4:1) - 0.1M Et_4NBF_4.
[b] An average of two experiments. The % and characterization is based mainly on GLC and GC/MS (EI or CI whenever necessary); DB-1 packed column, 30m long, 60-250°C/8°C/min. ^{1}H-, ^{13}C- and ^{29}Si-NMR are available for the major products.

A plausible mechanism which explains the formation of the variety of linear silicon products mentioned in Table 1 is described in Scheme 1. The initial anodic process leads to a cyclic (**A**) or linear (**B**) cation radical, followed by a chemical reaction with BF$_4^-$ to form a radical intermediate **C**. The latter could undergo hydrogen atom abstraction to form **D** and/or undergo additional electrochemical oxidation to generate cation **E** followed by various chemical reactions.

Scheme 1. Mechanism for formation of products in the presence of BF$_4^-$

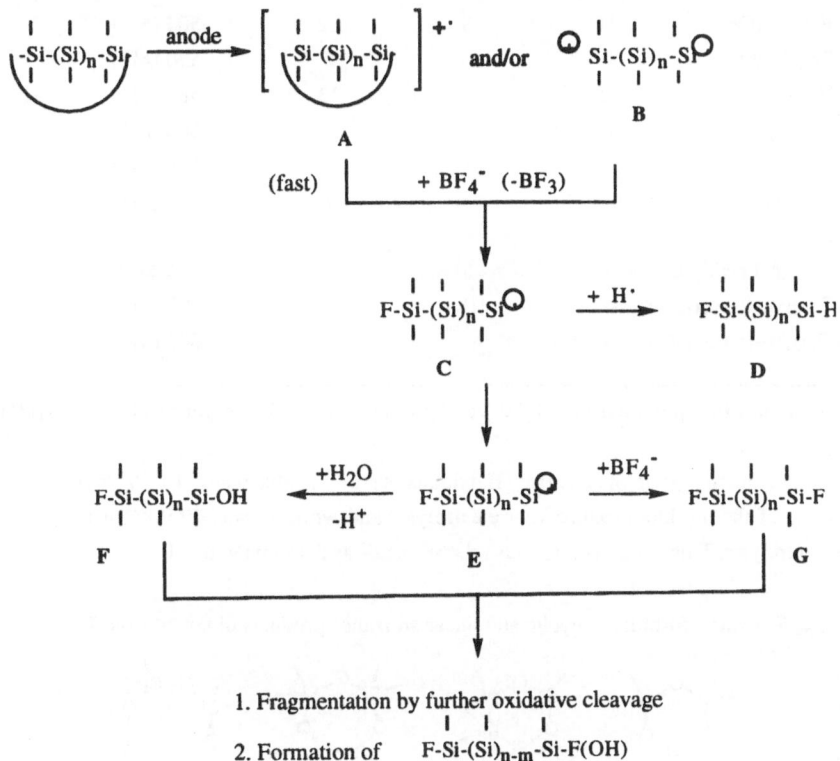

1. Fragmentation by further oxidative cleavage

2. Formation of $\quad$ F-Si-(Si)$_{n-m}$-Si-F(OH)

b) Electrolysis in the presence of ClO$_4^-$

Anodic oxidation of cyclic silanes **I-V** in the presence of Et$_4$NClO$_4$ undergo both oxygen insertion to form cyclic siloxans and ring opening to form linear siloxanes. In a typical example, Table 2 shows the results obtained by the oxidation of **III**, forming cyclic products (**1-8**) and linear ones (**9-11**), whose chemical structure is described in Scheme 2.

Table 1. Results from Anodic Oxidation^a of III^b on Pt in CH$_2$Cl$_2$-CH$_3$CN (4:1)-Et$_4$NClO$_4$ (0.1M)

Products	Yield (%)^c			Mass spectrum
	under air	under air	under N$_2$	(M$^+$)
	(2.2 F/mol)	(4 F/mol)	(4 F/mol)	
III	20	--	16	
Cyclic				
1, [(n-Pr)$_2$Si]$_3$O$_3$	2	25	10	390 (M$^+$)
2, [(n-Pr)$_2$Si]$_4$O$_2$	11	3	5	488 (M$^+$)
3, [(n-Pr)$_2$Si]$_4$O$_3$	16	8	12	504 (M$^+$)
4, [(n-Pr)$_2$Si]$_4$O$_4$	4	31	8	520 (M$^+$)
5, [(n-Pr)$_2$Si]$_5$O	25	--	13	586 (M$^+$)
6, [(n-Pr)$_2$Si]$_5$O$_2$	16	4	11	602 (M$^+$)
7, [(n-Pr)$_2$Si]$_5$O$_3$	--	--	7	618 (M$^+$)
8, [(n-Pr)$_2$Si]$_5$O$_4$	--	--	7	634 (M$^+$)
Linear				
9, (n-Pr)$_2$[(n-Pr)$_2$Si]$_4$O$_2$	--	5	--	574(M$^+$)
10, (n-Pr)$_2$[(n-Pr)$_2$Si]$_4$O$_3$	--	18	--	590 (M$^+$)
11, (n-Pr)$_2$[(n-Pr)$_2$Si]$_4$O$_4$	--		10	606 (M$^+$)

^{a} Applied controlled potential: 1.2-1.3V vs. Ag quasi-reversible reference electrode; [III]=6.6-11.7mM.

^{b} The cyclic voltammogram of substrate III exhibits one irreversible wave at 1.4V vs Ag/AgCl.

^{c} The mass of the product mixture after electrolysis and work-up was 70-90% of the mass of III before electrolysis. This column represents relative yields as determined by glc.

Scheme 2. Structural formula of cyclic and linear siloxanes products obtained from III

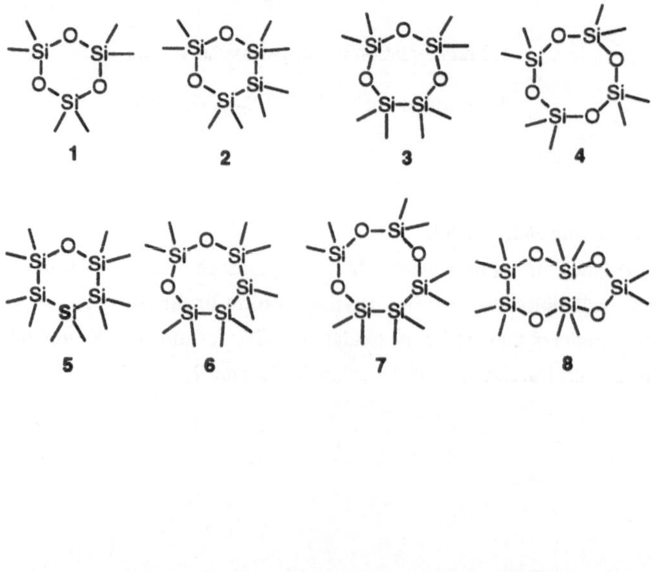

Scheme 2. (continued)

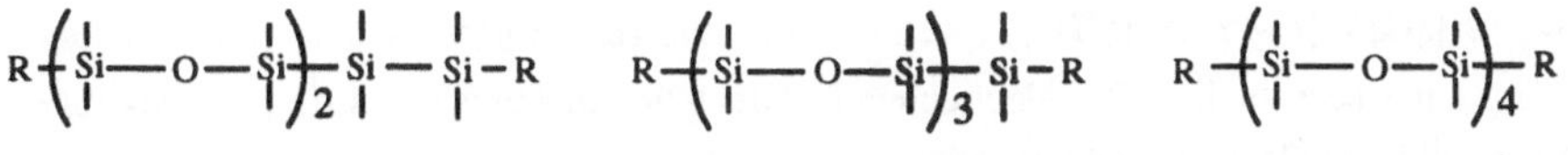

$$R \left(\overset{|}{\underset{|}{Si}} - O - \overset{|}{\underset{|}{Si}} \right)_2 \overset{|}{\underset{|}{Si}} - \overset{|}{\underset{|}{Si}} - R \qquad R \left(\overset{|}{\underset{|}{Si}} - O - \overset{|}{\underset{|}{Si}} \right)_3 \overset{|}{\underset{|}{Si}} - R \qquad R \left(\overset{|}{\underset{|}{Si}} - O - \overset{|}{\underset{|}{Si}} \right)_4 R$$

It is noteworthy that under different anodic experimental conditions [7], cyclic silanes yielded additional linear products of type:

$$HO \left(\overset{|}{\underset{|}{Si}} - O - \overset{|}{\underset{|}{Si}} \right)_n OH \qquad HO \left(\overset{|}{\underset{|}{Si}} - O - \overset{|}{\underset{|}{Si}} \right)_n H$$

Since the outcome in the presence of Et_4NBF_4 is different from that obtained in the presence of Et_4NClO_4, a different mechanism should exist to explain the different results. Scheme 3 describes such a possible mechanism in which perchlorate anions are involved.

Scheme 3. Mechanism for formation of products in the presence of ClO_4^-

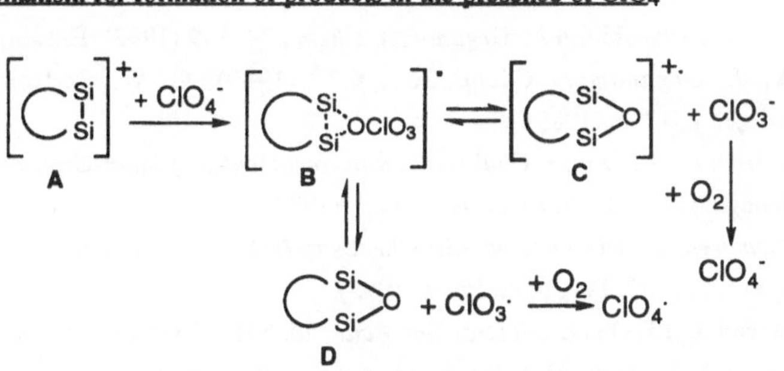

The electrogenerated cation radical 'A' co-ordinates with perchlorate anion to afford intermediate 'B'. The latter could be in unfavored equilibrium with 'C' and chlorate anion (or with 'D' and chlorate radical). As long as the chlorate species is not reoxidised to the corresponding perchlorate one, the further oxidation of siloxanes is inhibited. Obviously further work has to be done to investigate the mechanism of this reaction and support this hypothesis.

The fact that siloxanes are formed from **III** at various oxidation levels, while still unreacted **III** remains, indicates that the former are oxidised easier than **III**. This is a well established phenomenon for both cyclic and linear polysilanes. The observed [8] acceleration of the consecutive oxidation steps compared to the initial oxidation of the parent silane derivative, could stem from a n,σ-conjugation between the oxygen lone-pair and the Si-Si single bond, or intramolecular co-ordination due to interaction between the oxygen lone-pair and a vacant d-orbital of the silicon atom.

Conclusion

Electrochemical oxidation of cyclic polysilane derivatives, namely $(Mes_2Si)_3(I)$, [t-Bu(Me)Si]$_4$(II), [(n-Pr)$_2$Si]$_5$(III), $(Me_2Si)_6$(IV) and $(Et_2Si)_7$(V), under various experimental conditions, has been studied. The electrochemical oxidation of I underwent ring opening followed by further Si-Si bond cleavage and a reaction with the electrolyte to generate mostly linear products of type: X-$(Mes_2Si)_n$-Y (n=1,2; X,Y=mesityl, H, OH or F). Unlike I, the anodic oxidation of II led to oxygen insertion process and gave cyclic siloxanes as the major products. However, the electrochemical oxidation of III, IV and V provided both cyclic and linear siloxanes as major products, stemming from oxygen insertion and ring-opening parallel processes, respectively. The effects of ring-size, electrolyte, solvent, atmosphere, applied potential and anode material are discussed.

References

1. M. Biernaum and R. West, *Journal of Organometallic Chemistry*, **131**, 179 (1977); R. West, *J. Organomet. Chem.*, **300**, 327 (1986); R. West, *Pure Appl. Chem.*, **54**, 1041 (1982); V. F. Traven and S. Yu. Shapakin, *Adv. Organomet. Chem.*, **34**, 149 (1992); E. Hengge, M. Eibl and F. Schrank, *J. Organomet. Chem.*, **369**, C23 (1989); C. W. Carlson and R. West, *Organometallics*, **2**, 1792 (1983).

2. C. G. Pitt, in *Homoatomic Rings, Chains and Macromolecules of Main-Group Elements* (Edited by A. L. Rheingold), p. 203. Elsevier, New York (1977).

3. R. West in *Comprehensive Organometallic Chemistry II* (Edited by E. Abe, F.G.A. Stone and G. Wilkinson), Vol. 3, p. 77. Pergamon Press (1995).

4. E.M. Genies and F. EL Omar, *Electrochim. Acta*, **28**, 541 (1983) and references cited therein.

5. H. Watanabe and Y. Nagai in H. Sakurai, ed., *Organosilicon and Bioorganosilicon Chemistry*, Ellis Horwood Ltd., Chichester, U.K. p. 107-114 (1985); F. Shafiee and R. West, Silicon, *Germanium, Tin & Lead Compounds.*, **9**, 1-10 (1986).

6. J.Y. Becker, E. Shakkour and R. West, *Tetrahedron Lett.*, **33**, 5633-5636 (1992).

7. J.Y. Becker, M.-Q. Shen and R. West, *Electrochim. Acta*, **40**, 2775-2778 (1995); *Idem*, *J. Electroanal. Chem.*, **417**, 77-82 (1996); Z.-R. Zhang, J.Y. Becker and R. West, *Electrochim. Acta*, **42**, 1985-1992 (1997).

8. See a review article by G.A. Razuvaev, T.N. Brevnova and V.V. Semenov, *J. Organomet. Chem.*, 1984, **241**, 260-280.

SYNTHESIS, CHEMISTRY, AND ANODIC OXIDATION
OF 1,2-*BIS*(TRIALKYLSILYL)ETHANES

Albert J. FRY, John M. PORTER, Xiangyang XUAN, Burchelle
BLACKMAN, Brian MAYERS, Daniel HSU, and Shayna WHITE

Chemistry Department, Wesleyan University,
Middletown, CT 06459, U.S.A.

ABSTRACT

1,2-*Bis*[trialkyl-1,2-diarylethanes can be prepared by cathodic dimerization of *alpha*-trialkylsilyl benzyl bromides and metal-induced (Mg, HMPA; Li, THF) *bis*-silylation of substituted stilbenes. These substances are synthetic equivalents of both vicinal cations (by anodic oxidation in the presence of nucleophiles) and vicinal dianions (by fluoride-induced cleavage in the presence of electrophiles). A variety of experiments exploring the scope and limitations of this principle are described.

As part of a long-term study in our laboratory on the electrochemical reduction of alkyl halides, we had occasion to examine the electrochemical reduction of a series of substituted benzyl bromides.[1] Reduction affords radical dimers in high yield if electrolysis is carried out at the foot of the voltammetric wave; this is necessary because reduction of the radical to a carbanion occurs at a slightly more negative potential.[1] We noted that with some bromides the expected mixture of stereoisomeric (*meso* and *dl*) "head-to-head" dimers was accompanied by varying amounts of so-called "head-to-tail" isomers (eq 1). The relative proportion of head-to-head dimers in the mixture decreases roughly linearly as the conformational A-value of the

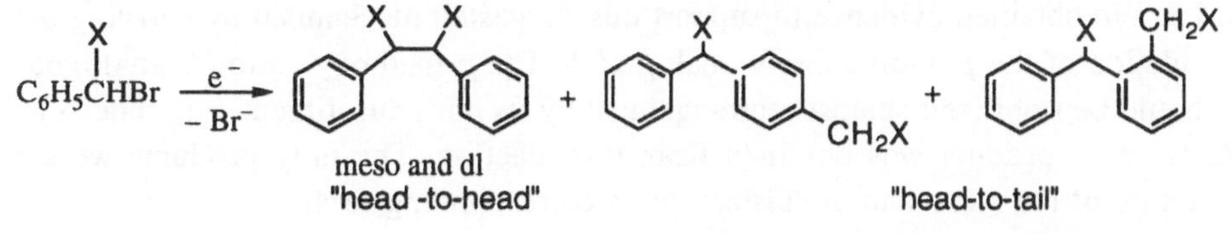

$$C_6H_5\overset{\overset{X}{|}}{C}HBr \xrightarrow[-Br^-]{e^-} \qquad \text{meso and dl} \quad + \qquad \text{"head-to-head"} \quad + \qquad \text{"head-to-tail"} \qquad (1)$$

substituent X increases, indicating that the reaction is sterically controlled; when X is large, undesirable non-bonded effects come into play in the head-to-head dimerization transition state. It appeared to us that the head-to-head dimers in which X is a trialkylsilyl group ought to be versatile synthetic intermediates. They may be

109

considered as doubly benzylic silanes. Benzyl silanes react with fluoride ion to afford benzyl anions, which can be trapped by electrophiles.[2,3] They can also be oxidized anodically to benzyl cations, which can be trapped by nucleophiles:[4,5] Substances **1** may therefore be regarded as synthons (synthetic equivalents) for both 1,2-dications and 1,2-dianions (eq 2). We are exploring these possibilities. A more convenient route to **1**, or more generally, **2**, involves treatment of an mono- or diarylsubstituted alkene with lithium or magnesium in the presence of a trialkylsilyl halide (eq 3).[6,7]

$$
\begin{array}{ccc}
\underset{\substack{|\\ \text{ArCHCHAr}\\ |\\ \text{E}}}{\overset{\text{E}}{|}} & \xleftarrow[\text{2 E}^{+}]{\text{F}^{-}} & \underset{\substack{|\\ \text{ArCHCHAr}\\ |\\ \text{SiR}_3}}{\overset{\text{SiR}_3}{|}} \xrightarrow[\text{2 Nu:}]{\text{[O]}} \underset{\substack{|\\ \text{ArCHCHAr}\\ |\\ \text{Nu}}}{\overset{\text{Nu}}{|}}
\end{array} \qquad (2)
$$

1

$$
\begin{array}{cc}
\text{ArCH=CHR'} & \xrightarrow[\text{or Li, R}_3\text{SiCl, THF}]{\text{Mg, R}_3\text{SiCl, HMPA}} \underset{\substack{|\\ \text{ArCHCHR'}\\ |\\ \text{SiR}_3}}{\overset{\text{SiR}_3}{|}} \\
(\text{R' = H or C}_6\text{H}_5)
\end{array} \qquad (3)
$$

2

Electrochemical oxidation of **1** (Ar = C_6H_5; R = Me) in methanol affords 1,2-dimethoxy-1,2-diphenylethane (**4**; *dl:meso* = 10:1) (65%) and diphenylacetaldehyde dimethylacetal (**5**) (35%) (Scheme 1). These products can be understood as arising from the common intermediate carbocation **6**, which can either react with methanol to afford **4** or undergo a 1,2-phenyl shift, followed by reaction with methanol, to afford **5**. We obtained evidence to support this suggested mechanism by carrying out the oxidation of the *p,p'*-dimethoxy analog of **1**. The *p*-methoxy cation **7** analogous to **6** should be stabilized against rearrangement by its *para* substituent, and indeed no rearrangement product was obtained from this reaction. The only products were a 1:1 mixture of the *meso* and *dl* diastereomers corresponding to **4.**

$$
\underset{\substack{+\\ \\ \textbf{7}}}{\overset{\text{OMe}}{\underset{|}{p\text{-MeOC}_6\text{H}_4\text{CHCHC}_6\text{H}_4\text{-}p\text{-OMe}}}}
$$

Scheme 1

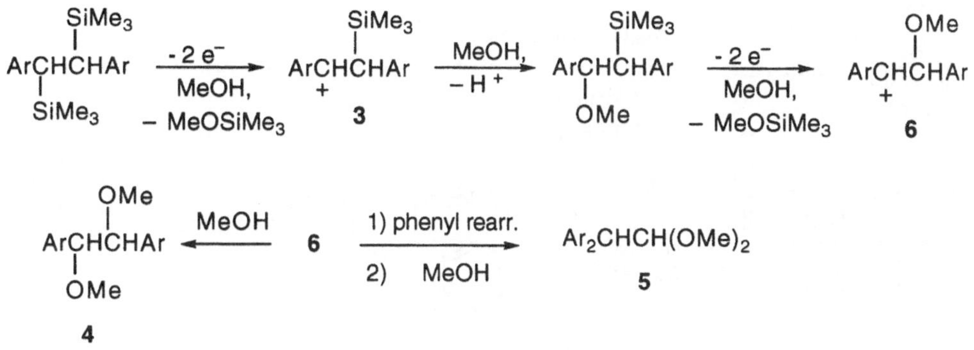

$$\underset{\overset{|}{\underset{SiMe_3}{ArCHCHAr}}}{\overset{SiMe_3}{|}} \xrightarrow[\substack{MeOH, \\ - MeOSiMe_3}]{-2e^-} \underset{\textbf{3}}{\overset{\overset{SiMe_3}{|}}{ArCHCHAr}} + \xrightarrow[-H^+]{MeOH,} \underset{\overset{|}{OMe}}{\overset{SiMe_3}{|}}{ArCHCHAr} \xrightarrow[\substack{MeOH, \\ - MeOSiMe_3}]{-2e^-} \underset{\textbf{6}}{\overset{\overset{OMe}{|}}{ArCHCHAr}} +$$

$$\underset{\textbf{4}}{\overset{\overset{OMe}{|}}{\underset{\overset{|}{OMe}}{ArCHCHAr}}} \xleftarrow{MeOH} \textbf{6} \xrightarrow[\substack{2) \quad MeOH}]{1) \; phenyl \; rearr.} \underset{\textbf{5}}{Ar_2CHCH(OMe)_2}$$

A side reaction with the initial *beta*-silyl cation **3** under some conditions is nucleophile-assisted loss of the silyl group to afford an alkene. We have been able to minimize this undesired side reaction using bulky silyl groups. For example, replacement of the trimethylsilyl groups in the following monoaryl disilane with t-butyldimethylsilyl groups almost completely eliminates nucleophilic reaction at silicon (Scheme 2).

Scheme 2

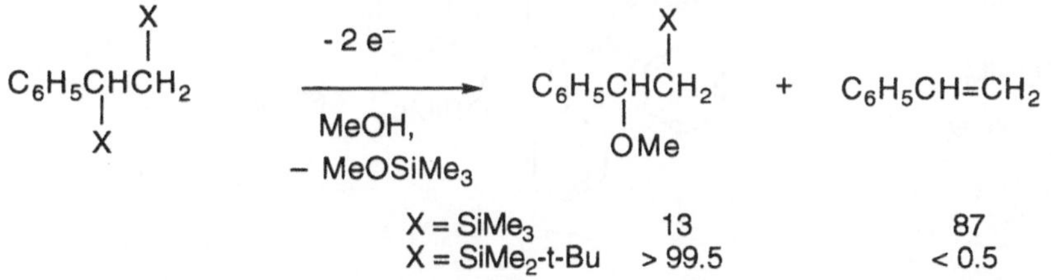

$$\underset{\overset{|}{X}}{\overset{X}{|}}{C_6H_5CHCH_2} \xrightarrow[\substack{MeOH, \\ - MeOSiMe_3}]{-2e^-} \underset{\overset{|}{OMe}}{\overset{X}{|}}{C_6H_5CHCH_2} + C_6H_5CH=CH_2$$

X = SiMe$_3$	13	87
X = SiMe$_2$-t-Bu	> 99.5	< 0.5

Fluoride-assisted cleavage of the carbon-silicon bond of **1** (Ar = C$_6$H$_5$; R = Me) and **2** (Ar = C$_6$H$_5$; R = Me, R' = H) with the aid of tetrabutylammonium difluorotriphenylsilicate [2,3] has been carried out in the presence of a variety of electrophiles, e.g., alkyl halides, aldehydes, acid derivatives, imines, and activated alkenes (Scheme 3). Only the benzylic silyl group of **2** is replaced in this manner, affording the substituted silanes **7-10**. Either one or both silyl groups of **1** are replaced, depending on the reaction conditions, affording adducts such as **11** and **12** (from reaction with benzaldehyde).

Scheme 3

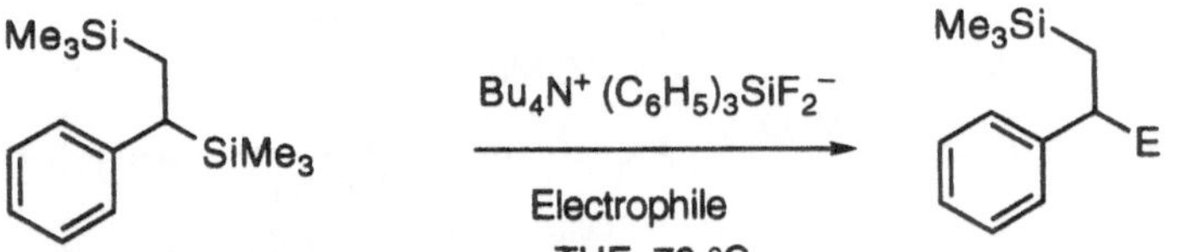
Me3Si
SiMe3
Bu4N+ (C6H5)3SiF2−
Electrophile
THF, 70 °C
Me3Si
E

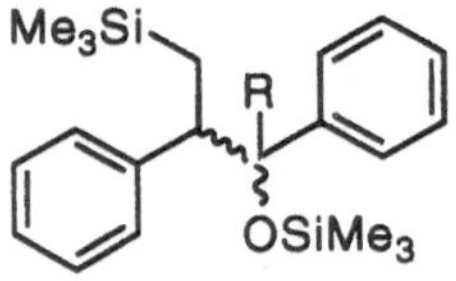
Me3Si
R
OSiMe3

R = H, 53%
R = Me, 48 %

7

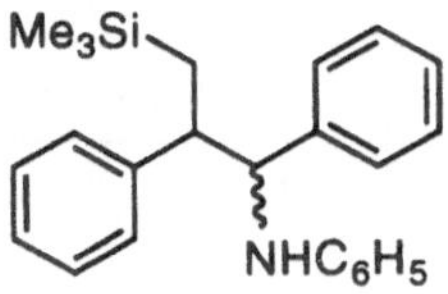
Me3Si
NHC6H5

52 %

8

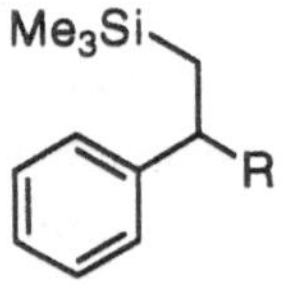
Me3Si
R

$R = C_6H_5CH_2$, 30 %
R = propargyl, 71 %
R = crotyl, 0 %

9

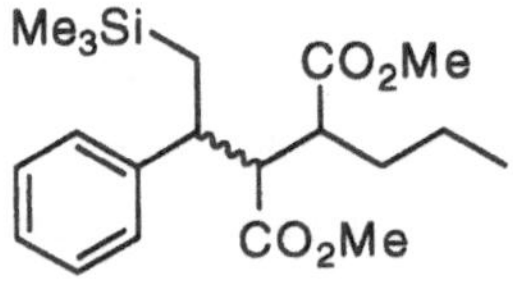
Me3Si
CO2Me
CO2Me

78%

10

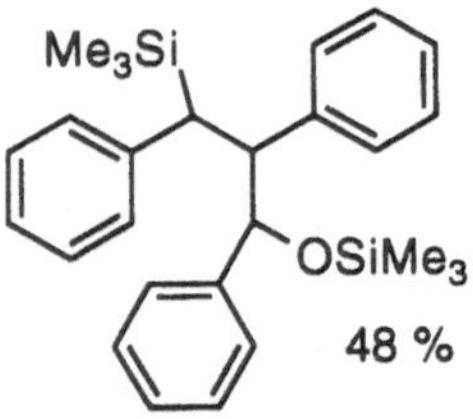
Me3Si
OSiMe3

48 %

11

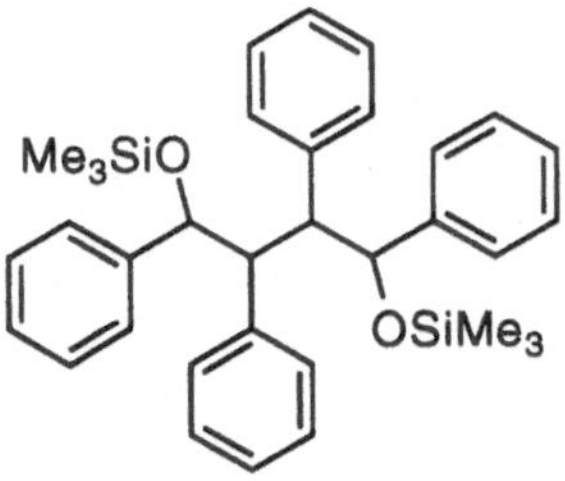
Me3SiO
OSiMe3

68 %

12

REFERENCES
(1) A. J. Fry, J. M. Porter, and P. F. Fry, *J. Org. Chem.*, **1996**, *61*, 3191.
(2) A. S. Pilcher, H. L. Ammon, and P. DeShong, *J. Am. Chem. Soc.*, **1995**, *117*, 5166.
(3) A. S. Pilcher and P. DeShong, *J. Org. Chem.*, **1996**, *61*, 6901.
(4) T. Koizumi, T. Fuchigami, and T. Nonaka, *Electrochim. Acta* , **1988**, *33*, 1635.
(5) J.-i. Yoshida, T. Murata, and S. Isoe, *Tetrahedron Lett.*, **1986**, *27*, 3373.
(6) J. Dunogues, R. Calas, N. Duffaut, P. Lapouyade, and J. Gerval, *J. Organomet. Chem.* **1969**, *20*, P20.
(7) D. R. Weyenberg, L. H. Toporcer, and A. E. Bey, *J. Org. Chem.*, **1965**, *30*, 4096.

SYNTHESIS OF OPTICALLY ACTIVE 1,3-DIENES FROM ALKENYLSTANNANES BY A COMBINATION OF ELECTRO-OXIDATION AND COPPER-MEDIATED HOMOCOUPLING REACTION

Toshiyuki ITOH,* Sachie EMOTO, Hideo TANAKA,† and Shigeru TORII†

Department of Chemistry, Faculty of Education, Okayama University, Okayama 700, Japan. †Department of Applied Chemistry, Faculty of Engineering, Okayama University, Okayama 700, Japan

The copper(II) salt mediated reaction of alkenylstannanes was reported to provide high stereospecificity for the homocoupling reaction to symmetrical dienes, though it required at least a stoichiometric amount of copper salt.[1] Because of its highly stereoselective nature, we refined the copper-mediated coupling reaction of alkenylstannanes using electro-oxidation; the stereocontrolled synthesis of optically active diene **2** from alkenylstannane **1** was accomplished using catalytic amounts of Cu(II)Cl$_2$ in combination with electro-oxidation (Eq. 1).[2]

$$ (1) $$

1aa: R$_1$=n-Pent, R$_2$=Ac **1aa**: R$_1$=n-Pent, R$_2$=Bz **1ba**: R$_1$=H, R$_2$=Ac
1bb : R$_1$=H, R$_2$=Bz **1bc** : R$_1$=H, R$_2$=Bn **1ca** : R$_1$=PhCH$_2$CH$_2$, R$_2$=Ac
1cb: R$_1$=PhCH$_2$CH$_2$ R$_2$=Bz **1db**: R$_1$=c-Hexyl, R$_2$=Bz **1eb**: R$_1$=i-Pr, R$_2$=Bz
1fb: R$_1$=n-Oct, R$_2$=Bz **1ga**: R$_1$=Ph, R$_2$=Ac **1ha**: R$_1$=4-MeO-C$_6$H$_4$, R$_2$=Ac
1ia: R$_1$=4-Cl-C$_6$H$_4$, R$_2$=Ac

The copper(II) ion catalyst is indispensable for completion of the coupling reaction, because the desired diene **2** was not obtained in the absence of a copper catalyst. Protection of the hydroxyl group at the 3-position as an acetate or a benzoate was necessary for providing the desired coupling product. The use of a divided cell was also essential for this reaction when the amount of CuCl$_2$ was reduced to catalytic amounts based on the substrate. Diene **2aa** was obtained in 63% yield in the presence of 20 mol% CuCl$_2$ using a divided cell. A better yield was observed under high dilution conditions than under a concentrated condition. Tetrabutylammonium perchlorate was the most suitable electrolyte among the tested ammonium salts. Better yield and quicker reaction were noted when the reactions were carried out in the presence of 1.0 eq. of the copper catalyst under electro-oxidative conditions. The best yield of 89% was recorded when (E)-**1db** was subjected to the reaction. Therefore, use of equivalent amounts of copper catalyst in this homo-coupling reaction is recommended from a practical viewpoint. It has been emphasized that the yield and product stereoselectivity of the copper(II)-mediated homocoupling reaction of the alkenylstannanes significantly varied with the reaction solvent. Isomerization of the double bond was observed when N-methyl-2-pyrolidone (NMP) was used as the solvent , and reaction under concentrated conditions when the solvent was 0.36 M DMF. It is noteworthy that the reaction proceeded with stereospecificity; (Z,Z)-**2aa** was obtained in 55% yield when (Z)-**1aa** was used as the substrate in 0.01 M DMF without isomerization in the presence of 20 mol% of $CuCl_2$. $CuCl_2$ gave better stereoselectivity for making dienes than $Cu(II)(NO_3)_2 \cdot 3H_2O$. We can also use Cu(I)Cl as an effective catlyst for this homocoupling reaction in combination with electro-oxidation, though the selectivity was slightly inferior to that of $CuCl_2$. To determine the scope and limitations, various types of vinylstannanes were subjected to the homo-coupling reaction. This reaction was significantly dependent on the nature and acyl moiety of the substrate. No desired diene was obtained when acetate **1ba** was subjected to this coupling reaction, whereas benzoate **1bb** and benzyl ether **1bc** provided the corresponding diene **2bb** and **2bc**, respectively. Although the diene was obtained with a slightly better yield from benzyl ether **1bc**, the diene part was isomerized. Alkenylstannanes with an aliphatic functional group at the β-position were thus essential to provide the homo-coupling products. It has recently been reported that the homocoupling reaction of the vinyl stannanes was achieved using excess amounts of copper(I) salts.[1b] In fact, we could also use Cu(I)Cl as an effective catlyst for this homocoupling reaction in combination with electro-oxidation, though the selectivity and chemical yield were slightly inferior to that of $CuCl_2$. GC-MS analysis revealed that Bu_3SnCl was produced by both reactions of Cu(I)Cl- and Cu(II)Cl_2-catalyzed homocoupling. VIS-UV spectroscopic analysis gave interesting results. The λmax of a solution of (E)-**1ab** and Cu(II)Cl_2 (1.0 eq.) in DMF was observed at 850 nm, then it gradually moved as the reaction proceeded and reached 697 nm when the starting material was entirely comsumed. Similarly, λmax that appeared at 760 nm for the solution of (E)-**1ab** and Cu(I)Cl moved to 660 nm. These results may suggest that a common intermediate is involved in the two reactions.

Optically active various types (S)-1-((E)- 2-tributylstannyl)-1-alken-3-yl acetates **1** were easily obtained by lipase-catalyzed optical resolution of the corresponding racemic acetate.[3] The copper-mediated oxidative coupling of the benzoate of alkenylstannanes proceeded with no racemization giving diens **2** in good yield. The synthesis of the four optically active types of dienes, **2ab**, **2cb**, **2eb**, and **2fb**, was thus accomplished by the combined method of the electro-oxidation and copper(II)-mediated coupling (Eq. 2); we succeeded in demonstrating that this methodology could be useful in the preparation of various types of optically active 2,4-diene-1,6-diol derivatives.

References
1) Recent examples see. (a) R. L. Beddos, T. Cheeseright, J. Wang, and P. Quayle, *Tetahedron Lett.*, **36**, 283 (1995). (b) E. Piers and M. A. Rowero, *J. Am. Chem. Soc.*, **118**, 1215 (1996).
2) T. Itoh, S. Emoto, M. Kondo, H. Ohara, H. Tanaka, and S. Torii, *Electrochimica Acta*, **42**, 2133 (1997).
3) (a) T. Itoh and T. Ohta, *Chem. Lett.* 217 (1991). (b) T. Itoh, Y. Takagi, and H. Tsukube, *J. Mol. Catalysis B; Enzymatic*, in press.

SELECTIVE ANODIC FLUORINATION OF HETEROCYCLES. POTENTIAL APPLICATION TOWARDS PHARMACEUTICALS

Toshio FUCHIGAMI
Department of Electronic Chemistry, Tokyo Institute of Technology
Nagatsuta, Yokohama 226, Japan

ABSTRACT

Various heterocyclic compounds having potentially biological activities were regioselectively monofluorinated by electrochemical fluorination using various supporting fluoride salts. Indirect anodic fluorodesulfurization of β-lactams using triarylamine mediators was also successful.

INTRODUCTION

Many heterocyclic compounds have unique biological activities. On the other hand, introduction of fluorine atom(s) into organic molecules sometimes enhances markedly their biological activities. Therefore, partially fluorinated heterocycles are focus of much biological interest. However, very few examples of selective anodic fluorination of heterocycles have been reported so far. From these view points, we have developed highly selective anodic fluorination of various heterocycles.[1-6]

In this work, we have successfully carried out regioselective anodic fluorination of biologically interesting nitrogen-containing heterocyclic compounds such as oxindoles and 3-oxo-1,2,3,4-tetrahydroisoquinolines. Furthermore, we have developed novel indirect anodic fluorodesulfurization of heterocycles such as β-phenylthio β-lactams using a mediator.

RESULTS AND DISCUSSION

Anodic Monofluorination of Oxindoles and 3-Oxo-1,2,3,4-tetrahydroisoquinolines

We have attempted anodic monofluorination of biologically interesting oxindoles 1.[7] As shown in Table 1, the fluorination was greatly affected by supporting fluoride salts, and $Me_4NF \cdot 4HF$ was the most effective among various fluoride salts. When

$Me_4NF \cdot 4HF$ was used, a carbon anode as well as a platinum anode was effective for this fluorination. This is quite important from a practical aspect. We extended this fluorination to 3-oxo-1,2,3,4-terahydroisoquinoline derivatives **3**.[7] $Et_4NF \cdot 3HF$ provided the desired fluorinated product **4** in good yields and with high current

Table 1. Anodic Monofluorination of 3-(Phenylthio)oxindole Derivative **1a**

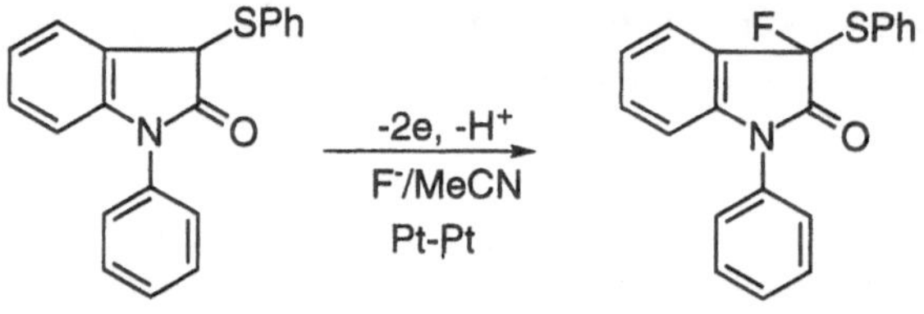
SPh
N
O
-2e, -H+
F-/MeCN
Pt-Pt
F
SPh
N
O

1a **2a**

Supporting Electrode	Charge Passed (F/mol)	Yield (%)
$Et_3N \cdot 3HF$	6	30
$Et_4NF \cdot 2HF$	4	41
$Et_4NF \cdot 3HF$	3.7	58
$Me_4NF \cdot 4HF$	3.5	64

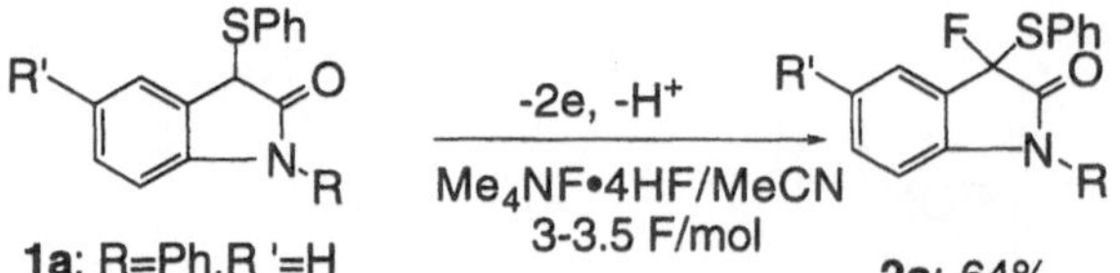
SPh
R'
O
N
R
-2e, -H+
Me4NF•4HF/MeCN
3-3.5 F/mol
F
SPh
R'
O
N
R

1a: R=Ph,R '=H
1b: R=p-Tol,R'=Me

2a: 64%
2b: 50%

[1]

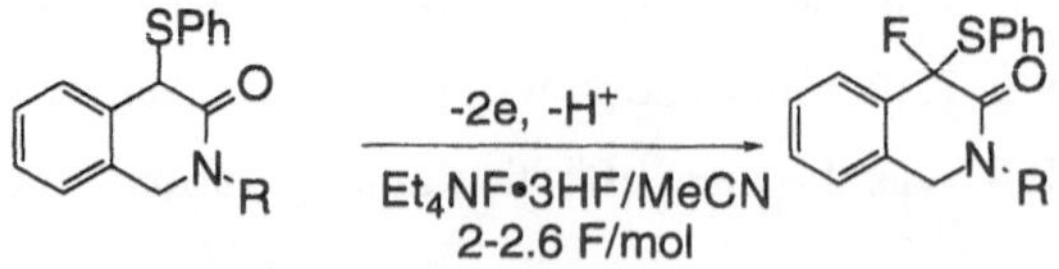
SPh
O
N
R
-2e, -H+
Et4NF•3HF/MeCN
2-2.6 F/mol
F
SPh
O
N
R

[2]

3a: R=H
3b: $R=CH_2Ph$

4a: 71%
4b: 70%

efficiencies as shown in Scheme 2.   Although **3b** has three kinds of benzylic carbons, the fluorination took place at the 4-position exclusively.   Therefore, it is noted that this anodic fluorination is highly regioselective.

In contrast to the cases of **1** and **3**, the corresponding oxindoles and 3-oxo-1,2,3,4-tetrahydroisoquinoline derivative devoid of a phenylthio group did not undergo selective anodic fluorination and complicated products owing to fluorination at the benzene rings were formed.   Therefore, the phenylthio group is essential in these selective fluorination.

Anodic Fluorodesulfurization of β-Lactams

Anodic monofluorination of α-phenylthio-β-lactams in Et$_3$N•3HF/MeCN was also successful to provide the corresponding α-fluorinated products in good yields and with high current efficiencies.[5]   In contrast, severe passivation of the anode took place during the anodic fluorination of β-phenylthio deivatives **5**.   In this case, pulse electrolysis was not so effective.   We found that indirect anodic oxidation of **5** in Et$_3$N•3HF/MeCN using triarylamine as a mediator was quite efficient for the fluorination as shown in Table 2.

Table 2. Anodic Fluorodesulfurization of β-Lactam **5a**[1]

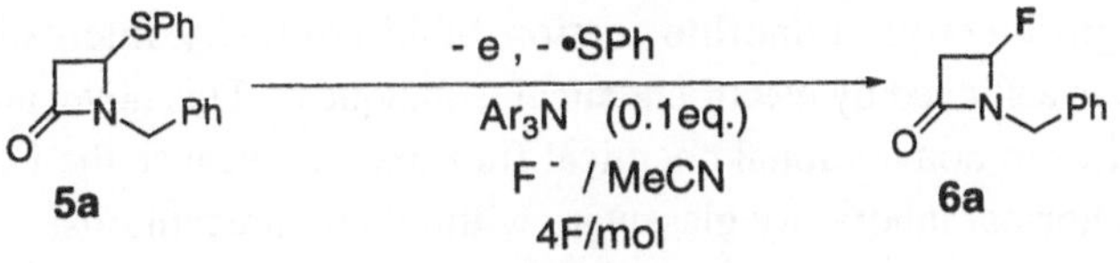

Run	Mediater (Ar$_3$N)	Supprting Electrolyte	Yield (%)
1	(4-BrC$_6$H$_4$)$_3$N	Et$_3$N•3HF	52
2	(2,4-Br$_2$C$_6$H$_3$)$_3$N	Et$_3$N•3HF	83
3	(2,4-Br$_2$C$_6$H$_3$)$_3$N	Et$_4$NF•3HF	43

1) Constant current electrolysis using a divided cell with an anion exchange membran

Tris(2,4-dibromophenyl)amine was more effective as a mediator than tris(2-bromophenyl)amine. It is noted that fluorodesulfurization took place efficiently without passivation of the anode by using such a triarylamine mediator.

Moreover, this mediatory system was also successfully applied to more complicated β-lactam derivative **5c** as shown in Scheme 4.

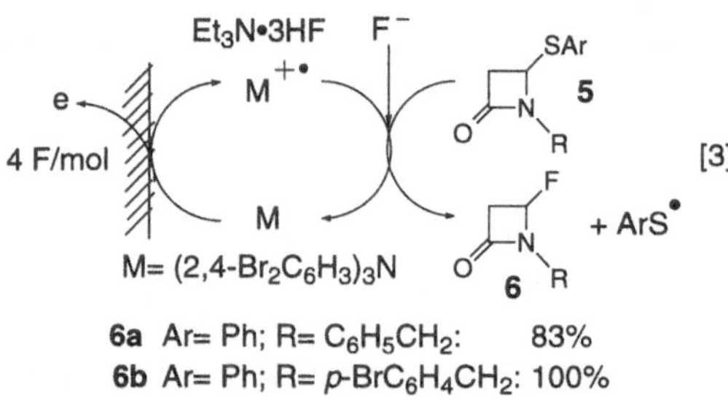

6a Ar= Ph; R= $C_6H_5CH_2$:   83%
6b Ar= Ph; R= p-$BrC_6H_4CH_2$: 100%

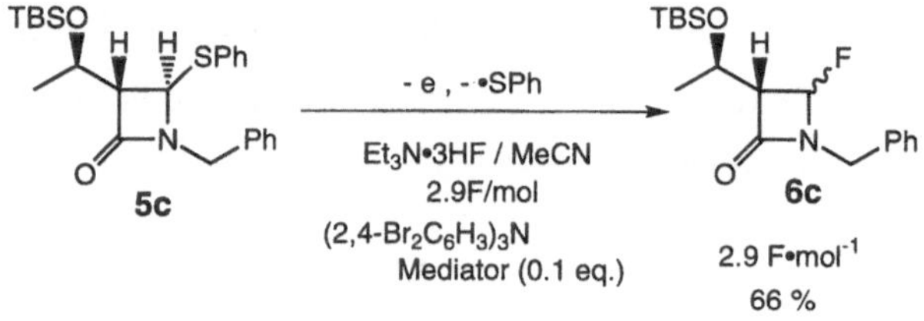

CONCLUSIONS

Highly regioselective monofluorination of biologically interesting heterocyclic compounds was achieved by electrochemical techniques.   This novel anodic fluorination is much superior to conventional chemical fluorination because the fluorination can be carried out in normal laboratory glassware without any precautions.

References
1) (a)T. Fuchigami, *Rev. Heteroatom Chem.*, **1994**, *10*, 155; (a) T. Fuchigami and A. Konno, *J. Synth. Org. Chem. Jpn.*, **1997**, *55*, 301; (b) T. Fuchigami and S. Nishiyama, *DENKI KAGAKU (J. Electrochem. Soc. Jpn.)*,**1997**, *65*, 626.
2) T. Fuchigami, S. Narizuka, and A. Konno, *J. Org. Chem.*, **1992**, *57*, 3755.
3) A. Konno, W. Naito, and T. Fuchigami, *Tetrahedron Lett.*, **1992**, *33*, 7017.
4) S. Narizuka and T. Fuchigami, *J. Org. Chem.*, **1993**, *58*, 4200.
5) S. Narizuka and T. Fuchigami, *Bioorg. Med. Chem. Lett.*, **1993**, *5*, 1293.
6) A. Konno, M. Shimojo, and T. Fuchigami, *J. Fluorine Chem.*, in press.
7) (a) Y. Hou, S. Higashiya, and T. Fuchigami, *SYNLETT.*, **1997**, 655;
   (b) Y. Hou, S. Higashiya, and T. Fuchigami, *J. Org. Chem.*, in press.

ELECTROCHEMICAL PARTIAL FLUORINATION OF α,β-UNSATURATED ESTERS

Norihiko YONEDA

Division of Molecular Chemistry, Graduate School of Engineering, Hokkaido University, Sapporo 060, Japan

ABSTRACT

A novel electrochemical partial fluorination of α,β-unsaturated esters efficiently took place accompanied by the rearrangement of the β-substituted alkyl groups to the α-position to afford gem-β,β-difluoroesters in good yields under the pulse electrolysis conditions using Et$_3$N-5HF electrolyte.

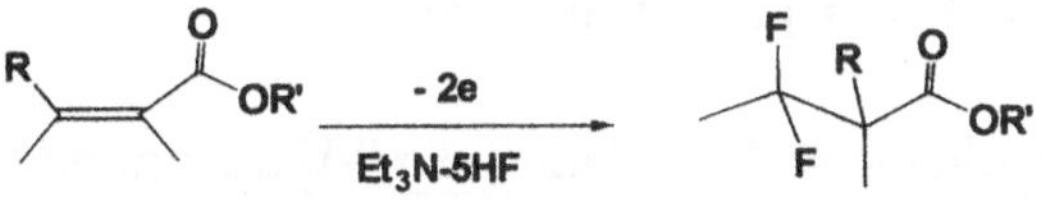

INTRODUCTION

Selective fluorination of organic compounds is of considerable interest in medicinal and materials chemistry. HF, which is the key material in fluorine chemistry, is an economical fluorine source for the fluorination of organic compounds. However, HF is an extremely hazardous chemical due to its low boiling point, high acidity and toxicity. In order to overcome these difficulties, HF combined with a Lewis organic base is widely employed as a convenient fluorination reagent[1]. Under these circumstances, electrochemical syntheses began to attract much attention among synthetic chemists due to their high energy efficiency and cleanliness, and electrochemical partial fluorination of organic compounds using a conventional electrolyte such as Et$_3$N-3HF or Et$_4$NF-nHF (n=3-4) which has been quickly developed over the past few years[2]. We have also developed the electrochemical fluorination of carbonyl derivatives using a novel Et$_3$N-5HF supporting electrolyte. Notably, Et$_3$N-5HF has been found to be electrochemically highly stable and an excellent electrolyte for the electrochemical fluorination of aldehydes and ketones to produce the corresponding acylfluorides and alkylfluorides in good yields[3]. In this paper, we report a novel electrochemical partial fluorination of α,β-unsaturated esters to produce gem- β,β-difluoro esters. The reaction mechanism will be also discussed for this unusual electrochemical partial oxidative fluorination of α,β-unsaturated esters.

RESULTS AND DISCUSSION

The electrochemical reaction of alkenes such as styrene, stilbene, pinene, butadiene and so on in conventional Et_3N-3HF electrolyte has been reported to give complex mixture of unidentified products or vic-difluoroalkanes in low yields.[4] On the contrary, as shown in Table 1, the gem-difluorination took place exclusively during the electrochemical oxidation of α,β-unsaturated esters in Et_3N-5HF electrolyte to afford β,β-gem-difluorocarboxylic esters as the major products accompanying the migration of β-alkyl groups to the α-position in the substrates in fairly good yields. Interestingly, under similar conditions, the electrochemical fluorination of simple alkenes such as dodecene-1 did not take place at all but exclusively gave a reductive product, dodecane.

Table 1.   Electrochemical Fluorination of α,β-Unsaturated Esters[a].

Substrate	Potential (V vs Ag/Ag⁺)	Q (F/mol)	Product	Yield/%
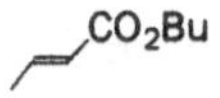 CO₂Bu	2.8	4.0	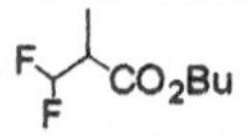 F₂C–CO₂Bu	50
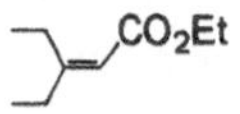 CO₂Et	2.5	4.0	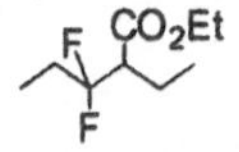 F,F CO₂Et	47 (40)[b]
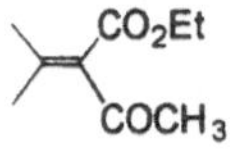 CO₂Et, COCH₃	2.5	3.5	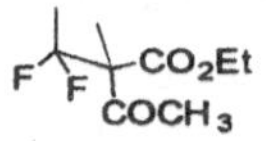 F,F CO₂Et, COCH₃	37
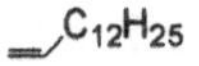 C₁₂H₂₅	2.4	2.0	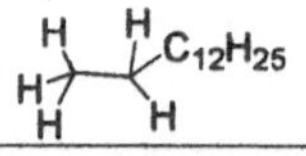 C₁₂H₂₅ (dodecane)	60

a) Conditions: -20°C, Supporting electrolyte; Et_3N-5HF.   b) Isolated yield.

Cycloalkylidenacetates(**1**), on the other hand, exhibited much more conspicuous results during this novel electrochemical partial fluorination of α,β-unsaturated esters to produce fluoroesters(**2-5**) with or without ring expansion as shown in the following equation:

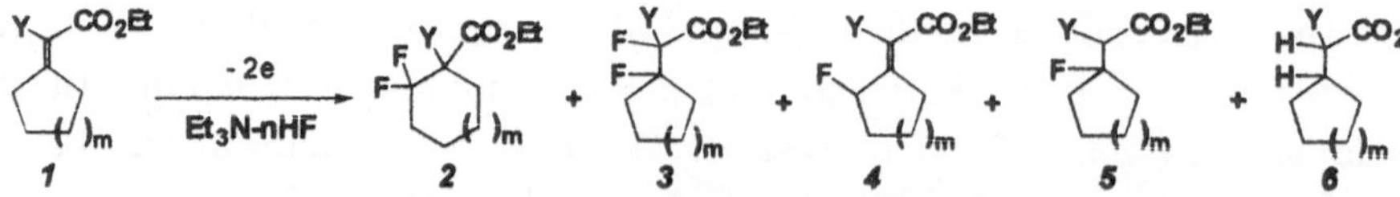

As shown in Table 2, product distribution is greatly influenced by the ring size and the substituents Y located at the α-position in **1**. Namely, by employing the Et_3N-5HF electrolyte, substrate **1** having 5- and 6-membered rings caused ring

Table 2. Electrochemical Fluorination of Cycloalkylidenacetates(1) [a]

m	Y	Conditions Q F / mol	V (vs Ag/Ag⁺)	Temp °C	Yield %	Product Distribution / % 2	3	4	5	6
1	H	4.0	2.4	-20	56	~99	0	0	0	Trace
1	CO₂Et	7.0	2.4	-20	51	100	0	0	0	0
2[b]	H	4.0	2.3	-20	20	92	0	0	0	8
2	H	4.0	2.3	-20	76	93	0	0	0	7
2	H	2.5	2.2	50	71	49	0	0	0	51
2	CH₃	2.8	1.8	-20	83	73	20	0	7	0
2	CO₂Et	4.0	2.4	-20	52	100	0	0	0	0
3	H	4.2	2.4	-40	67	48	0	52	0	0
3	CO₂Et	5.0	2.4	-40	22	100	0	0	0	0

a) The reaction was carried out in Et₃N-5HF under the pulse electrolysis conditions
with exchanging electrode using undivided Cell.   b) Electrolyte: Et₃N-3HF.

expansion with the incorporation of two fluorines in a ring to exclusively afford the 6-
and 7-membered di-fluorocycloalkanecarboxylic esters(2) in good yields, respectively.
On the other hand, curiously, during the reaction of cycloheptylideneacetate (7
membered ring), a simple mono-fluorination at the γ-position in 1 competitively took
place affording γ-fluoro-cycloheptylideneacetate (4) as major product together with
the formation of the ring expanded di-fluorocyclooctanecarboxylic ester 2. However,
the presence of other electron withdrawing groups such as an ester group at α-
position in 1 brought about the exclusive formation of 2.   On the other hand, when
Y is a methyl group in the starting esters, the correspondin α,β-di-fluoro (3) and β-
mono-fluoro (5) esters were produced together with   the   formation   of   the   ring

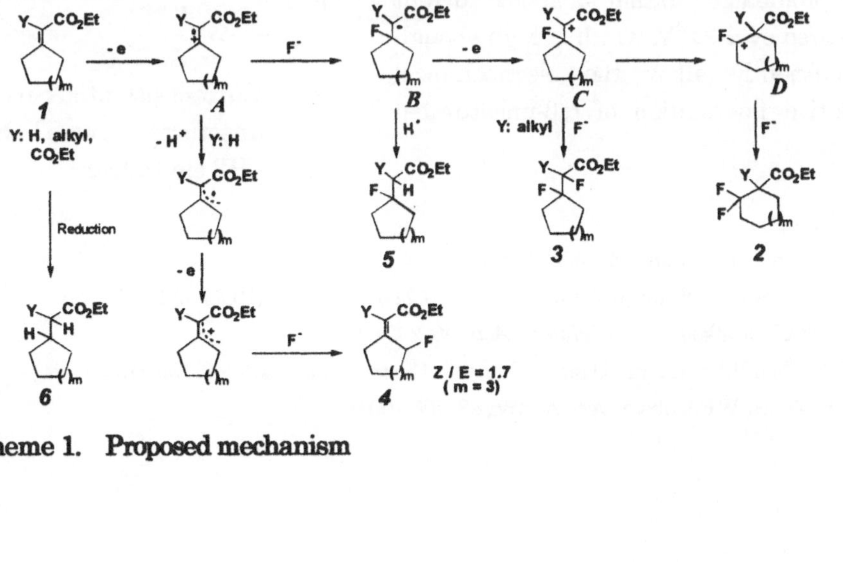

Scheme 1.   Proposed mechanism

expanded α-methyl-β,β-di-fluoro cycloheptane carboxylic ester *2* as a major product.

The reaction probably proceeds as follows. Mono-electron oxidation and fluorination of the substrate *1* gives a β-fluoroalkyl radical (*B*). The subsequent oxidation of *B* may yield the unstable β-fluorocarbocation (*C*), which readily rearranges to the more stable intermediate (*D*) accompanied by ring expansion and eventually the final product *2* by reaction with fluoride ion. When intermediate *C* is stabilized by the *Y* (= methyl) group in *1*, subsequent fluorination brings about the formation of *3.* The γ-fluorinated product *4* may come from the initially formed intermediate (*A*) by its subsequent deprotonation, oxidation, and fluorination sequence. Products *5* and *6* may be formed by electroreduction at cathode.

Replacement of Et₃N-5HF by commercially available Et₃N-3HF resulted in a significant decrease in the yields of the corresponding products. Judging from the current-potential curve of the substrates, as shown in Fig. 1, the half-wave potential of the substrates is observed to be 1.9-2.6 V. The anodic oxidation of Et₃N-3HF started at 2.0 V, of which the electrochemical stability is not sufficiently high enough for an electrolyte to perform the desired electrochemical oxidative fluorination of α,β-unsaturated esters. On the contrary, the Et₃N molecule in Et₃N-5HF solution is fully protonated[3] so that its anodic current appeared over 2.8 V, which is high enough to efficiently allow the electrochmical oxidative fluorination of α,β-unsaturated esters.

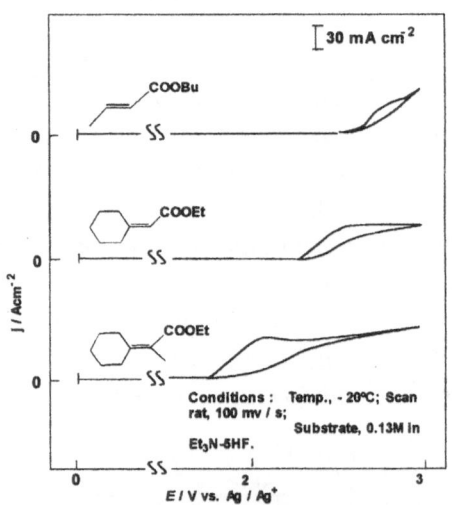

Fig. 1. Current-potential curves of unsaturated esters in Et3N-5HF electrolyte

References

1) N.Yoneda, *Tetrahedron*, **47**, 5329 (1991).

2a) T.Fuchigami, M.Shimojo, A.Konno, *J. Org. Chem.*, **60**, 3495 (1995). b) K. Momota, K. Kato, M. Morita, Y. Matsuda, *Electrochimica Acta*, **40**, 233 (1995).

3) S.-Q.Chen, T.Fukuhara, S.Hara, N.Yoneda, *Electrochemica Acta*, **42**, 1951(1997).

4) J.H.Meurs, W.Eilenberg, *Tetrahedron*, **47**, 705(1991).

ELECTRO-OXIDATION OF ENAMINES USING IODIDE ION AS A MEDIATOR

Toshiro CHIBA, Isao SAITHO, and Mitsuhiro OKIMOTO
Department of Applied Chemistry, Kitami Institute of Technology, Kitami, Japan 090

ABSTRACT

A series of pyrrolidinoenamines of an alicyclic ketone (**1**) was subjected to the indirect electrooxidation in NaCN-MeOH using iodide ion as a mediator. It was found that the electrolysis products were dependent on the ring size of the ketone component. For example, the cyclohexanone and cycloheptanone enamines exclusively provided a ring fused cyclopropane aminonitrile (**2**), whereas the cyclooctanone enamine yielded only the corresponding β-cyanoenamine (**4**).

INTRODUCTION

Previously, we reported that the anodic oxidation of enamines in the presence of organic anions derived from β-dicarbonyl compounds gives rise to nucleophilic substitution with these anions at the carbon β to the nitrogen atom.[1] As an extension of this reaction, we attempted analogous electrolysis in the presence of cyanide ion as a nucleophile and found that indirect electrolysis using iodide ion as the mediator gave a unique product, such as cyclopropane aminonitriles (**2, 3**), depending on the structure of the enamine. Of particular interest is that the present reaction proceeds stereoselectively with cyclohexanone and cycloheptanone enamines(**1a, 1b**). In these cases, only *exo*-nitriles (**2a, 2b**) are formed.

Scheme 1

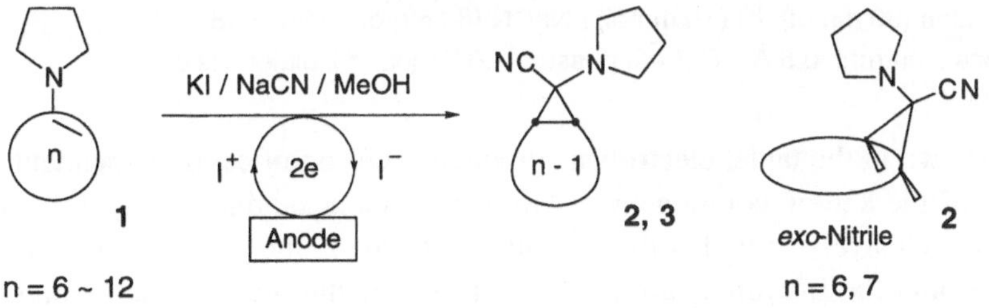

n = 6 ~ 12

n = 6, 7

RESULTS AND DISCUSSION

Preparative electrolyses were performed in a divided cell using a platinum anode. The anolyte, consisting of the enamine, a 3-fold excess of sodium cyanide, and a catalytic amount of KI in methanol, was electrolyzed at a constant current at room temperature. After 2 F/mol of charge had been passed through the cell, the electrolyzed solution was treated in the usual manner, and the product was isolated by distillation or column chromatography.

The representative results are summarized in Table 1.

Table 1. Electrooxidation of Pyrrolidine Enamines in NaCN / KI / MeOH[a]

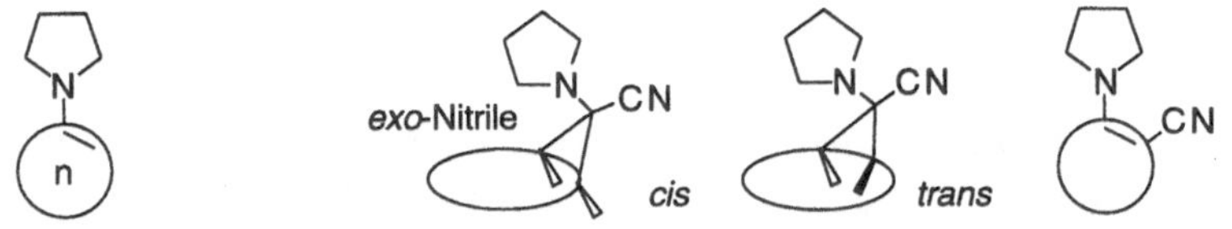

enamine		exo-Nitrile	cis	trans	
		2	**3**	**4**	
1a	n = 6	**2a** 63			0
1b	7	**2b** 60			0
1c	8	0		**4c**	61
1d	9	**2d** 14	**3d** 18	**4d**	33
1e	10	**2e** 12	**3e** 20	**4e**	36
1f	11	**2f** 31	**3f** 22		0
1g	12	**2g** 31	**3g** 17		0

[a] Enamine (30 mmol), KI (10 mmol) / NaCN (80 mmol) / MeOH (80 mL).
Constant current, 0.5 A. Current passed, 2.0 F/mol. [b] Isolated yield.

As shown in the table, electrolysis products were significantly dependent on the ring size of the ketone component. Although cyclohexanone and cycloheptanone enamines exclusively provided the *exo*-nitrile of type **2**, the cyclooctanone enamine (**1c**) gave only the β-cyanoenamine (**4c**) instead of the bicyclic compound. The cyclononanone and cyclodecanone enamines (**1d, 1e**) also produced the corresponding cyanated enamine (**4d, 4e**) together with the bicyclic compounds which consist of a mixture of *cis* and *trans* isomers (**2, 3**). The cyanated enamines

4 were no longer produced from the cycloundecanone and cyclododecanone enamines(**1f, 1g**).   In these cases, the predominant products were mixtures of the *cis* and *trans* isomers of the bicyclic compounds **2** and **3**.

The structures of the electrolysis products were confirmed by elemental analysis, IR, NMR, Mass spectra, and by their reactions.

For example, in the case of norcarane aminonitrile (**2a**), the IR spectrum showed a characteristic absorption arising from the cyano group near 2216 cm[-1], but no C=C bond.   The ^{1}H NMR spectrum of **2a** was very complex, and therefore, we could not employ a difference NOE technique to assign the structure.   In contrast to this, the ^{13}C NMR spectrum was quite simple.   In this case, only seven carbon signals were observed, including the quaternary and nitrile carbon signals, due to the symmetry of the molecule. The presence of the cyclopropane in a fused-ring system was confirmed by converting it into the known 7-*endo*-pyrrolidino-norcarane.[2)]   The reductive decyanation was successfully performed using sodium-liquid ammonia. The stereochemical configuration of **2a** was determined as follows.

At first, we found that the aminonitrile **2a** undergoes thermal isomerization to give the stereoisomer (**2a'**) upon heating in ethylene glycol at 200 °C.   Probably, the isomerization proceeds through a transition state or intermediate such as the iminium cation, and it may adopt a more energetically favorable conformation in which the pyrrolidine ring is *trans* to the six-membered ring.   The two isomers were then converted into the methyl ester (**6a, 6a'**), respectively, which were compared with an authentic sample of the 7-*endo*-methoxycarbonyl-7-*exo*-pyrrolidino-norcarane (**9**) that had been independently prepared.

Scheme 2

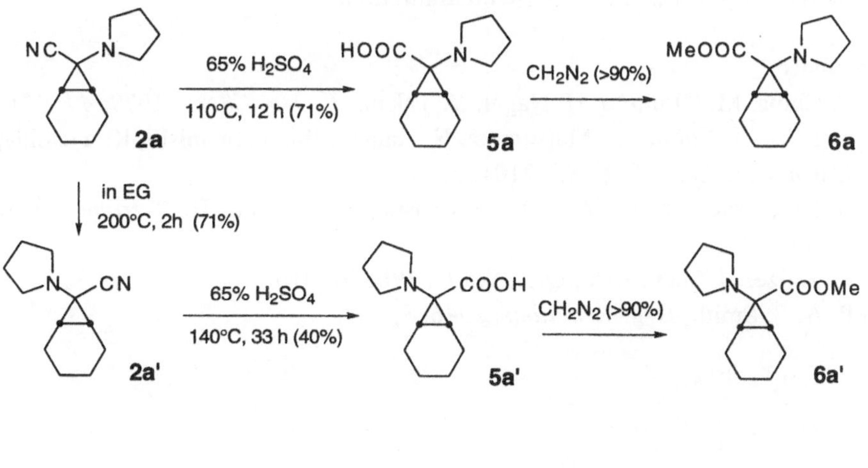

The *endo*-ester **9** was prepared from 7-*exo*-carboxyl-7-*endo*-methoxycarbonyl norcarane (**8**) by employing the Curtius rearrangement according to the following scheme.[3]

Scheme 3

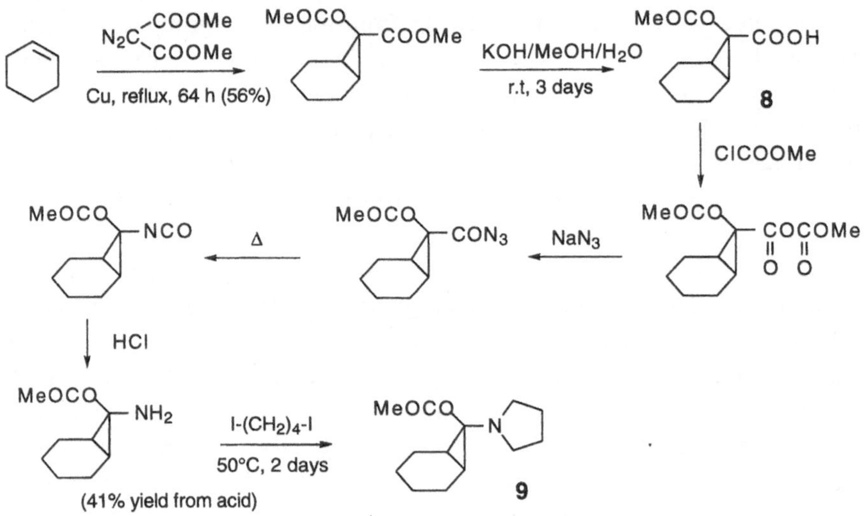

N₂C(COOMe)(COOMe), Cu, reflux, 64 h (56%) → MeOCO–COOMe → KOH/MeOH/H₂O, r.t, 3 days → MeOCO–COOH (**8**)

8 → ClCOOMe → MeOCO–COCOMe (O O) → NaN₃ → MeOCO–CON₃ → Δ → MeOCO–NCO

MeOCO–NCO → HCl → MeOCO–NH₂ (41% yield from acid) → I-(CH₂)₄-I, 50°C, 2 days → MeOCO–N(pyrrolidine) (**9**)

The *endo*-ester **9** prepared in this way was identical to ester **6a'** that was derived from isomer **2a'**, therefore, the electrolysis product **2a** was confirmed to be the *exo*-nitrile, since it has been knoun that the Curtius rearrangement occurs with retention of optical and geometric configuration.[4]

References
1) T. Chiba, M. Okimoto, H. Nagai, Y. Takata, *J. Org. Chem.*, **1979**, *44*, 3519.
   see also; T. Shono, Y. Matsumura, H. Hamaguchi, T. Imanishi, K. Yoshida, *Bull. Chem. Soc. Jpn.*, **1978**, *86*, 2101.
2) J. Szmuszkovicz, E. Cerda, M. F. Grostic, J. F. Zieserl, Jr., *Tetrahedron Lett.*, **1967**, 3969.
3) C. Kaizer, J. Weinstock, *Org. Synth.*, **1988**, *6*, 910.
4) P. A. S. Smith, *Org. Reactions*, **1946**, *3*, 337.

A CONVENIENT ELECTROCHEMICAL O-GLYCOSYLATION OF THIOGLYCOSIDE USING CATALYTIC Br

Junzo NOKAMI, Masahiro OSAFUNE, and Shin-ichi SUMIDA

Department of Applied Chemistry, Faculty of Engineering, Okayama University of Science, Ridai, Okayama 700, Japan

O-Glycosylation of thioglycosides **1** with alcohols **3-5** was successfully performed by electro-oxidation in the presence of catalytic amount of ammonium or metal bromide to afford O-glycosides **2**.

Intensive efforts have been devoted to the synthesis of oligosaccharides because of their numerous and diverse biological functions. Thioglycosides have been widely utilized for O-glycosylations in the presence of soft electrophiles or heavy metal salts. On the other hand, Balavoine[1] and Amatore[2] performed electrochemical O-glycosylations of arylthioglycosides although excess amounts of electrolyte and electricity were required to complete the reactions. We, therefore, sought more efficient method of O-glycosylation, and found that electrolysis using of catalytic amount of ammonium or metal bromide as electrolyte accomplished the O-glycosylation in high current efficiency. Herein, we describe that the electrooxidative O-glycosylation of phenylthioglycosides **1** with alcohols **3-5** in the presence of catalytic amount of ammonium or metal bromide, leading to O-glycosides **2** (Scheme 1).

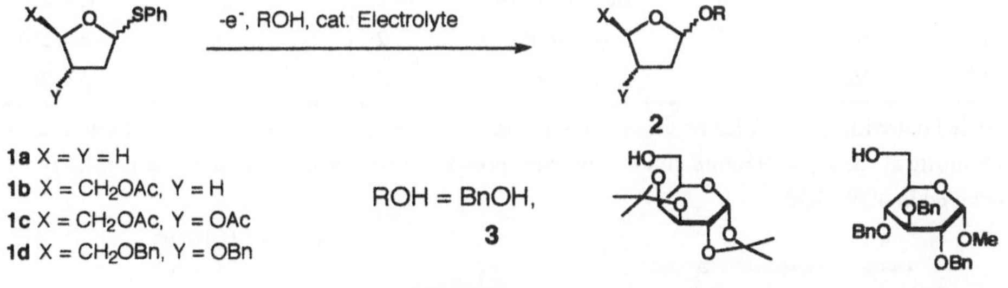

1a X = Y = H
1b X = CH₂OAc, Y = H
1c X = CH₂OAc, Y = OAc
1d X = CH₂OBn, Y = OBn

ROH = BnOH, **3**

Scheme 1   **4**   **5**

A typical procedure is as followed: electrolysis of **1** was carried out in an undivided cell fitted with platinum electrodes at room temperature. Thus, a mixture of **1a**, benzyl alcohol **3** (1.5 equiv.), and sodium bromide (0.1 equiv.) in acetonitrile was electrolyzed under a constant current (5 mA/cm²). After most of starting material **1a** was consumed (1 F/mol), work up of the electrolytes afforded **2a** (89%) (Table 1, entry 1). Notably, the yields of **2a** are depending on the electrolyte, decreasing in the following order: NaBr (89%)>KBr (86%)>MgBr₂•6H₂O (77%) (entries 1-3). Next, the electrochemical O-glycosylation of phenylthioglycosides **1a-d** with alcohols **4** and **5**, derived from sugars, was investigated. Treatment of **1a** with **4** in the presence of NaBr gave **2b** in only 31% yield (entry 4) while

Bu_4NBr effectively worked, resulting in increase of the yield of **2b** (52%) (entry 5). Therefore, the latter reagent was employed for the synthesis of the *O*-glycosides **2c-f**. The electrolysis of **1b-d** with **4** or **5** in the presence of Bu_4NBr (0.1 or 0.01 equiv.) proceeded smoothly to afford **2c** (63%), **2d** (15%), **2e** (54%) and **2f** (62%), respectively (entries 6-9).

Although the reaction mechanism is not clear at present, it is likely that $Br^+/Br^•$ redox system may promote the *O*-glycosylation of the phenylthioglycosides **1** with alcohols **3-5**. The active species, "Br^+", would be initially formed from "Br^-" and regenerated from "$Br^•$" in the electrooxidative media. (Scheme 2).

Table 1. *O*-Glycosylation of Phenylthioglycosides 1 by Electrochemical Oxidation[a]

Entry	Substrate	ROH	Electrolyte (equiv.)	Current (mA/cm^2)	Electricity (F/mol)	Yield of **2** (%) (Product)
1	**1a**	**3**	NaBr (0.1)	5	1	89 (**2a**)
2	**1a**	**3**	KBr (0.1)	5	1	86 (**2a**)
3	**1a**	**3**	$MgBr_2 \cdot 6H_2O$ (0.1)	5	1	77 (**2a**)
4	**1a**	**4**	NaBr (0.1)	5	1	31 (**2b**)
5	**1a**	**4**	Bu_4NBr (0.1)	5	1	52 (**2b**)
6	**1b**	**4**	Bu_4NBr (0.1)	2.5	1.5	63 (**2c**)
7[b]	**1c**	**4**	Bu_4NBr (0.01)	2	1.5	15(**2d**)[c]
8[b]	**1d**	**4**	Bu_4NBr (0.01)	2	1	54 (**2e**)
9[b]	**1d**	**5**	Bu_4NBr (0.01)	2	1.5	62 (**2f**)

[a]Carried out with alcohol (1.5 equiv.) in an undivided cell using platinum electrodes (1 cm^2 x 2) in acetonitrile at room temperature unless otherwise noted. [b]Propionitrile was used as solvent. [c]**1c** was recovered in 49% yield.

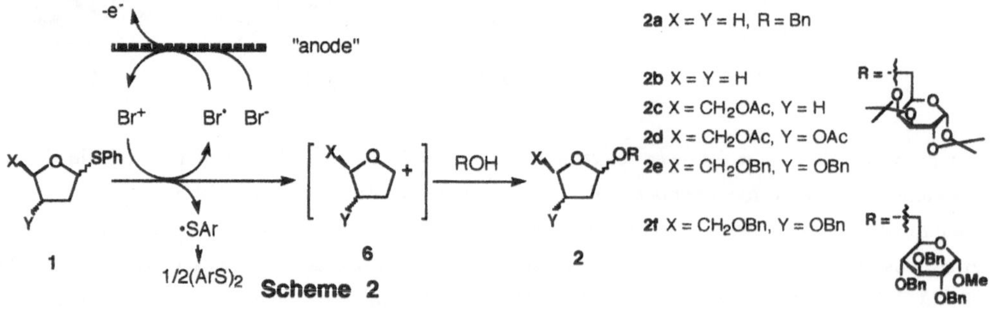

-e⁻
"anode"
Br+
Br• Br⁻
SPh
•SAr
1/2(ArS)2
ROH
OR
1
6
2
2a X = Y = H, R = Bn
2b X = Y = H
2c X = CH2OAc, Y = H
2d X = CH2OAc, Y = OAc
2e X = CH2OBn, Y = OBn
2f X = CH2OBn, Y = OBn
R =
OBn
OMe

Scheme 2

References

1) G. Balavoine, S. Berteina, A. Gref, J.-C. Fischer, and A. Lubineau, *J. Carbohydr. Chem.* **1995**, *14*, 1217, *ibid.* **1995**, *14*, 1237.

2) C. Amatore, A. Jutand, G. Meyer, P. Bourhis, F. Machetto, J.-M. Mallet, P. Sinaÿ, C. Tabeur, Y. M. Zhang, *J. Applied Elctrochem.* **1994**, *24*, 725 and the references cited therein.

INDIRECT ANODIC FLUORODESULFURIZATION USING TRIARYLAMINE MEDIATORS

Toshio FUCHIGAMI
Department of Electronic Chemistry, Tokyo Institute of Technology
Nagatsuta, Yokohama 226, Japan

ABSTRACT

Indirect anodic *gem*-difluorodesulfurization of dithioacetals and monofluoro-desulfurization of β-phenylthio-β-lactams were successfuly carried out by using triarylamine mediators.

INTRODUCTION

In contrast to electrochemical perfluorination, electrochemical partial fluorination has not been develped well.   One of main reasons is passivation of the anode.   Pulse electrolysis is effective in some cases, however electrolysis is very often impossible due to the inevitable severe passivation.   From these viewpoints, we have developed indirect electrochermical partial fluorination using various mediators.[1-3]

In this work, indirect anodic *gem*-difluorodesulfurization of dithioacetals and monofluorodesulfurization of β-phenylthio-β-lactams were attempted using triarylamines as a mediator.

RESULTS AND DISCUSSION

Indirect Anodic *gem*-Difluorodesulfurization of Dithioacetals Using Triarylamine Mediators.

An enhanced catalytic oxidation current was observed in the cyclic voltammogram in the presence of dithioacetal **1** and triarylamine **3** in $Et_4NF \cdot 3HF/CH_2Cl_2$.

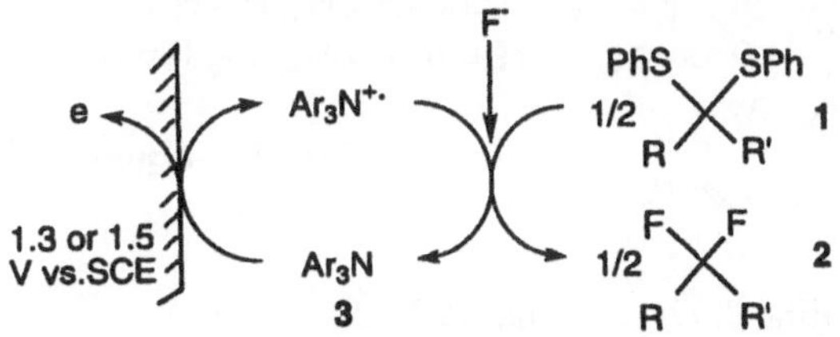

Fig. 1

Macro-scale electrolysis of **1** was carried out at a constant potential where the triarylamine **3** was selectively oxidized to provide the desired *gem*-difluoro product **2** in moderate to good yields. As a supporting electrolyte, $Et_3N \cdot 3HF$ gave better results than $Et_4NF \cdot 3HF$.

Table 1. Indirect Anodic Difluorodesulfurization of Dithioacetals

Run	Dithioketal		Charge Passed	Ar_3N	Yield of 2 [a]
	R	R'	(x96480C / mol)	Ar	(%)
1	Ph	Ph	4	$p\text{-}BrC_6H_4$	58
2	$p\text{-}FC_6H_4$	Ph	5	$p\text{-}BrC_6H_4$	83
3	$p\text{-}FC_6H_4$	$p\text{-}FC_6H_4$	5	$p\text{-}BrC_6H_4$	58
4[b]	$p\text{-}ClC_6H_4$	$p\text{-}ClC_6H_4$	5	$o,p\text{-}Br_2C_6H_3$	74
5	PhS SPh 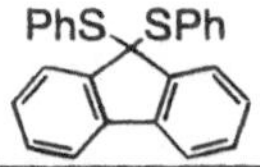		5	$p\text{-}BrC_6H_4$	61
6[b]			5	$o,p\text{-}Br_2C_6H_3$	76

a) Determined by ^{19}F NMR   b) 1.5 V vs SCE

Indirect Anodic Fluorodesulfurization of β-Phenylthio-β-lactams

We have alreadyreported that anodic monofluorination of α-phenylthio-β-lactams in $Et_3N \cdot 3HF$ provided the corresponding α-fluorinated β-lactams in good yields and with high current efficiencies.[4] In contrast, severe passivation of the anode took place during the anodic fluorination of β-phenylthio deivatives **4**. In this case, pulse electrolysis was not so effective. We found that indirect anodic oxidation of **4** in $Et_3N \cdot 3HF/MeCN$ using triarylamine (0.1 equiv.) as a mediator was quite efficient for the fluorination as shown in Figure 2. It is noted that fluorodesulfurization proceeded efficiently without passivation of the anode in this case.

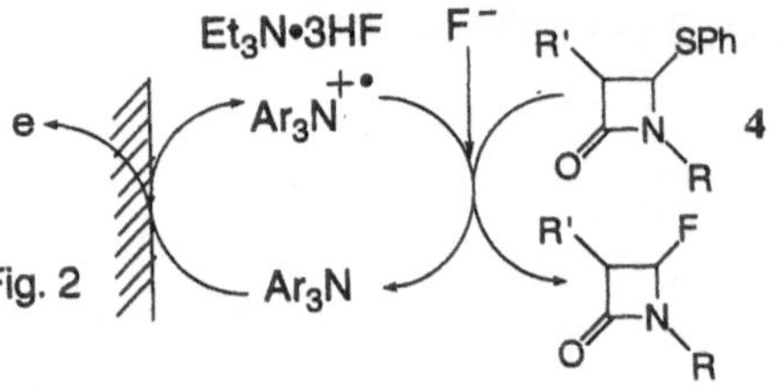

Fig. 2

$Ar= 2,4\text{-}Br_2C_6H_3$
$R= C_6H_5CH_2$; R'= H: 83%
$R= p\text{-}BrC_6H_4CH_2$; R'= H: 100%
$R= C_6H_5CH_2$; R'= $t\text{-}BuMe_2SiOCH\text{-}$: 66%

References
1) T. Fuchigami and T. Fujita, *J. Org. Chem.*, **1994**, *51*, 7190.
2) T. Fujita and T. Fuchigami, *Tetrahedron Lett.*, **1995**, *37*, 4725.
3) T. Fuchigami and M. Sano, *J. Electroanal Chem.*, **1996**, *414*, 81.
4) S. Narizuka and T. Fuchigami, *J. Org. Chem.*, **1993**, *8*, 4200.

ELECTROOXIDATION OF CYCLOHEXANOL TO CYCLOHEXANONE IN THE PRESENCE OF A MEDIATORY SYSTEM

Maria OLEA, Ioan IOSUB[*], and George SEMENESCU[*]

Faculty of Chemistry and Chemical Engineering, "Babes-Bolyai" University, 3400 Cluj, Romania, [*]University of Pitesti, Faculty of Sciences, 0300 Pitesti, Romania

ABSTRACT

The paper deals with the study of both the catalytical and the electrocatalytical oxidation of cyclohexanol to cyclohexanone, because, sometimes, the electrochemical option, can be much better than the chemical one. However, the rate of the direct electrooxidation of cyclohexanol is lower than the rate of catalytic reaction hence, the use of a mediatory system is proposed.

INTRODUCTION

It is very well known that the cyclohexanone is the main intermediate in the adipic acid synthesis, an acid used most frequently and with good results in obtaining the polyamide fibers. Cyclohexanone also has an important utility as a solvent for lacquers and synthetic paints. In Romania cyclohexanone is used on a large scale and it is obtained by a solid – catalyzed reaction.

Though at its beginnings in my country, the electroorganic synthesis has offered solutions that are more suitable for this subject, especially in the production of some aromatic amines by the electroreduction of the suitable nitrocompounds (1).

That's why we have considered the study of the electrooxidation process of cyclohexanol more appropriate.

SOLID – CATALYZED REACTION

In our laboratories we have studied the oxidation reaction of cyclohexanol to cyclohexanone in a gas phase, in the presence of a Zn – Cr catalyst. We used a fixed bed 40 cm^3 reactor; the catalyst particles had a 3 – 5 mm diameter.

As we know this oxidation reaction is in fact a heterogeneous process, its rate could be controlled both by diffusion and by electrode kinetics, because the following stages could occur in its unfolding:

- the reactants' diffusion from the main body of the fluid to the exterior surface of the catalyst;
- surface phenomena (adsorption, reaction, and desorption);

- the products' diffusion from the catalyst into the main fluid stream.

We studied the influence of temperature as well as space–velocity on the rate of the process, in order to find out the rate determining step (rds).

A linearly increase of the reaction yield with space–time is obtained, for a given temperature. That means that the gas film resistance is important. Also, for a given space – time, the conversion depends on temperature. Thus the importance of the surface phenomena is confirmed.

The analysis of the reaction products has been performed by gas – chromatografic methods as well as by chemical ones.

Practically, for the space–times longer than one hour the gas film resistance becomes less important while the surface steps are more important. Because all the three surface steps are influenced by temperature, the surface reaction was chosen as a rds, because the reaction yield was not influenced by the cyclohexanol concentration.

Therefore, the rate expression could be derived from the LHHW theory (2) leading to:

$$r_p = \frac{kKC_{ol}}{1 + KC_{ol}} \qquad (1)$$

Where r_p is the rate of process, k is the reaction rate constant, K is the adsorption constant, and C_{ol} is the cyclohexanol concentration. One can determine the K value from adsorption measurements and the expression of k from experimental data, by an integral analysis. Our values were in good agreement with the ones in the specific literature (3).

DIRECT ELECTROOXIDATION

By its heterogeneous feature and by the fact that the same steps occur in its development the electrooxidation resembles the solid – catalyzed oxidation.

The direct electrochemical oxidation of alcohols has been surveyed by many scientists (4 – 7) and their conclusion was that the chemical yield is very high (8). But, unfortunately, as the current efficiency is very low in all cases, its industrial application is not efficient. This is revealed by the experiments we performed on a PAR Potentiostat Galvanostat.

First, we obtained the polarization curves, in a divided electrolytic cell, on platinum electrodes, in both acidic (H_2SO_4 - 0.07 M) and basic media (NaOH – 0.07M) and 5M cyclohexanol concentration, and then we obtained the cyclic voltammograms and the chronocoulometric plots, using an undivided cell, on platinum electrodes, in an acidic medium only. The electrooxidation reaction was not efficient either in an acidic medium or in a basic one, only a maximum $5A/m^2$ current density was obtained. However an improvement of the rate of electrochemical reaction is obtained when the stirring is present. Consequently, we can say that both diffusion and kinetics control the rate of the process (9). This statement is in a good agreement with the CV results.

The chronocoulometric experiments have proved that cyclohexanol, in its first reaction step, is adsorbed on the electrode surface, so that the conclusion drawn before that the two oxidation processes (chemical and electrochemical) look alike, is confirmed once again. Thus, by the CV, we have studied the influence of cyclohexanol concentration, cyclohexanol-water ratio and pH on the reaction rate. At the beginning, we found out that the anodic peak at 2.4V *vs nhe* is caused by the irreversible oxidation of cyclohexanol to cyclohexanone (in fact it is a quasireversible oxidation because the plot i_{peak} versus scan rate has had a maximum at the 20mV/s)(10).

The anodic peak current is not influenced by the C_{ol} (we made experiments at a few concentration values, such as: 1.2M, 2.4M, 3.6M, and 4.8M) and so, we can say that the adsorption doesn't control the reaction rate. Then, because the value of i_{peak} is low and because the ratio $i_{peak}/v^{1/2}$ (where v is the scan rate) is not constant, the mixed control for the reaction is proved. But, when the mass transfer is improved, the rds becomes the surface reaction and the expressions of the reaction rate of electrochemical and chemical process are the same.

INDIRECT ELECTROOOOXIDATION

The presence of a mediatory system improves the value of the reaction rate. For example, Shono and co-workers (11-13) have mentioned that electrooxidation of secondary alcohols to ketones, in the presence of KI occurred with best yields; the active intermediate was thought to be I^{+}. In Savéant's terminology (14) this process is called "chemical catalysis with electrochemical regeneration", Shono uses the expression "heteromediatory system"(8) while Simonet (15) says that this indirect oxidation is an inner sphere process".

As mediatory systems we have used: I_2/KI, I_2, NaI, KI, and CsI.

The CV studies have confirmed the catalytical effect of all these substances, which, in a small concentrations (about 0.01M) and in their oxidation states could increase the oxidation rate. We have here an EC scheme reaction (indeed, in all these cases, in CV the return peak is smaller or absent, that means the reaction half life is much lower than the scan duration). The indirect electrooxidation has a great advantage because as the chemical step does not depend on the potential, the potential need is 950mV lower than in the case of direct oxidation. The highest rates of reaction have been obtained in the presence of I_2/KI system, due to the fact that both I_2 and I_3^{-} (I_2 + KI <=> I_3^{-}) have oxidation potentials and the reaction occurs both in a chemical and in an indirect electrochemical manner (C_1EC_2). The reaction rate decreases with the scan rate that means that the C_2 reaction rate is lower than the E-rate. At this time, the CV experiments show that the direct electrooxidation of cyclohexanol didn't occur for concentrations of cyclohexanol smaller than 5M. When I_2 was used as a mediatory system, the reaction rate wasn't so big as in the case of I_2 /KI system, but it was bigger than in the other cases.

Let us consider now the case of iodides. As we expected, the best results were obtained in the presence of CsI (the mobility of Cs^+ is bigger than K^+ and Na^+ ones). Both the CV experiments and chronocoulometric plots show a mixed mechanism (namely indirect and direct electrooxidation), even at the C_{ol} smaller than 5M. From the CV we have determined the optimum pH value (such is 2.8). The rds is the chemical one, therefore the reaction kinetics could be a homogeneous one. Neverthless, further experiments are needed to prove that.

CONCLUSIONS

The CV and chronocoulometric experiments allow us to choose the optimum conditions of pH and Col, and the mediatory system that made the indirect electrooxidation of cyclohexanol more suitable than the solid-catalyzed one.

Another advantage is the fact that the indirect electrooxidation occurs at room temperature.

References
1) M. Olea, L. Oniciu, and G. Petran, *Proceedings of Chemical and Chemical Engineering Conference,*Bucharest, Roumania, **1995**, 270.
2) O. Levenspiel, *Chemical Reaction Engineering,* John Wiley & Sons, NY, **1972**, Chapter 14.
3) B.M. Mirovskoya, A.S. Bedrian, *USSR Khim. Prom. St.,* **1974,** (8), CA **1974**, 81, 160055x.
4) G. Horany, P. Konig, and I. Telcs, *Acta Chim. Acad. Sci. Hung.,* **1972**, 72(2), 165.
5) P.C. Scholl, S.E. Lentsch, and M.R. Van de Mark, *Tetrahedron*, **1976**, 32, 303.
6) J.E. Leonard, P.C. Scholl, T.P. Steckel, S.E. Lentsch, and M.R. Van de Mark, *Tetrahedron Lett.*, **1980**, 21, 4695.
7) J. Kaulen, and H.J. Schafer, *Tetrahedron ,* **1982,** 32, 3299.
8) T. Shono, *Electroorganic Chemistry as a New Tool in Organic Synthesis*, Springer - Verlag, Tokyo, **1984**, Chapter 2.3 and 2.9.
9) K. Koster, and H. Wendt, *Electro - Organic Synthesis,* Chapter 4 in Comprehensive Treatice of Electrochemistry, Eds. J.O'M. Bockris, B.E. Conway, and E. Yager, **1981.**
10) A. J. Bare, L.R. Faulkner, *Electrochemical Principles. Methods and Applications*, Mason, Paris, **1983**, Chapter 6.
11) T. Shono, Y. Matsumura, J. Hayashi, and M. Mizoguchi, *Tetrahedron Lett.*, **1979**, 165.
12) T. Shono, Y. Matsumura, J. Hayashi, and M. Mizoguchi , *Tetrahedron Lett.*, **1979**, 3861.
13) T. Shono, Y. Matsumura, J. Kayashi, and M. Mizoguchi, *Tetrahedron Lett.*, **1980**, 21, 1867.
14) C.P. Andrieux, J.M. Dumas - Bouchiat, and J. M. Savéant, *J. Electroanal. Chem. Interfacial Electrochem.*, **1978**, 87, 39.
15) J. Simonet, *Electrogenerated Reagents in Organic Electrochemistry. An Introduction and a Guide*, Eds. H. Lund and M.M. Baizer, Marcel Dekker Inc., NY, **1991**, Chapter 29.

Electrochemical Oxidation of Diols Mediated by Organotin Compounds

Toshihide MAKI, Kazuhiro FUKAE, Hitomi HARASAWA, Takahiro OHISHI, and Yoshihiro MATSUMURA*

Faculty of Pharmaceutical Sciences, Nagasaki University
1-14 Bunkyo-machi, Nagasaki 852, Japan

ABSTRACT

A new diol selective electrochemical oxidation method has been developed utilzing dibutyltin oxide and bromide anion as mediators. The oxidation proceeded effectively at 0°C under nutral condition. Cyclohexane diol was selectivly oxidized even in the presence of excess primary or secondary alcohols.

INTRODUCTION

It has been known that dibutylstannylenes **A** are oxidized with bromine to keto-alcohols **C** (eq 1).

This reaction was first reported by S. David et al. [1] and has been developed by Y. Tsuda et al. as a versatile oxidation method of a variety of sugars to oxo-sugars without protection-deprotection procedure.[2] In this method, however, dibutylstannylenes **A** are prepared from dibutyltin oxide with diols prior to brominolysis and therefore usually more than two equivalent of dibutyltin oxide is required to complete the reaction.[2] We wish to present here a high diol selective electrochemical oxidation mediated by dibutyltin oxide.

Table Electorochemical Oxidation of cis-1,2-cyclohexanediol(*cis*-1)

Run	Catalyst(equiv.)	Temperature	Current(mA)	electlicity(F/mol)	Yield of C_1 (%)*
1	$Bu_2Sn=O$ (0.1)	r. t.	100	2.0	58
2	$Bu_2Sn=O$ (0.1)	r. t.	100	3.5	41
3	$Bu_2Sn=O$ (0.1)	r. t.	50	2.0	57
4	$Bu_2Sn=O$ (0.1)	reflux	100	2.0	36
5	non	0°C	100	2.0	11
6	$Bu_2Sn=O$ (0.1)	0°C	100	2.0	95
7	$Bu_2Sn=O$ (0.1)	-20°C	100	2.0	84
8	$Bu_2Sn=O$ (0.05)	0°C	100	2.0	96
9	$Bu_2Sn=O$ (0.02)	0°C	100	2.8	94

* Determined by GLC.

135

METHOD

Electrolysis of *cis*-1 in the presence of catalytic amount (0.1 to 0.02 equiv.) of dibutyltin oxide in methanol selectively afforded 2-hydroxycyclohexanone (C_1) (see runs 6, 8, 9 in Table). The result was summarized in the Table. Considerably lower yield of C_1 was observed if the electrolysis was carried out without dibutyltin oxide (see run 5). These facts suggest that the oxidation was mediated by dibutyltin oxide. The yield of C_1 was dramatically changed with the reaction temperature (see runs 1, 4, 6, 7) and it was found that the electrolysis at 0°C afforded the best result.

A typical procedure was as follows; A methanol solution (10ml) containing *cis*-1 (1mmol) and dibutyltin oxide (0.1mmol) was refluxed for 30min prior to electrolysis to form dibutylstannylene. The resulting solution was placed in a glass cell equipped with Pt anode and cathode (1cm x 2cm), and then subjected to constant current electrolysis (100mA) in the presence of tetraethylammonium bromide (1mmol) as a supporting electrolyte.

The electrolysis proceeded with high chemoselectivities. For example, when the electrolysis was carried out in the presence of cyclohexanol (5 equiv.), C_1 was obtained in 84% yield with a 99% recovery of cyclohexanol after 2F/mol of electricity was passed (eq.2).

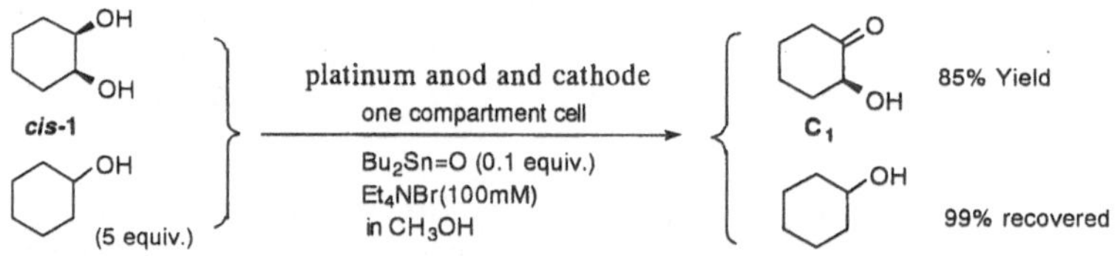

The mechanism of the process is considered as the Scheme depicted below.

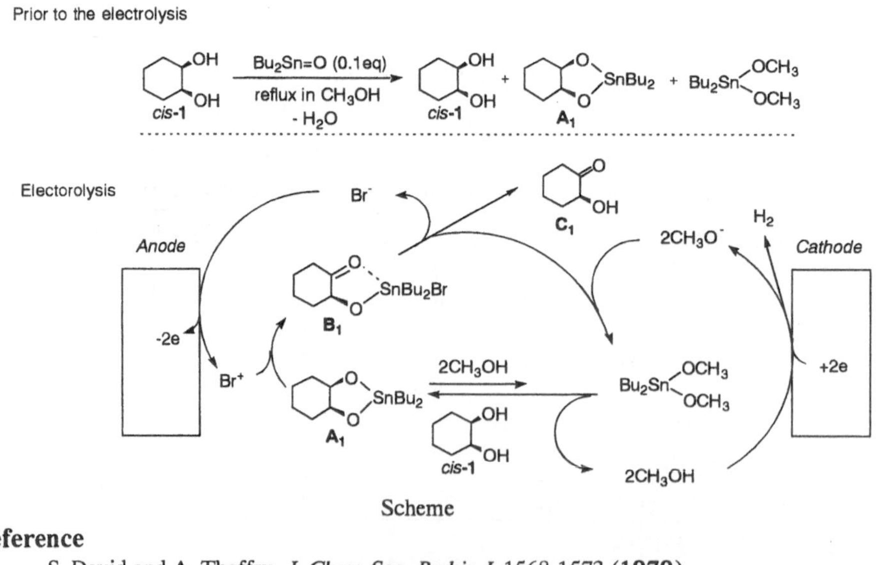

Scheme

reference

1.   S. David and A. Theffry, *J. Chem Soc. Perkin I*, 1568-1573 (**1979**).
2.   Y. Tsuda, M. Hanajima, and K. Yoshimoto, *Chem. Pharm. Bull.*, **31**, 3778-3788 (1983).

PHOTOINDUCED ELECTROCHEMICAL OXIDATION OF ALCOHOLS UTILIZING FLAVIN ANALOG MEDIATORS IN ORGANIC ELECTROLYTES

Masashi ISHIKAWA, Kazunori WATANABE, and Masayuki MORITA
Department of Applied Chemistry and Chemical Engineering, Faculty of Engineering, Yamaguchi University, 2557 Tokiwadai, Ube 755, Japan

ABSTRACT

Flavinmononucleotide, a dopant anion, in polypyrrole derivatives as electrode matrices acts as an efficient mediator in photoinduced electrochemical oxidation of benzyl alcohol to benzaldehyde in MeCN containing $HClO_4$ under visible-light irradiation.

INTRODUCTION

Many workers have applied flavins to electron donors and acceptors for substrates and to electrochemical mediators in aqueous systems. However, little has been known about application of flavins to electrochemical mediators in organic electrolyte systems[1]. We previously reported the photoinduced electrochemical oxidation of benzyl alcohol to benzaldehyde in the presence of riboflavin-2',3',4',5'-tetraacetate (RF) as a mediator in MeCN containing $HClO_4$[2,3]. The present study reports that flavinmononucleotide (FMN), a dopant anion, in polypyrrole derivatives acts as a mediator in the photoinduced electrochemical oxidation of benzyl alcohol in MeCN containing $HClO_4$.

PREPARATION OF ELECTRODES

FMN-doped polypyrrole (FMN-PPy) and FMN-doped poly-N-methylpyrrole (FMN-PMePy) electrodes were prepared on a Pt electrode by a constant-current oxidation method (current density: 1 mA/cm^2) in a water solution containing pyrrole or N-methylpyrrole (0.1 M) and sodium salt of FMN (0.01 M).

RESULTS AND DISCUSSION

Cyclic voltammetry measurements revealed that both the FMN-PPy and FMN-PMePy electrodes had electrochemical activity in MeCN in the presence of appropriate amount of $HClO_4$. The activity was ascribed to the redox of incorporated FMN as well as the polypyrrole derivatives. The electrochemical oxidation of benzyl alcohol (1 x 10^{-2} M) was performed under visible light irradiation (> 310 nm) and an argon atmosphere in MeCN containing $HClO_4$ (5 x 10^{-2} M) using the FMN-PPy or FMN-PMePy electrodes. The constant potential (0.75 V vs. Ag/AgCl) polarization of the FMN-PPy electrode under the irradiation yielded ca. 80 % conversion of benzyl alcohol into benzaldehyde in 4 h as shown in Fig. 1. Although the FMN-PPy electrode showed high activity within middle reaction period, the electrode degraded after 4 h polarization, resulting in about 80 % final conversion. On the other hand, the FMN-PMePy electrode showed no degradation and high activity throughout the electrolysis; the final conversion of benzyl alcohol into benzaldehyde reached to about 95 % in 9 h as given in Fig. 1. The FMN-PMePy system showed high activity in the photoinduced electrochemical oxidation in comparison with our previous homogeneous mediator system where riboflavin-tetraacetate (RF) acts as a mediator soluble itself in MeCN[3], as shown in Fig. 2. The electrochemical oxidation of benzyl alcohol (1.0×10^{-2} M) proceeded at a GC electrode polarized at 0.5 V vs. Ag/AgCl in the presence of RF (3.3×10^{-3} M) and $HClO_4$ (6.7×10^{-3} M) in MeCN to yield benzaldehyde when the cell system under an argon atmosphere was irradiated with the visible light (> 310 nm)[3]. The conversion into benzaldehyde reached to about 90% in 25 h in the homogeneous system [3].

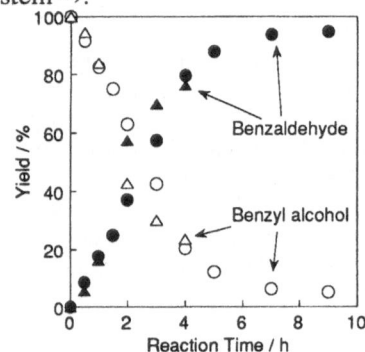

Fig.1 Oxidation of benzyl alcohol at FMN-PPy ($\triangle$,$\blacktriangle$) and FMN-PMePy ($\bigcirc$,$\bullet$) electrodes in MeCN containing benzyl alcohol (1×10^{-2} M), TEAP (0.1M) and $HClO_4$ under irradiation (> 310 nm), applied voltage: 0.75V(vs. Ag/AgCl).

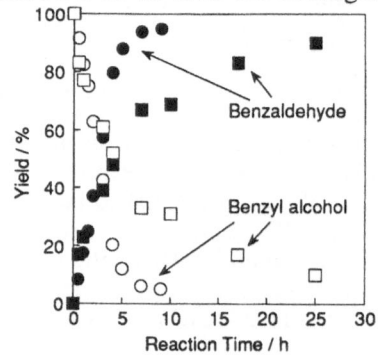

Fig.2 Oxidation of benzyl alcohol at FMN-PMePy electrode ($\bigcirc$,$\bullet$) and GC electrode with RF mediator[3] ($\square$,$\blacksquare$) in MeCN containing benzyl alcohol (1×10^{-2} M), TEAP (0.1 M) and $HClO_4$ under irradiation (> 310 nm), applied voltage: 0.75V(for FMN-PMepy), 0.5V(for GC electrode with RF).

It should be emphasized that the present FMN-modified electrode systems are excellently efficient mediatory systems; although the amount of FMN as a mediator in the present electrodes is much less than that of the RF mediator in our previous homogeneous system (the amount of RF in the previous system is about 10^4-fold excess over that of FMN in the present systems), the activity against the oxidation of benzyl alcohol in the present systems is much higher than that in our previous system.

References
1) M. Ishikawa, Y. Takahashi, M. Morita, and Y. Matsuda, *Denki Kagaku*, **1994**, *62*,1227.
2) M. Ishikawa, H. Okimoto, M. Morita, and Y. Matsuda, "*Novel Trends in Electroorganic Synthesis*" ed. by S. Torii, Kodansha, Tokyo, p. 389 (1995).
3) M. Ishikawa, H. Okimito, M. Morita, and Y. Matsuda, *Chem. Lett.*, **1996**, 953.

APPLICATION OF TRANSITION METAL SUBSTITUTED α-KEGGIN SILICON POLYOXOTUNGSTATES AS OXIDATION ELECTROCATALYST

Masahiro Sadakane and Eberhard Steckhan*

Kekulé-Institut für Organische Chemie und Biochemie der Universität Bonn

Gerhard-Dormagk-Str. 1, 53121, Bonn, Germany

ABSTRACT

Transition-Metal (Mn, Co, Fe, and Cr) substituted silicon polyoxotungstates ($SiW_{11}O_{39}M^{n-}$) can be used successfully as electrocatalyst for alcohol oxidation. The behavior of a manganese substituted silicon polyoxotungstate ($K_6SiW_{11}O_{39}Mn(II)(H_2O)$) under electrochemical oxidation conditions was investigated by means of cyclic voltammetry and the formation of the active complex $SiW_{11}O_{39}Mn(OH)$ was confirmed.

INTRODUCTION

Considerable attention has been directed towards transition metal substituted α-Keggin heteropolyanions as analogues of metalloporphyrins[1]. Similar to metalloporphyrins, transition metal substituted α-Keggin heteropolyanions are able to oxidize various organic substrates. The heteropolyanion ligands are robust under the oxidation conditions and thus have an important advantage over metalloporphyrin systems, which decompose under these conditions.

Here, we want to present the electrochemical oxidation of alcohols catalyzed by a transition metal substituted silicon tungstate and its electrochemical behavior.

ANODIC OXIDATION OF ALCOHOLS CATALYZED BY TRANSITION-METAL SUBSTITUTED SILICON POLYOXOTUNGSTATES[2, 3]

Electrochemical catalytic activity of several different transition metal substituted silicon polyoxotungstates was examined by using 1-phenyl-ethanol as substrate (Table 1). All heteropolyanions show catalytic activity and the material yields and the current yields are higher than in the blank experiment (Entry 1) for all heteropolyanions examined (Entry 2-5). The best catalyst in terms of material yield is the manganese substituted silicon polyoxotungstate. Constant potential electrolysis of 1-phenyl-ethanol at 1.25 V in the presence of 5 mol% of $K_6SiW_{11}O_{39}Mn(II)$ in pH 6.0 phosphate buffer produced acetophenone in 61 % material at a current yield of 52 %. No formation of side products was observed and 38 % of the starting material were recovered (Entry 2). Under this condition, other alcohols as cyclohexenol and benzylalcohol derivatives could be oxidized to the corresponding carbonyl compounds.

Table 1. Anodic oxidation of 1-phenyl-ethanol catalyzed by heteropolyanions

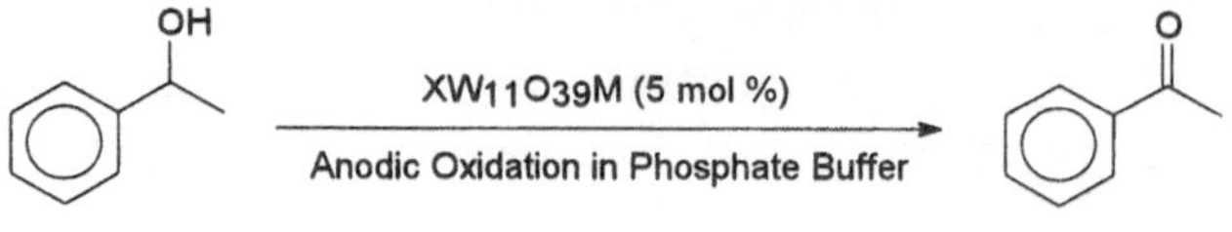

Entry	Catalyst	Yield	Current Yield
1	-	2.9 %	6.4 %
2	$K_6SiW_{11}O_{39}Mn(II)$	61.0 %	52.4 %
3	$K_6SiW_{11}O_{39}Co(II)$	19.8 %	30.2 %
4	$K_5SiW_{11}O_{39}Fe(III)$	10.2 %	27.1 %
5	$K_5SiW_{11}O_{39}Cr(III)$	34.4 %	65.0 %

Condition : Alcohol (50 mM), Phosphate Buffer ($[K^+]$ = 0.5 M, pH = 6.0), Divided Cell, Carbon (Workingelectrode), Pt (Counterelectrode), 1.25 V v.s. Ag/AgCl (3 M KCl), r.t., 4 h..

CYCLIC VOLTAMMETRY OF α-$SiW_{11}O_{39}Mn$ COMPLEX[2)]

The cyclic voltammetry of the complex was carried out in pH 6.0 phosphate buffer (Fig. 1). With continuous scanning, the first oxidation peak shifted to a more negative potential, and the second oxidation peak developed with increasing number of scans. Additionally, all peak currents increased. With continuous cycling, all peaks grew with the number of scans and finally reached a stable value (consolidated peak).

By measuring the effect of pH and the concentration of the potassium cation of the consolidated peak (Fig. 2) we have confirmed that the consolidated peaks corresponds to the following redox reaction (eq. (1) and (2)).

$$K_nSiW_{11}O_{39}Mn(II)(H_2O)^{(6-n)-} = K_{n-1}SiW_{11}O_{39}Mn(III)(H_2O)^{(5-n)-} + e^- + K^+ \qquad (1)$$

$$K_{n-1}SiW_{11}O_{39}Mn(III)(H_2O)^{(5-n)-} = K_{n-1}SiW_{11}O_{39}Mn(IV)(OH)^{(5-n)-} + e^- + H^+ \qquad (2)$$

The shape-change of the CV was explained as follows:

1. A counter anion (L) of the electrolyte substituted the aquo ligand.

$$K_nSiW_{11}O_{39}Mn(II)(H_2O)^{(6-n)-} + L = K_nSiW_{11}O_{39}Mn(II)(L)^{(6-n)-} + H_2O \qquad (3)$$

2. The Mn(II)(L) complex is oxidized to Mn(IV)(L) at the eletrode surface in two steps (eq. (4) and (5)).

$$K_nSiW_{11}O_{39}Mn(II)(L)^{(6-n)-} = K_{n-1}SiW_{11}O_{39}Mn(III)(L)^{(5-n)-} + e^- + K^+ \qquad (4)$$

$$K_{n-1}SiW_{11}O_{39}Mn(III)(H_2O)^{(5-n)-} = K_{n-1}SiW_{11}O_{39}Mn(IV)(H_2O)^{(4-n)-} + e^- \tag{5}$$

3. At the Mn(IV) oxidation state, the counter anion is replaced by a hydroxy anion (eq. (6)).

$$K_{n-1}SiW_{11}O_{39}Mn(IV)(L)^{(4-n)-} + HO^- = K_{n-1}SiW_{11}O_{39}Mn(IV)(OH)^{(5-n)-} + L \tag{6}$$

The counter anion L is supposed to be the HPO_4^{2-} of the buffer solution.

By continuous cycling of CV, the Mn(L) complex is converted to the manganese aquo complex at the electrode surface. The substitution of the aquo ligand at lower potentials on the reverse potential sweep is slow on the time-scale of cyclic voltammetry. Therefore, although in the bulk solution the aquo ligand is replaced by an electrolyte anion, upon repeated cycling, the aquo complex dominates in the double layer, leading to formation of the Mn(IV)(OH) complex upon oxidation as explained above. The Mn(IV)(OH) complex is formed at the potential of 1.25 V v.s. Ag/AgCl, which corresponds to the potential of the preparative electrolysis.

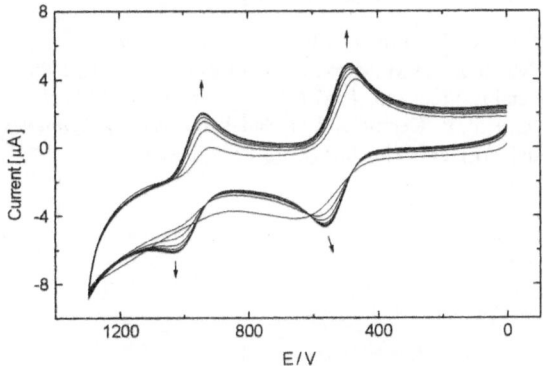

Fig. 1 Continuous cyclic voltammograms at a glassy carbon electrode for a 1 mM of $SiW_{11}O_{39}Mn$ in pH 6.0 phosphate buffer. Supporting electrolyte : 0.5 M phosphate solution. Scan rate = 20 mVs^{-1}. The initial potential was 0 V and the initial scan direction was towards more positive potential. The arrows indicate the direction of peak changes in the course of continuous scanning.

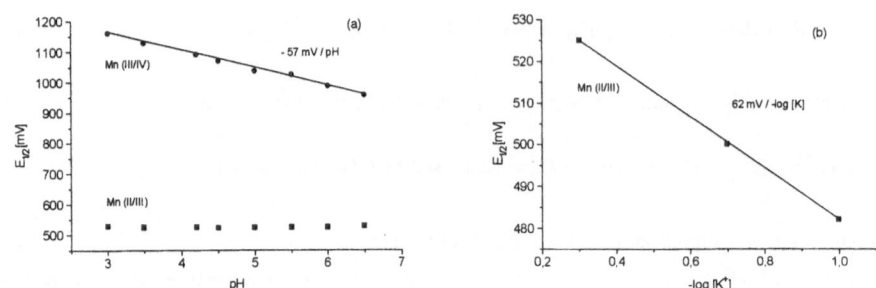

Fig. 2 (a) The pH effect and (b) the effect of the potassium cation concentration of the redox potential of the consolidated peaks

CONCLUSION

The preparative electrochemical oxidation of alcohols has been successfully performed. For the $SiW_{11}O_{39}Mn$, the most active complex, the activated Mn(IV)(OH) complex could be confirmed by using cyclic voltammetric analysis.

References

1) a) C. L. Hill and C. M. Prosser-McCartha, *Coord. Chem. Rev.*, **1995**, *143*, 407. b) T. Okuhara, N. Mizuno, and M. Misono, *Advances in Catalysis*, **1996**, *41*, 113.

2) M. Sadakane and E. Steckhan, *J. Mol. Cat. A: Chemical*, **1996**, *114*, 221. '

3) E. Steckhan, G. Hilt, R. Kempf, and M. Sadakane, *Organic Synthesis via Organometallics* (OSM5, Heidelberg), G. Helmchen (Ed.), **1997**, p 253.

NEW CONDUCTING POLYPYRROLES
FUSED WITH AROMATIC RINGS

Department of Chemistry, Faculty of Science, Ehime University, Matsuyama, Japan

Noboru Ono

ABSTRACT

Pyrroles fused with aromatic rings are prepared by the reaction of aromatic nitro compounds with ethyl isocyanoacetate followed by deethoxycarbonylation on treatment with KOH in ethylene glycol at 170 °C. Such pyrroles are good precursors for conducting polypyrroles and their optical and electrical properties are well controlled by the kinds of fused aromatic rings and substituents of aromatic rings.

INTRODUCTION

Chemical or electrochemical oxidation of numerous resonance stabilized aromatic molecules including pyrrole, thiophene, aniline, furan, carbazole, azulene and indole produces novel electronically conducting polymers.[1] Conducting polymers which are obtained by electrochemical or chemical polymerization promise many applications such as sensors, display-devices. In order to apply conducting polymer for electronic devices, we need well defined polymers. Molecular engineering of conducting polymers with the desired properties involves the design of the monomer by synthetic chemist, the design of polymer structure, the arrangement of polymer chain and the assembly of conducting polymers. The properties of such conducting polymers can be controlled by the monomer prior to polymerization.

Functionalization of polythiophenes or polyphenylenes has been extensively studied so far, but functionalization of polypyrroles has been less studied. This is due to the difficulty of synthesis of the requisite monomers with the required functionality. Recently we have found that Barton-Zard pyrrole synthesis[2] based on the reaction of nitroalkenes with ethyl isocyanoacetate is the excellent method for functionalization of pyrroles. It affords pyrroles with long alkyl groups, aryl groups or sugar molecules, which may be used as functionalized conducting polymers.[3] Furthermore the Barton-Zard reaction can be extended to aromatic nitro compounds to afford pyrroles fused with aromatic rings.[4] We report herein scope and limitation of this pyrrole synthesis from aromatic nitro compounds and application to the synthesis of new types of conducting polymers which are fused with aromatic rings.

RESULTS AND DISCUSSION

Various aromatic nitro compounds (**1**) including heteroaromatics react with ethyl isocyanoacetate to give the corresponding pyrroles (**2**). Polycyclic nitro aromatics were converted into **2** by the reaction in THF using DBU as base (procedure A).[5] However, monocyclic nitro compounds or some nitro heteroaromatics did not give **2** under these conditions. The use of more stronger nonionic iminophosphorane base in THF was required for the conversion of such nitro compounds into **2** (procedure B).[6] The ester group of **2** can be removed by heating with KOH in ethylene glycol to give α-free pyrrole **3** as a monomer for polypyrroles **4** as shown in Scheme 1. Thus, pyrroles fused with various aromatic rings are readily prepared from aromatic nitro compounds. They are good precursors for conducting materials, in fact, they are converted into polymers either by anodic oxidation or chemical oxidation. The conductivities of new polymers are about 0.01-4 Scm^{-1}.

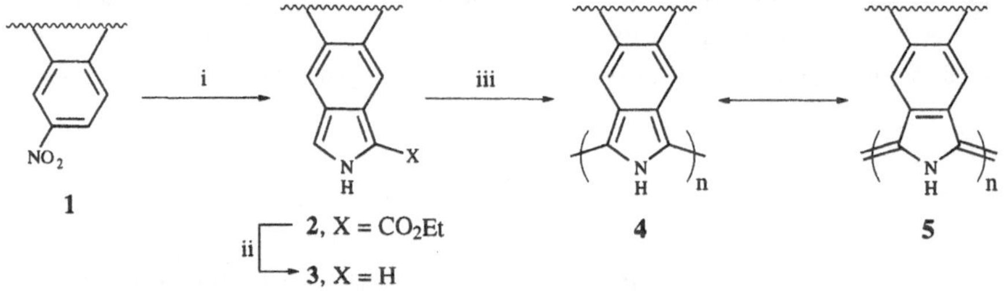

Scheme 1 : i, Procedure A : $CNCH_2CO_2Et$, DBU, THF, RT. Procedure B : $CNCH_2CO_2Et$, BTPP, THF. ii, KOH, $HOCH_2CH_2OH$, 170°C. iii, Anodic Oxidation, TBAP, MeCN.

Table 1. Preparation of Fused Pyrrole **3**.

Pyrrole	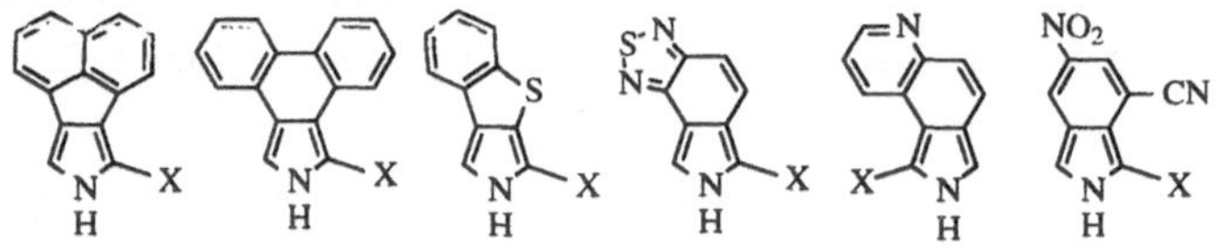					
Procedure	A	A	A	A	B	A
Yield (%) **2**	70	87	60	33	53	64
Yield (%) **3**	72	80	60	24	65	9

Electronic properties of pyrrole **3** and polymer **4** were determined by the CV method. Oxidation potentials of polymers refer to the top of the valence band. The band gap was determined by the absorption edge of the neutral polymer. So the bottom of the conduction band is estimated by these values. The dependency of these values on fused aromatic rings is shown in Fig. 1. Thus, fused pyrroles covers the band levels between polypyrroles and polythiophenes. As fused polycyclic systems such as **4** maintain conjugation between the polymer chain and aromatic rings fused with the pyrrole ring, the band gap, ionization potential (IP), electron affinity (EA) of the polymer can be effectively controlled by the fused aromatic rings.

Furthermore, the electronic properties of polypyrroles can be controlled by the change of substituents of fused aromatic rings or by copolymerization of different monomers. Introduction of electron withdrawing groups such as Cl or CN into the acenaphthene ring lowers an oxidation potential (valence band). Interestingly, the cyano substituted polymers lowers the band gap to be 1.90 eV. Copolymerization with thiophenes is also effective to lower their band gap. The pyrroles with long alkyl groups such as two hexyl groups are soluble in THF and their molecular weight is measured by GPC to be 13100.

Table 2. Electronic Properties of Polypyrroles.

Polymer	Oxidation Potential (V)	Band Gap (eV)
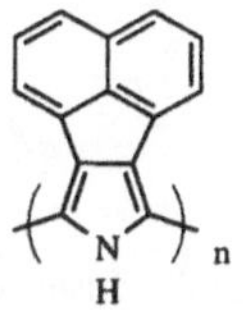	0.19	2.25
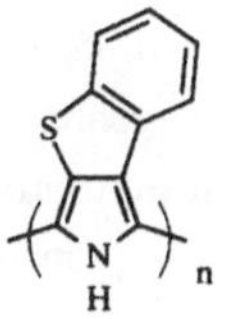	0.29	2.35
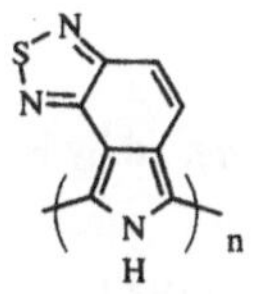	0.30	2.20
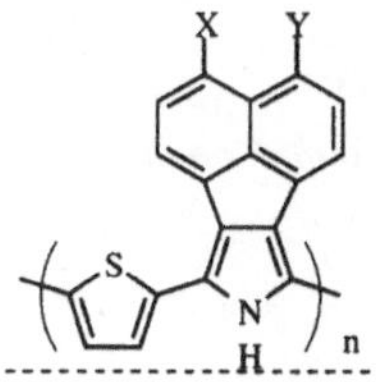 X,Y = H,H	0.29	1.80
X,Y = n-Hex,n-Hex	0.27	1.86
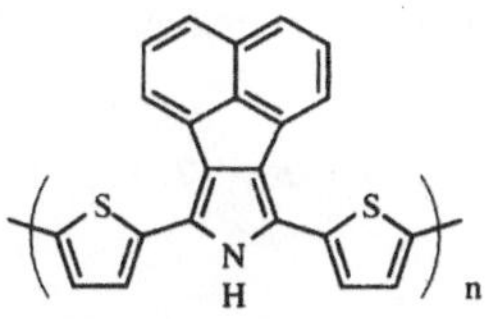	0.32	1.79

146

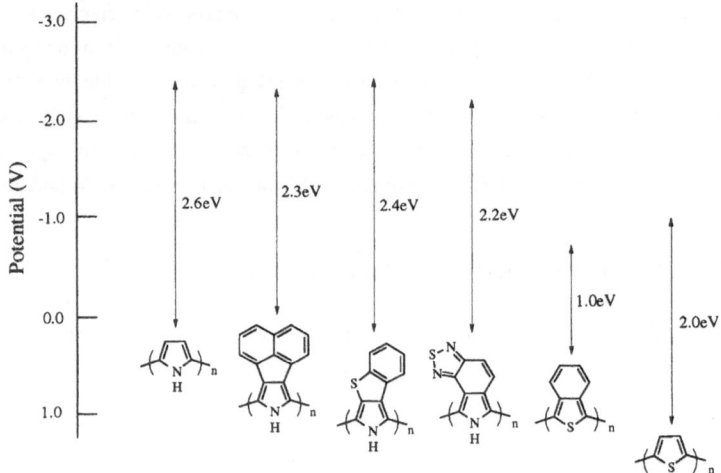

Fig. 1 Band Structure of Conductive Polymers

In summary, we have now established a new synthetic method for pyrroles fused with aromatic rings which are used as monomers for conducting polymers. The band structure which is main factor for the control of electrical properties of polymers is nicely controlled by the choice of the fused aromatic rings. The substituents on aromatic rings and the copolymerization with other monomers control the band structure very effectively. As the fused aromatic rings are derived from nitro aromatics, the present method provides a very effective way to control optical and electrical properties of conducting polymers.

REFERENCES

1) C. P. Evans, in *Advances in Electrochemical Science and Engineering*, ed by H. Gerischer and C. W. Tobias, Vol 1, VCH, Weiheim, 1-74 (1990)

2) D. H. R. Barton, J. Kervagoret, S. Z. Zard, *Tetrahedron*, **46**, 7587 (1990).

3) N. Ono, K. Maruyama, *Bull. Chem. Soc Jpn.*, **61**, 4470 (1980);
   N. Ono, M. Bougauchi, K. Maruyama, *Tetrahedron Lett.*, **33**, 1629 (1992);
   N. Ono, H. Kawamura, M. Bougauchi, K. Maruyama, *J. Chem. Soc., Chem. Commun.*, 1580 (1989).

4) N. Ono, H. Hironaga, K. Shimizu, K. Ono, K. Kuwano, T. Ogawa,
   J. Chem. Soc., Chem. Commun., 1019 (1994).

5) N. Ono, H. Hironaga, K. Ono, S. Kaneko, T. Murashima, T. Ueda, C. Tsukamura, T. Ogawa,
   J. Chem. Soc., Perkin Trans 1, 417 (1996).

6) T. Murashima, R. Tamai, K. Fujita, H. Uno, N. Ono, *Tetrahedron Lett.*, **37**, 8391 (1996).

ELECTROCHEMICAL CYCLOOLIGOMERIZATION OF 1,4-BIS(1,4-DITHIAFUL-VEN-6-YL)BENZENE AND ANALOGUES

Dominique LORCY, Philippe HASCOAT, Roger CARLIER, André TALLEC and Albert ROBERT

Laboratoire de synthèse et électrosynthèse organiques, UMR 6510, Université de Rennes, Campus de Beaulieu, 35042 Rennes, France

ABSTRACT

Intermolecular coupling of 1,4-dithiafulvenes and intramolecular coupling of bis(1,4-dithiafulvenes) upon electrochemical oxidation have been studied. In addition we report that 1,4-bis(1,4-dithiafulven-6-yl) benzene and a cyclic analogue undergo the same electrooligomerization process upon oxidation to afford a new kind of cyclophanes.

INTRODUCTION

The search for novel organic materials exhibiting new properties has represented an active field these past few years.[1] Within this frame, cage molecules incorporating electroactive units such as tetrathiafulvalenes (TTF) or analogues have received a growing attention. These molecules could be, for example, specific hosts for acceptor molecules and strong interactions are expected between the electroactive units. We recently reported the electrochemical synthesis of TTF vinylogues **2** starting from 1,4-dithiafulvenes **1**.[2] This oxidative dimerization of dithiafulvenes is very convenient and appears to be very general.[3] We have now extended it to the electrochemical synthesis of cage molecules incorporating TTF analogues.

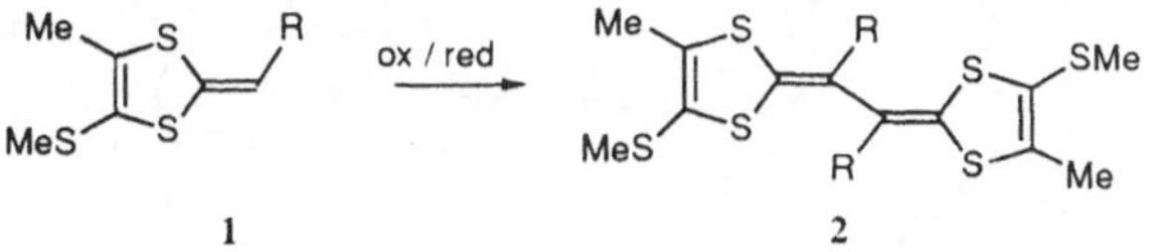

RESULTS AND DISCUSSION

Similar redox behavior was observed for 1,4-dithiafulvenes **1** and bis(1,4-dithiafulvenes) **3**.[4] The electroanalytical investigations were carried out at a platinum disk electrode (A = 1 mm^2) in a 10^{-3} M solution of **1** or **3** in acetonitrile, containing 1 M tetrabutylammonium hexafluorophosphate as the supporting electrolyte. The cyclic voltammogram of compound **3**, recorded at 0.1 V. s^{-1}, is reported in Figure 1 ; very similar results were obtained for **1**. As can be seen in Figure 1 on the first anodic scan, an irreversible oxidation peak (a) is observed. Then the new redox (b/c) system

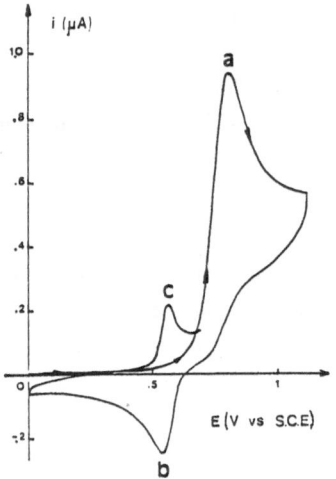

Figure 1 Voltammetric behavior of **3** (R = C$_6$H$_4$CN, A = (CH$_2$)$_4$) in CH$_3$CN, 1 M nBu$_4$NPF$_6$, platinum anode, scan speed : 0.1Vs^{-1}.

which appears upon successive scans is associated with the redox behavior of the formed cage **4**.[4] It is noteworthy that even with a short bisthioalkyl chain between the two dithiafulvene moieties the shape of the voltammograms remains unchanged (Figure 1). Mechanism of the oxidative dimerization of dithiafulvenes **1** to form TTF vinylogues **2** has been investigated by cyclic voltammetry at low and high scan rates for a series of substituted dithiafulvenes.[5] In this ECE process several reaction steps are involved. The first electron transfer corresponds to the formation of the cation radical. Then the C-C bond formation results from a fast irreversible dimerization of the generated cation radical into a protonated dimer. This dication slowly deprotonates (k=0.5-1 s^{-1}) to give the final dimer. This dimer is then oxidized at a lower potential than the starting 1,4-dithiafulvenes. The dimerization rate constant was found to be in the range of k$_{dim}$=(2 - 4)10^8 l. mol^{-1}. s^{-1} and does not vary much with the nature of the substituent.

Therefore we performed macroscale electrolyses in a divided cell, the working electrode being a platinum grid. In a typical experiment, 50 ml of an acetonitrile solution containing 2 mmol of **1** or 1 mmol of **3** and 1 M of tetrabutylammonium hexafluorophosphate are introduced in the working compartment. The solution is oxidized under controlled potential (0.6 V to 0.8 V vs aqueous SCE, depending on R). The highly coloured solution is then, without any treatment, reduced at -0.2 V SCE. Work up of the solution leads to neutral dimers **2** or **4**. The redox behaviour of these new dimers **2** and **4** were studied by cyclic voltammetry, the results are collected in Table I.

Table I Cyclic voltammetry data of dithiafulvenes **1**, bis(dithiafulvenes) **3** and TTF vinylogues **2** and **4**, E in V vs SCE, Pt working electrode with 1 M n-Bu$_4$NPF$_6$ 100 mV/s in CH$_3$CN

R	Ep1	E$_0$ 2	R	A	Ep 3	E$_0$ 4
C$_6$H$_4$NO$_2$	0.74	0.48	C$_6$H$_4$CN	(CH$_2$)$_3$	0.79	0.48
C$_6$H$_4$CN	0.69	0.44	C$_6$H$_4$CN	(CH$_2$)$_4$	0.77	0.54
C$_6$H$_5$	0.56	0.34	C$_6$H$_4$CN	(CH$_2$)$_3$	0.71	0.49
C$_6$H$_4$OMe	0.47	0.32	MeOC$_6$H$_4$	(CH$_2$)$_4$	0.50	0.42
C$_6$H$_4$NMe$_2$	0.28	0.23				
thiophene	0.5	0.35				
(CH$_2$)$_2$CH$_3$	0.58	0.32				

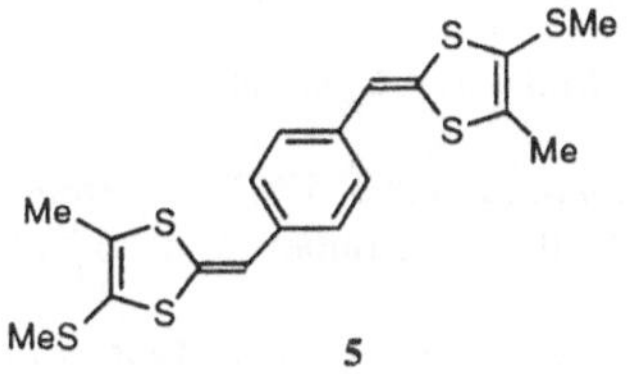

We also analyzed the electrochemical oxidation of 1,4-bis(1,4-dithiafulven-6-yl) benzene **5** and related analogue **6** where the two dithiole rings are linked together by a bisthioalkyl chain. The redox behavior of **5** and **6** was first studied by cyclic voltammetry, but what occured with this compounds was not as obvious as in the case of dithiafulvenes **1** or bis(dithiafulvenes) **3**.[4] In both cases, on the first anodic scan two irreversible peaks are observed and upon successive scans the second irreversible oxidation peak tends to decrease. We performed a preparative electrolyses of **5** and **6** (oxidation-reduction) under the same experimental conditions described for **1** and **3**. Work up of the reaction mixture and ^{1}H NMR as well as the FAB spectra provided evidence of the presence of macrocyclophanes including 4 and 5 bis(dithiafulvenyl) benzene units. In this case bis(dithiafulvenyl) benzenes upon oxidation undergo cyclooligomerization leading to novel macrocyclophane compounds with multi 1,3-dithiole rings.

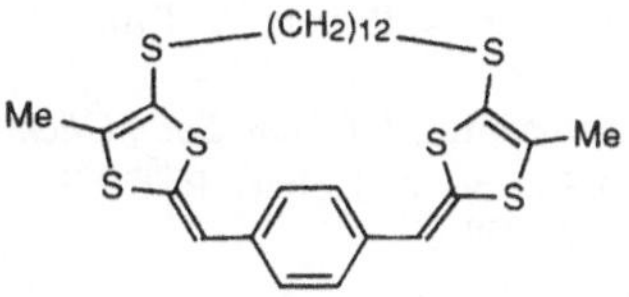

5 Ox / Red | $R^1 = CH_3$

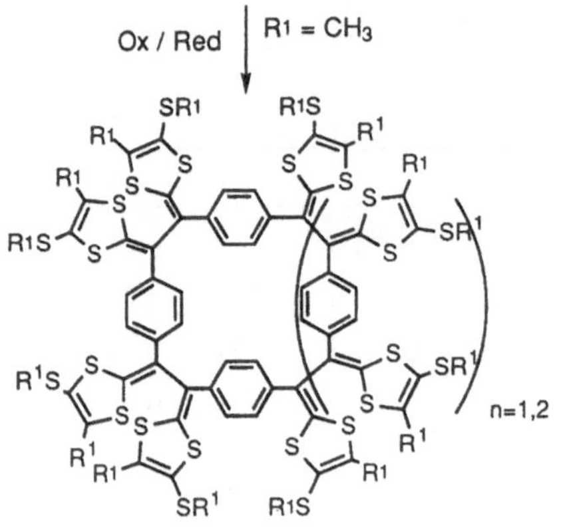

6 Ox / Red | $R^1 = CH_3$, $A = (CH_2)_{12}$

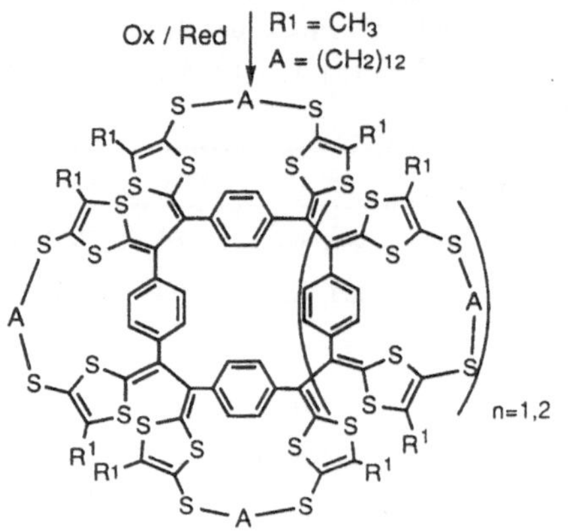

References

1) T.Otsubo, Y. Aso, K. Takimiya, *Adv. Mater.,* **1996**, *8*, 203. See also, M.Adam, K. Mullen, *Adv Mater.,* **1994**, *6*, 439.

2) D. Lorcy, R. Carlier, A. Robert, A. Tallec, P. Le Maguerès, L. Ouahab, *J. Org. Chem.,* **1995**, *60.*, 2443.

3) A. Ohta, Y. Yamashita, *J. Chem. Soc., Chem. Commun.,* **1995**, 1761 and references therein.

4) P. Hascoat, D. Lorcy, A. Robert, R. Carlier, A. Tallec, K. Boubekeur, P. Batail, *J. Org. Chem.,* **1997**, *62*, 6086.

5) P. Hapiot, D. Lorcy, R. Carlier, A. Tallec, A. Robert, *J. Phys. Chem.,* **1996**, *100.*, 14823.

RECTIFYING AND PHOTOVOLTAIC EFFECTS OF ORGANIC HETERO-JUNCTION FILMS BASED ON FLUORENONE DERIVATIVES / POLY (3-METHYLTHIOPHENE) PREPARED BY ELECTROCHEMICAL POLYMERIZATION AND UNDOPING

Takeshi MIKAYAMA[a], Yonggu SHIM[a], Nobuyuki YAMAMOTO[b], and Kaku UEHARA*[a]

[a]*Research Institute for Advanced Science and Technology, Osaka Prefecture University, Gakuen – cho, Sakai, Osaka, 599 – 8570 Japan*
[b]*College of Engineering, Osaka Prefecture University, Gakuen–cho, Sakai, Osaka, 599–8531 Japan*

1. Introduction

The eletrooxidative polymerization of 3-methylthiophene is a convenient method to synthesize an electroconducting polymer poly(3-methylthiophene) (PMeT) in form of a thin solid film incorporated by the counter anion from the supporting electrolytes as a dopant. Electropolymerized PMeT is konwn to be very stable toward oxygen, moisture and temperature in both their doped and undoped state. The thickness of these films, the amount of dopant and the properties of the films can be easily controlled electrochemically[1]. We have proposed dry organic hetero-junction devices based on electron transfer reaction between an organic dye molecule and PMeT [2]. Wet devices based on electron transfer reaction have been already accomplished by many reseahers, but these devices have complex structure. Our dry device is very simple. We expect that these devices should not behave like a conventional semiconductor device base on band theory and that they might be a next coming origanic molecular devices.

2. Experimental

3000-Å-thick PMeT was grown onto Au electrode in a classical three-electrode electrochemical cell equipped with a potentiostat (Hokuto Denko HA-501). Au electrode maintained at +1.4 V vs. Ag/AgCl(saturated KCl). The obtained polymer was undoped electrochemically at the -1.0 V vs. Ag/AgCl(saturated KCl). The amount of electric charge was monitored by the digital electrometer (Advantest TR8652) during undoping. A 300-Å-thick fluorenone derivatives were vacuum-deposited onto the top of PMeT under 5×10^{-5} Torr at deposition rate 5 Åsec^{-1}. And finally a 300-Å-thick Al was evaporated onto the fluorenone layer at deposition rate 1Åsec^{-1}. Current - voltage and current - time measurements were carried out by Keithley model 263 calibrator/source and Advantest TR8652 digital electrometer. These measurements carried out automatically by a personal computer. A halogen-tungsten lamp was used as a light-source. Fig. 1 shows (a): chemical structures of 2-amino-9-fluorenone (AF) and 2, 4, 7,-trinitrofluorenylidenemalononitrile (TNFM), (b): chemical doping and undoping of PMeT, and (c): Au/PMeT/ 9-fluorenone derivatives/Al sandwich cell.

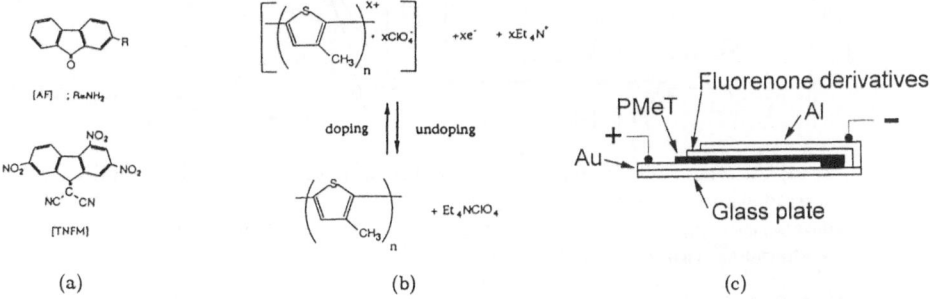

(a)　　　　　　　　　(b)　　　　　　　　　(c)

Fig. 1. (a) Chemical structures of 9-fluorenone derivatives. (b) Doping and undoping reactions of PMeT. (c) Schematic diagram of the Al/9-fluorenones/PMeT/Au sandwich cell

3. Results and Discussion

As shown in Fig. 2(a), normal rectification effect was observed for AF diode : a very small backward current was observed when the Au electrode was negatively biased with respect to the Al electrode. On the other hand, anomalous conduction effect was observed for TNFM diode : a significantly large backward current was observed when the Au electrode was negatively biased with respect to the Al electrode. To be interesting, TNFM diode showed that the backward current in the dark was markedly increased upon irradiation of light (Fig. 2(b) and (c)). The short-circuit photocurrent J_{sc} and open-circuit photovoltage V_{oc} for TNFM diode, increased with increase in undoping charge of PMeT (Fig. 2(d)). It was suggested that a redox reaction in the interface region of PMeT and TNFM [2] and a high photoconductivity for undoped PMeT [3] might contribute to the observed effects.

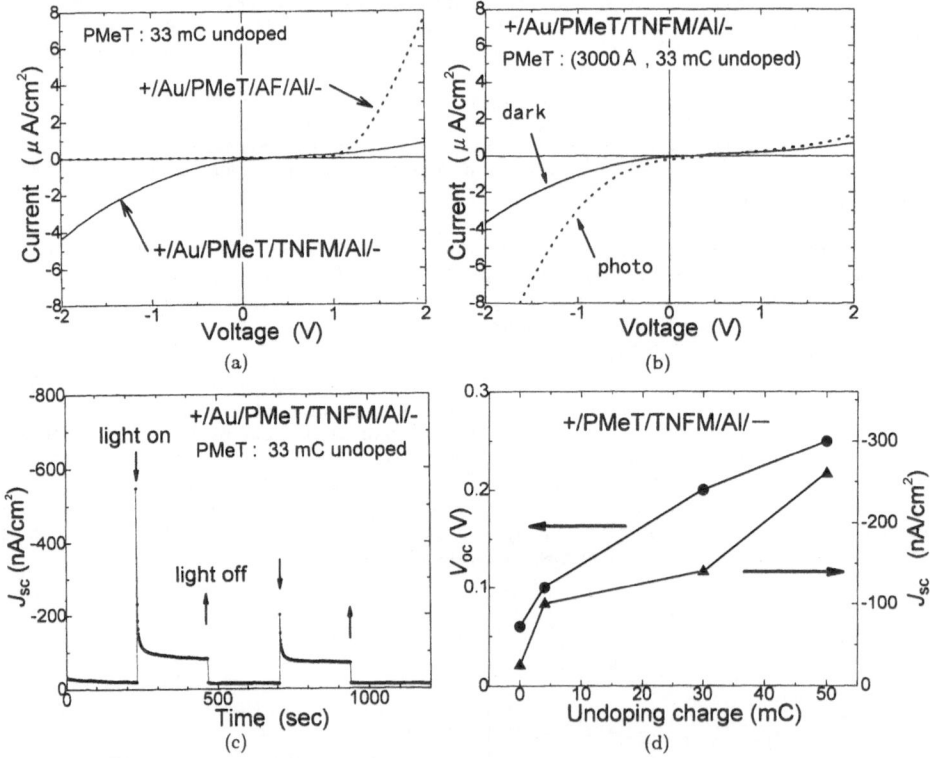

Fig. 2. (a) Current - voltage characteristics in the dark for AF and TNFM diode. (b) Effect of photo-irradiation on the current - voltage characteristics for TNFM diode. (c) Current - time characteristics upon illumination for TNFM diode. (d) Undoping effect on the current - voltage characteristics for TNFM diode

4. References

[1] G. Tourillon and F. Garnier, *J. Electrochem. Soc.* , **130**, 2042 (1983).

[2] K. Uehara, T. Ichikawa, T. Serikawa, N. Nishiyama, and M. Tsunooka, *Thin Solid Films*. **266**. 263 (1995).

[3] S. Glenis, G. Tourillon, and F. Garnier, *Thin Solid Films*. **139**, 221 (1986)

ELECTROCHEMICAL SYNTHESIS AND PROPERTIES OF POLY (3-METHYLTHIOPHENE) DOPED WITH PENTACHLOROSTANNATE ANION

Dan SINGH and Ram A MISRA

Departemnt of Chemistry, Faculty of Science, Banaras Hindu University, Varanasi- 221 005, INDIA

The preparation and properties of organic conducting polymers have been a rapidly developing area over the last few years. Special attention of researchers is directed at polymer based on aniline and five membered heterocyclic compounds such as pyrrole and thiophene. Conducting polymers are an interesting class of synthetic metals which find applications in a wide variety of technological avenues as electroactive meterials.

Poly(3-methylthiophene) doped with pentachlorostannate anion has been synthesized by electrochemical polymerization on platinum electrode in different solvent using tetra-n-butyl ammonium pentachlorostannate as supporting electrolyte cum dopant. The effect of dopant concentration, current densities, reaction temperature and nature of solvents on polymer yields and polymer conductivities have been investigated. The cyclic voltammetry of 3-methylthiophene and poly (3-methylthiophene) film deposited on platinum electrode was examined in 1,2-dichloroethane containing $n-Bu_4NSnCl_5$ in presence and in absence of monomer. Morphological properties of the polymer was obtained by scanning electron micrograph. Thermal analysis of polymer sampel was investigated through thermogravimetric analysis (TGA).

Both the yield and conductivity of the polymer was found to increase with the increase of current densities. The effect of temperature was observed on the yield and conductivity of the polymer, both the yield and conductivity decreased with the increase of temperature. Further, when the dopant concentration was increased the yield and conductivity of the polymer increased.

The cyclic voltammogram (Fig.1) of 3-Methylthiophene (0.1M) in 1,2-dichloroethane containing tetra-n-butyl ammonium pentachlorostannate

(0.02M) on a platinum microelectrode at scan rate (a) 20 (b) 50 (c) 100 and (d) 200 mVs^{-1} shows that the redox peak current increases with the increase of scan rate. The anodic peaks are sharp and higehr showing the polymer film is in conductive state which is slightly shifted towards positive potential. From Fig. 1 it is clear that the first anodic peak appears due to the formation of oligomers which disappears at higher scan rate. The linear plot of anodic peak curent (data from voltammetric curve fig. 1) suggesting that the redox reaction is reversible (Fig. 2).

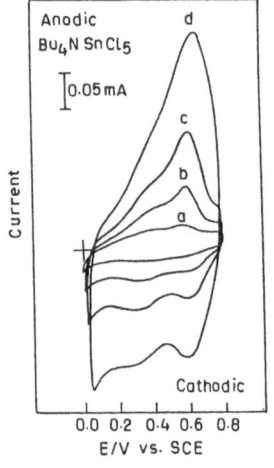

Fig.1 Cyclic voltammograms of 3-methyl thiophene

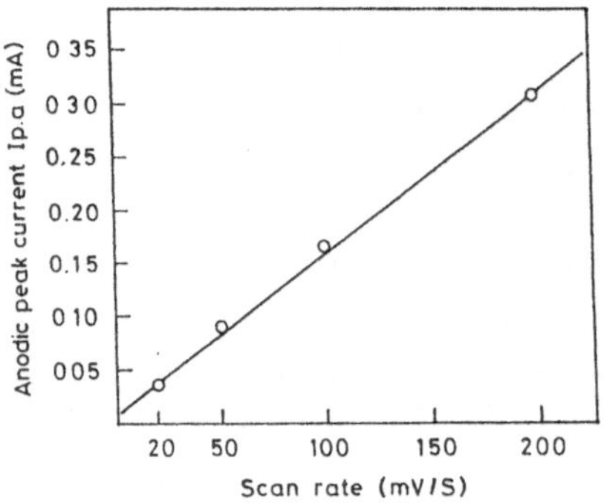

Fig. 2 Plot of anodic peak current against scan rate.

The cyclic voltammogram of poly (3-methylthiophene) shows that the polymer film is fairly stable without any severe degradation. The regular deposition globular morphology was observed from SEM photographs of the polymer film. The poly (3-methylthiophene) $SnCl_5^-$ undergo a three step weight loss process in heating cycle. The polymer is apparently stable up to 300°C. The polymerization is believed to proceed via the cation radical intermediate.

A NEW METHOD FOR CHEMICAL MODIFICATION OF
CONDUCTING POLYPYRROLES

Noboru ONO, Chikanori TSUKAMURA, Youta NOMURA, Takanori YAMAMOTO,
Takashi MURASHIMA and Takuji OGAWA

Department of Chemistry, Faculty of Science, Ehime University, Matsuyama, Japan

ABSTRACT

Chemical modification of polypyrroles derived from acenaphtho[1,2-c]pyrroles provides a new strategy for functionalization of conducting polymers without destroying their conductivity. By this method, a new soluble and low band gap polypyrrole was obtained, which may be useful as a stable material for electrochromic or light emitting devices.

INTRODUCTION

Electronically conducting aromatic polymers have been the object of great amount of research in recent years. Among them, poly(pyrroles) and poly(thiophenes) are the most attractive due to their relative high conductivity, stability and ready modification. The functionalization of these polymers in the 3-position has been shown to be a very powerful method for controlling the properties of them. However, the introduction of large groups into the pyrrole or thiophene rings at the N- or 3-position brings about a dramatic drop in the conductivity and an increase of the band gap due to the steric hindrance to planarity of the conjugated system in the corresponding polymer. Thus, it is not easy to modify the conducting polymers in a desired way without losing their conductivity or other properties. Recently we have found a new strategy to control the band gap of poly(pyrroles), namely , isoindoles prepared by the reaction of aromatic nitro compounds with ethyl isocyanoacetate serves as good precursor molecules for conducting polymers.[1] The band gap of these polymers depends on the fused aromatic rings which are introduced from the starting aromatic nitro compounds. Another merit of this method lies in the easy chemical modification of the controlling polymer, which can be done by the use of aromatic nitro compounds with desired properties.[2]

RESULTS AND DISCUSSION

The reaction of 1-nitroacenaphthylene with various substituents at 5,6-positions with ethyl isocyanoacetate in the presence of DBU was carried out to give the fused pyrrole 2. The subsequent deethoxycarbonylation by heating with KOH in ethylene glycol gave 3, which was converted into polymer by anodic oxidation as shown in Scheme 1. The substituents X and Y in 3 are chosen to control the electronic and physical properties of the conducting polymer. Anodic oxidation of 3 affords the corresponding polymer 4, and electronic and optical properties are summarized in Table 1. The

155

substituents X and Y can control the HOMO and LUMO band of the polymer **4** effectively and improve the physical properties of the polymer without destroying their conductivity. The conductivities are in the range of 10^{-2}~4 Scm^{-1}, which are independent on X and Y. As polymer **4b** is soluble in organic solvents, the molecular weight is measured to be 13100 by GPC. Polymer **4** is highly conjugated through the fused ring and the polymer backbone. Thus, the substituents X and Y in **4** can control the electronic properties such as ionization potential (IP), electron affinity (Ea) and band gap (Eg) of the polymer effectively. It is noteworthy that the strong electron withdrawing group such as CN lowers Eg effectively.

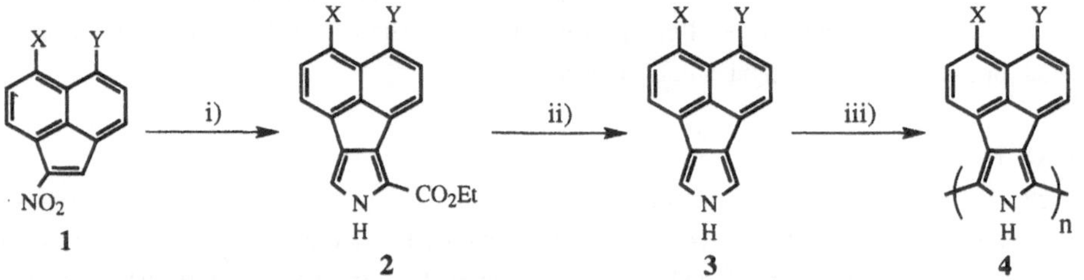

Scheme 1. i) $CNCH_2CO_2Et$, DBU / THF; ii) KOH / $HOCH_2CH_2OH$; iii) Anodic Oxidation.

Table 1. Optical and electronic property of **4**.

Entry	X	Y	$E_{pa}(V)$ **4**	Absorption Edge (nm)	Eg (eV)	Conductivity (Scm^{-1})
a	H	H	0.19	552	2.25	4
b	C_6H_{13}	C_6H_{13}	0.04	574	2.16	2
c	Cl	Cl	0.29	586	2.12	0.3
d	Cl	CN	0.49	652	1.90	——

E_{pa} was measured by CV in a 0.1M solution of TBAP in MeCN cast on a platinum electrode (potentials relative to Ag / Ag^+).

REFERENCES

1) N. Ono, H. Hironaga, K.Shimizu, K. Ono, K. Kuwano, T. Ogawa, *J. Chem. Soc. Chem. Commun.*, 1019 (1994).

2) N. Ono, C. Tsukamura, Y. Nomura, H. Hironaga, T. Murashima, T. Ogawa, *Adv. Mater.*, **9**, 149 (1997)

MODIFICATION OF GLASSY CARBON SURFACES BY ANODIC OXIDATION IN 1-ALKANOLS

Hatsuo MAEDA, Kazunori KATAYAMA, and Hidenobu OHMORI
Faculty of Pharmaceutical Sciences, Osaka University,
1-6 Yamada-oka, Suita, Osaka 565, Japan

ABSTRACT

Electrochemical oxidation of a glassy carbon (GC) electrode in a 1-alkanol leads to the covalent attachment of the alkanol molecules to the electrode surface. Measurements of contact angle, capacitance, electrochemical performance, and/or adsorption property of bovine serum albumin allowed surface characterization of the electrodes anodized in a wide variety of 1-alkanols. Anodization in 1,ω-diols turned out to have potentiality as a convenient starting point for chemical modifications of carbon surfaces.

INTRODUCTION

Recently, we have found that the anodic treatment of a glassy carbon (GC) electrode in a 1-alkanol with constant or cycled potentials allows the alkanol molecules to be fixed on the electrode surface *via* an ether-linkage (Scheme 1).[1] The resulting membrane on a GC electrode has turned to be not as densely packed as that on a gold electrode modified by the self-assembly technique, which permitted the modified electrodes to find some applications to electrochemical analysis.[2,3] In order to further develop the utility of the modified GC electrodes in electro-organic synthesis as well as electrochemical analysis, our attention has been directed to evaluating more detailed surface conditions of the modified electrodes. Here, we describe the surface characterization of GC electrodes anodized in a wide variety of 1-alkanols (Scheme 1). Some attempts to prepare catalytic electrodes are also described, to demonstrate the possibility of anodization in 1,ω-diols as a simple tool to facilitate chemical modification of carbon surfaces.

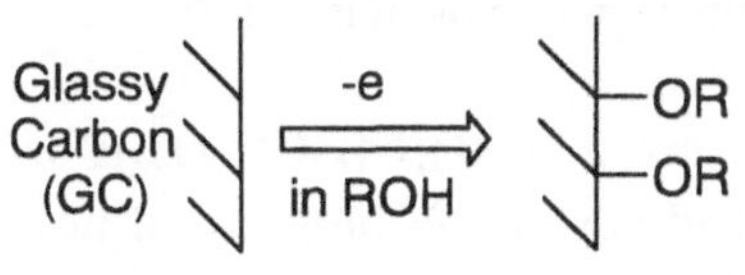

Glassy Carbon (GC) $\xrightarrow[\text{in ROH}]{-e}$ —OR / —OR

ROH: $n\text{-}C_nH_{2n+1}OH$ (n=2, 4, 6, 8)
$HO(CH_2)_nOH$ (n=2~5)
$HO(CH_2CH_2O)_nR'$ (n=1~4; R'=H, CH_3)

Scheme 1

RESULTS AND DISCUSSION

Surface characterization

The anodic modification was carried out by controlled-potential electrolysis at 2.0 V *vs.* Ag wire in a 1-alkanol (0.1 M $LiClO_4$) in one compartment cell equipped with a Pt foil cathode, where 5 mC of electricity had been allowed to be consumed. The surface conditions of GC electrodes anodized in $C_nH_{2n+1}OH$ and $HO(CH_2)_nOH$ were explored by measuring the wettability and capacitance as well as the voltammetric response of $Fe(CN)_6^{3-}$: wetting properties were evaluated by the contact angle (θ) formed by a drop of water; the ratio (C_{mod}/C_{bare}) between the values of capacitance obtained before and after the anodic modification was used instead of the capacitance itself. The results are summarized in Table 1. The data have revealed that the surface conditions of the modified electrodes reflect the identities of the modifiers remarkably: electrodes anodized in n-$C_nH_{2n+1}OH$ have surface membranes with hydrophobicity and thickness correlated to the length of carbon chains of the modifiers; the surfaces of GC anodized in $HO(CH_2)_nOH$ are hydrophilic.[4]

Table 1. Contact Angle (θ) of Water, Relative Capacitance (C_{mod}/C_{bare}), and Voltammetric Response (E_{pc} and I_{pc}) of $K_3Fe(CN)_6$ at Bare and ROH-Modified GC Electrodes

ROH	θ (°)	C_{mod}/C_{bare}	E_{pc} (V)	I_{pc} (µA)
-	64	-	0.122	71.5
C_2H_5OH	74	1.11	0.093	63.7
n-C_4H_9OH	85	0.86	0.040	50.7
n-$C_6H_{13}OH$	91	0.78	-[a]	41.7[b]
n-$C_8H_{17}OH$	95	0.60	-[a]	36.0[b]
$HO(CH_2)_2OH$	50	0.99	0.103	71.3
$HO(CH_2)_3OH$	50	0.91	-[a]	18.5[b]
$HO(CH_2)_4OH$	56	0.95	-0.137	45.0
$HO(CH_2)_5OH$	64	0.63	-[a]	3.0[b]

a) No cathodic peak was observed. b) Cathodic response obtained at -0.2 V.

The surface conditions of GC electrodes anodized in $HO(CH_2CH_2O)_nR'$ were examined by measurements of the wettability and adsorption property of bovine serum albumin (BSA): the latter was evaluated by comparing anodic responses of caffeic acid (3,4-dihydroxycinnamic acid) on cyclic voltammograms at GC electrodes before and after a treatment with a buffer solution (pH 7.0) of BSA

Table 2. Voltammetric Data[a] of Caffeic Acid at Bare and $HO(CH_2CH_2O)_nR'$ Modified GC Electrodes Before and After a Treatment with a BSA Solution, and Contact Angle (θ) of Water at the Electrodes

$HO(CH_2CH_2O)_nR'$ n	R'	E_{pa} (V)	I_{pa} (µA)	I_{pa}^{BSA} / I_{pa}	θ (°)
-	-	0.23	38	0.68	66
1	H	0.30	31	0.84	54
2	H	0.46	26	0.96	50
3	H	0.48	25	1.00	47
4	H	_[b]	15[c]	-	40
1	CH_3	0.24	39	0.79	72
2	CH_3	0.31	34	0.91	62
3	CH_3	0.43	28	1.00	50

a) E_{pa} coupled with I_{pa} denote the potential and the currrent of the anodic peak observed at the electrodes before the treatment, and I_{pa}^{BSA} stands for the current after the treatment.

b) No peak was observed.   c) Obtained as anodic current at 0.8 V.

(0.01 w/v %) for 5 min.   The results are summarized in Table 2.   GC electrodes modified with triethylene glycol and its monomethyl ether have proved to most effectively resist surface fouling due to BSA adsorption as well as to show satisfactory electrochemical performance, which allowed the electrodes to find a useful application to HPLC analysis of samples containing proteins.[5]

Preparation of catalytic electrodes

On cyclic voltammetry in pH 7.0 phosphate buffer at GC electrodes anodized in $HO(CH_2)_nOH$, anodic oxidation of ascorbic acid is totally suppressed, while dopamine shows a well-defined anodic peak.[2]   The observation that the electrochemical reaction of the anionic compound was retarded at the modified electrodes can be explained by invoking a mechanism depicted in Scheme 2 for the anodic modification of carbon surfaces with 1,ω-diols: the anodization of GC initially confines the diol molecules to the surface *via* an ether-linkage, and then the terminal hydroxy groups of the attached molecules seem to be oxidized to carboxyl groups during the modification procedure.   Thus, it can be expected that electrochemical mediators such as catechol and TEMPO are immobilized on GC surfaces through anodization in a 1,ω-diol followed by a coupling reaction with **1** or **2** (Scheme 2).   In fact, it was found that immobilization of **1** on a GC surface is achieved by the sequence of the anodization in triethylene glycol and a treatment

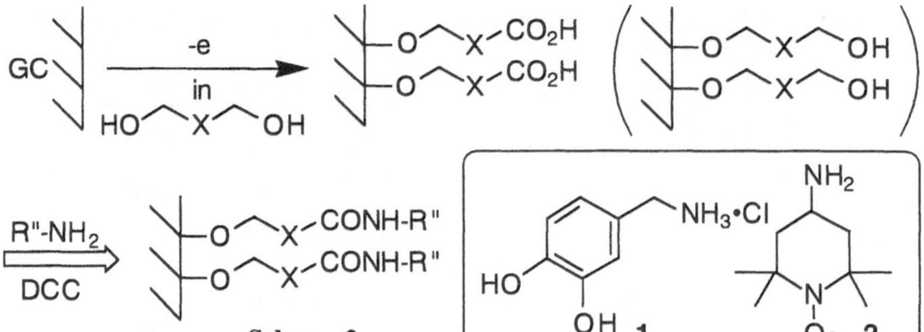

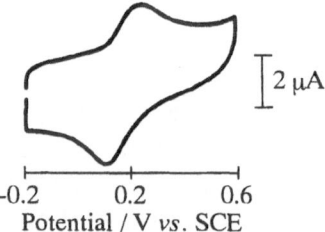

Scheme 2

with a CH_2Cl_2 solution of **1** in the presence of DCC and Et_3N for 24 h. As shown in Fig. 1, this modified electrode in pH 7.0 phosphate buffer shows a redox wave due to the electrochemical process of the catechol moiety. A similar procedure with **2** yielded a GC electrode modified with TEMPO. Although the electrochemical performance of these modified electrodes as catalytic electrodes is not examined in detail, the results have demonstrated that anodization in 1,ω-diols has potentiality as a convenient starting point for chemical modifications to introduce preselected functionalities on carbon surfaces.

Figure 1. Cyclic Voltammogram of a GC Electrode Modified with **1** in pH 7.0 Phosphate Buffer (0.1 M): Sweep Rate, 0.1 V/s.

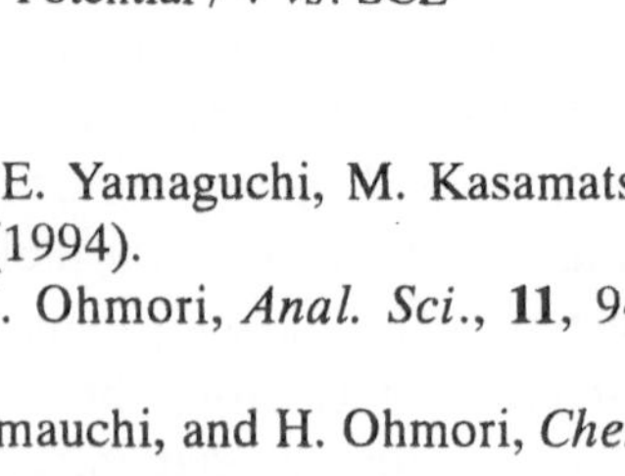

Potential / V *vs*. SCE

References

1) H. Maeda, Y. Yamauchi, M. Hosoe, T.-X. Li, E. Yamaguchi, M. Kasamatsu, and H. Ohmori, *Chem. Pharm. Bull.*, **42**, 1870 (1994).

2) H. Maeda, Y. Yamauchi, M. Yoshida, and H. Ohmori, *Anal. Sci.*, **11**, 947 (1995).

3) H. Maeda, M. Hosoe, T.-X. Li, M. Itami, Y. Yamauchi, and H. Ohmori, *Chem. Pharm. Bull.*, **44**, 559 (1996).

4) H. Maeda, M. Itami, Y. Yamauchi, and H. Ohmori, *Chem. Pharm. Bull.*, **44**, 2294 (1996).

5) H. Maeda, M. Itami, K. Katayama, Y. Yamauchi, and H. Ohmori, *Anal. Sci.*, in press.

MODIFIED GRAPHITE FELT ELECTRODE. APPLICATION TO ELECTRO-CATALYSIS.

Patricia JEGO and Claude MOINET

Laboratoire "Electrochimie et Organométalliques", UMR CNRS 6509, Université de Rennes I, Campus de Beaulieu, 35042 - Rennes Cédex, France

ABSTRACT

Homogeneously electrooxidized graphite felt electrode was modified by various organo-metallic compounds. The homogeneity of the τ-bonding inside the porous electrode was controlled by cyclic voltammetry and the volume density of attached redox centers was obtained by coulometric measurements. Catalytic properties of such modified electrodes were characterized by cyclic voltammetry and by electrolyses in a flow cell fitted with these electrodes. As an application, styrene oxide was produced from styrene at a graphite felt cathode modified by a manganese (III) porphyrin.

INTRODUCTION

Chemically modified electrode is well known since more than 20 years [1-2]. Modification at the surface of the electrode can be obtained by covalently attached monolayer [3-5], electro-polymerization [6] or chimisorption [7]. Generally, several steps are necessary to do a τ-bonding between an electrode and a substrate. For example, graphite electrodes have been previously oxidized [9-11] in order to produce an high surface density of carboxylic groups. Then a modified carbon electrode is obtained after activation of the carboxylic group by thionyl chlorid and reaction with an amino derivative (equation 1).

$$C \equiv \xrightarrow{\text{oxidation}} C \equiv -C\!\!\nwarrow^{O}_{OH} \xrightarrow{\text{SOCl}_2} C \equiv -C\!\!\nwarrow^{O}_{Cl} \xrightarrow{\text{RNH}_2} C \equiv -C\!\!\nwarrow^{O}_{NHR} \quad (1)$$

In order to adapt modified porous electrodes to our flow cells we have treated a graphite felt according to the reaction (1) in which the oxidation step occured only electrochemically [11].

RESULTS

Oxidation in the presence of air of graphite felt (Le Carbone Lorraine) by heating at 450°C failed. Electrooxidation of graphite felt was obtained in a flow cell equiped with a disc of this material (diameter 5.2 cm, thickness 9 mm) located between two cathodes. An aqueous solution of 2M KNO_3 was pumped through the porous anode (flow rate 15 $cm^3.mn^{-1}$). To reach an high volume density of carboxylic centers at the surface of graphite, the current intensity for electrolysis was optimized to 2.4 A for 30 mn. For higher intensities and higher times, a dramatic damage of felt occured.

After oxidation, a successive treatment of graphite felt by $SOCl_2$ in toluene for 3 h at 30-40°C and by various organometallic compounds bearing either an amino group (substituted ferrocenes, ruthenium (II) complex or manganese (III) porphyrins) in toluene or an activated methylen group ($[\eta^5$-cyclopentadienyl η^6-arene iron (II)] cations) in DMF lead to modified electrode.

Cyclic voltammetry in 0,5 M H_2SO_4 of samples (diameter 8 mm, length 9 mm) of graphite felt modified by α-aminoethylferrocene was used to control the homogeneity of the τ-bonding inside the porous electrode (Fig. 1).

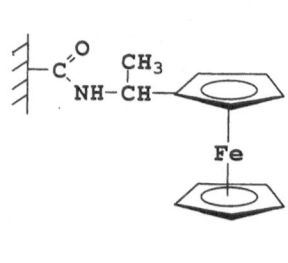
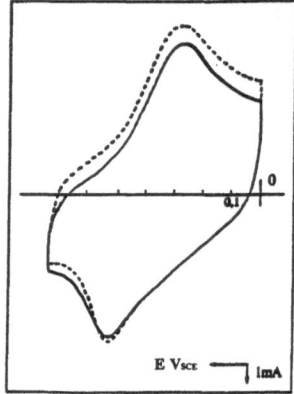

Fig. 1 - Voltammograms of a modified graphite felt by α-aminoethylferrocene ; 0,5 M H_2SO_4 ; scan rate :10 mV.s^{-1}
—— external part of felt   ----- internal part of felt

The volume density of the same attached redox center was obtained from a series of coulometric measurements after successive oxidations and reductions at controlled potential (respectively 0.5 V_{SCE} and - 0.15 V_{SCE}) of a sample (diameter 14 mm, thickness 9 mm) of a modified felt in 0.5 M H_2SO_4, until reproducible results. The volume density was approximatively 2 to 8x10^{-6} mol per cm^3 of felt according to oxidation conditions. We observed a good correlation between coulometric data and global peak intensities on the cyclic voltammograms.

When complexes were substituted by two amino groups only one was bonded to graphite. This was observed with 1,1'-di(α-aminoethyl) ferrocene or Ru (bpy)$_2$ (3-py NH$_2$)$_2$ (PF$_6$)$_2$. Hence, the second group could be functionalyzed by another substrate.

The reduction of ferrocenium cation by ascorbic acid is well known and currently used in our laboratory during synthesis of ([η^5-cyclopentadienyl η^6-arene iron (II)] cations. Catalytic oxidation of ascorbic acid by a ferrocenium cation can be shown by cyclic voltammetry. For example, the reversibility observed for the oxidation of N,N'-dimethylaminomethylferrocene (Ep$_a$ = 0.39 V_{SCE}, Ep$_c$ = 0.31 V_{SCE}) in aqueous phosphat buffer (pH 2.2) totally disappeared with an increasing of anodic peak after addition of ascorbic acid.

We observed also an increasing of the anodic peak at a graphite felt electrode modified by α-aminoethylferrocene (Fig. 2). Ascorbic acid contained inside the porous electrode was totally oxidized during the first anodic scan.

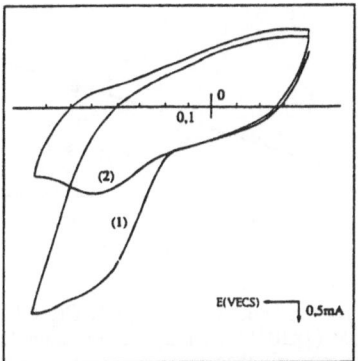

Fig. 2 - Voltammogram of a graphite felt modified by α-aminoethylferrocene first immersed in phosphat buffer (pH 2.2) containing ascorbic acid ; phosphat buffer (pH 2.2) ; scan rate : 10 mV.s⁻¹
(1) first scan   (2) second scan.

Electrocatalytic oxidation of ascorbic acid into dehydroascorbic acid was performed in a flow cell fitted with a graphite felt anode (diameter 4 cm, thickness 9 mm) modified by α-aminoethyl-ferrocene (3.6×10^{-6} mol per cm^3 of felt). The porous electrode was percolated (flow rate : 4.6 $cm^3.mn^{-1}$) by a solution of ascorbic acid (5×10^{-2} M) in aqueous phosphat buffer (pH 2.2). The current intensity calculated for 2F per mole of ascorbic acid was 0.74 A. HPLC analyses of the electrolyzed solution showed that 80% of ascorbic acid were oxidized after a single passage. We controlled that voltammograms of modified felt before and after electrolysis (turn over 1500) were similar showing a stability of the organometallic catalyst.

The electrochemical reduction, in the presence of oxygen, of metallated porphyrin in solution [12-14] or bonded to an electrodeposited polymer [15] lead to an superoxo intermediate used for epoxidation reactions according to a catalytic cycle [12-15]. To verify the catalytic efficiency of a graphite felt modified by the manganese (4-aminophenyl)-triphenyl-porphyrin diacetate, 100 cm^3 of a solution containing styrene (0.1 M), 1-methylimidazole (2×10^{-2} M) as axial base, benzoic anhydride (0.1 M) as activator and Bu_4NBF_4 (0.2 M) as electrolyte in dichloromethane flowed through the porous modified cathode (diameter 4 mm, thickness 9 mm). The solution was continuously saturated by air or pure oxygen. Because of a slow reaction, the flow rate was low (1.5 $cm^3.mn^{-1}$) and the current intensity (90 mA) lower than theoritical intensity (480 mA) ; thus, it was necessary to recycle the solution through the flow cell. The yield in styrene oxide continuously increased after each electrolysis (Fig. 3).

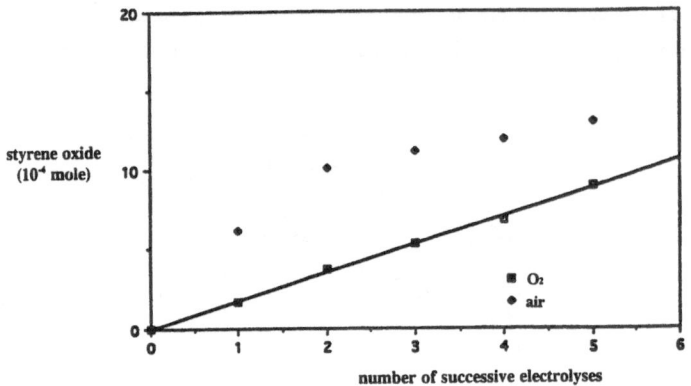

Fig. 3 - Electrocatalytic epoxidation of styrene (0.1 M) in CH_2Cl_2 + Bu_4NBF_4 (0.2 M), $(C_6H_5CO)_2O$ (0.1 M), 1-methylimidazole (2×10^{-2} M) ; graphite felt cathode modified by manganese (III) (4-aminophenyl) triphenylporphyrin diacetate.

CONCLUSION

In summary, we have shown that it is possible to modify homogeneously a graphite felt by redox organometallic compounds. A good stability of such electrodes was observed and several applications in electrocatalysis using a flow cell with modified porous electrode can be considered.

References

1) A.F. Diaz, in *Organic Electrochemistry*, H. Lund and M. Baizer Ed., 3rd edition, **1982**, p. 1363.
2) R.W. Murray, A.G. Ewing and R.A. Durst, *Anal. Chem.*, **1987**, *59*, 379 A.
3) B.F. Watkins, J.R. Behling, E. Kariv and L.L. Miller, *J. Am. Chem. Soc.*, **1975**, *97*, 3549.
4) J.R. Lenhard and R.W. Murray, *J. Electroanal. Chem.*, **1977**, *78*, 195.
5) A. Diaz, *J. Am. Chem. Soc.*, **1977**, *99*, 5838.
6) A. Deronzier and J.C. Moutet, *Accts Chemical Research*, **1989**, *22*, 249.
7) A.P. Brown and F.C. Anson, *J. Electroanal. Chem.*, **1977**, *83*, 203.
8) A.P. Brown, C. Koval and F.C. Anson, *J. Electroanal. Chem.*, **1976**, *72*, 379.
9) C.M. Elliot and R.W. Murray, *Anal. Chem.*, **1976**, *48*, 1247.
10) J.C. Lennox and R.W. Murray, *J. Electroanal. Chem.*, **1977**, *78*, 395.
11) H.M. Wu, R. Olier, N. Jaffrezic-Renault, P. Clechet, A. Nyamsi and C. Martelet, *Electrochim. Acta*, **1994**, *39*, 327.
12) S.E. Greager, S.A. Raybuck and R.W. Murray, *J. Am. Chem. Soc.*, **1986**, *108*, 4225.
13) G. Balavoine, D.H.R. Barton, J. Boivin, A. Gref, N. Ozbalik and H. Rivière, *Tetrahedron Lett.*, **1986**, *27*, 2849.
14) P. Leduc, P. Battioni, J.F. Bartoli and D. Mansuy, *Tetrahedron Lett.*, **1988**, *29*, 205.
15) G. Cauquis, S. Cosnier, A. Deronzier, B. Galland, D. Limousin, J.C. Moutet, J. Bizot, D. Deprez and J.P. Pulicani, *J. Electroanal. Chem.*, **1993**, *352*, 181.

ANODIC OXIDATION OF ORGANICS ON OXIDE ANODES

Olivier SIMOND, Christos COMNINELLIS, György FOTI

Institute of Chemical Engineering, Swiss Federal Institute of Technology,
CH-1015 Lausanne, Switzerland

ABSTRACT

The electrochemical oxidation (or combustion) of organics with simultaneous oxygen evolution has been investigated using different oxide electrodes. A simplified mechanism for the electrochemical oxidation of organics is presented, according to which selective oxidation occurs with oxide anodes (MO_x) for which the formation of higher oxides (e.g.: MO_{x+1}) is possible. Combustion occurs at electrodes with oxidatively saturated surface. Detection of $^{\cdot}OH$ radicals formed by water discharge at different anodes using N,N dimethyl-p-nitrosoaniline (RNO) as a spin trap and preparative electrolysis confirmed the proposed mechanism.

INTRODUCTION

There is an increasing interest in the use of the electrochemical oxidation of organic compounds for synthesis (electrochemical conversion) and for the treatment of toxic waste water (electrochemical combustion).

The electrode material is clearly an important parameter when optimizing such processes since the mechanism and the products of several anodic reactions are known to depend on the anode material. For example, anodic oxidation of phenol yields hydroquinone and benzoquinone at Ti/IrO_2 anodes and mainly carbon dioxide at $Ti/SnO_2-Sb_2O_5$ anodes. [1]

Theoretically the electrochemical oxidation (or combustion) of organics is possible before oxygen evolution (due to H_2O discharge) but in practice the oxidation reaction is very slow as a consequence of kinetic rather than thermodynamic limitations.

To increase the oxidation rate electrocatalytic noble-metal anodes have been proposed (Pt,Pd,...) but the main problem during oxidation of organics at a fixed anodic potential, before oxygen evolution, is the decrease of the anode activity as the consequence of poison formation at the anode surface. [2]

Dimensionally stable anodes (DSAs), which consists of a titanium base metal covered by a thin layer (few microns) of oxide or mixed oxides, have been commercialized for chlorine production (Ti/RuO_2-TiO_2) and oxygen evolution ($Ti/IrO_2-Ta_2O_5$) but the application to electroorganic oxidation has remained relatively rare.

In this paper the electrochemical oxidation of organics has been studied using DSA type anodes and a simplified mechanism for the electrochemical oxidation-combustion of organics is presented. [1,3,4]

$$IrO_3 + R \xrightarrow{k_1} IrO_2 + RO$$
$$IrO_3 \xrightarrow{k_2} IrO_2 + \tfrac{1}{2}O_2$$

where k_1 is the rate constant of the organics oxidation reaction ($m^3 \cdot mol^{-1} \cdot s^{-1}$) and k_2 the rate constant of oxygen evolution reaction (s^{-1}).
Neglecting concentration polarization, a simple relation (eq. 1) has been derived for the experimental determination of the k_2/k_1 ratio [4].

$$\frac{1}{\eta} = 1 + \frac{k_2}{k_1} \cdot \frac{1}{C_R} \tag{1}$$

Thus for a given anode, the plot of the inverse of the current efficiency (η) for the oxidation of an organic compound as a function of the inverse of its concentration (C_R) gives a straight line with a slope equal to the k_2/k_1 ratio. If the concentration polarization is considered, the following relation can be derived [4] :

$$\frac{1}{\eta_d} = 1 + \frac{k_2}{k_1}(1+\phi) \cdot \frac{1}{C_R}$$

in this case, the slope of the straight line increases by a factor $(1+\phi)$. The ϕ parameter depends strongly on the configuration of the anodic current collector.

In order to get an experimental verification of the model in a simple case, oxidation of i-Propanol in 1 M H_2SO_4 at 20 to 60° C was investigated on Ti/IrO2 electrodes. Under the chosen experimental conditions (high organic concentration, low conversion) no important diffusion limitation is expected ($\phi \approx 0$). The applied current density ranged from 30 to 60 mA/cm^2. The oxidation of i-Propanol is selective yielding acetone as oxidation product:

$$(CH_3)_2CHOH \quad \rightarrow \quad (CH_3)_2CO \; + \quad H_2O \; + \quad 2H^+ \; + \quad 2e$$

Even at i-Propanol conversions higher than 90 %, the selectivity of acetone production is better than 80 %. When the inverse of current efficiency is plotted as a function of the inverse of bulk i-Propanol concentration, a straight line is obtained with an intercept close to unity, as predicted by the kinetic model (eq. 1). It is also found that up to 50 mA/cm^2, the current efficiency is fairly independent of the current density, as it should be when diffusion limitation is negligible. The experimental results have shown that the slope of the $1/\eta$ vs. $1/C_R$ plot, which equals to the ratio k_2/k_1, decreases with increasing temperature. This means that selective oxidation of i-Propanol to acetone by the intermediation of IrO_3 active sites has higher activation energy than that of oxygen evolution at the same sites. All these results are in good agreement with the proposed kinetic model and confirm its applicability for the case of negligible diffusion limitation.

OXIDATION OF ORGANICS ON OXIDE ANODES

Cyclic voltammetry measurements of organics in $1M$ H_2SO_4 at oxide anodes (IrO_2) has shown that they are inactive for the oxidation of most organics (Formic acid, Oxalic acid, Maleic acid , Methanol, Ethanol, n-Propanol, i-Propanol, t-Butanol) before oxygen evolution and only simple electron transfer reactions can occur before oxygen evolution, a typical example is the oxidation of 1,4 hydroquinone to 1,4 benzoquinone (Fig. 1).

1,4 Hydroquinone =   1,4 Benzoquinone + $2H^+$ + $2\,e^-$

Preparative electrolysis under conditions of simultaneous oxygen evolution has shown that depending on the oxide anode used the organic compound can be oxidized selectively (Ti/IrO_2 anode) or oxidized mainly to CO_2 (Ti/SnO_2-Sb_2O_5 anode).

To explain the catalytic action of simultaneous oxygen evolution toward organic oxidation or combustion, and to understand the influence of oxide anode (MO_x) on the reaction products, a generalized mechanism for the electrochemical conversion/combustion of organics on oxide anodes is proposed (Fig 2). In the first step, H_2O in acid solution is discharged at the anode to produce adsorbed hydroxyl radicals (eq.1 in Fig 2).In a second step, the adsorbed hydroxyl radicals may interact with the oxide anode forming the higher oxide (MO_{x+1}) (eq.2 in Fig 2).

Thus we can consider that at the oxide anode surface two states of "active oxygen" can be present:
i) Physisorbed "active oxygen" (adsorbed hydroxyl radicals, $^{\cdot}OH$)
ii) Chemisorbed "active oxygen" (Oxygen in the oxide lattice MO_{x+1})

In the absence of organics the physisorbed and the chemisorbed "active oxygen" produce dioxygen (eq 3 and 4 in fig 2).

In the presence of organics we speculate that the physisorbed "active oxygen" ($^{\cdot}OH$) should cause predominantly the combustion of organics (eq 5 in fig 1) and chemisorbed "active oxygen" (MO_{x+1}) participate in the selective oxidation of organics (eq 6 in fig 2)

Thus oxides forming the higher oxide (MO_{x+1}) at potentials above the thermodynamic potential for O_2 evolution (IrO_2,RuO_2..) can favour selective oxidation («active» electrodes) and oxides whose surface is oxidatively saturated (PbO_2,SnO_2 ...) can favour combustion of organics (« non-active » electrodes).

KINETIC MODEL OF OXIDATION OF ORGANICS ON « ACTIVE » ELECTRODES.

The mathematical model for the anodic oxidation of organics on « active » oxide electrodes with simultaneous oxygen evolution has been described in a previous paper [4]. In this model water is discharged in a first step forming hydroxyl radicals which interact with the oxide anode (in case IrO_2) forming the higher oxide IrO_3 [4]. This reactive intermediate is in competition between oxidation of organics (R) and oxygen evolution.

References

1) Ch. Comninellis and A. De Battisti, J.Chim.Phys. **1996**, *93*, 673
2) C. Lamy, Electrochim. Acta, **1984**, *29*, 1581
3) Ch. Comninellis, Electrochim. Acta, **1994**, *39*, 1857
4) O. Simond, V. Schaller and Ch. Comninellis, Electrochim. Acta, **1997**, *42*, 2009

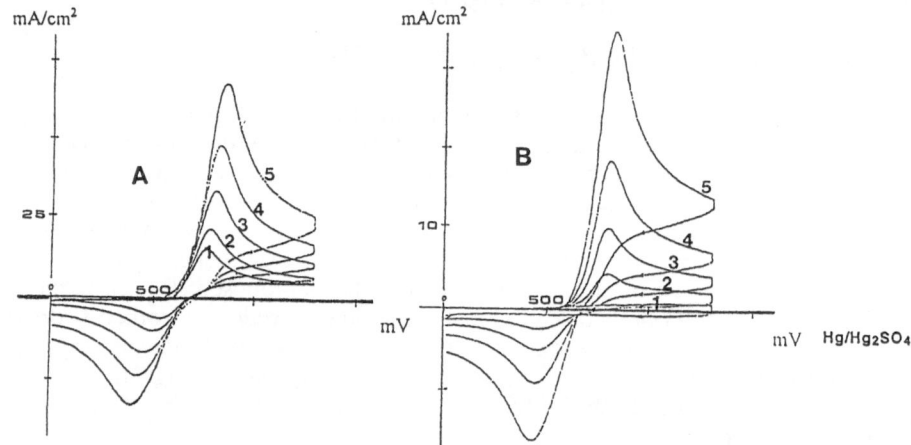

Figure 1: Cyclovoltammograms of 1,4 hydroquinone/1,4 benzoquinone redox couple at Ti/IrO_2 anodes obtained in 1 M H_2SO_4. T = 25° C.
(A) Influence of 1,4 hydroquinone concentration: 1 to 5: 0,00M/ 0,01M/ 0,05M/ 0,1M (Scan rate: 50 mV/s); (B) Influence of scan rate: 1 to 5: 10, 20, 50, 100, 200 mV/s (conc. 0,1 M)

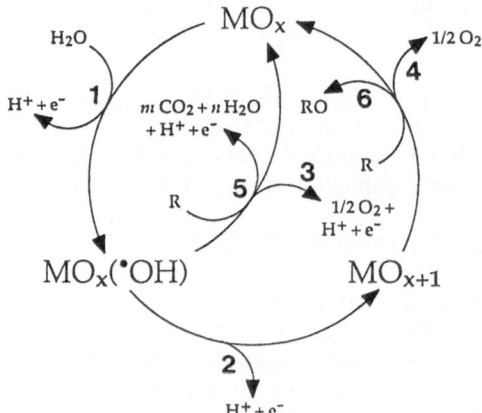

Figure 2: Generalized schema of the electrochemical conversion (6)/ combustion (5) of organics with simultaneous oxygen evolution (3), (4). (1) H_2O discharge, (2) Transition of O from 'OH to the lattice of the oxide anode.

ELECTROCATALYRIC OXIDATION ON AN (S)-ENANTIOSELECTIVE CHIRAL TEMPO-MODIFIED ELECTRODE

Yoshitomo KASHIWAGI, Futoshi KURASHIMA, Chikara KIKUCHI, Jun-ichi ANZAI, and Testuo OSA

Faculty of Pharmaceutical Sciences, Tohoku University, Aobayama, Aoba-ku, Sendai 980-77, Japan

ABSTRACT

(6S,7R,10R)-1-aza-4-amino-2,2,7-trimethyl-10-isopropylspiro[5.5]undecane-1-yloxyl   (Chiral TEMPO)-modified electrode was prepared by attaching this mediator chemically to the carboxyl groups of thin poly(acrylic acid) (PAA) layer coated on graphite felt (GF).   Preparative electrolysis of racemic *sec*-alcohols and diols on Chiral TEMPO-modified GF electrode proceeded electrocatalytically in the presence of achiral base such as 2,6-lutidine, and yielded the corresponding products with high current efficiency and (S)-enantioselectivity.

INTRODUCTION

The use of mediator-modified electrode has many advantages over that of bare electrode for organic electrosynthesis.   However, the modified electrode must stand high current (A/dm^2 level), high stability, long life, etc.   Such a modified electrode has scarcely been reported,[1] though many modified electrodes have been studied for sensors, memories and displays.

Recently, our group has succeeded in preparation and applications of mediator-modified electrode using poly(acrylic acid) (PAA)-coated graphite felt (GF).[2]   Especially, the attractive results which are only attained in modified electrode are enantioselective oxidation of racemic alcohols   (electrochemical resolution of racemic alcohols),[3] diols (direct synthesis of optically pure lactones)[4] and naphthols (direct synthesis of optically pure binaphthyls)[5] on 2,2,6,6-tetramethylpiperidiny-1-yloxy (TEMPO)-modified PAA-GF electrode in the presence of (-)-sparteine. Connecting with the above, the similar enantioselective oxidation was succeeded in the use of a (6S,7R,10R)-1-aza-4-amino-2,2,7-trimethyl-10-isopropylspiro[5.5]undecane-1-yloxyl (Chiral TEMPO, 3)[6]-modified PAA-GF electrode in the presence of an achiral base of 2,6-lutidine.

1 : (6R,7R,10R)
2 : (6R,7S,10R)
3 : (6S,7R,10R)
4 : (6S,7S,10R)

Chiral TEMPO

METHOD

Preparation of modified electrode

A tipycal Chiral TEMPO-modified PAA-GF electrode was prepared as shown in Scheme 1. GF piece (National Electric Carbon Corp., WDF, surface area 0.7 m^2/g) first coated with a thin PAA layer, and then Chiral TEMPO was immobilized, hexamethylenediamine was crosslinked to the PAA layer by

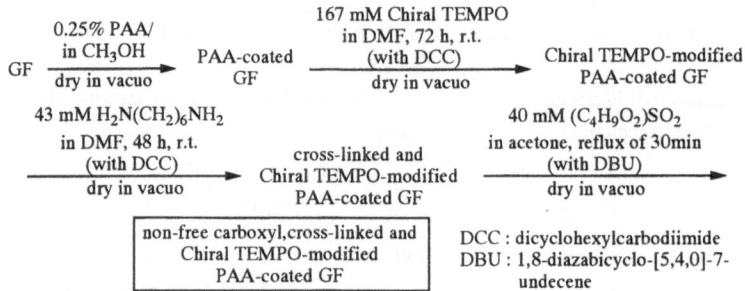

Scheme 1. Preparation of Chiral TEMPO-modified PAA-GF electrode for macroelectrosynthesis.

Table 1. Composition of PAA Layer of Chiral TEMPO-modified PAA-GF Electrode

Thickness of PAA layer	Chiral TEMPO-modified (%)	Cross-linked (%)	Butylated (%)	Density of Chiral TEMPO (μmol/cm^3)
ca. 40 nm	29.8	54.3	15.9	14.2

amido bond, and finally the remained free carboxyl groups of the PAA layer wsa esterified. Some analytical data of the surface of **3**-modified PAA-GF electrode are demonstrated in Table 1. Each values were estimated base on the reacted ratio of carboxylate groups of PAA layer. The density of **3** in the PAA layer was also indicated in Table 1 (14.2 μmol·cm^3).

Macroelectrolysis

Preparative potential-controlled electrolysis was performed in CH$_3$CN solution, using an 'H' type divided cell separated by cationic exchange membrane (Nafion 117). The anolyte contained 1.2 mmol of substrate (diol; 0.3 mmol), 1 mmol of tetralin as a chromatographic standard, 1.2 mmol of 2,6-lutidine as a base and 1 mmol of NaClO$_4$ as a supporting electrolyte in a total volume of 5 ml. The catholyte was 5 ml of CH$_3$CN solution containing 1 mmol of NaClO$_4$. Controlled potential electrolysis was carried out at + 0.80 V (*vs*. Ag/AgCl) under nitrogen atmosphere. The size of the modified anode was 1.0 x 1.0 x 0.5 cm. During electrolysis, the substrates and products were analyzed by gas chromatography (GC) or high performance liquid chromatography (HPLC). After the electrolysis was over, the anolyte was evaporated, dissolved in 30 cm^3 ethyl acetate, and the mixture was washed with 0.1 M HCl and H$_2$O, dried with sodium sulfate and concentrated. Then, for example, the reaction mixture from 1-phenylethanol thus obtained was fed onto a silica gel column (Wako Gel C-200, 3 cm ϕ x 50 cm) and eluted with a hexane-ethyl acetate mixture (9:1 v/v). The eluted solution was evaporated and distilled. The product was indentified by conventional methods. The reaction mixtures of the other substrates were similarly treated.

RESULTS AND DISCUSSION

Enantioselective electrocatalytic oxidation of racemic *sec*-alcohols

The cyclic voltammograms of (*R*)-(+)- and (*S*)-(-)-1-phenylethanol   (*R*-**5** and *S*-**5**, respectively)

presence of 2,6-lutidine are shown in Figure 1. The anodic peak current for R-5 was slightly larger than that for blank (the electrode itself), even if 2,6-lutidine in present in the electrolyte solution. On the other hand, the anodic peak current for S-5 was 6.1 fold larger than that for R-5 and no cathodic peak was observed on the reverse scan, showing that S-5 was oxidized electrocatalytically.

The results of enantioselective electrocatalytic oxidation of racemic sec-alcohols are shown in Table 2. In the case of $(\pm)$-5, S-5 was completely oxidized to acetophenone whereas 98.6% of R-5 remained unreacted. The current efficiency for the oxidation, the ee of the unreacted alcohol and the turnover number based on TEMPO were 99.1%, 99.6% and 74 respectively. All racemic sec-alcohols were also S-enantioselectively oxidized to the corresponding

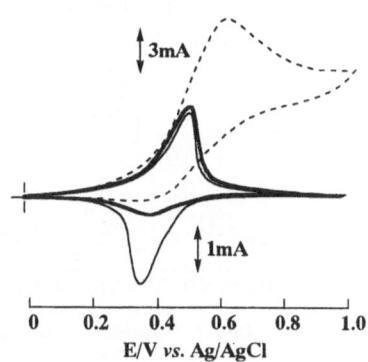

Figure 1 Cyclic voltammograms of 3-modified PAA-GF electrode (1.0 x 1.0 x 0.5 cm) in 0.2 M $NaClO_4$ / CH_3CN at the scan rate of 10 mV/s. ———— : in the presence of 10 mM R-5 and 10 mM 2,6-lutidine, - - - - : in the presence of 10 mM S-5 and 10 mM 2,6-lutidine, ———— : blank.

Table 2 The Results of Enantioselective Electrocatalytic Oxidation of Racemic sec-Alcohols on 3-modified PAA-GF Electrode [a)]

Racemic alcohol	Product	Main remained alcohol	Charge passed / C	Oxidized alcohol / %	Unreacted alcohol / %	η / %	T. N.	R : S	% ee
HO—CH₃	O—CH₃	HO—CH₃	117.8	50.4 (98.6)	49.3	99.1	74	99.8 : 0.2	99.6
—OH	—O	—OH	123.5	51.6 (96.2)	48.1	96.8	76	97.8 : 2.2	95.4
OH	O	OH	118.1	50.5 (97.8)	48.9	98.8	74	99.3 : 0.7	98.6
OH	O	OH	120.9	51.1 (97.0)	48.5	97.8	75	98.9 : 1.1	97.8

a) 5 ml of 0.2 M $NaClO_4$ / CH_3CN containing 0.30 mmol racemic alcohols and 0.30 mmol of 2,6-lutidine.

ketones and the unreacted alcohols contained more than 96% of R-isomers. The current efficiency and selectivity for the formation of ketones were also high. On the contrary, the electrooxidation of $(\pm)$-2-phenylpropanol, which has a chiral center at β-position to hydroxy group, was non-enantioselective. This fact means that that an α-hydrogen of chiral center adjacent to the hydroxy group will be necessary to attain an enantioselective oxidation in the present electrochemical method. This enantioselective electrolysis can be applied to an optical resolution process of racemic alcohols.

Electrocatalytic oxidation of diols to optically active lactones

Diols were electrocatalytically oxidized to optically active lactones (Table 3). For instance, 3-

Table 3 The Results of Stereoselective Electrocatalytic Lactonization of Diols on 3-modified PAA-GF Electrode [a]

Substrate	Product	Charge passed / C	Reaction time / hr	η / %	Isolated yield / %	R : S	% ee	T. N.
(structure)	(structure)	115.2	10.5	97.0	96.5	0.3 : 99.0	99.4	70.6
(structure)	(structure)	118.9	28	89.3	91.7	0.7 : 99.3	98.6	67.1
(structure)	(structure)	113.8	9	96.5	94.8	0.1 : 99.9	99.8	69.4

a) 5 ml of 0.2 M $NaClO_4$ / CH_3CN containing 0.30 mmol diols and 1.2 mmol of 2,6-lutidine.

Table 4 Stereoselective Electrocatalytic Lactonization of 3-Methylpentane-1,5-diol to 3-Methyl-δ-valerolactone [a]

Method	Charge passed / C	η / %	Isolated yield / %	R : S	% ee	T. N. [d]
electrocatalysis on 3-GF	115.2	97.0	96.5	0.3 : 99.7	99.4	70.4
electrocatalysis on 4-amino-TEMPO-GF	116.8	96.4	97.2	50 : 50	0	47.8
electrocatalysis on bare GF [b]	127.9	78.5	86.7	31.2 : 68.8	37.6	28.8
reagent oxidation [c]	———	———	91.3	35.1 : 64.9	29.8	1

a) the presence of 0.30 mmol 3-methylpentane-1,5-diol and 1.2 mmol 2,6-lutidine in each reaction, b) 0.04 mmol 4-acetylamino-Chiral TEMPO(3) in 5 ml of 0.2 M $NaClO_4$ / CH_3CN, c) 1.2 mmol of oxoammonium tosylate of 4-acetylamino-Chiral TEMPO (3) in 5 ml CH_3CN, 48 hrs, d) calculated from 3-methyl-δ-valerolactone (mol) / TEMPO (mol).

methylpentane-1,5-diol was oxidized to (S)-3-methyl-δ-valerolactone in 92.6% current efficiency and 93.8% isolated yield with 98.0% R enantiomeric excess. All diols were also S-stereoselectively oxidized to the corresponding lactones.

This stereoselective lactonization was achieved only on the Chiral TEMPO-modified electrode in the presence of achiral base (Table 4), because a low enantioselective oxidation proceeded not only on a bare GF with chiral TEMPO in solution but also in a homogeneous chemical system under the similar conditions.

References

1) For instance, A. Deronzier, D. Limosin, and J-C. Moutet, *Electrochimica Acta*, **1987**, *32*, 1643.

2) Y. Kashiwagi, H. Ono, and T. Osa., *Chem. Lett.*, **1993**, 257.

3) Y. Kashiwagi, Y. Yanagisawa, F. Kurashima, J. Anzai, T. Osa, and J. M. Bobbitt, *Chem. Commun.*, **1996**, 2745.

4) Y. Yanagisawa, Y. Kashiwagi, F. Kurashima, J. Anzai, T. Osa, and J. M. Bobbitt, *Chem. Lett.*, **1996**, 1043.

5) T. Osa, Y. Kashiwagi, Y. Yanagisawa, and J. M. Bobbitt, *J. Chem. Soc., Chem. Commun.*, **1994**, 2535.

6)   Z. Ma, G. Huang, and J. M. Bobbitt, *J. Org. Chem.*, **1993**, *58*, 4837.

ENANTIOSELECTIVE, ELECTROCATALYTIC OXIDATION OF 1-PHENYLETHYL-AMINE ON A CHIRAL TEMPO-MODIFIED ELECTRODE

Yoshitomo KASHIWAGI, Futoshi KURASIMA, Chikara KIKUCHI, Jun-ichi ANZAI, and Testuo OSA

Faculty of Pharmaceutical Sciences, Tohoku University, Aobayama, Aoba-ku, Sendai 980-77, Japan

ABSTRACT

Preparative enantioselective electrocatalytic oxidation of 1-phenylethylamine was succeeded achived on a poly (acrylic acid) (PAA)-coated graphite felt (GF) electrode immobilized $(6R,7R,10R)$-1-aza-4-amino-2,2,7-trimethyl-10-isopropylspiro[5.5]undecane-1-yloxyl (Chiral TEMPO) in high current efficiency and (R)-enantioselectivity.

INTRODUCTION

For long years, much attention for enantioselective electrosynthesis has been paid by many electroorganic chemists. We have reported that *sec*-alcohols and diols were enantioselectively oxidized to ketones and lactones, respectively, on a graphite felt (GF) electrode coated with a thin poly (acrylic acid) (PAA) layer immobilizing $(6R,7R,10R)$-1-aza-4-amino-2,2,7-trimethyl-10-isopropylspiro[5.5]undecane-1-yloxyl (Chiral TEMPO, **1**)[1], in high yield, current efficiency, (R)-enantioselectivity at a constant potential of + 0.80 V *vs.* Ag/AgCl.[2] Connecting with the above, the enantioselective electrocatalytic oxidation of 1-phenylethyl-amine (**5**) was succeeded in the use of a **1**-modified PAA-GF electrode in the presence of 2,6-lutidine.

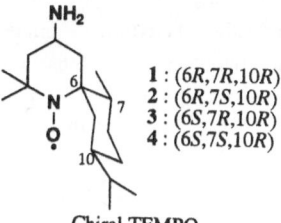

1 : $(6R,7R,10R)$
2 : $(6R,7S,10R)$
3 : $(6S,7R,10R)$
4 : $(6S,7S,10R)$

Chiral TEMPO

METHOD

A PAA-GF modified by **1** introducing 3 chiral centers was prepared in a similar method as the PAA-GF modified by achiral 4-amino-TEMPO.[3] Preparative potential-controlled electrolysis was performed in CH_3CN solution, using an 'H' type divided cell separated by cationic exchange membrane (Nafion 117). The anolyte contained 1.2 mmol of **5**, 1 mmol of tetralin as a chromatographic standard, 4.8 mmol of 2,6-lutidine as a base and 1 mmol of $NaClO_4$ as a supporting electrolyte in a total volume of 5 ml. The catholyte was 5 ml of CH_3CN solution containing 1 mmol of $NaClO_4$. Controlled potential electrolysis was carried out at + 0.50 V (*vs.* Ag/AgCl) under nitrogen atmosphere. The size of the modified anode was 1.0 x 1.0 x 0.5 cm. During electrolysis, the substrates and products were analyzed by gas chromatography (GC) or high performance liquid chromatography (HPLC).

REAULT AND DISSCUSSION

This chiral TEMPO-modified PAA-GF was deactivated by cyclic voltammetric sweeps, but restored completely of its electrocatalytic activity treated by a *m*-chloroperbenzoic acid ether solution probably due to formation of a different structure of chiral TEMPO (1'), and the changed electrode was used for preparative electrolysis. Most of the research has been devoted to identify a changed stabler structure of chiral TEMPO. The cyclic voltammograms of (R)-(+)- and (S)-(-)-1-phenylethylamine (R-5 and S-5, respectively) presence of 2,6-lutidine are shown in Figure 1. The anodic peak current for S-5 was slightly larger than that for blank (the electrode itself), even if 2,6-lutidine in present in the electrolyte solution. On the other hand, the anodic peak current for R-5 was 2.2 fold larger than that for S-5 and no cathodic peak was observed on the reverse scan, showing that R-5 was oxidized electrocatalytically.

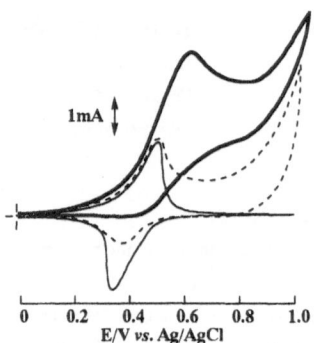

Figure 1 Cyclic voltammograms of 1'-modified electrode (1.0 x 1.0 x 0.5 cm) in 0.2 M NaClO₄/CH₃CN at the scan rate of 10 mV/s. ——: in the presence of 10 mM R-5 and 10 mM 2,6-lutidine, - - - -: in the presence of 10 mM S-5 and 10 mM 2,6-lutidine, ——: blank.

Base on the cyclic voltammetry results, apreparative and controlled potential electrolysis racemic **5** was perfomed at +0.5 V vs. Ag/AgCl. For racemic amine such as **5** which possesses a chiral center at the α-position to the amino group, the R-isomer were only oxidized probably to the corresponding imine (**6**) and the S-isomer remained unreacted on the re-activated electrode. Imine **6** is the expected oxidation intermediate, which can be easily hydrolyzed to acetophenone (**7**); **6** has not actually been detected.

The enantioselectivity of the remaining S-**5** was 98% ee. The current efficiency for the produced **7** was 90.1%. On the other hand, **5** was only oxidized to **7** both in 92.6% isolated yield and 94.3% current efficiency on 4-amino-TEMPO-modified PAA-GF electrode in place of chiral TEMPO-modified PAA-GF electrode. Thus, this enantioselective electrolysis can be useful for an optical resolution process of racemic amines.

References

1) Z. Ma, G. Huang, and J. M. Bobbitt, *J. Org. Chem.*, **1993**, *58*, 4837.
2) Y. Kashiwagi, F. Kurashima, Y. Yanagisawa, J. Anzai, T. Osa, and J. M. Bobbitt, *Proceedings of 7th International Symposium on Electrochemistry* (Taiwan), **1996**, 58.
3) Y. Kashiwagi, H. Ono, and T. Osa, *Chem. Lett.*, **1993**, 257.

MODIFICATION OF THIOLS ONTO SINGLE CRYSTAL SURFACES OF GOLD ELECTRODES FOR METALLOPORTEIN ELECTROCHEMISTRY

Isao TANIGUCHI, Soichiro YOSHIMOTO, Masahito YOSHIDA,
Yasuhiro MIE, Kazuhiko KUDO, and Katsuhiko NISHIYAMA
*Department of Applied Chemistry and Biochemistry, Faculty of
Engineering, Kumamoto University, 2–39–1, Kurokami, Kumamoto 860,
Japan*

ABSTRACT

Adsorption behavior of pyridine thiols (PySH) and corresponding disulfides (PySSPy) onto gold single crystal surfaces has been examined to understand in detail the suitable surface structures of the modified electrodes for cytochrome c electrochemistry. On the Au(111) surface PySH was rather easily decomposed with the C-S bond cleavage during modification of thiols from an ethanolic PySH solution, while PySSPy was more stable under the same conditions. Very interestingly, cytochrome c electrochemistry depended very much on the modifier structure at single crystal surfaces.

INTRODUCTION

Recently, functional surfaces modified with thiols have been studied extensively. We have reported[1] that electron transfer reactions of metalloproteins can be accelerated on the gold surfaces modified with thiols. In order to understand such surface functions at the molecular level, use of well–defined surfaces in atomic level would be preferable. In the present study, we have prepared Au(111), Au(100) and Au(110) surfaces to modify with thiols, and the dependence of the surface functions on the single crystal faces was examined by cyclic voltammetry (CV) using both the reductive desorption of the adsorbed thiols in an alkaline solution and cytochrome c electrochemistry as a monitor reaction in a neutral solution.

EXPERIMENTAL

Gold single crystals were prepared by the method developed by Clavilier et al.[2] (so–called the flame–annealing–quenching method). Surface modification of the electrode was carried out by contacting with 4,4'-dithiodipyridine (4,4'-PySSPy), 4-mercaptopyridine (4-PySH), and other related modifiers by the meniscus method. The electrochemical reductive desorption of thiols from each single crystal surface was performed in a 0.1 M KOH solution.[3]

RESULTS AND DISCUSSION

Au(111) surface should be used with careful attention when pyridine thiols (PySH) are modified onto the gold surface, because, although 4-PySH acts as an effective electron-transfer promoter for cytochrome c, decomposition of PySH itself has been reported[3,4] to make the structure of the modified surface complicated. This was found not only due to the surface reaction such as the pyridine-S bond cleavage but also the decomposition in solution. The carbon-sulfur bond cleavage occurred even in an ethanolic solution, and thus use of an old solution of the modifier, for example, the modifier solution is no longer a simple solution but various chemical species must be inside: This was confirmed by the MS measurements. Also, during modification of 4-PySH onto the Au(111) surface the pyridine-S bond cleavage took place. Again, the C-S bond cleavage was faster in ethanol than in water. The reaction was slower when the corresponding disulfide instead of PySH was used as a modifier. Also, the reaction occurred much faster on Au(111) than on Au(110) and (100) surfaces.[4] After examination of this C-S bond cleavage in some detail, when an aqueous solution with a low concentration (< 50 μM) of modifier, and preferably using the disulfide but not thiol, one can obtain a suitable promoter-modified surface even on Au(111) electrode, where rapid direct electron transfer of cytochrome c takes place. Promoter-modified Au(110) and Au(100) single crystal electrodes were obtained more easily than on the Au(111) surface.

Using atomically flat single crystal surfaces, clear effect of the structure of the promoter molecule on cytochrome c electrochemistry has been observed: At Au(110) and Au(100) surfaces, only 4,4'-PySSPy and 4-PySH modified electrodes gave well-defined voltammetric responses of cytochrome c, but no response was seen at all on 2,2'-PySSPy and 2-PySH modified electrodes. No response was also observed from diphenyldisulfide (PhSSPh) and mercaptophenol modified electrodes.[5] The present results suggest that an interaction between cytochrome c and the modifier is essential for the rapid electron transfer reaction of cytochrome c at the modified electrodes.

The poor electrochemical response at Au(111) electrode modified with 4,4'-PySSPy can be explained in terms of various reasons depending on the experimental conditions: 1) Pyridine-S bond cleavage, 2) desorption of the promoter molecule from the PyS-Au modified electrode, and 3) difference in structure of the promoter-modified electrode surfaces.

References
1) (a) I. Taniguchi, K. Toyosawa, H. Yamaguchi, and K. Yasukouchi, *J. Chem Soc., Chem. Commun.*, **1982**, 1032; (b) I. Taniguchi, K. Toyosawa, H. Yamaguchi, and K. Yasukouchi, *J. Electroanal. Chem.*, **1982**, *140*, 187; (c) F. M. Hawkridge and I. Taniguchi, *Comments on Inorg. Chem.*, **1995**, 17, 163.
2) J. Clavilier, R. Faure, G. Guinet and R. Durand, *J. Electroanal. Chem.*, **1980**, *107*, 205.
3) B. E. Lamp, D. Hobara, M.D. Porter, K. Niki and T. M. Cotton, *Langmuir*, **1997**, *13*, 736.
4) I. Taniguchi, S. Yoshimoto, K. Nishiyama, *Chem. Lett.*, **1997**, 353.
5) I. Taniguchi, S. Yoshimoto, M. Yoshida, *Langmuir*, in press.

DIELS-ALDER REACTION OF IN SITU GENERATED QUINONES ON A LITHIUM PERCHLORATE / POLYETHYLENEGLYCOL MODIFIED ELECTRODE

Kazuhiro CHIBA, Shokaku KIM and Masahiro TADA

Laboratory of Bio-organic Chemistry, Tokyo University of Agriculture andTechnology, 3-5-8 Saiwai-cho, Fuchu, Tokyo 183, Japan

ABSTRACT

Lithium perchlorate was highly concentrated on the polyethyleneglycol modified electrode and promoted Diels-Alder reactions. By using the modified electrode in acetonitrile, hydroquinones were efficiently oxidized to corresponding quinones which were immediately trapped by dienes to give varied cycloadducts.

INTRODUCTION

The solvent system which possesses internal solvent pressures provides comparable rate accelerations of intermolecular cycloaddition process.[1] We obtained varied cycloadducts *via* electrochemically generated intermediates in lithium perchlorate / nitroalkane solution which was found to accelerate cycloaddition reactions.[2] In this reaction system, PTFE-fiber coated electrode efficiently oxidized hydroquinones even in the presence of easily oxidizable dienes and desired cycloadducts of quinones and dienes were obtained in high yields. Furthermore, we report our findings, which shows that the cycloaddition reaction was markedly promoted on a

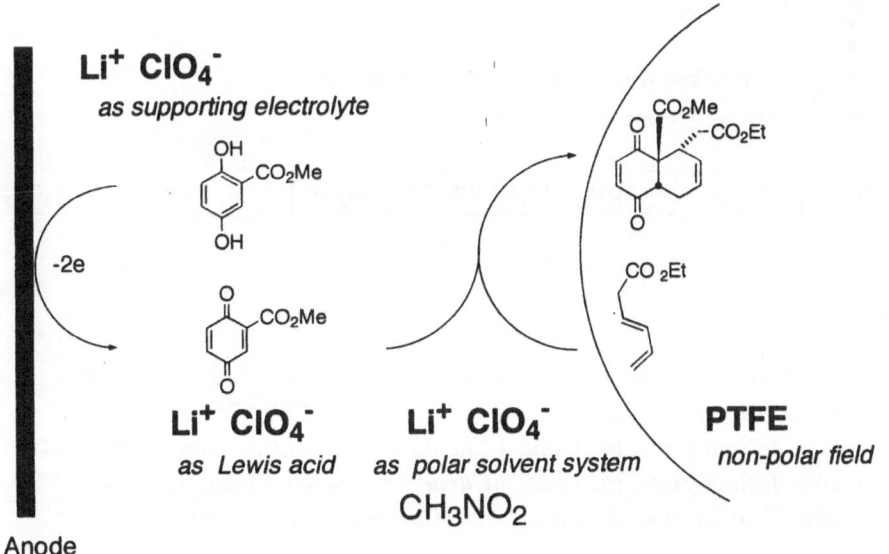

polyethyleneglycol coated electrode in the presence of lithium perchlorate in acetonitrile.

METHOD

In a preliminary study, acceleration property of Diels-Alder reaction by poly-ethers in the presence of lithium perchlorate was examined. Diels-Alder reaction of *p*-benzoquinone and isoprene was scarcely promoted in saturated lithium perchlorate in dry acetonitrile. On the other hand, desired cycloadduct was obtained in high yield in lithium perchlorate / acetonitrile and diglyme (1:1 v/v). In this solution, lithium perchlorate was dissolved in higher concentration. The results suggested that poly-ether compounds in acetonitrile promoted Diels-Alder reactions. Accordingly, we envisioned that Diels-Alder reaction of *in situ* generated dienophiles with dienes could be readily accessible on the polyethyleneglycol coated electrode. The modification was performed by esterification of polyethyleneglycol monoalkylethers on a poly (acrylic acid) modified electrode. By using the electrode, varied hydroquinones were successfully oxidized to corresponding quinones which was trapped by dienes.

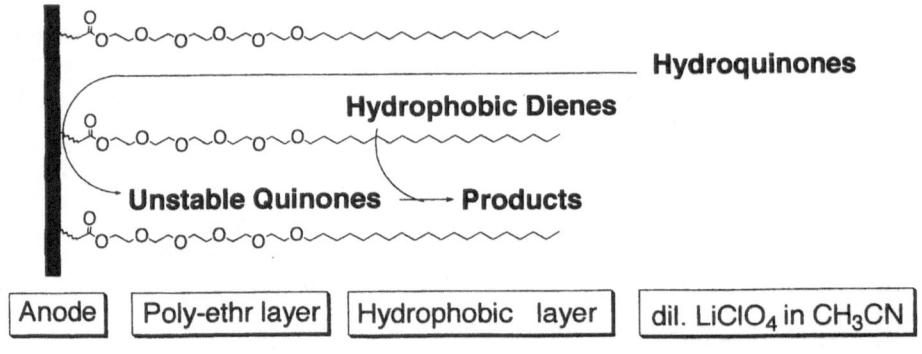

References
1) P. A. Grieco, J. J. Nunes and M. D. Gaul, *J. Am. Chem. Soc.*, **1990**,4595
2) K. Chiba, M. Tada, *J.Chem.Soc., Chem.Commun.*, **1994**,2485
3) K. Chiba, J. Sonoyama, M. Tada, *J.Chem.Soc., Chem.Commun.*, **1995**,1381
4) K. Chiba, J. Sonoyama, M. Tada, *J.Chem.Soc., Perkin Trans. 1*, **1996**,1435
5) K. Chiba, T .Arakawa, M. Tada, *Chem.Commun.*, **1996**,1763

ELECTROCHEMICAL OXIDATION OF BIOLOGICAL AMINES ON THE ELECTRODES MODIFIED WITH HYDROPHOBIC MONOLAYERS

Hiroaki SHINOHARA and Masahiko SISIDO

Department of Bioscience and Biotechnology, Faculty of Engineering,
Okayama University, Tsushima-Naka, Okayama 700, Japan

ABSTRACT

The electrodes modified with hydrophobic monolayers were applied to detect electrochemically an indicator hormone for biological rhythm, melatonin, without separation preocesses. It was demonstrated that octanol-modified glassy carbon electrode could be available to oxidize melatonin selectively to a certain degree against the interfering substances and that the modified layer was stable under the high positive potential application for the detection of melatonin.

INTRODUCTION

In this paper, the surface structure of the electrodes was designed to perform selective electrochemical detection of an important biological amine, melatonine, without column separation process. Melatonin (N-acetyl-5-methoxytriptamine) has been considered as an indicator hormone for biological rhythm. Its levels *in vivo* are photoperiod-dependent and exhibit circadian rhythm under the constant dark conditions. The rapid and simple detection systems for melatonin has been required in deed to diagnose the biological rhythms of human and other animals. To realize the separation-free quick determination of melatonin, we paid much attention to the point that melatonin is hydrophobic while most interfering substances are hydrophilic. We expected that hydrophobic monolayer-modified electrodes could be available to oxidize melatonin selectively. Therefore self assembly of n-alkanethiol on gold electrodes and electrochemical modification of n-alkanol on glassy carbon electrodes were applied to make hydrophobic monolayers onto electrode surfaces and to detect melatonin selectively by electrochemical oxidation. The stability of the modified electrodes for high potential application was also investigated to evaluate the reusability.

METHOD

Alumina-polished gold(Au) disk electrodes were immersed into the ethanol solutions containing 20mM of n-alkanethiols($C_mH_{2m+1}SH$, m=4,8,12,16) for 20 min to make alkanethiol self-assembled monolayer on Au surfaces. After rinse the modified Au electrodes were set into the 0.1M phosphate buffer solutions (pH 7.0), in which samples were dissolved. The electrochemical oxidation of melatonin and interfering substances on a bare Au and modified Au electrodes was characterized by differential pulse voltammetry with an electrochemical analyzer.

The modification of n-Alkanol monolayer onto glassy carbon (GC) electrodes was carried out by electrochemical oxidation of n-alkanol as reported by Prof. Maeda[1]. Cyclic voltammetry of a polished GC electrode was done in the potential range from 0V to 2.0V vs. Ag/AgCl in an alkanol solution containing H_2SO_4 as an electrolyte. After rinse, electrochemical measurements for various biological amines with the modified GC electrodes were carried out in the same manner as the alkanethiol-modified gold electrodes.

RESULTS AND DISCUSSION

At a bare Au electrode, oxidation peaks for melatonin and dopamine were observed clearly at potentials of 0.68V and 0.16V vs. Ag/AgCl, respectively. The oxidation waves for dopamine at octanethiol-modified Au electrodes declined significantly, whereas the oxidation waves for melatonin at the modified electrode declined only slightly as shown in Figure 1. The oxidation of serotonin, noradrenaline and ascorbic acid was also suppressed at the octanethiol-modified Au electrodes. However, at dodecanethiol and hexadecanethiol-modified Au electrode, the oxidation of melatonin was also disturbed.

Oxidation current for melatonin on an octanol-modified GC electrode was suppressed at a half as compared with that on an unmodified electrode, on the other side the oxidation currents for interfering substances such as dopamine declined significantly on the modified GC electrode. The suppression effects of octanol-modification for the electrochemical oxidation of interfering substances were arranged in Figure 2. The dependency of suppression effect on alkyl chain length was almost similar to the case of alkanethiol-modified Au electrodes.

These data support that electrodes modified with hydrophobic monolayers are available to detect melatonin selectively to a certain degree due to hydrophobic interaction and that octyl group-modification is most suitable for electrochemical sensing of melatonin.

The stability of the monolayer-modified electrodes for oxidation measurements was evaluated by voltammetry of interfering substances after the high potential application at 0.9V vs. Ag/AgCl. The oxidation suppression for interfering substances on the octanethiol-modified Au electrode was unfortunately disappeared after the high potential application, however the suppression effect on the octanol-modified GC electrodes was maintained after the high potential application. It was demonstrated that alkanol-modifed GC electrodes are promising to detect melatonin selectively and stably.

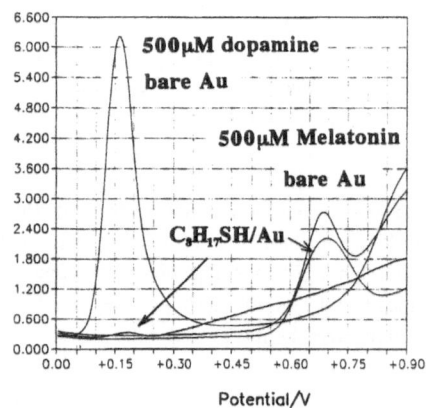

Fig.1 DPV waves of melatonin and dipamine at a bare and an octanethiol-modified Au electrode.

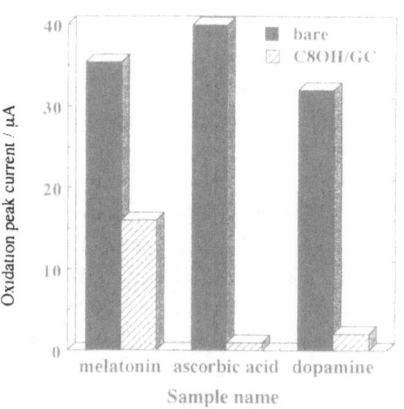

Fig.2 Peak currents in DPV waves with a bare and an octanol-modified GC electrode. Sample conc.:500μM.

Reference
1) H.Maeda et al., Chem.Pharm.Bull., 42, 1870-1873 (1994).

STUDY ON ANODIC OXIDATION OF ESTERS ON TRANSITION METAL OXIDE ELECTRODES

Kiyoshi Kanamura, Masahiro Ohashi, Zen-ichiro Takehara, and Zempachi Ogumi

Department of Energy and Hydrocarbon Chemistry, Graduate School of Engineering, Kyoto University, Yoshida-honmachi, Sakyo-ku, Kyoto 606-01, Japan

ABSTRACT

An electrochemical oxidation of propylene carbonate (PC) and diethyl carbonate (DEC) with 1.0 mol dm^{-3} $LiClO_4$ on the $LiCoO_2$ electrode was investigated using *in situ* Fourier Transform Infrared (FTIR) spectroscopy. The obtained spectra showed that C-O-C bonds in both solvents were broken at the initial stage of the electrochemical oxidation process to form radical species which have a similar chemical structure to carboxylic compounds. These radical intermediates were changed by a further electrochemical oxidation and/or a following chemical reactions. In the case of PC, the following chemical reaction proceeded to form carboxylic anhydrous adsorbed on the $LiCoO_2$ surface. In the case of DEC, such adsorbates were not observed.

INTRODUCTION

Rechargeable lithium batteries have been developed as portable power sources. One of battery failure processes is related to the decomposition of nonaqueous electrolytes, so that the stability of nonaqueous electrolytes in batteries is very important. [1, 2] Since, in batteries, cathode materials have the largest surface area, nonaqueous electrolytes may be mostly oxidized on cathode materials. Cathode materials are usually transition metal oxides which may have a high catalytic activity for the electrochemical oxidation of nonaqueous electrolytes. Therefore, the study on the electrochemical oxidation of nonaqueous electrolytes on cathode materials is extremely important. In this study, the electrochemical oxidation of some ester electrolytes on transition metal oxide electrodes was investigated using an *in situ* Fourier Transform Infrared (FTIR) spectroscopy.

EXPERIMENTAL

Transition metal thin film oxide electrodes were prepared by a rf-sputtering method (13.56 MHz). The target electrode was made of $LiCoO_2$ disc pellet which was prepared by pressing $LiCoO_2$ fine powder. A stainless steal cylinder was used as the substrate. A mixture gas of oxygen and argon (2:1) was used for the preparation of the thin film. The thickness of the prepared thin film was estimated to be 1 μm with a scanning electron microscope.

Lithium metal was used as the reference electrode, and Ni wire was used as the counter electrode. Propylene carbonate (PC) or ethylene carbonate (DEC) containing 1.0 mol dm^{-3} $LiClO_4$ was used as the electrolyte. The electrode potential of $LiCoO_2$ was swept to 4.0 V at 10 mV min⁻¹ and then was kept at 4.0 V vs. Li/Li⁺ until the Subtractively Normalized Interfacial FTIR (SNIFTIR) spectrum became stable (until no peaks were observed). Then, the *in situ* FTIR measurement was started from 4.0 V vs. Li/Li⁺ with a potential step method. The potential width of each step was 100 mV. Reflectance spectra were obtained before and after the potential step. SNIFTIR spectra were calculated from these spectra at each potential step. Upward peaks in SNIFTIR spectra show disappearance of chemical bonds corresponding to wavenumbers, and downward ones correspond formation of new chemical bonds corresponding to wavenumbers.

RESULTS AND DISCUSSION

Figures 1 and 2 shows the SNIFTIR spectra for the electrochemical oxidation of PC and DEC containing 1.0 mol dm^{-3} LiClO$_4$ on the LiCoO$_2$ thin film electrode. A lot of upward and downward peaks were observed in these spectra. Peaks were already observed at 4.1 V vs. Li/Li$^+$, so that the on-set potential for the electrochemical oxidation was seemed to be more cathodic than 4.1 V vs. Li/Li$^+$. The downward peaks in these spectra correspond to the formation of new compounds and the upward ones are due to the disappearance of some compounds existing near the LiCoO$_2$ thin film electrode surface. These peaks were independent of the electrode potential, showing to that the chemical composition of decomposition products does not change in this potential range. All upward peaks observed in Figures 1 and 2 correspond to PC and DEC, respectively, indicating that these solvents are adsorbed on the LiCoO$_2$ electrode before the electrochemical oxidation and then oxidized to some new species. Most of downward peaks in both spectra were assigned to species having carboxylic groups. On the other hand, a peak observed at 1850 cm^{-1} was only observed in Figure 1, corresponding to C=O in carboxylic anhydrous compounds which is formed by a following chemical reaction of intermediate species generated from PC.

REFERENCES
1) D.Aurbach, M. Daroux, P. Faguy and E. Yeager, *J. Electrochem. Soc.*, **1991**, 297, 225.
2) A. Campbell, C. Bowes and R. S. McMillan, *J. Electroanal. Chem.*, **1990**, 284, 195.

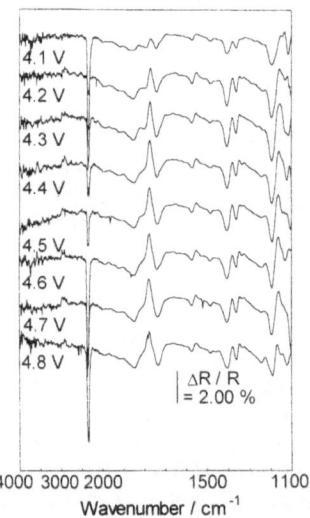

Figure 1 SNIFTIR spectra during the anodic polarization of LiCoO$_2$ electrode in PC with LiClO$_4$.

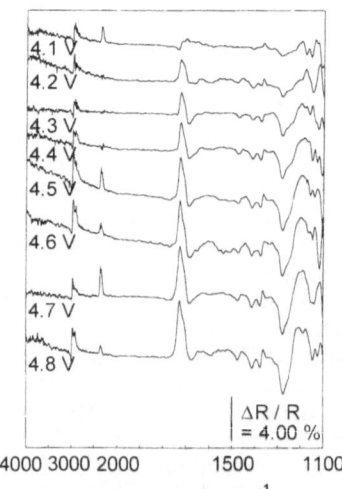

Figure 2 SNIFTIR spectra during the anodic polarization of LiCoO$_2$ electrode in DEC with LiClO$_4$.

UNUSUAL REARRANGEMENT BY ANODIC OXIDATION OF CYCLOALKANONES IN THE PRESENCE OF TRIFLUOROACETIC ACID

Ikuzo NISHIGUCHI

*Department of Chemistry, Nagaoka University of Technology,
1603-1, Kamitomioka-cho, Nagaoka, Niigata 940-21, JAPAN*

ABSTRACT

Anodic oxidation of cyclopentanone and cyclohexanones in a mixed solvent of CF_3COOH and CH_2Cl_2 containing triethylamine led to unusual rearrangement to give selectively γ-valerolactone and a mixture of γ- and δ-lactones, respectively.

Anodic cleavage of a bond between the carbonyl carbon and the adjacent carbon of cycloalkanones followed by Wagner-Meerwein rearrangement and re-cyclization may be proposed as one of the most plausible reaction mechanisms.

INTRODUCTION

Trifluroacetic acid (CF_3COOH) has been known to be a unique solvent possessing specific characters such as strong acidity, small nucleophilicity, high anodic discharge potential(ca. 2.4 V vs. SCE) and satisfactory affinity with other organic solvents. Usefulness of this solvent has been increased, especially in the study of anodic oxidation. Thus, it was shown that anodic oxidation procceded smoothly in the presence of trifluoroacetic acid for the substrates which are difficult to oxidize by conventional methods, such as cycloalkanes[1] and aromatic compounds possessing electron-withdrawing groups and benzene itself[2], possibly because of the enhanced stabilizing effect of this organic strong acid on the radical cations generating electrochemically.

RESULTS AND DISCUSSION

In this study, we wish to report a unique transformation of cyclic ketones to lactones or lactams[3], different from ones formed by Baeyer-Villiger oxidation, through tandem oxidative ring-opening / rearrangement / re-cyclization by anodic oxidation[4] in the presence of trifluoroacetic acid.

The reaction was generally carried out in a mixed solvent of CF_3COOH and CH_2Cl_2(volume ratio; 2:1) containing triethylamine or lithium perchlorate as a supporting electrolyte using a divided cell equipped with platinum plates as anode and cathode, and a ceramic cylinder as a diaphragm at room temperature under the constant current conditions (current density: 22 mA/cm2) until 3F/mol of electricity passed through the system.

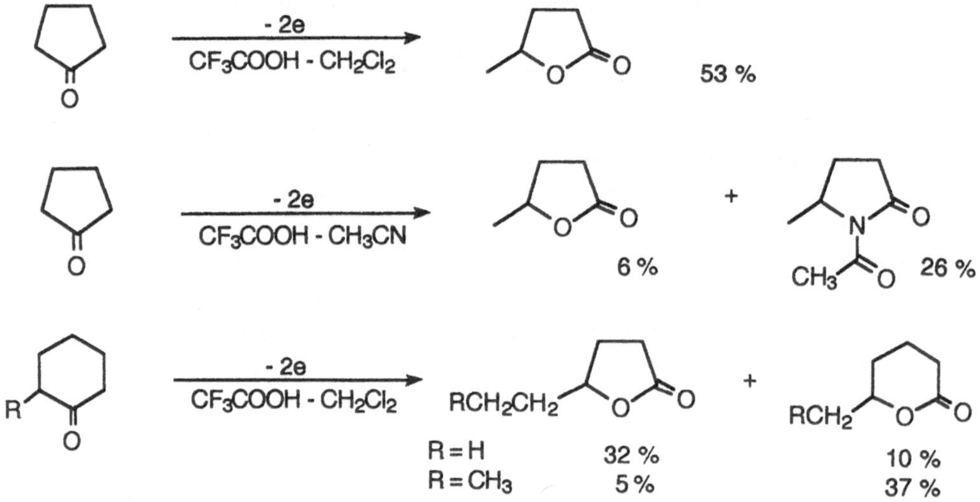

γ–Valerolactone was obtained as a sole product from the reaction of cyclopentanone in a 53% yield, while anodic oxidation of cyclohexanone gave a mixture of γ-hexanolactone and δ-hexanolactone in 32 and 10% yields, respectively[5]. It may be interesting that the introduction of a methyl group into the 2-position of cyclohexanone led to preferential formation [yield: 37%] of the δ-lactone to that[yield: 5%] of the γ-lactone. Use of acetonitrile instead of methylene chloride as a co-solvent brought about formation of N-acetyl-5-methyl-2-pyrrolidone(yield: 26%) along with γ-valerolactone[yield: 6%].

The presence of a methyl group on the 4-positions of a cyclohexanone ring resulted in the selective formation of γ-ethyl-γ-methylbutyrolactone [yield=45%] while 4-tert-butylcyclohexanone was not electrochemically oxidized under the same conditions, suggesting much influences of steric effect on the present anodic oxidation.

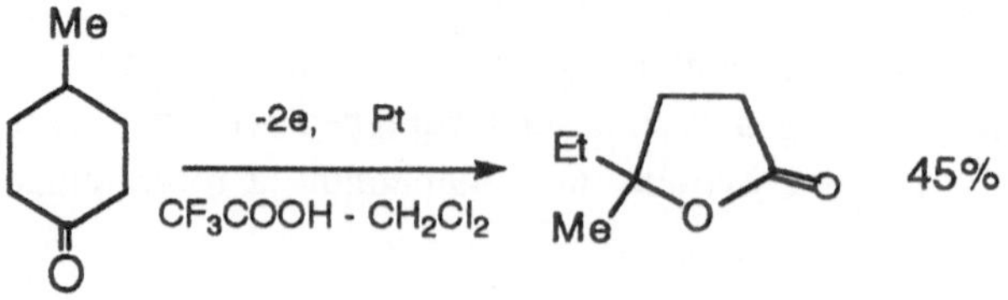

As one of the most plausible reaction mechanism, the following scheme may be proposed, in which an initial electron-transfer is
involved from the unpaired electrons of the oxygen atom of cycloalka-nones followed by ring-opening to give a radical cation **A**.   Nucleo-philic attack of the solvent to the acyl cationic center of **A** followed by the second electron transfer may give a primary carbocation **B**, which is subjected to Wagner-Meerwein rearrangement to generate a more stable secondary carbocation , subsequently re-cyclizing to the final product, lactones.

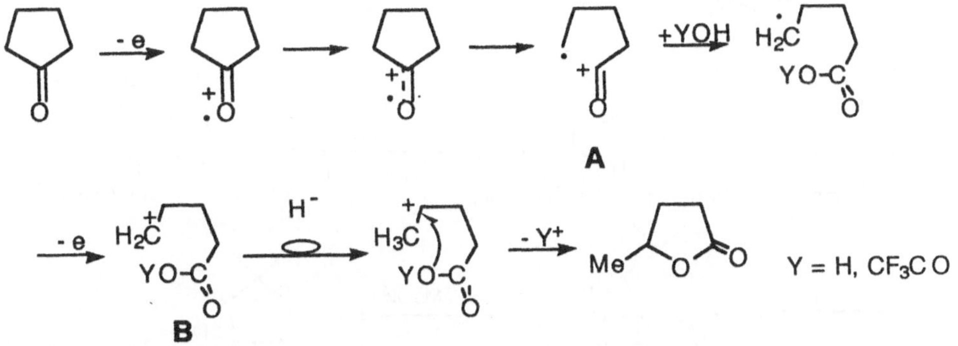

Furthermore, it was found that anodic oxidation of cyclohexa-1,3-diones in the presence of trifluoroacetic acxid also led to unusual rearrangement of the carbon skeleton to give the ring -contraction products, α-hydroxy-α-carbomethoxycyclo-pentanones as almost sole products.   Thus, anodic oxidation of dimedone in a mixed solvent of trifluoruacetic acid and methanol(volume ratio=2:1) containing a small amunt of triethylamine was carried out using a divided cell equipped with platinum plates as anode and cathode and a ceramic cylinder as diaphragm under the constant current conditions(current density=2.0 mA/cm^2) until 4.0 F/mol of electricity passed through the system.   After usual esterification of the reaction mixture in refluxing

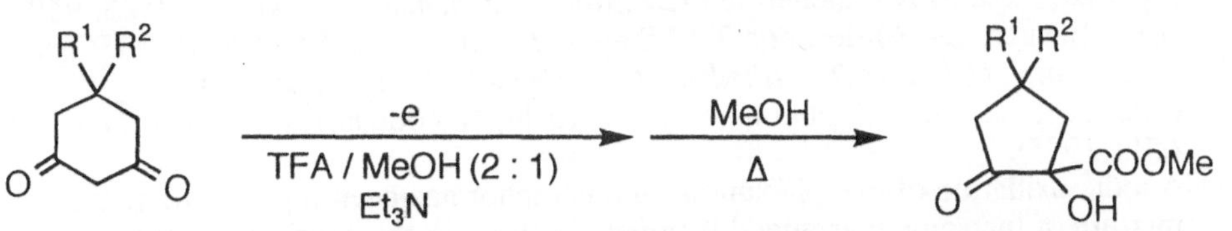

$R^1=R^2=CH_3$ Isolated Yield= 47%

$R^1=H, R^2=CH(CH_3)_2$ Isolated Yield= 36%

$R^1=H, R^2=CH_3$ Isolated Yield= 23%

methamol, γ–dimethyl-α-hydroxy-α-carbomethoxycyclo-pentanone was obtained in a 47% yield as almost a sole product accompanying much tarric products.   Similarl electrochemical oxidation of 5-methyland 5-isopropropylcyclohexa-1,3-diones gave the corresponding ring-contraction products in a 36 and 23% yields,respectively.

This unusual rearrangement may be initiated by electron transfer from initially formed 3-methoxycyclohexenone as shown in the following scheme.

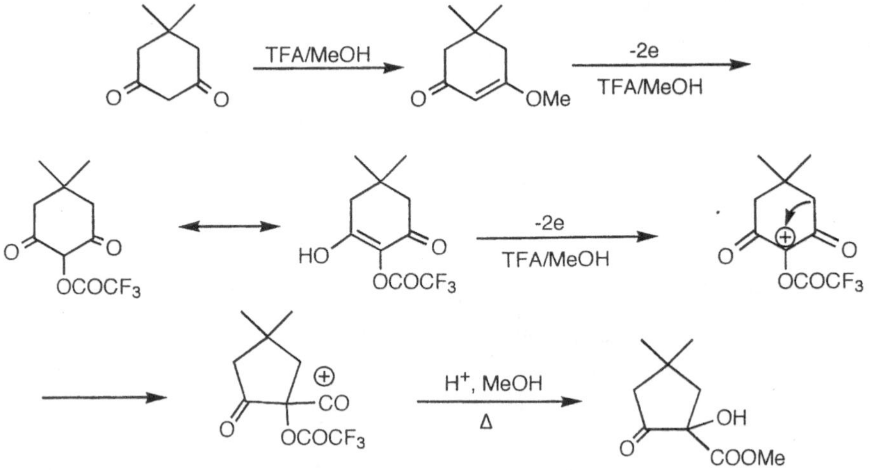

REFERENCES AND NOTES

[1]H.J.Schafer, E.Cramer, A.Hembrock, and G.Matusczyk,"*Electroorganic Synthesis-Festschrift for Manuel M. Baizer*",   R.D.Little and N.L.Weinberg Ed., Maecel Dekker Inc.,New York, **1991**, pp 169-180, and others are cited therein.

[2]a)K.Fujimoto, H.Maekawa, Y.Tokuda, Y.Matsubara, T.Mizuno,and I.Nishiguchi, *Synlett.*, **1995**, 661.   b) K.Fujimoto, Y.Mataubara, H.Maekawa, and I.Nishiguchi., *Tetrahedron*, **5 2**(11), 3838(1996).

[3]All of the products were characterized by spectroscopic (^{1}H-, ^{13}C NMR, MASS and IR spectra) and elemental analyses.

[4]Some works on anodic oxidation of ketones under various conditions has been reported.   See a) D.Pletcher and C.Z.Smith, *J.Chem.Soc.Parkin I* , **1975**, 948 b)J.Y.Becker, L.L.Miller, and T.M.Siegel, *J. Am. Chem. Soc.*,**9 7**, 849(1975). c)M.Oyama, M.Ohono,*Tetrahedron Lett.*,**1966**,5201. d)S,Hara, S.Q. Chen,T,Hatakeyama, T.Fukuhara,M.Sekiguchi, N.Yoneda,*Tetrahedron Lett.*, **3 6**, 6511(1995).

[5]Anodic oxidation of cyclohexanone and camphor has been reported to give a mixture of the same rearranged lactones, as those in this work, as well as ε– caprolactone, a Baeyer Villiger type product. See a)F.Barba, A.Guidrado, M.L. Serura, *An. Quim.,1̃7 5*,404, 967(1979). b)B. Wermeckes, F,Beck, *DECHEMA- Monogr.***112**, 1(1989). c)S.Ye, F.Beck, *Tetrahedron*, **47**, 5463(1991).d)F,Beck, B.Wermeckes, *DECHEMA- Monogr.*, **125**, 823(1992).

CONVERSION OF BIOMASS DERIVED PRODUCTS BY ANODIC ACTIVATION

Hans J. SCHÄFER, Silke KRATSCHMER, Andreas WEIPER, Elisabeth KLOCKE, Mark PLATE, and Reinhard MALETZ

Organisch-Chemisches Institut der Universität Münster
Corrensstraße 40, D-48149 Münster, Germany,
Fax +49 251 83 39772, E-Mail: schafeh@uni-muenster.de

ABSTRACT

Fatty acids and carbohydrate carboxylic acids are homo- and heterocoupled to intermediates for polyesters, new oleochemicals and potential enzyme inhibitors. - L-Ketogulonic acid is decarboxylated quantitatively to L-xylonolactone. - To conjugated linoleic acid nucleophiles can be added anodically to afford disubstituted oleates. - Enone fatty acid methylesters are hydrodimerized to dimer fatty acids and dimethyl aconitate is catalytically cyclized to hexamethyl cyclopentanehexacarboxylate.

Anodic Decarboxylation of fatty acids and carbohydrate carboxylic acids

Carboxylic acids can be decarboxylated at the anode to form either radicals or carbocations depending on their structure[1]. Biomass derived carboxylic acids are suitable substrates to be converted by this reaction to higher value compounds. Radical homo- and heterocouplings of e.g. dimethyl citrate[2] (eq. 1,2), carboxylic acids derived from fatty acids (eq. 3)[3] or carbohydrate carboxylic acids (eq. 4)[4] afford a short access to monomers for polyesters, to cosmetics, fragances, new oleochemicals or potential enzyme inhibitors.

(1)

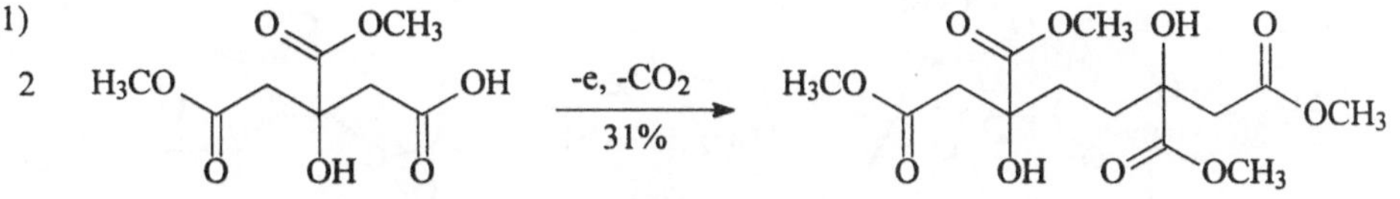

(2)

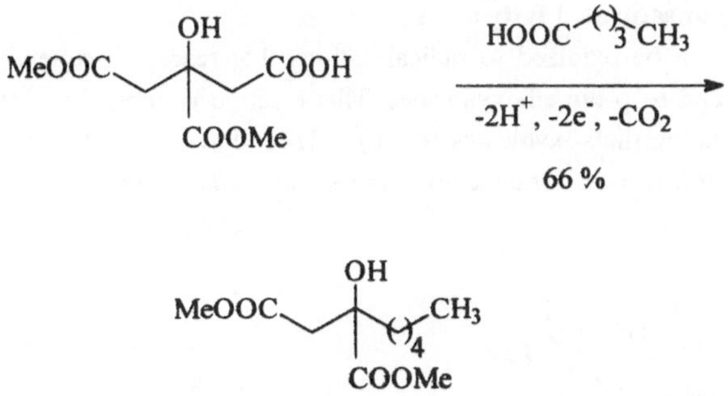

(3)

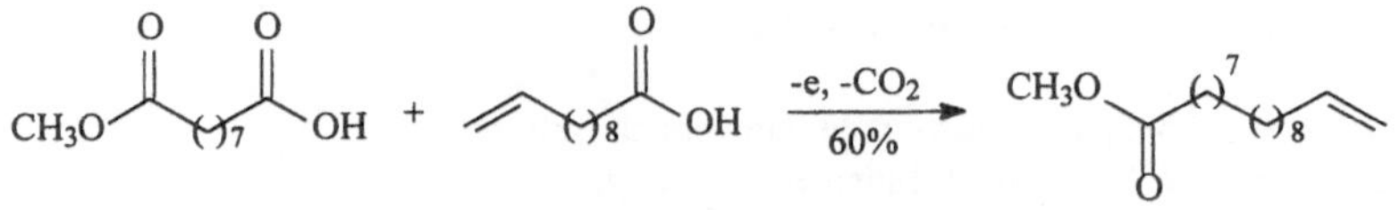

$$\text{CH}_3\text{O}-\overset{O}{C}-(\,)_7-\overset{O}{C}-\text{OH} \;+\; \diagup\diagdown(\,)_8-\overset{O}{C}-\text{OH} \xrightarrow[60\%]{-e,\,-CO_2} \text{CH}_3\text{O}-\overset{O}{C}-(\,)_8^{\,7}\diagdown\diagup$$

(4)

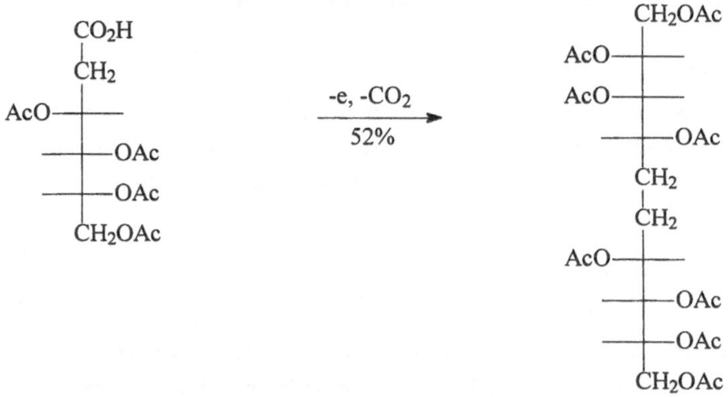

Carboxylic acids with electron donating substituents in α-position are anodically decarboxylated to carbocations that can undergo β-cleavage[1]. Anodic oxidation of diacetone-2-keto-L-gulonic acid, an intermediate in the technical vitamin C synthesis, yields quantitatively L-xylonolactone (eq. 5)[5] an otherwise only difficult accessible carbohydrate, which can be further converted to e.g. L-xylo-C-glycosides.

(5)

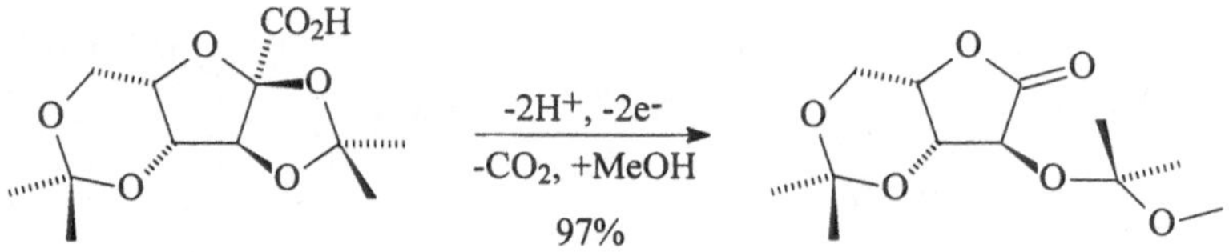

$$\xrightarrow[97\%]{\substack{-2H^+,\,-2e^- \\ -CO_2,\,+MeOH}}$$

Anodic addition to unsaturated fatty acids

Conjugated dienes can be oxidized to radical cations that react subsequently with the solvent, e.g. methanol, to α, δ- and α, β-dimethoxyalkenes. This reaction is applied to conjugated linoleic acid to afford the corresponding dialkoxyoleates (eq. 6)[6]. These have interesting properties as tensides; the diacetates are key intermediates for the conversion to other oleochemicals.

(6)

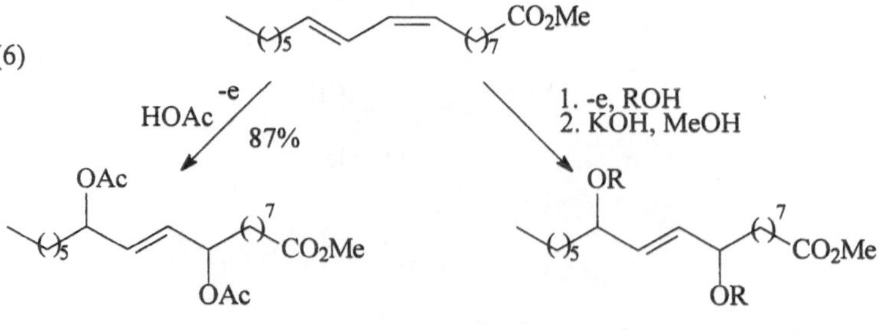

Cathodic dimerization of unsaturated fatty acids and dimethyl aconitate

Cathodic reduction of conjugated enones leads to radical anions, that can couple to hydrodimers. Application of this cathodic dimerization to enones derived from fatty acids opens a short route to dimer fatty acids (eq. 7)[7] that are of interest for the conversion to *gemini*-tensides, lubricants or as monomers for polyesters.

(7)

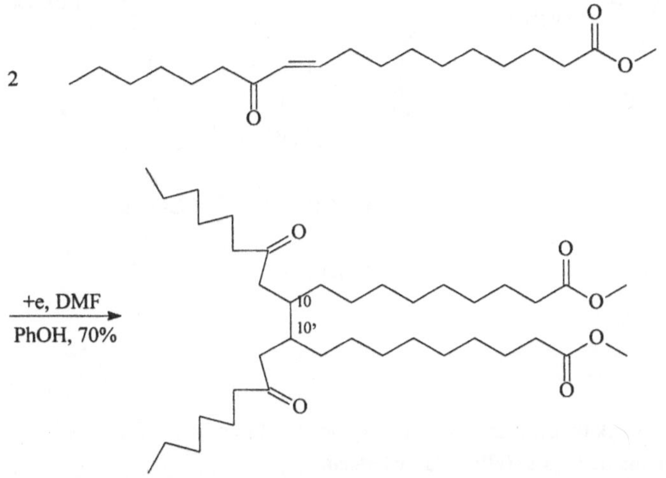

Dimethylaconitate (**4**), which is available by dehydration of citric acid, can be cathodically subjected to a cyclodimerization to afford one diastereomer of hexamethyl cyclopentanehexacarboxylate (**6**) (eq. 8). As mechanism we propose a non-electrochemical cycloaddition, proceeding in a chain reaction, which is initiated cathodically. Thereby **4** is reduced to the allylanion **7**, having a W-shaped conformation, which undergoes an anionic [4+2]-cycloaddition to afford anion **8**, which continues the chain by deprotonation of **4** yielding the product **6** (eq. 9).

(8)

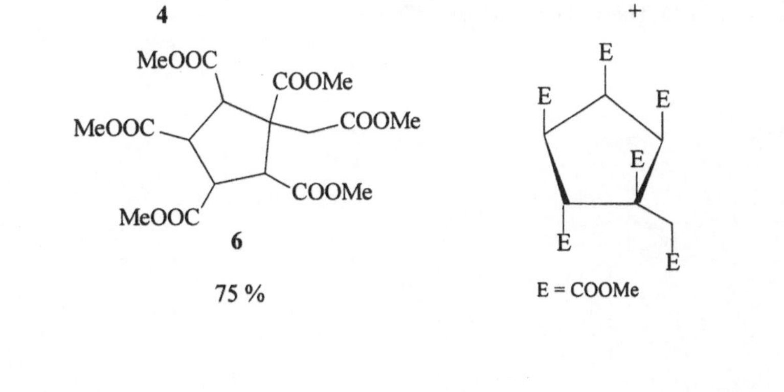

(9)

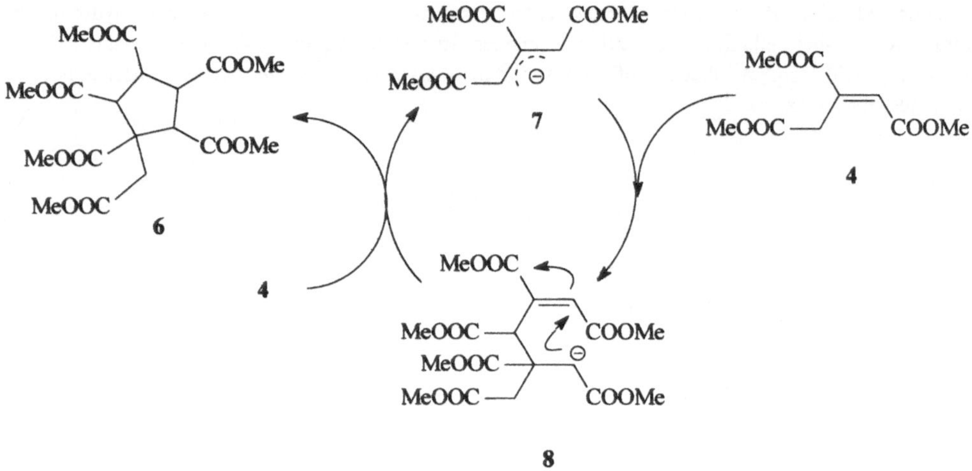

6

7

4

4

8

Support of this work by the minister of agriculture and forestry, the Bayer AG and the Fonds der chemischen Industrie is gratefully acknowledged.

References

1) H.J. Schäfer, *Top. Curr. Chem.* **1990**, *152*, 91; S. Torii and H. Tanaka, In: H. Lund, M.M. Baizer, Eds. *Organic Electrochemistry*, 3rd. ed. M. Dekker, New York, **1991**, Chap. 14.

2) S. Kratschmer, H.J. Schäfer, 20. Sandbjerg Meeting 1997 on Organic Electrochemistry.

3) (a) H.J. Schäfer, A. Weiper, M. aus dem Kahmen, A. Matzeit, in *Nachwachsende Rohstoffe*, Ed. M. Eggersdorfer, S. Warwel, G. Wulff, Verlag Chemie **1993**, 97 - 108; (b) A. Weiper-Idelmann, M. aus dem Kahmen, H.J. Schäfer and M. Gockeln, *Acta Chem. Scand.*, in press.

4) M. Harenbrock, A. Matzeit, H.J. Schäfer, *Liebigs Ann. Chem.* **1996**, 55-62.

5) E. Klocke, Ph.D. Thesis, University of Münster, **1992**.

6) M. Plate, 19. Sandbjerg Meeting 1996 on Organic Electrochemistry, Ph.D. Thesis, Univ. Münster **1997**.

7) R. Maletz, Ph.D. Thesis, Univ. Münster **1994**.

Part II
Electroreduction

ELECTROCHEMICAL ACTIVATION OF CARBON DIOXIDE. SYNTHESIS OF ORGANIC CARBONATES AND CARBAMATES

Marta FEROCI,[a] Achille INESI,[b] Leucio ROSSI[b]

[a]*Department I.C.M.M.P.M. University of Rome "La Sapienza"*
[b]*Department of Chemistry, Chemical Engineering and Materials, University of L'Aquila, Monteluco di Roio, I-67040; L'Aquila, Italy*

ABSTRACT
Electrochemically activated CO_2 reacts, under mild conditions, with primary and secondary amines and alcohols bearing a leaving group at the α or β position, affording cyclic carbamates and carbonates in high yields. After addition of EtI, unsubstituted amines and alcohols are converted into the corresponding ethyl carbamates and carbonates in good yields. In this paper, the expression *electrochemical activation of carbon dioxide* refers to some changes in the chemical properties of solutions of Et_4NClO_4 in aprotic solvents saturated with CO_2 or an O_2/CO_2 mixture, induced by electrochemical methodologies. Owing to these changes, the solutions acquire an interesting carboxylating power versus amines and alcohols. Our method of synthesis of organic carbamates and carbonates relies on this carboxylating power.

Organic carbamates[1] and carbonates[2] are compounds of growing interest because of their widespread applications in the synthesis of pharmaceutical and agricultural chemicals, and in chemical industry as starting materials and versatile intermediates. The most important syntheses of carbamates and carbonates require the utilization of toxic and corrosive reagents such as phosgene and isocyanates, with all the considerable drawbacks that these harmful reagents involve. Therefore, the development of alternative and effective synthetic routes that avoid the use of dangerous starting materials has received significant consideration. The structures of carbamate and carbonate esters suggest that the reaction of CO_2 with amines and alcohols represents a potential route to these classes of compounds and pave the way for the utilization of carbon dioxide as a safe substitute of phosgene.

The utilization of CO_2 as source of carbon in organic synthesis has been proposed by several authors in the last decade. Hovever, the thermodynamic stability and the relative kinetic inertness of CO_2 require its preliminary activation. As regards the electrochemical techniques, carbon dioxide has been activated either by direct or indirect reduction[3]. The activation of CO_2, carried out by direct electrochemical reduction, has been described by Savéant *et al.*[4] (eq. 1-4 as a monoelectronic process yielding the radical anion $CO_2^{\cdot-}$ which, in aprotic solvents, evolves to oxalate, carbon monoxide and carbonate. The oxalate/carbon monoxide ratio is affected by several factors including the nature of the electrodic material, the solvent and the supporting electrolyte, the CO_2 concentration, the temperature and the current density. Formate is produced only in the presence of residual water[5].

$$CO_2 + e \longrightarrow CO_2^{\cdot -} \tag{1}$$

$$CO_2^{\cdot -} + CO_2^{\cdot -} \longrightarrow C_2O_4^{2-} \tag{2}$$

$$CO_2^{\cdot -} + CO_2 \longrightarrow O=\overset{\cdot}{C}-O-C\overset{O}{\underset{O^-}{\lessgtr}} \xrightarrow{+e} CO + CO_3^{2-} \tag{3}$$

$$CO_2^{\cdot -} + H_2O + e \longrightarrow HCOO^- + OH^- \tag{4}$$

The problems concerning the synthesis of organic carbamates and carbonates via addition of alkyl halides to solutions of amines or alcohols saturated with CO_2 (i.e., the difficulty of inserting the CO_2 molecule into these substrates) have been recently reviewed and discussed[1a,2]. As regards the carbamates, these problems do not derive from a peculiar inertness of CO_2 with respect to amines, but rather from the following transfer of the carbamate group to alkyl halides. In fact, even at room temperature, CO_2 reacts with primary and secondary aliphatic amines giving the corresponding alkylammonium alkyl carbamates. Unfortunately, when in solution, these substances have a poor chemical stability, readily decomposing to amines and CO_2 via carbamic acid. Actually, carbamate anion behaves as an ambident nucleophile. In the presence of alkyl halides, the reaction of N-alkylation (to alkyl amines and CO_2) is selective with respect to the O-alkylation reaction (to carbamic esters). Accordingly, N-alkylation products and CO_2 were isolated after addition of alkyl halides to the CO_2-amine systems, while organic carbamates were never isolated.

Moreover, considering that organic carbamates have been isolated by addition of amines and alkyl halides to solutions of tetraethylammonium (hydrogen) carbonate[6], it seems reasonable that:

- i - a possible route to the carbamate ion is the direct reaction of carbonate anion with the amine;

- ii - the presence of tetraethylammonium cation modifies the reactivity of carbamate anion, making the O-alkylation reaction competitive with respect to the N-alkylation.

Furthemore, solutions containing carbonate and tetraethylammonium ions are easily obtained by electrolysis of aprotic solutions of Et_4NClO_4 (TEAP) saturated with CO_2 (E=-2.1 V vs SCE). We have verified that these electrolysed solutions are possible reagents in the process of carbon dioxide fixation to give carbamate or carbonate. In fact, carbamates[7a] or carbonates[7b] have been isolated, in good to excellent yields, from these solutions after addition of amines or alcohols, respectively, and, subsequently, of alkyl halides.

We have also studied the reactivity of aprotic solutions of TEAP saturated with an O_2/CO_2 mixture, previously electrolysed at a potential (E=-1.1 V vs SCE) negative enough to cause the selective reduction of O_2 to $O_2^{\cdot -}$. The behaviour of these last solutions versus amines and alcohols is similar to the previous ones. In fact, addition of amines and alcohols allows obtaining organic carbamates[1b] and carbonates[8]. The reaction product of $O_2^{\cdot -}$ with CO_2 has been identified by Sawyer et $al.$[9] as the peroxydicarbonate anion $C_2O_6^{2-}$ (eq. 5-6). According to this hypothesis, the reactivity of $C_2O_6^{2-}$ anion versus amines and alcohols in the presence of tetraethylammonium cations is like that of CO_3^{2-} anion.

$$O_2 + e \longrightarrow O_2^{\cdot -} \tag{5}$$

$$2\,O_2^{\cdot -} + 2CO_2 \longrightarrow C_2O_6^{2-} + O_2 \tag{6}$$

In summary, two simple and safe electrochemical processes for the synthesis of linear and cyclic carbamates and carbonates starting from the corresponding amines and alcohols have been established.

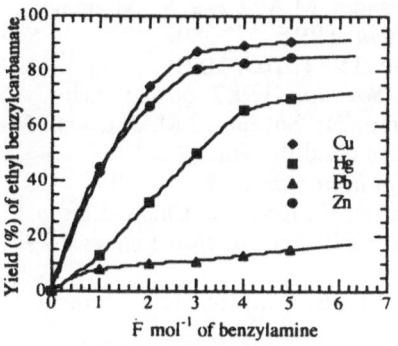

These processes rely upon either the direct electrochemical reduction of carbon dioxide (method A) or the reaction of CO_2 with the electrogenerated superoxide ion $O_2^{\cdot -}$ (method B). Both methodologies have the advantage of operating under mild conditions, without catalysts, under atmospheric pressure and at room temperature.

The nature of the electrode and the amount of electricity supplied to the system have remarkable effects on the yields. As regards method A, Cu and Zn electrodes gave better results than Pb or Hg ones (see figure). In a typical procedure, a solution of MeCN-TEAP (0.1 mol dm^{-3}) kept under continuous CO_2 bubbling, was electrolysed (divided cell, Pt anode, 0°C) over a Cu cathode at a potential (E = -2.1 V vs SCE) that allows the reduction of CO_2 to $CO_2^{\cdot -}$ (method A). Alternatively, the

MeCN-TEAP solution, under continuous O_2/CO_2 mixture bubbling, was electrolysed (divided cell, Pt anode, room temperature) over a Hg cathode at a potential (E = -1.1 V vs SCE) that allows the selective reduction of O_2 to $O_2^{\div}$ (method B). At the end of the electrolysis, the substrate was added and the reaction mixture was allowed to stand at room temperature for the appropriate time before adding an excess of alkyl halide, when necessary. The "ex-cell" features of both methods greatly enhance their usefulness, widening their application also to substrates bearing additional electroactive groups. As an example, aliphatic amines and alcohols bearing a reducible leaving group (halide, tosylate) at the α or β position were readily converted into the corresponding cyclic carbamates and carbonates, in high yields.

In evaluating the best synthetic conditions, it should be considered that the potential allowing the reduction of O_2 is less negative (E=-1.1 V vs SCE) than the potential involving the reduction of CO_2 (E=-2.1 V vs SCE). Moreover, a weak oxidative power of the electrolysed solutions used in method B has been recently evidenced.[10] Consequently, the choice of method A is advisable in the carboxylation of more complex aminic and alcoholic structures containing easily oxidizable functional groups.

References
1) (a) Aresta, M.; Quaranta, M. *Proceedings of the International Conference on Carbon Dioxide Utilization* ; Bari: Italy,1993, p.63 and references cited therein. (b) Casadei, M.A.; Micheletti Moracci, F.; Zappia, G.; Inesi, A.; Rossi, L. *J. Org. Chem.* **1997**, *62*, 0000 and references cited therein
2) Shaikh, A.G.; Sivaram, S. *Chem. Rev.* **1996**, 96, 951.
3) (a) Aresta, M.; Forti, G. *Carbon Dioxide as a Source of Carbon;* Nato Asi Ser. C; Reidel: Dordrecht, 1987; Vol. 206. (b) Sullivan, B.P.;Krist, K.; Guard, H.E. *Electrochemical and Electrocatalytic Reactions of Carbon Dioxide*; Elsevier Science Publishers: Amsterdam, 1993. (c) Casadei, M.A.; Cesa, S.; Micheletti Moracci, F.; Inesi, A.; Feroci, M. *J. Org. Chem.* **1996**, *61*, 380.
4) Amatore, C.; Savéant, J.M. *J. Am. Chem. Soc.* **1981**, *103*, 5021.
5) (a) Ikeda, S.; Takagi, T.; Ito, K. *Bull. Chem. Soc. Jpn.* **1987**, *60*, 2517.(b) Gennaro, A.; Isse, A.A.; Severin, M.G.; Bhugun, I;. Savéant, J.M. *J. Chem. Soc., Faraday Trans.* **1996**, *92*, 3963 and references cited therein.
6) Mucciante, V.; Inesi, A.; Rossi, L. *Manuscript in preparation.*
7) (a) Casadei, A.M.; Inesi, A.; Micheletti Moracci, F.; Rossi, L. *Chem. Comm.* **1996**, 2575. (b) Casadei, M.A.. Inesi, A.; Rossi, L; *Tetrahedron Letters* **1997**, *38*, 3565;
8) Casadei, M.A.; Cesa, S.; Feroci, M.; Inesi, A.; Rossi, L.; Micheletti Moracci, F. *Tetrahedron,* **1997**, *53*, 167.
9) Roberts, J.L.,Jr; Calderwood, T.S.; Sawyer, D.T. *J. Am. Chem. Soc.* **1984**, *106*, 4667.
10)(a) Nagano, T.; Yamamoto, H.; Hirobe, H. *J. Am. Chem. Soc.* **1990**, *112*, 3529. (b) Casadei, M.A.; Mazzei, P; *Atti delle Giornate Dell'Elettrochimica Italiana GEI-97*, Belgirate: Italy, 1997.

ELECTROCATALYTIC REDUCTION OF CO_2 ON THE DUAL-FILM ELECTRODES MODIFIED WITH VARIOUS METAL COMPLEXES

Kotaro OGURA

Department of Applied Chemistry, Faculty of Engineering, Yamaguchi University, Tokiwadai 2557, Ube 755, Japan

INTRODUCTION

A highly negative overpotential is required to electrochemically reduce CO_2 owing to its chemical inertness. A key to solve this problem is how to activate CO_2 at electrode, and the development of electrocatalytic process with an electrode mediator is very important. We have proposed a catalytic process of CO_2 reduction at the modified electrode having two laminated films consisting of an inorganic conductor and a conducting polymer. As the inorganic conductor and the conducting polymer, Prussian blue (PB) and polyaniline (PAn) were used, respectively. In the present study, various metal complexes were fixed onto the conducting polymer, and the activation process of CO_2 reduction was monitored with *in situ* Fourier transform infrared spectroscopic (FTIR) method.[1, 2]

EXPERIMENTAL

The PB film was first deposited from an aqueous ferric ferricyanate solution, and the PAn film was then deposited on the PB film electrode by repeated potential cycling in hydrochloric acid solution (pH 1) containing 0.1 M aniline. The prepared PAn/PB/Pt electrode was immersed in the phosphate buffer solution (pH 7) for 30 min to release the anions incorporated into the PAn film during the electrodeposition. After the dual-film electrode was rinsed thoroughly with distilled water, it was put in solutions containing various metal complexes. The complexes used were iron complexes of chromotropic acid, chromotrope 2R, nitroso-R-salt and 1-nitroso-2-naphtol-4-sulfonic acid. The structures of iron(II)-nitroso-R-salt (Fe-L(N)) and iron(II)-chromotropic acid (Fe-L(C)) are shown by the following scheme. These were immobilized onto PAn by the electrochemical method.

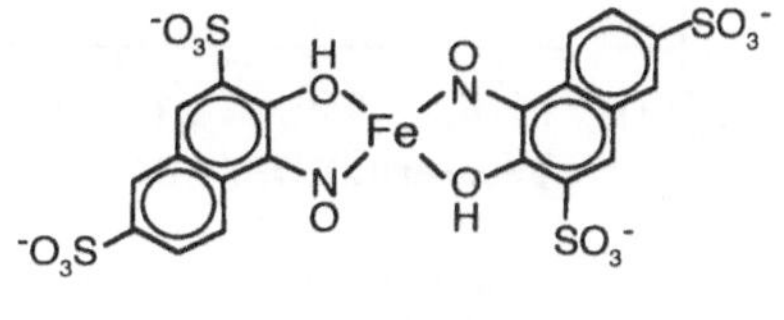

Fe-L(N) 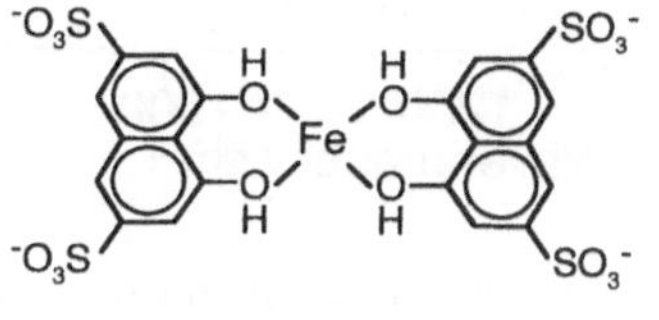   Fe-L(C)

The electroreduction of CO_2 was performed with the prepared electrode in a gas tight H-type electrolysis cell. The electrolyte was 0.5 M KCl solution of pH 3.0. The products were analyzed with a Shimadzu organic acid analyzer and an Okura steam chromatograph. The properties of the dual-film electrode were characterized by *in situ* Fourier transform infrared (FTIR) reflection absorption spectroscopy. The FTIR spectrometer used was a Shimadzu FTIR-8100M equipped with a wide-band mercury cadmium teluride (MCT) detector cooled with liquid nitrogen. *In situ* FTIR measurements were carried out in a spectroelectrochemical cell in which the modified electrode was pushed against an IR transparent silicon window to form a thin layer of solution. A total of 100 interferometric scans was accumulated with the electrode polarized at a given potential.

RESULTS AND DISCUSSION

The Fe-L(N)-immobilized PAn/PB/Pt electrode was first polarized at -0.8 V vs Ag/AgCl for 1 min in N_2-saturated KCl solution. After the base spectrum was taken at -0.8 V, the potential was subjected to anodic step. Fig. 1 shows the resulting spectrum in the wavenumber region between 1400 and 1000 cm^{-1}. In this figure,

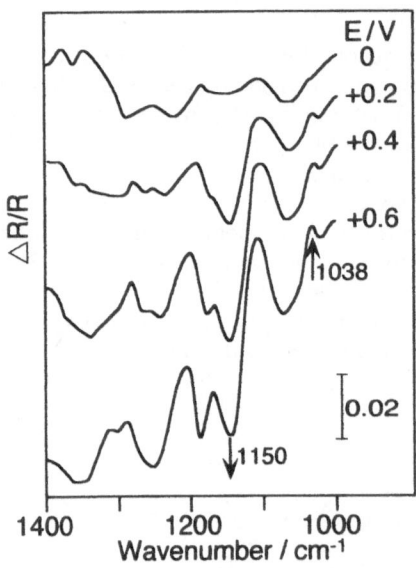

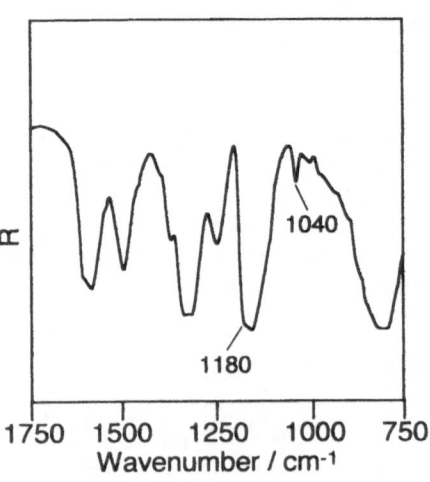

Fig. 1 *In situ* FTIR spectra of PAn doped with Fe-L(C) in 0.1 M KCl solution of pH 3. The potential was stepped from -0.8 V (base) to 0 to +0.6 V vs Ag/AgCl.

Fig. 2 *Ex situ* FTIR spectra of PAn doped with Fe-L(N).

the IR spectrum is expressed by $\triangle R/R$ in the normalized form according to the formula : $\triangle R/R = [R(E_s)-R(E_b)] / R(E_b)$ where $R(E_s)$ and $R(E_b)$ are the reflected intensities measured at the sample and base potentials, respectively. As seen from this figure, a distinct and downward peak is observed at the wavenumber of 1150 cm^{-1}, which is attributable to B-NH$^+$=Q vibration where B and Q mean the benzenoid and quinoid rings, respectively.[3] The enhancement of this peak with shifting the potential to anodic side indicates that PAn becomes more conductive at more positive potentials. Another distinct peak is observed upward at 1038 cm^{-1}. This peak can be assigned to the symmetric stretching vibration of the SO_2 group of SO_3^- belonging to the ligand of the Fe-L(N) complex in comparison with the *ex situ* spectrum of PAn doped with Fe-L(N). In the *ex situ* spectrum (Fig. 2), the symmetric vibration due to SO_3^- was found at 1040 cm^{-1} with the asymmetric mode at 1180 cm^{-1} that was unobservable in the *in situ* spectrum. The upward peak at

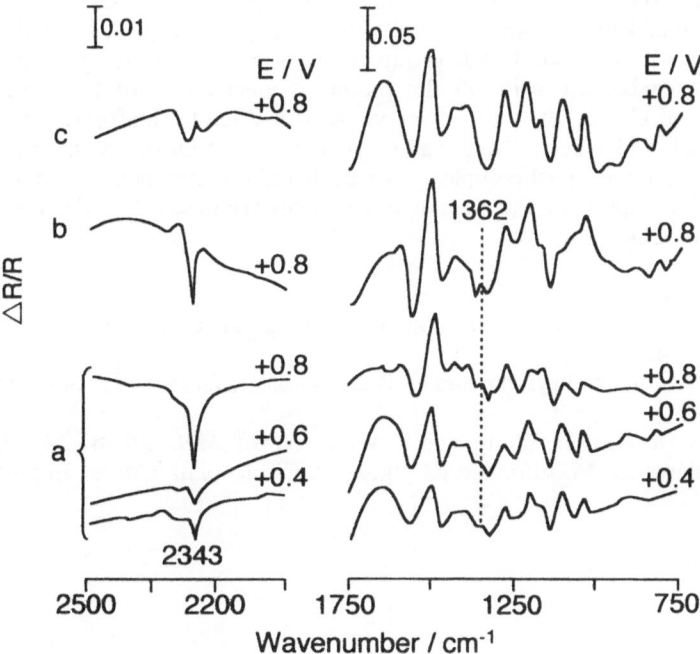

Fig. 3 *In situ* FTIR spectra of a Pt/PB/PAn/Fe-L(C) electrode in 0.1 M KCl solutions saturated with CO_2 (a and b) and N_2 (c). The spectra were obtained at the noted potentials. The electrode was potentiostatically polarized at -0.8 V vs Ag/AgCl (base potential) for 1 min (a and c) and 7 min (b).

1038 cm^{-1} does not mean the decrease in concentration of the SO$_3^-$ ligand because there was no detachment of the anionic metal complex in the anodic polarization but is related to the molecular orientation of the ligand on electrode surface.[4] In the oxidized state, PAn is positively charged, and the SO$_3^-$ may lie flat by associating with the positive PAn, leading to relative decrease in number of the dipoles perpendicular to the surface. This results in the enhancement of intensity of the upward peak at high positive potentials.

In situ FTIR spectra of a Fe-L(C)/PAn/PB/Pt electrode in KCl solutions saturated with CO$_2$ and N$_2$ are shown in Fig. 3 where the electrode was polarized at -0.8 V for 1 min before it was shifted to each potential. A well-defined downward peak appears at 2343 cm^{-1} in the CO$_2$-saturated solution, and the peak intensity increases with an increase of potential. This band can be assigned to the CO$_2$ generated in the reoxidation of the reduced species of CO$_2$ because no band is observed at the same wavenumber with the modified electrode treated in the N$_2$-saturated solution. Furthermore, an upward peak was observed at 1362 cm^{-1} in the presence of CO$_2$ but not in the solution with N$_2$. This peak is attributable to the asymmetric vibration of COO$^-$ group in carboxylic acid, suggesting that CO$_2$ is reduced to carboxylic acid on the modified electrode. In fact, the prolonged electrolysis of CO$_2$ on the same electrode at -0.8 V led to the formation of lactic and formic acids.[1] However, there was no reduction of CO$_2$ on the modified electrode without the anionic metal complex. It was therefore concluded that the existence of such metal complex in the mediated films are requisite for the electrocatalytic reduction of CO$_2$.

REFERENCES
1) K. Ogura, N. Endo, M. Nakayama, and Y. Ootsuka, *J. Electrochem, Soc.,* **1995,** 142, 4026.
2) K. Ogura, M. Nakayama, and C. Kusumoto, *J. Electrochem. Soc.,* **1996,** 143, 3506.
3) J. Tang, X. Jing, B. Wang, and F. Wang, *Synth. Met.,* **1988,** 24, 231.
4) M. Nakayama, M. Iino, and K. Ogura, *J. Electroanal. Chem.,* in press.

CONTROL OF PRODUCT DISTRIBUTION BY USE OF SURFACTANTS IN CATHODIC REDUCTION OF ACETOPHENONE

Sotaro Ito, Yorimitsu Kodama, Akira Kitani, and Kazuo Sasaki
Department of Applied Physics and Chemistry, Faculty of Engineering,
Hiroshima University, Higashi Hiroshima 739, Japan

ABSTRACT

In the cathodic reduction of acetophenone in aqueous Na_2SO_4 solutions on Pb electrode, 1-phenyl-ethanol was selectively formed by use of cationic surfactant.

The control of product distribution is one of the important subject in electro-organic syntheses. A large number of studies has been reported, describing the effect of electrolytic conditions, such as electrode potential, current density, solvent, electrode materials, and supporting electrolyte, on the product distribution. In this study, the effect of surfactants added to the electrolyte on the product distribution in cathodic reduction of acetophenone (**1**) was studied.

The first step of the cathodic reduction of acetophenone is one-electron reduction to give a radical anion , which will dimerize chemically to give diols. The radical anion can also undergoes further one-electron reduction to give 1-phenyl-ethanol (**2**), as illustrated in Scheme 1. The question in this study is if a cationic surfactants can stabilize the radical anion and change the product distribution, the molar ratio of 1-phenyl-ethanol to 2,3-diphenyl-2,3-butanediol (**3**).

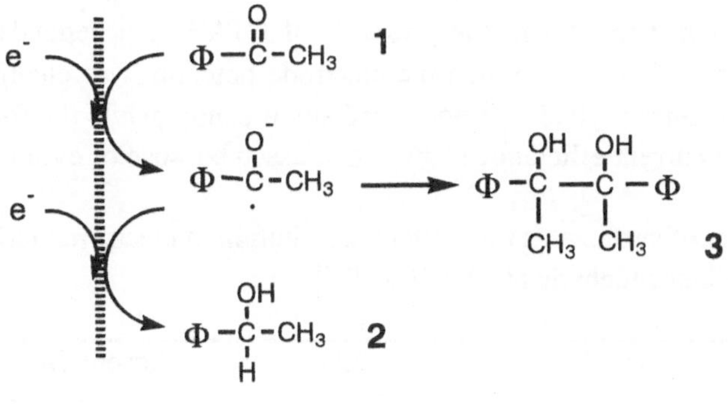

Scheme 1   Cathodic reduction of acetophenone

Table 1 Effect of surfactants on product distribution in electroreduction
of acetophenone at –2.1 V vs SCE

Additive	Yield [a]/ %		Molar ratio	Current yield / %		
	2	3	2 / 3	2	3	Total
CTAB	40	2.3	17	67	4	71
SDS	23	7.4	3.1	40	13	53
PELE	23	6.7	3.4	40	1212	52
None	19	18	1.1	33	30	63

a) Based on acetophenone added. The electricity passed was 1.16 F/mol.

The constant-potential electrolysis (at -2.1 V vs SCE) of acetophenone(4.3 mmol) in 100 ml of 50 mM sodium sulfate on Pb cathode (12 cm^2) in H-type cell at room temperature gave 1-phenyl-ethanol (2) and 2,3-diphenyl-2,3-butanediol (3) in a moderate yield[1]. In the absence of surfactant added, the current yields of 2 and 3 are comparable, as shown in Table 1. The electricity passed was 5 mF. The addition of 2.7 mmol of a cationic surfactant (cetyltrimethylammonium bromide, CTAB) increased the yield of 2 and decreased the yield of 3, increasing the molar ratio of 2/3 up to 17.   Neither anionic (sodium dodecylsulfate, SDS) nor nonionic surfactants (polyethylene-glycol monolauryl ether, PELE) showed a significant effect on the selectivity of 2.   In the electroreduction of benzaldehyde in place of acetophenone, essentially the same result was observed, as shown in Table 2.

The product distribution in the presence of CTAB also depends on the cathode potential, as shown Figure 1.   When the electrode potential was changed from –1.5 to 2,1 V vs SCE, the current yield of 3 decreased significantly, while the formation of 2 was favored. The total current efficiency slightly decreased because of evolution of hydrogen

Table 2 Effect of surfactants on product distribution in electroreduction
of benzaldehyde at –2.1 V vs SCE

Additive	Yield / %		Molar ratio	Current yield / %		
	2	3	2 / 3	2	3	Total
CTAB	48	3	16	82	5	87
SDS	36	10	3.4	61	18	79
PELE	9.5	22	0.44	16	38	54
None	40	13	3.0	69	23	92

2: benzylalcohol,   3: 1,2-diphenyl-1,2-ethanediol

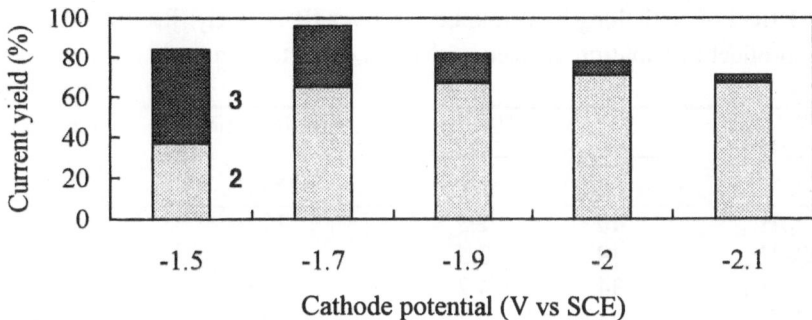

Figure 1　Effect of cathode potential on the product distribution
2: 1-phenyl-ethanol,　**3**: 2,3-diphenyl-2,3-butanediol

gas.　Without the cationic surfactant added (not shown in Figure 1) the yields of both **2** and **3** did not change so much. The result in Figure 1 can be explained by the increase of the concentration of the surfactant near the electrode surface, which is charged more negatively at –2.1 V vs SCE.　Therefore, the effect of bulk concentration of CTAB on the product distribution was examined and shown in Figure 2. The effect of the concentration in Figure 2 was similar to the result in Figure 1, suggesting that the effect of the cathode potential is due to the increase in the concentration of the cationic surfactant in the vicinity of the cathode.

　　　Alkyltrimethyammonium bromides with C_{16}, C_{12} and C_{10}-chains increased the selectivity of **2**, as shown Table 3, while the addition of　C_8 and C_6-trimethylammonium

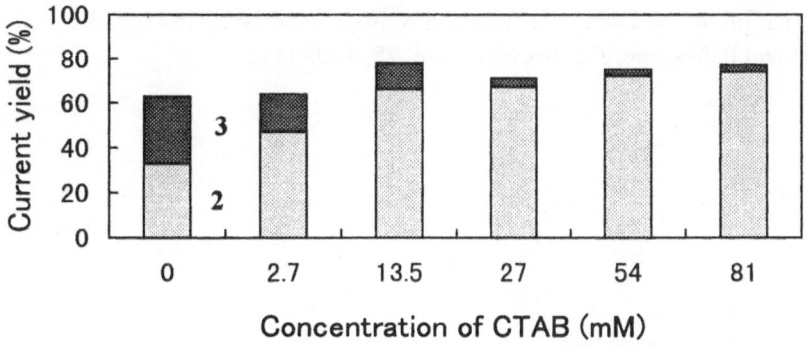

Figure 2　Effect of concentration of CTAB on product distribution
2: 1-phenyl-ethanol,　**3**: 2,3-diphenyl-2,3-butanediol

Table 3   Effect of chain length in cationic additives $[R-N^+(CH_3)_3]Br^-$ on product distribution in electroreduction of acetophenone

R	Yield / %		Molar ratio
	2	**3**	**2 / 3**
$C_{16}H_{33}$	40	2.3	17
$C_{12}H_{25}$	39	3.5	11
$C_{10}H_{21}$	34	4.7	7.6
C_8H_{17}	28	6.7	4.2
C_6H_{13}	23	9.5	2.4
CH_3	23	15	1.6
None	19	18	1.1

bromide slightly accelerated the formation of **2**.

All of these experimental results can be explained in terms of the stabilization of an intermediate anion (derived from reduction of acetophenone as seen in Scheme 1) by the formation of the ion-pair with a micelle of cationic surfactant, favoring the further one-electron reduction of the radical anion.   The effect of CTAB on the electroreduction of acetophenone was also studied by Honnorat and Martinet[2]. The result was, however, quite different from the present work: the formation of the dimeric product was favored in the presence of CTAB at higher pH. They used a Hg cathode and the cathode potential applied seems to be lower than that of present study.

References
1) Y. Kodama, M. Imoto, N. Ohta, A. Kitani and S. Ito, *Chem. Lett.*, **1997**, 337.
2) A. Honnorat and P. Martine, *Electrochim. Acta*, **28**, 1703 (1983).

ELECTROORGANIC SYNTHESIS OF OPTICALLY ACTIVE ALCOHOLS FROM KETONE AND ALDEHYDE DERIVATIVES USING ALCOHOL DEHYDROGENASE AS AN ELECTROCATALYST

Susumu KUWABATA, Ruo YUAN, Shozo WATANABE,
and Hiroshi YONEYAMA
Department of Applied Chemistry, Faculty of Engineering, Osaka University
Yamada-oka 2-1, Suita, Osaka 565, Japan

ABSTRACT

Asymmetric electroreduction of ketone and aldehyde derivatives was examined for two electrochemical reduction systems using alcohol dehydrogenase (ADH as an electrocatalyst. The reaction system (A) allowed asymmetric reduction of acetophenone, propiophenone, phenoxy-2-propanone, pyruvic acid and 2-phenylpropionaldehyde to corresponding optically active alcohols with the enantiomer excesses (ee) close to 100% and the current efficiencies larger than 92%, and the reaction system (B) gave 100% ee for the reduction of propiophenone, phenoxy-2-propanone, and pyruvic acid.

INTRODUCTION

Many kinds of enzymes having high catalytic activities and high selectivities are commercially available nowadays, and their use as catalysts is current topic in organic syntheses. Especially, the extremely high selectivity that enzymes possess is greatly useful to proceed asymmetric reactions with high enantiomer excesses (ee). In the field of electrochemistry, although the fabrication of biosensors with use of enzymes are eagerly

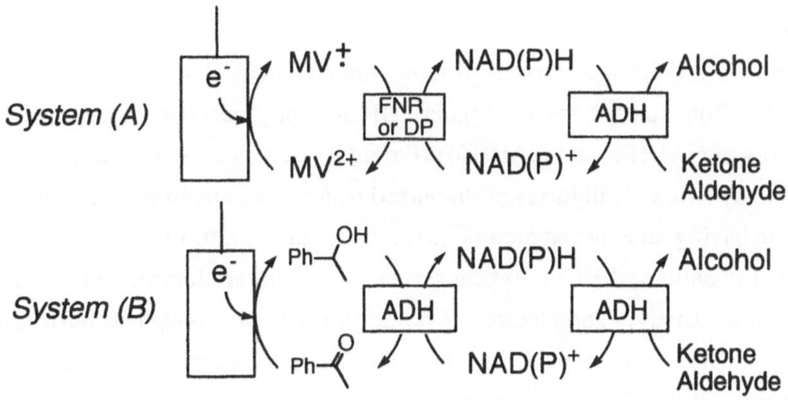

Fig. 1. Two electrochemical reduction systems using alcohol dehydrogenase as an electrocatalyst.

studied,[1-3] there have been published few reports concerning electroorganic sysntheses by utilizing enzymes as electrocatalysts. We have developed some kinds of electrochemical reaction systems using enzymes as electrocatalysts: reductive fixation of carbon dioxide in organic molecules, such as α-oxoglutaric acid[4] and pyruvic acid,[5] and electrochemical reduction of carbon dioxide to methanol.[6,7] In this study, we have developed two kinds of electrochemical reaction systems with the use of alcohol dehydrogenase (ADH) as a electrocatalyst, schematically shown in Figure 1. In the reaction system A, NAD(P)H is regenerated from NAD(P)$^+$ using either ferredoxin-NADP$^+$ reductase (FNR) for NADPH or diaphorase (DP) for NADH. Methyl viologen (MV^{2+}) is used as an electron mediator.[8] The other method of the system B uses ADH as the sole electrocatalyst which catalyzes both regeneration of NAD(P)H and reduction of reaction substrates.[9] It was found in our previous study[9,10] that phenethyl alcohol is oxidized by ADH accompanied by reduction of NADP$^+$ to NADPH and its oxidation product, acetophenone, is reduced electrochemically at a glassy carbon cathode. By combining this electrochemical regeneration reaction of NAD(P)H with the reduction of substrates catalyzed by ADH, the system B is fabricated. The purpose of this study is to examine asymmetric electroorganic syntheses using the reaction systems shown in Fig. 1.

METHOD

ADH(EC 1.1.1.2) from *T. brockii* (Sigma) and ADH (EC 1.1.1.1) from equine liver (Sigma), FNR from spinach leaves (Sigma), DP from pig heart (Boehringer Mannheim) and cofactors of NAD(P)$^+$ and NAD(P)H (Oriental yeast were used as received. Water was purified by twice distillations of deionized water. The electrode used was a glassy carbon plate having an exposed area of 2.0 cm^2. A platinum foil having 5 cm^2 area and an Ag/AgCl in saturated KCl solution served as a counter electrode and a reference electrode, respectively. The electrolyte solution used was phosphate buffer, pH 7.1, containing enzymes and chemical species desired for each electrolysis. An H-type two compartment cell divided by a cation exchange membrane (Nafion) was used as an electrolytic cell, and all electrolysis experiments were initiated after both catholyte and anolyte were bubbled with N$_2$ for at least 1 h.

RESULTS AND DISCUSSION

The electrochemical reduction of several ketone and aldehyde derivatives was attempted using the reaction systems A and B, and the results obtained by the electrolysis for 30 h are summarized in Table 1. In the case of using the reaction system A, expected products were obtained with the current efficiencies ranging from 89% to 100%, suggesting that the reaction selectivity of ADH was maintained for all cases. The enantiomer excess greater than 98% was obtained except for the electrochemical reduction of benzoylformic acid and 2-phenylbuthylaldehyde. The configuration of each product indicated that the reaction obeys stereoselective rule proposed for the reactions catalyzed by ADH (EC 1.1.1.2) from *T. brockii* and ADH (EC 1.1.1.1) from equine liver. Judging from the result obtained by the electrochemical reduction of benzoylformic acid where racemic product was obtained, it seems difficult for ADH (EC 1.1.1.2) to recognize the difference in the sterical sizes between phenyl and carboxylic groups. In the case of the reduction of 2-phenylpropionaldehyde, both substrate and product have the chiral center. Although the racemic substrate was used for the electrolysis, the product of *S*-configuration alone

Table 1. Electroreduction of Ketone and Aledehyde Derivatives with Use of the Electrochemical Reaction Systems shown in Fig. 1[a]

Substrate	Product	ADH enzyme No.	Reaction system (A)[a]			Reaction system (B)[b]		
			Configuration and enantiomer excess (%ee)	Amount of product (μmol)	Current efficiency (%)	Configuration and enantiomer excess (%ee)	Amount of product (μmol)	Current efficiency (%)
Ph–C(=O)CH₃	Ph–CH(OH)CH₃	1.1.1.2	R / 98	18.2	92			
Ph–C(=O)–	Ph–CH(OH)–	1.1.1.2	R / 100	12.5	97	R / 100	3.4	91
Ph–O–C(=O)–	Ph–O–CH(OH)–	1.1.1.2	S / 100	14.5	100	S / 100	2.1	83
O=C–COOH	HO–CH–COOH	1.1.1.2	R / 100	8.3	89	R / 100	3.6	86
Ph–C(=O)COOH	Ph–CH(OH)COOH	1.1.1.2	RS / ~0	5.5	93	RS / ~0	> 0.3	76
Ph–CH–CHO	Ph–CH–CH₂OH	1.1.1.1	S / 98	9.6	96	RS / ~0	4.7	90
Ph–CH–CHO	Ph–CH–CH₂OH	1.1.1.1	RS / ~0	3.2	100			

[a] The electrolyte solution used was 10 mL of phosphate buffer (pH 7.1) containing 3.0 mmol dm⁻³ reaction substrate, 10 units ADH, 0.1 mmol dm⁻³ NAD(P)⁺, 0.1 mmol dm⁻³ MV⁺, 3.0 vol% t-butanol, and 0.5 units FNR or 10 units DP. Applied potential chosen was -0.65 V vs. Ag/AgCl.   [b] The electrolyte solution used was 10 mL of phosphate buffer (pH 7.1) containing 3.0 mmol dm⁻³ reaction substrate, 0.1 mmol dm⁻³ acetophenone, 20 units ADH, 0.1 mmol dm⁻³ NAD(P)⁺, and 3 vol% t-butanol. Applied potential chosen was -0.85 V vs. Ag/AgCl.

was obtained. As well known, 2-phenylpropionaldehyde exhibits keto-enol tautomerism. If the enzyme recognizes selectively the C=C in the enol form instead of C=O of the formyl group as double bonds to be hydrated, then the 2-phenylpropanol of *S*-configuration should be produced.

The electrolysis using the system B gave the enantiomer excess of 100% with the current efficiencies higher than 83% for the reduction of phenoxy-2-propanone, indicating that ADH can work as a bifunctional catalyst without any decrease in the reaction selectivity. However, racemic product was obtained for the reduction of 2-phenylpropionaldehyde, being in marked contrast to the case of using the system A which gave the enantiomer excess of as high as 98%. It was revealed by the electrolysis of 2-phenylpropionaldehyde in the absence of ADH and acetophenone that this substrate was directly reduced to racemic products on the electrode with reaction rate which was a little higher than that of the electrochemical reduction of acetophenone.

References
1) C. Iwakura, Y. Kajiya, and H. Yoneyama, *J. Chem. Soc., Chem. Commun.*, **1988**, 1019.
2) Y. Kajiya, H. Sugai, C. Iwakura, and H. Yoneyama,*Anal. Chem.* **1991**, *63*, 49
3) S. Kuwabata, T. Okamoto, Y. Kajiya, and H. Yoneyama, *Anal. Chem.*, **1995**, *67*, 1684.
4) K. Sugimura, S. Kuwabata, and H. Yoneyama, *J. Am. Chem. Soc.*, **1989**, *111*, 2361.
5) K. Sugimura, S. Kuwabata, and H. Yoneyama, *Bioelectrochem. Bioenerg.*, **1990**, *24*, 241.
6) S. Kuwabata, R. Tsuda, K. Nishida, and H. Yoneyama, *Chem. Lett.*, **1993**, 1631.
7) S. Kuwabata, R. Tsuda, H. Yoneyama, *J. Am. Chem. Soc.*, **1994**, *116*, 5437.
8) S. Kuwabata, K. Nishida, and H. Yoneyama, *Chem. Lett.*, 407 (1994).
9) R. Yuan, S. Kuwabata, and H. Yoneyama, *Chem. Lett.*, 137 (1996).
10) R. Yuan, S. Watanabe, S. Kuwabata, and H. Yoneyama, *J. Org. Chem.*, **62**, 2494 (1997).

ELECTROCHEMISTRY OF DIPHENYLCYCLOPROPENONE IN APROTIC SOLVENTS: ELECTROCHEMICAL SYNTHESIS OF OLIGOMERS AND ELECTROCHEMICALLY DRIVEN DECARBONYLATION REACTION

Leonardo MATTIELLO and Liliana RAMPAZZO
Dept. of ICMMPM - University of Rome "La Sapienza"
Via del Castro Laurenziano 7 - 00161 Rome Italy

We have shown that electrochemistry of Diphenylcyclopropenone (**1**) in aprotic solvents presents some interesting and peculiar aspects.[1] For instance, (a) the voltammetric curves clearly show that the irreversible reduction of (**1**) in DMF at a G.C. electrode is followed by a polymerization-dimerization process, by which a new substance is formed; this species shows reversible behaviour at more positive potentials (Figure 2); (b) a preparative electrolysis in the same solvent system and at a moderate concentration of (**1**) originates monomeric decarbonylation and fragmentation-rearrangement products, as for example Diphenylacetylene (tolan) (**2**), Benzylphenylketone (**3**), Benzophenone (**4**) (all isolated and identified), together with an oligomeric-derived mixture of (**5**) and (**6**) (GC, EI.MS).[2] Now, we report that the preparative electrochemistry of (**1**), in repeated experiments,(DMF-TEAP 0.1 M and R.V.C. electrode, concentration of (**1**) 10^{-2} M), produces the new substances (**7**), (**8**) and (**9**), all isolated and identified by ^{1}H-NMR, ^{13}C-NMR (decoupled and DEPT) (Figure 1), and FABMS. To this purpose, work-up was carried out in two stages: firstly, the oligomeric fractions were separated from a polymeric mixture, which was not further investigated. The oligomeric mixture was then processed, thus obtaining products (**7**)-(**9**). The simplest way to reconcile the above mentioned results is as follows: anion radical of A=(**1**), A$^{\cdot-}$, once formed at the electrode, may decarbonylate to (**2**), rearrange, or dimerize-polymerize. Three different types of polymers may be anticipated, and correspondingly, all the types of the above mentioned oligomeric products follow.

For example, as for (**7**), (**8**) and (**9**), they clearly originate from a trimerization-decarbonylation process.

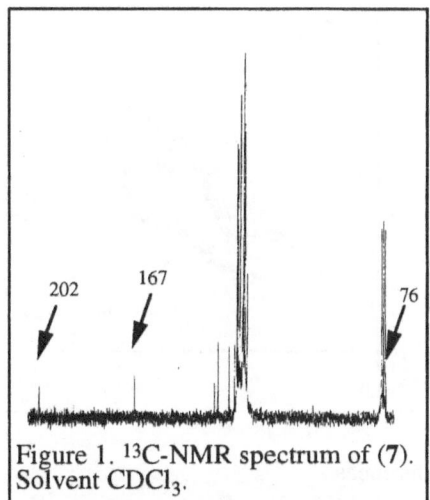

Figure 1. ^{13}C-NMR spectrum of (**7**). Solvent CDCl$_3$.

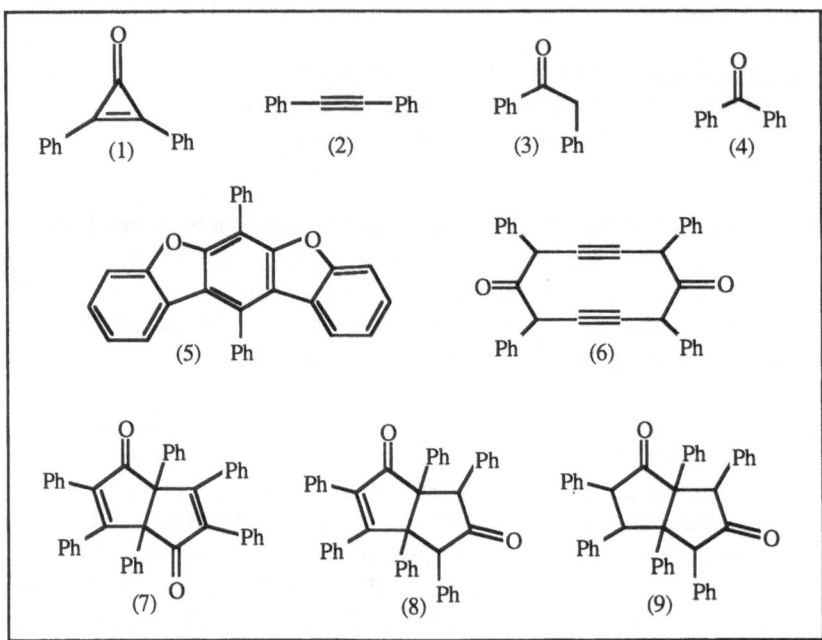

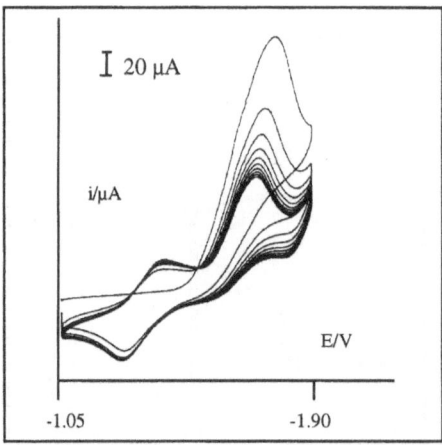

Figure 2. Cyclic voltammogram for diphenylcyclopropenone (**1**). C=0.02 M. Glassy-carbon cathode, E/V *vs*. SCE. Solvent system: DMF-TEAP (0.1 M). Sweep rate V = 0.2 V/s. Ten cycles. From top to bottom, first to tenth cycle.

References
1) Mattiello L., Rampazzo L. Resumes des Communications Journees d'Electrochimie **1995**, Strasbourg, France, May 1995, CO 8-9.
2) Mattiello L., Rampazzo L. Poster Communication N° 1302; **1997** Joint International Meeting, ECS-ISE, Paris, France, September 1997.

REMARKABLE EFFECT OF SUPPORTING ELECTROLYTE ON THE ELECTROREDUCTIVE CYCLIZATION OF CYCLIC ENONES

Akinori KONNO, Heinrich BODE, and R. Daniel LITTLE

Faculty of Engineering, Shizuoka University, 3-5-1 Johoku, Hamamatsu 432, Japan
Department of Chemistry, University of California, Santa Barbara, CA 93106

ABSTRACT

The stereoselectivity of electroreductive cyclization of butenolide 1 changes significantly depending upon the choice of supporting electrolyte. Metal coordination to the anionic intermediate of electroreductive cyclization is postulated.

INTRODUCTION

Controlling stereochemistry of organic electrochemical reactions has been still challenging object. A number of efforts have been done to control stereochemistry of electrochemical reactions.[1] For example, electroreductive cyclizations afford stereoisomeric mixtures. The stereoselectivity of these reactions could be controlled by almost all factors concerning in electrochemistry.[2] In this paper, we describe a remarkable effect of supporting electrolytes on the stereoselectivity of the electroreductive cyclization of butenolide.

$$Z= O, NH, CHEWG, C(EWG)_2, etc.$$

RESULTS AND DISCUSSION

Electroreductive cyclization of butenolide 1,[3] which is interesting precursor to pentalenolactone E, proceeded successfully to afford a mixture of the cis-anti-cis and cis-syn-cis linearly fused lactones 2 and 3. An interesting feature of the electrochemistry of 1 can be discerned from the data presented in Table 1. The stereoselectivity changes significantly depending upon the choice of supporting electrolyte, varying from zero (i.e., a 1/1 mixture of stereoisomers) when n-Bu$_4$NBr is used to as high as 11/1 in the presence of Mg(ClO$_4$)$_2$. It is noted that the major product corresponds to the cis-anti-cis adduct 2, as required in the construction of pentalenolactone E-methyl ester. Chemical reduction of 1

211

using SmI$_2$ does not afford cyclized product at all.   Therefore this electrochemical method is superior to the chemical methods.

We suggest that the stereochemistry may have its origin in the ability of lithium and magnesium to serve as coordinating metals.   In the case of 1 undergoing cyclization in the presence of magnesium perchlorate, it seems reasonable to postulate the existence of an intermediate, where the metal is associated with the butenolide as well as the reduced alkylidene malononitrile.

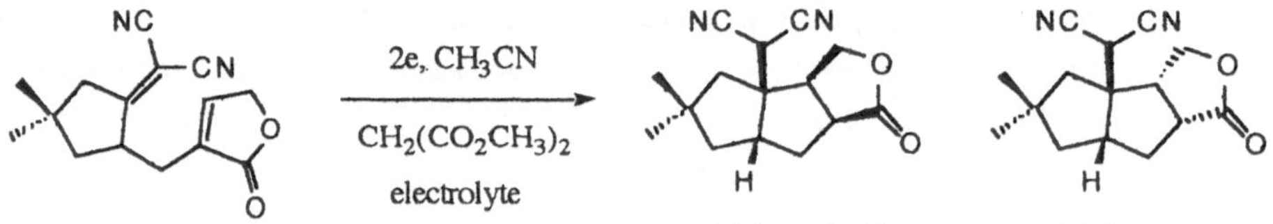

1   2 (cis-anti-cis)   3 (cis-syn-cis)

Table 1

potential (V vs SCE)	combined yield (%)	ratio 2/3	electrolyte
-1.6	23	1/1	n-Bu$_4$NBr
-1.7	77	3/1	LiClO$_4$
-1.6	62	11/1	Mg(ClO$_4$)$_2$
a)	0	--	SmI$_2$

a) Chemical reduction in THF.

References

1. Nonaka, T. In *Organic Electrochemistry*, Lund, H. and Baizer, M. M. eds., Dekker: New York, 1991; chapter 7.
2. Little, R. D.; Schwaebe, M. K. *Top. Curr. Chem.*, **1997**, *185*, 1.
3. Bode, H. E.; Sowell, C. G.; Little, R. D. *Tetrahedron Lett.* **1990**, *31*, 2525.

INFLUENCE OF ELECTROGENERATED BASES ON THE ELECTROREDUCTION OF SOME CARBONYL DERIVATIVES

Henning LUND, Benoit SOUCAZE-GUILLOUS and Zi-rong ZHENG

Department of Organic Chemistry, University of Aarhus, DK-8000 Aarhus C, Denmark

ABSTRACT

The influence of bases generated during an electrolytic reduction may influence the reduction route of compounds having a removable hydrogen in different ways. In this communication the reduction of some carbonyl derivatives in DMF has been investigated. The presence of a labile hydrogen may lead to reactions which can be classified as follows.

1.   Evolution of hydrogen

2.   Formation of anion

3.   Elimination to nitrile

4.   Rearrangement.

RESULTS

In protic media hydrazones and oximes are reduced in a four-electron reaction, and the cleavage of the nitrogen-hetero bond precedes the saturation of the carbon-nitrogen double bond.[1] In aprotic media the first reaction, after the uptake of an electron, of a hydrazone with no labile hydrogen, such as N,N-dimethylhydrazones, is a cleavage of the hetero bond. The rate of the cleavage of the radical anion is linearly dependent (fig. 1) on the reversible reduction potential of the compound.[2] This is analogous to the reductive cleavage of aryl[3] and vinyl halides[4] in which the electron is accepted in the π-system and later transferred to the σ-bond in a dissociative electron transfer reaction.[3]

$$ArRC=NNR'R'' + e \rightleftarrows [ArRC=NNR'R'']^{-\cdot} \rightarrow ArRC=N\cdot + R'R''N^- \qquad (1)$$

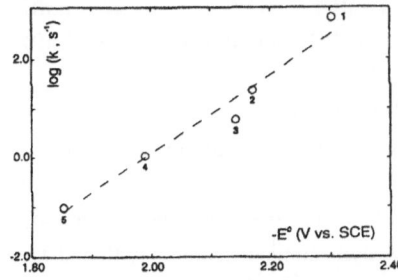

Fig. 1. Rate of cleavage of radical anions of N,N-dimethylhydrazones vs the reversible reduction potential of the hydrazones. N,N-dimethyl-hydrazones of (1) benzophenone, (2) 1-(2-naphthyl)ethanone, (3) 1-(1-naphthyl)ethanone, (4) 4-pyridinealdehyde, (5) 4-cyanobenzaldehyd.

When the labile hydrogen is sufficiently acidic the reduction at a platinum electrode may induce evolution of molecular hydrogen. This is the case in the reduction of fluorenone N-tosylhydrazone.[2,5]

$$Fl=NNHTs + e \rightleftarrows [Fl=NNHTs]^{-\cdot} \rightarrow Fl=NN^-Ts + 1/2\ H_2 \qquad (2)$$

In the more general case the electrogenerated base (EGB) deprotonates the compound to an anion which is reducible at a considerably more negative potential, often outside the available potential window. The EGB may be the radical anion formed (a "father/son" reaction)[6] as observed for acetophenone N-tosylhydrazone[2] or the EGB may be a base formed on further reduction of the cleavage products. This is observed for many ketoximes[7] and phenylhydrazones of ketones.[2]

$$Ar(CH_3)C=NNHTs + e \rightleftarrows [Ar(CH_3)C=NNHTs]^{-\cdot} \qquad (3)$$
$$Ar(CH_3)C=NNHTs + Ar(CH_3)C=NNHTs]^{-\cdot} \rightarrow [Ar(CH_3)C=NNTs]^- + [Ar(CH_3)C=NNTs]^{\cdot} \quad (4)$$
$$[Ar(CH_3)C=NNTs]^{\cdot} \rightarrow products \qquad (5)$$

Derivatives of aldehydes have a further possibility, the elimination of water (oximes) or amine (hydrazones) to a nitrile.[7,8] For hydrazones the reaction has been suggested to be a base-induced E2 elimination of an amine from the N,N-disubstituted hydrazones to a nitrile,[8] (reaction 9). The base consumed in the attack on the carbon-hydrogen bond is recovered on protonation of the amine anion thus making the elimination a catalytic circle. The initial cleavage of the radical anion to the imine anion and the amine radical might be considered as an alternative to reaction 7 as well as a hydrogen atom abstraction from the solvent followed by reduction of the radicals to anions. In all cases bases are produced which can initiate the elimination reaction.

$$PhCH=NNMe_2 + e^- \rightleftarrows PhCH=NNMe_2^{-\cdot} \qquad (6)$$
$$PhCH=NNMe_2^{-\cdot} \rightarrow PhCH=N\cdot + Me_2N^- (+ H_2O \rightarrow Me_2NH + OH^-) \qquad (7)$$
$$PhCH=N\cdot + e^- \rightarrow PhCH=N^- (+ H_2O \rightarrow PhCH=NH + OH^-) \qquad (8)$$
$$PhCH=NNMe_2 + OH^- \rightarrow PhCN + H_2O + Me_2N^- \quad (\rightarrow OH^- + HNMe_2) \qquad (9)$$

The imine formed in reaction 8 may be hydrolyzed or further reduced.

For oximes there exists, besides a simple E2 elimination, the possibility of a catalytic base-induced abstraction of a proton from the imine radical to the radical anion of the nitrile (reaction 12). The imine radical is formed by cleavage of the radical anion of the oxime. The nitrile radical anion then reduces the incoming oxime[7] thus forming a catalytic circle, reactions 11, 12 and 13. A simple base-induced E2 elimination of water from the oxime is unlikely in view of the fact that benzaldoxime is prepared by reaction of the aldehyde with hydroxylamine in 25 % aqueous NaOH. A base would attack the proton at the OH-group faster than it would attack the carbon bonded hydrogen; a loss of a proton from the hydroxyl group makes it a very poor leaving group, so an elimination is disfavored. The N,N-substituted hydrazones do not have the possibility to lose a hydrogen from the leaving group, so a base-induced elimination is more likely for such compounds. An elimination akin to that suggested below for the oximes may, however, be a possibility also for the N,N-dimethylhydrazones.

$$PhCH=NOH + e^- \quad\rightleftarrows\quad PhCH=NOH^{-\cdot} \tag{10}$$

$$PhCH=NOH^{-\cdot} \quad\rightarrow\quad PhCH=N\cdot + OH^- \tag{11}$$

$$PhCH=N\cdot + OH^- \quad\rightarrow\quad PhCN^{-\cdot} + H_2O \tag{12}$$

$$PhCN^{-\cdot} + PhCH=NOH \quad\rightleftarrows\quad PhCN + PhCH=NOH^{-\cdot} \tag{13}$$

$$PhCN + e^- \quad\rightleftarrows\quad PhCN^{-\cdot} \tag{14}$$

A further type EGB-induced reaction is found during the reduction of derivatives of acylated aldehyde cyanohydrins. The scheme shows the transformation of O-benzoyl benzaldehyde cyanohydrin (**1**) to benzil. In this case the base may be the anion of benzylcyanide[9] formed on reduction of **1**, which on electron uptake cleaves to benzoate and the further reducible cyanobenzyl radical. The reactions are written as reversible reactions as the electrochemical reduction of benzoylcyanide, C_6H_5COCN, in anhydrous acetonitrile forms **1** through the dicyanohydrin of benzil in a reaction akin to the reverse of the transformation of **1** to benzil.[10] In the rearrangement of **1** the equilibria are driven to the right by the continous removal of benzil by reduction.

Scheme 1

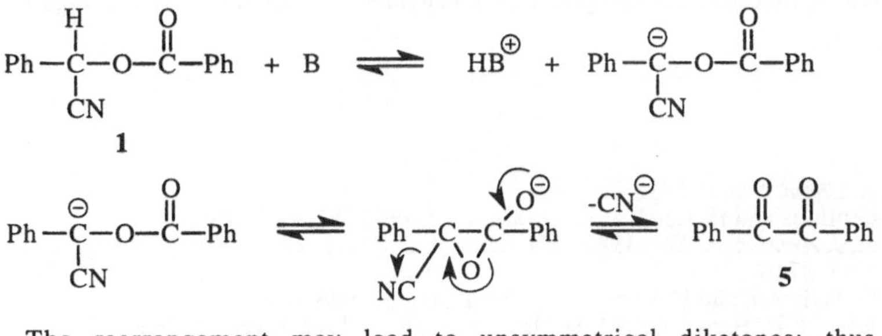

The rearrangement may lead to unsymmetrical diketones; thus O-benzoyl 4-methoxybenzaldehyde cyanohydrin rearranges to 4-methoxybenzil and O-acetyl benzaldehyde cyanohydrin to 1-phenylpropane-1,2-dione. The rearrangement is, however, not useful for a synthesis of unsymmetric diketones using *in situ* electrogenerated bases, as the diketone formed is either reduced further or reacts with the EGB in an addition reaction. The reduction of O-benzoyl benzaldehyde cyanohydrin in DMF in the presence of 2,6-di-*tert*-butylphenol gives mainly 1,2-

diphenylethanone and benzylcyanide, besides benzoic acid. The reduction of benzil to 1,2-diphenylethanone probably follows the route in Scheme 2.[11]

Scheme 2

$$Ph\text{-}CO\text{-}CO\text{-}Ph \ + \ 2\ e^- \ \rightarrow \ \underset{\underset{O^-\ \ O^-}{|\quad |}}{Ph\text{-}C=C\text{-}Ph} \ \ (+2\ [H^+]) \rightarrow PhCHOH\text{-}COPh \tag{15}$$

$$\underset{\underset{OH}{|}}{Ph\text{-}CH\text{-}CO\text{-}Ph} \ + \ e^- \ \rightarrow \ \underset{\underset{HO\ \ O^-}{|\quad |}}{Ph\text{-}CH\text{-}C\text{·}\text{-}Ph} \tag{16}$$

$$\underset{\underset{HO\ \ O^-}{|\quad |}}{Ph\text{-}CH\text{-}C\text{·}\text{-}Ph} + [H^+] + e^- \ \rightarrow \ \underset{\underset{HO\ \ OH}{|\quad |}}{Ph\text{-}CH\text{-}C^-\text{-}Ph} \ \rightarrow \ \underset{\underset{OH}{|}}{Ph\text{-}CH=C\text{-}Ph} \ + OH^- \tag{17}$$

$$\underset{\underset{OH}{|}}{Ph\text{-}CH=C\text{-}Ph} \ \rightarrow \ Ph\text{-}CH_2\text{-}CO\text{-}Ph \tag{18}$$

Benzoin is also reduced to 1,2-diphenylethanone in DMF; polarography in aqueous ethanol of benzoin gives a single two-electron wave ($E_{1/2} = -0.78$ V (SCE)) and 1,2-diphenylethanone gives one with $E_{1/2} = -1.06$ V (SCE). Although the latter is the product from the reduction of benzoin the reduction of it does not show on the polarographic wave of benzoin. This indicates a slow step in the reaction which probably is the elimination of the hydroxyl group or the tautomerization to the ketone of the enol formed in the elimination.

An alternative route to 1,2-diphenylethanone through a condensation of the benzylcyanide anion with 1 followed by a cleavage to 3-oxo-2,3-diphenylpropionitrile is not feasible as the reduction of 3-oxo-2,3-diphenylpropionitrile does not give 1,2-diphenylethanone by reductive cleavage of the cyano group.

References

1) H. Lund, *Acta Chem. Scand.* **1959**, *13*, 249.
2) B. Soucaze-Guillous and H. Lund, *J. Electroanal. Chem.*, **1997**, *423*, 109.
3) J.-M. Savéant, *J. Am. Chem. Soc.* **1987**, *109*, 6788; *Acc. Chem. Res.*, **1993**, *26*, 455; *J. Phys. Chem.*, **1994**, *98*, 3716.
4) N. Gatti, S. U. Pedersen and H. Lund, *Acta Chem. Scand.* **1988**, *B 42*, 11.
5) D. A. Van Galen, J. H. Barnes and M. D. Hawley, *J. Org. Chem.* **1986**, *51*, 2544.
6) M. C. Arévalo, G. Farna, M. G. Severin and E. Vianello, *J. Electroanal. Chem.* **1987**, *220*, 201; F. Maran, S. Roffia, M.G.Severin and E Vianello, *Electrochim. Acta*, **1990**, *35*, 81.
7) B. Soucaze-Guillous and H. Lund, *Acta Chem. Scand.* in press.
8) Yu. M. Kargin, M. Yu. Kitaeva, V. Z. Latypova, R. M. Zaripova and A. V. Il'yasov. *Izv. Akad.Nauk SSSR, Ser Khim.*, **1988**, 607.
9) Z.-R. Zheng, N. T. Kjær and H. Lund, *Acta Chem. Scand.* in press.
10) M. Okimoto, T. Itoh and T. Chiba, *J. Org. Chem.* **1996**, *61*, 4835.
11) Z.-R. Zheng and H. Lund, *J. Electroanal. Chem.* in press.

ELECTROREDUCTIVE INTERMOLECULAR COUPLING OF UNSATURATED COMPOUNDS

Naoki KISE, Syun-ichiro MASHIBA, and Nasuo UEDA
Department of Biochemistry, Faculty of Engineering, Tottori University, Tottori 680, Japan

ABSTRACT

The stereoselectivities of electroreductive hydrocoupling of chiral *N-trans*-cinnamoyl-2-oxazolidones are significantly influenced by supporting electrolyte.

INTRODUCTION

It has been well known that the electroreduction of cinnamic acid esters in aprotic solution gave cyclized products of hydrodimers and they were obtained as all-trans isomers stereospecifically.[1] These results prompted us to investigate enantioselective hydrocoupling of cinnamic acid derivatives. We have already been reported the enantioselective synthesis of 3,4-diphenyladipic acid utilizing electroreductive intermolecular hydrocoupling of chiral *N-trans*-cinnamoyl-2-oxazolidones **1** (eq 1).[2] We wish to report the effects of supporting electrolyte on the product yields and stereoselectivities in the hydrocoupling of **1**.

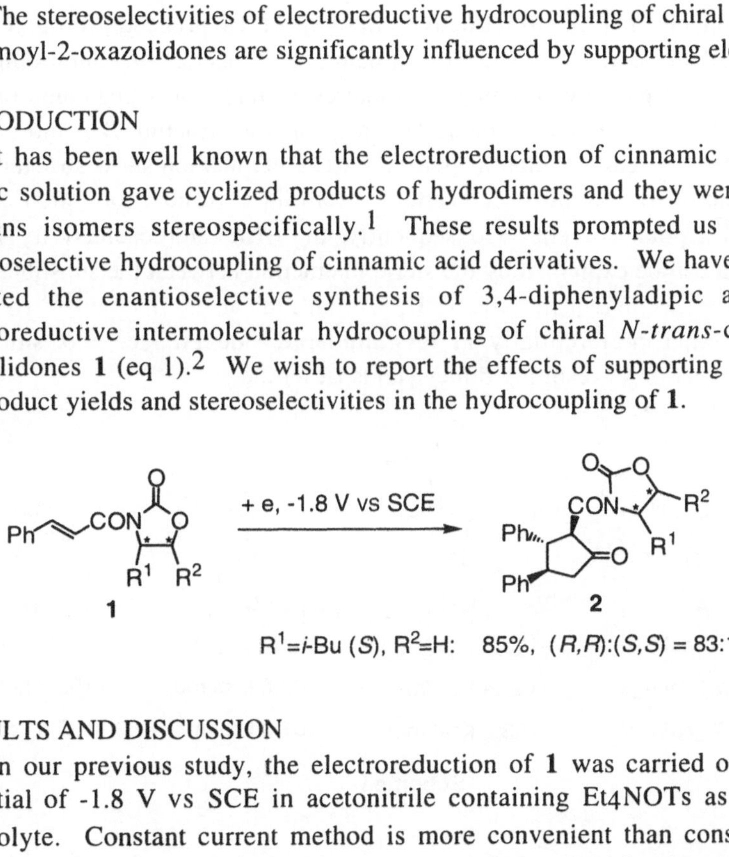

$$+ e, -1.8 \text{ V vs SCE} \quad (1)$$

R^1=*i*-Bu (*S*), R^2=H: 85%, (*R,R*):(*S,S*) = 83:17

RESULTS AND DISCUSSION

In our previous study, the electroreduction of **1** was carried out at constant potential of -1.8 V vs SCE in acetonitrile containing Et$_4$NOTs as a supporting electrolyte. Constant current method is more convenient than constant potential method. Therefore, we employed constant current method for the electroreduction of optically active *N-trans*-cinnamoyl-2-oxazolidone **1a**. It was found that constant current method was also effective, since hydrocoupling product **2a** was obtained in 72% yield and (*R,R*):(*S,S*) = 85:15 selectivity at constant current of 0.1 A (eq 2)

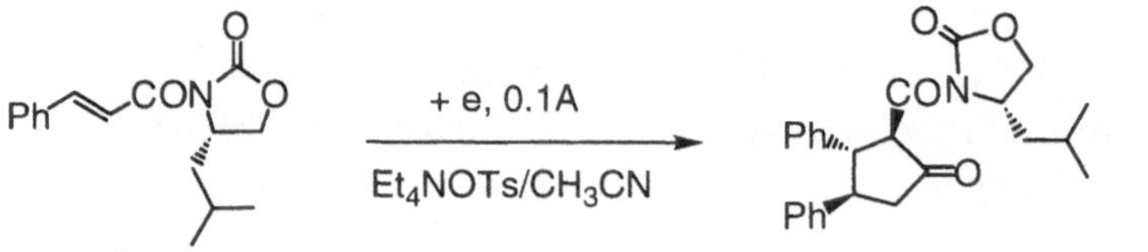

$$\text{(2)} \quad \textbf{2a} \ 72\%, \ (R,R):(S,S) = 85:15$$

It is reasonable that electroreductive coupling of **1a** proceeds through anion radical intermediate similarly to that of cinnamic acid esters.[1a,f] Semi-empirical calculations (PM3)[3] gave two optimized structures **A** and **B** for naked anion radicals of *N*-cinnamoyl-2-oxazolidone (Scheme I). In gas phase structure **B** is more stable than structure **A**, whereas calculations with the COSMO method show structure **A** is more stable than **B** in acetonitrile or H_2O.[4] From these results, formation of **A** is more likely in a polar solvent. Consequently, the (R,R)-stereoselectivity in the formation of **2a** can be explained by the steric interaction between the substituents on oxazolidone rings as shown in Scheme II. Anion radical intermediate **E** (**A** type) couples each other preferentially at less hindered side (*si* face = α side) and subsequent cyclization of resulting dimer yields (R,R)-**2a**.

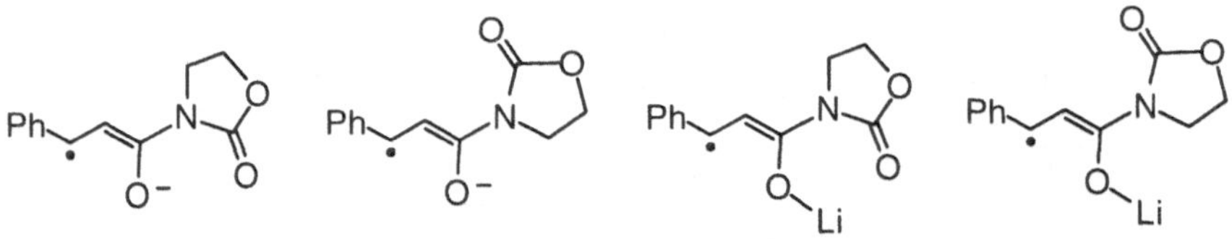

	A	B	C	D
gas	-112.3 kcal/mol	-113.3 kcal/mol	-79.0 kcal/mol	-66.9 kcal/mol
in AN	-182.7 kcal/mol	-181.4 kcal/mol	-100.5 kcal/mol	-103.8 kcal/mol
in H_2O	-184.7 kcal/mol	-183.2 kcal/mol	-101.3 kcal/mol	-105.0 kcal/mol

Scheme I

On the other hand, PM3 calculations of lithiated anion radicals suggest that structure **D** (corresponding to **B**) is more stable than structure **C** in a polar solvent (Scheme I). According to these results, it is expected that lithiated anion radical **F** (**D** type) is formed and couples preferentially at *re* face (β side) to give (S,S)-**2a** when the supporting electrolyte is a lithium salt (Scheme II). We therefore examined the electroreduction of **1a** using $LiClO_4$ as a supporting electrolyte in place of Et_4NOTs.

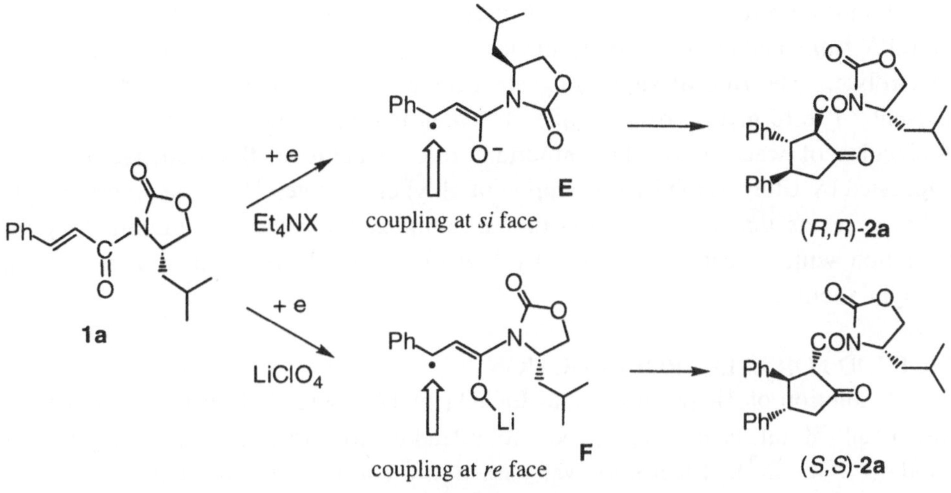

Scheme II

The electroreduction using LiClO4 was carried out in THF and afforded hydrodimers **3a** with a small amount of cyclized dimers **2a**. The mixture of the products was transformed to dimethyl ester **4** and the stereoselectivity was determined as shown in Table I. As we would expect, (S,S)-selectivity increased with an increase in concentration of LiClO4. However, significant amounts of (R,S)-isomers of **3a** were also formed.

Table I Electroreduction of **1a** Using LiClO$_4$ as a Supporting Electrolyte

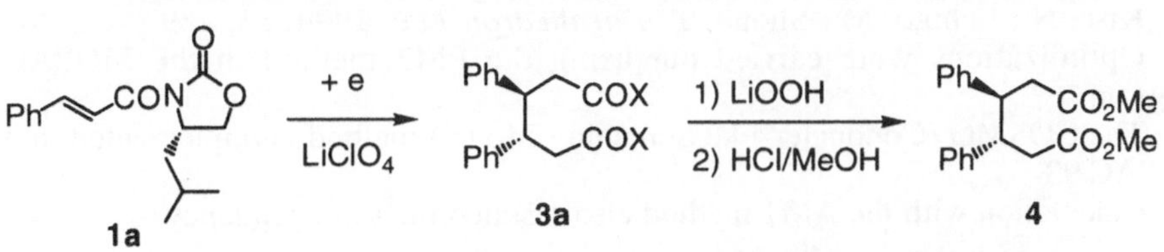

LiClO$_4$ (M)	Yield of 4 (%)[a]	dl:meso of 4[b]	RR:SS of dl-4[c]
0.3	63	50:50	46:54
1.1	45	50:50	24:76
1.7	44	48:52	11:89

a. Isolated yields from **1a**. b. Determined by ^{1}H NMR analyses.
c. Determined by ^{1}H NMR analyses with Eu(hfc)$_3$.

From these results, the stereoselectivities of electroreductive hydrocoupling of chiral *N-trans*-cinnamoyl-2-oxazolidones are significantly influenced by supporting electrolyte. The role of supporting electrolyte on stereoselectivity is not clear at present. The formation of a complex between two anion radicals and one or more molecules of water may play important role to achieve dl-stereospecificity, as suggested by Utley in the hydrocoupling of alkyl cinnamates.[1f] In the presence of Li salts, it seems that the formation of the complex with water is inhibited by the chelation with Li cation and thereby low dl-meso selectivity in dimerization is brought about.

METHOD FOR ELECTROREDUCTION

A solution of **1a** (2 mmol) and $LiClO_4$ (1.12 g) in dry THF (35 mL) was put into a cell (50 mL beaker) equipped with a lead cathode (5 X 10 cm^2) and platinum anode (2 X 2 cm^2). Electricity was passed at a constant current of 0.2 A at 25 °C until almost all of **1a** was consumed (ca. 400 c). The reaction mixture was poured into water (100 mL) and extracted with CH_2Cl_2. The products **3a** and **2a** were isolated as a mixture of diastereomers by column chromatography on silica gel.

References and Notes
1. (a) Klemm, L. H.; Olson, D. R. *J. Org. Chem.* **1973,** *58,* 3390. (b) Kanetsuna, H.; Nonaka, T. *Denki Kagaku* **1981**, *49*, 526. (c) Smith, C. Z.; Utley, H. P. *J. Chem. Soc. Chem. Commun.* **1981**, 492. (d) Nishiguchi, I.; Hirashima, T. *Angew. Chem. Int. Ed. Engl.* **1983**, *22*, 52. (e) Utley, J. H. P.; Güllü, M.; Motevalli, M. *J. Chem. Soc. Perkin Trans. 1,* **1995**, 1961. (f) Fussing, I.; Güllü, M.; Hammerrich, O.; Hussain, A.; Nielsen, M. F.; Utley, *J. M. P. J. Chem. Soc. Perkin Trans. 2,* **1996**, 649.
2. Kise, N.; Echigo, M.; Shono, T. *Tetrahedron Lett.* **1994,** *35*, 189.
3. Optimizations were carried out using the PM3 method in the MOPAC 93 program.
4. The COSMO (Conducter-like Screening Model) method is implemented into the MOPAC 93.
5. Calculation with the AM1 method also resulted the same tendency.

ELECTROREDUCTIVE CYCLIZATION REACTIONS. STUDIES DIRECTED TOWARD THE PHORBOL ESTERS AND BIOACTIVE DITERPENES

José I. LOZANO, Georgia L. CARROLL, and R. Daniel LITTLE

Department of Chemistry, University of California, Santa Barbara,
Santa Barbara, CA 93106

ABSTRACT

An approach to a phorbol ester analog and four bioactive tetracyclic diterpenes is presented. The electroreductive cyclization (ERC) reaction plays a key role in the successful implementation of each pathway.

INTRODUCTION

The phorbol esters have long been recognized for their ability to act as co-carcinogens.[1] Current interest focuses upon (a) their ability to activate protein kinase C (PKC), an enzyme that mediates cellular signal transduction initiated by hormones, neurotransmitters, growth factors, and oncogenes, and (b) upon using the esters to probe in detail the mechanism of the activation process and the sequence of events which follow. This is particularly so in relation to the formulation of a mechanism for carcinogenesis at the molecular level, and the development of chemotherapeutic agents for the treatment of cancer.[2] Significant effort is being directed toward the development of new agents that either block or stimulate PKC activity.

As illustrated in Scheme 1, phorbol (7) and its esters are structurally quite complex. One of the consequences of efforts to synthesize these materials has been a further clarification of the structural features needed for an expression of biological activity.[3] However, because of the structural complexity of the phorbols as well as their instability toward a wide variety of common reagents and conditions, it has not been possible to make extensive structural or functional group modifications. The development of more readily modifiable ring systems active at the phorbol ester binding site of protein kinase C would be of utility in exploring and identifying structure activity relationships and may facilitate the development of antitumor agents.

221

METHOD

Our approach to a bioactive phorbol analog, portrayed in Scheme 1, capitalizes on at least two electroreductive cyclizations[4] to assemble the basic ring system in a relatively few steps. The chemistry features (a) an enantioselective closure of enoate **1** to afford **2**, a system with much of the C-ring of the target structures in place; (b) electroreductive cyclization onto a lactone (**4** to **5**); (c) capitalization upon the ability of the electroreductive cyclization to temporarily form a bicyclo[3.2.1] ring system that promises to undergo ring opening *in situ* to afford the seven-membered B-ring with functionality nicely positioned for elaboration; (d) a novel introduction of the C_4 hydroxyl group through the silylation and subsequent Rubottom oxidation of the intermediate formed upon conducting electroreductive cyclization in the presence of silylating agents.

Scheme 1. Synthetic approach to phorbol analogs.

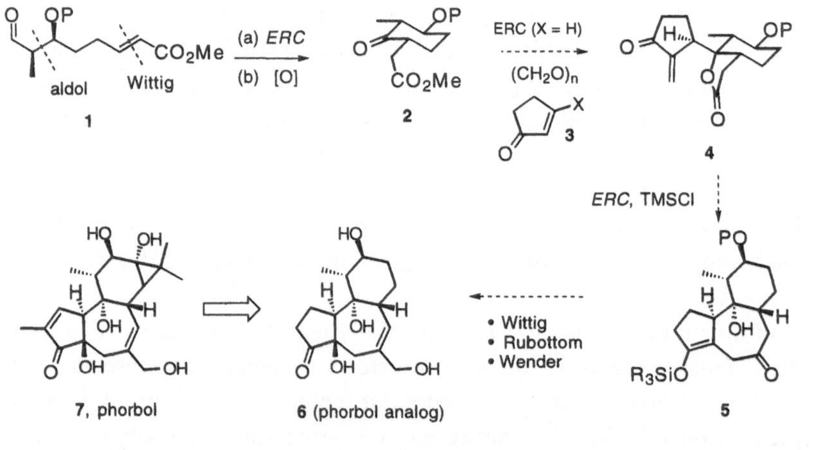

The diterpenes **8-11** display the bioactivity indicated below their structural formulations; in particular, they are active against breast and prostrate cancer, as well as HIV protease.[5] Our approach to **12**, a substance that promises to serve as a common intermediate *en route* to each of the target structures is illustrated in Scheme 2. It features the use of two electroreductive cyclizations, the first to assemble the CD-ring system (**17** to **16**), the second establishing the B-ring and completing the tetracyclic skeleton (**14** to **13**). In preparation for these reactions, we note the use of a Trost-Chan cycloaddition to construct what ultimately becomes the D ring,[6] and a Stille coupling[7] (**16** to **14**) to link the A-ring and the bicyclo[3.2.1] core that is formed after the first electrochemical step. This sets the stage for the second reductive cyclization. Overall, the route is convergent, exceptionally direct, and promises to showcase the electrochemical methodology.

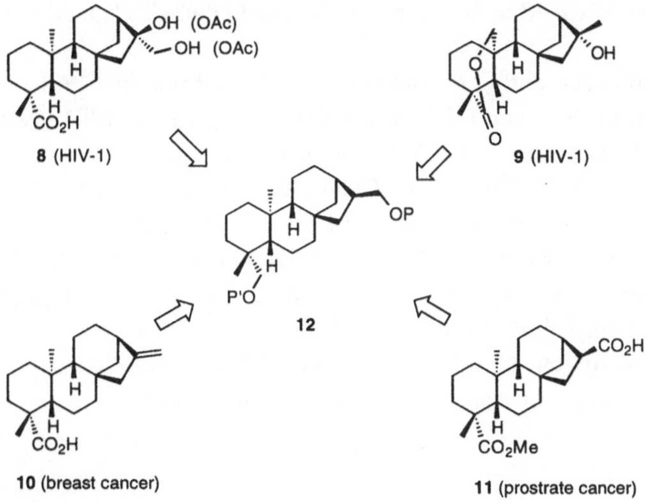

Scheme 2. Possible route to bioactive tetracyclic diterpenes.

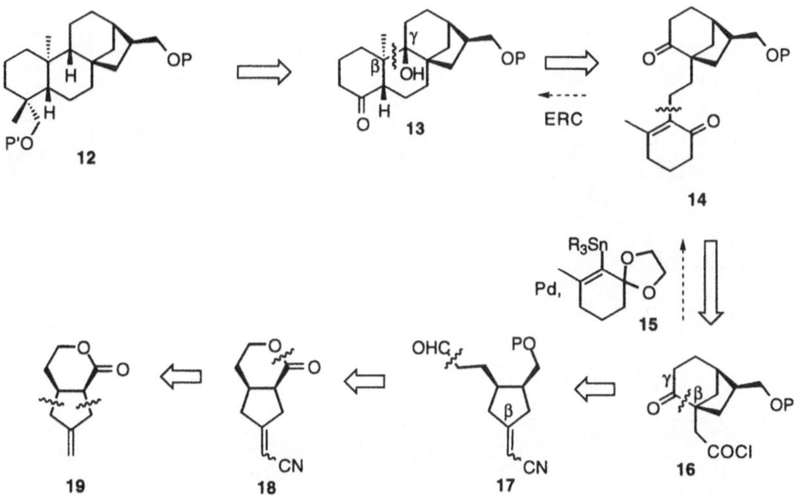

References

1) I. Berenblum, In *Risk Factors and Multiple Cancer*; B. Stoll, Ed.; John Wiley and Sons: New York, 1984.

2) P. M. Blumberg, "Protein Kinase C as the Receptor for the Phorbol Ester Tumor Promoters: Sixth Rhoads Memorial Award Lecture", *Cancer Research,* **1988**, *48*, 1.

3) (a) P. A. Wender, and F. E. McDonald, *J. Am. Chem. Soc.* **1990**, *112,* 4956. (b) P. A. Wender, P.A. and J. L. Mascareñas, *J. Org. Chem.* **1991**, *56*, 6267.

4) A. J. Fry, R. D. Little, and J. A. Leonetti, *J. Org. Chem.*, **1994**, *59*, 5017.

5) (a) Y-C. Wu, Y-C. Hung, F-R. Chang, M. Cosentino, H-K. Wang, and K-H. Lee, *J. Nat. Prod.*, **1996**, *59*, 635; (b) M. O. Fatope, O. T. Audu, Y. Takeda, L. Zeng, G. Shi, H. Shimada, and J. L. McLaughlin, *J. Nat. Prod.*, **1996**, *59*, 301; (c) K. Chen, Q. Shi, T. Fujioka, D-C. Zhang, C-Q. Hu, J-Q. Jin, R. E. Kilkuskie, and K-H. Lee, *J. Nat. Prod.*, **1992**, *55*, 88.

6) B. M. Trost, *Angew. Chem. Int. Ed. Engl.*, **1986**, *25*, 1.

7) J. K. Stille, *Angew. Chem. Int. Ed. Engl.*, **1986**, *25*, 508.

SELECTIVE ELECTROORGANIC REDUCTIONS IN UNDIVIDED CELLS

N. L. WEINBERG, J. D. GENDERS, E. A. GEORGE, P. M. KENDALL, and
D. J. MAZUR

The Electrosynthesis Company, Inc., 72 Ward Road, Lancaster, NY 14086, USA

ABSTRACT

Three selective electroorganic reductions have been performed successfully in undivided cells. New process developments include the sacrificial oxidation of acetic acid in the reductive deacetoxylation of a steroid intermediate, and the use of gas diffusion anodes in the reductive carboxylation of chlorobenzene and the reduction of 3-hydroxybenzoic acid.

INTRODUCTION

Significant advantages in synthetic electroorganic processes are often realized by avoiding the use of an ion-exchange membrane or separator. A number of approaches have been studied in order to prevent the possibility of anodic oxidation of the substrate or product in electroreduction reactions performed in undivided cells. These include the use of dissolving metal anodes[1], the addition of oxidizable salts such as oxalate or formate[2], and the use of aqueous/nonaqueous mixtures[3]. A potential alternative is the utilization of a gas diffusion anode. A hydrogen consuming anode is expected to offer several advantages for electroorganic reductions. These include: a low potential for the hydrogen oxidation reaction which is expected to prevent oxidation of organics; the generation of protons needed for the cathode reduction chemistry; and the possibility for allowing gaseous byproducts to exit the cell through the gas diffusion electrode. Gas diffusion anodes have been successfully utilized in the synthesis of inorganic reagents such as hydrogen peroxide[4] and titanium (III)[5], but to the best of our knowledge have not been used in electroorganic reactions.

RESULTS AND DISCUSSION

This paper is concerned with three selective electroorganic reduction reactions, all performed in undivided cells[6]. The first example is a reductive deacetoxylation of a steroid intermediate, 11-

225

ketorockogenin diacetate (1), to provide 11-ketotigogenin (2). This reaction has been accomplished with calcium in liquid ammonia[8], but due to the low temperature required for the reaction, as well as safety and environmental factors, an alternative route was desired. The electrochemical reaction utilizing a mercury cathode in aqueous ethanol provided the product with the ketone over-reduced to the alcohol[9]. A more selective route was desired.

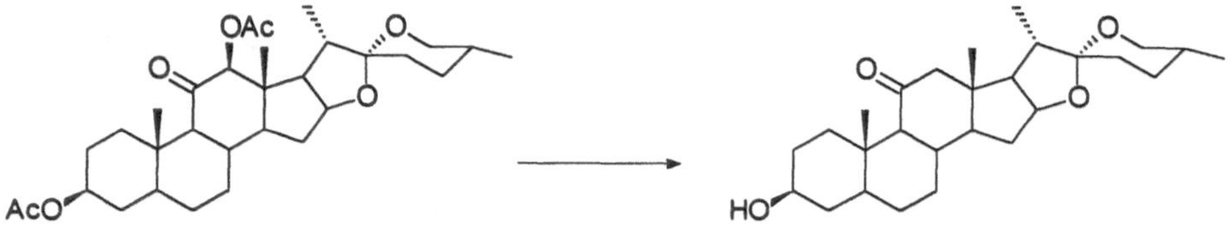

1   **2**

The standard divided cell reaction in aprotic solvent was compared to undivided cell reactions with a dissolving metal anode, and alternatively, with the sacrificial anodic oxidation of acetic acid. The best results were achieved in the undivided cell at a graphite felt cathode with 0.3 M Bu_4NOAc in N-methyl-2-pyrrolidinone/tetrahydrofuran (1:1). An isolated yield of 75% of the desired product was recovered (compared to 64% when utilizing a dissolving magnesium anode) at a current density of 25 mA/cm^2. The use of acetic acid provides a highly efficient and cost-effective synthesis. Two anodic processes are possible.

$$2AcO^- \rightarrow 2CO_2 + CH_3CH_3 + 2e^-$$
$$2AcO^- \rightarrow CO_2 + CH_3OAc + 2e^-$$

The next two examples utilized an undivided cell with a gas diffusion anode. The reductive carboxylation of chlorobenzene to benzoic acid was compared to the similar synthesis utilizing a dissolving metal anode[1]. A small flow cell (Micro Flow Cell, ElectroCell Systems, Täby, Sweden) was equipped with a nickel cathode and an ECFG gas diffusion anode containing 1 mg/cm^2 of platinum dispersed throughout the electrode (E-Tek, Inc., Massachusetts). The chlorobenzene was dissolved in dimethyl formamide with tetrabutylammonium bromide in the presence of carbon dioxide.

A current density of 50 mA/cm^2 was applied. It was necessary to add calcium oxide to the system to neutralize acid formed at the anode. The current efficiency for formation of benzoic acid was only 19%, and was believed due to the inability of the gas diffusion electrode to allow hydrogen chloride to exit from the system through the back side of the electrode.

The third reaction studied was the electroreduction of 3-hydroxybenzoic acid (3) to 3-hydroxybenzyl alcohol (4) which has been successfully reported in a divided cell[7]. A small flow cell (Micro Flow Cell, ElectroCell Systems, Täby, Sweden) was equipped with a lead cathode and an ELAT gas diffusion anode containing 1 mg/cm^2 of platinum as a thin surface coating (E-Tek, Inc., Massachusetts). The anode was covered with a prewetted piece of Nafion® 117 cation exchange membrane to provide a barrier layer. The process was operated under similar operating conditions as those utilized in the divided cell, including 160 mA/cm^2 current density, addition of the starting material in portions to the aqueous 17.5% sulfuric acid solution, and a temperature of 50°C. A 0.5 M solution of the desired product was obtained with a chemical yield of 89% at a current efficiency of 40%. These results are very similar to those reported for the divided cell. The cell voltage utilized is about 40% less for the undivided cell due to a smaller interelectrode gap in the undivided cell system.

3	4

CONCLUSIONS

The results obtained with the three reductions offer different levels of success. The use of an inexpensive sacrificial reagent in a nonaqueous system provided a unique solution to allow for an economical scale up of the electrochemical process. The complexity of a gas diffusion electrode or a dissolving metal anode was not required. In the case of chlorobenzene, the use of a gas diffusion anode in a nonaqueous solvent system was not as successful. The performance of a gas diffusion electrode is strongly influenced by factors such as permeability, hydrophobicity, and the nature of the catalyst used. These types of electrodes have not been optimized for reactions in nonaqueous solvents.

In the case of 3-hydroxybenzoic acid, the use of a gas diffusion electrode in an aqueous-organic solvent mixture was more successful. We believe that through systematic study and modification of gas diffusion electrode properties, a wide range of new applications can be realized.

ACKNOWLEDGEMENT

The authors are grateful to the National Science Foundation, the Electric Power Research Institute, and to Pfizer, Inc. for financial support for this work.

REFERENCES

1) J. Chaussard, J. C. Folest, J. Y. Nedelec, J. Perichon, S. Sibille and M. Troupel, *Synthesis*, **1990**, 369 .

2) R. Engels, C. J. Smit and W. J. M. Van Tilbourg, *Angew. Chem. Suppl..*, **1980**, 691.

3) D. J. Mazur and N. L. Weinberg, *U. S. Patent 4,968,393*, **1990**.

4) D. F. Dong and A. L. Clifford, *U. S. Patent 4,891,107*, **1990**.

5) P. C. Foller, C. W. Tobias and M. L. Goodwin, *U. S. Patent 4,375,395*, **1982**.

6) Portions of this work have been reported previously: (a) D. J. Mazur, P. M. Kendall, C. W. Murtiashaw, P. Dunn, S. L. Pezzullo, S. W. Walinsky, and J. B. Zung, *J. Organic Chem.*, **1996**, *61*, 405 ; (b) N. L. Weinberg, J. D. Genders, E. A. George, P. M. Kendall, D. J. Mazur and G. D. Zappi, "Fundamentals and Potential Applications of Electrochemical Synthesis", Eds. R. D. Weaver, F. Fisher, F. R. Kalhammer, and D. Mazur; The Electrochemical Society, Inc.: Pennington, NJ, 1997; pp 69-72.

7) R. Oi, C. Shimakawa, Y. Shimokawa and S. Takenaka, *Bull. Chem. Soc. Japan*, **1987**, *60*, 4193.

8) J. H. Chapman, J. Elks, G. H. Phillips and L. J. Wyman, *J. Chem. Soc.*, **1956**, 4344.

9) P. Kabasakalian, J. McGlotten, A. Basch and M. D. Yudis, *J. Org.*, **1961**, *26*, 1738.

ENANTIOSELECTIVE CATHODIC HYDROGENATION OF PROSTEREOGENIC ACTIVATED DOUBLE BONDS

M.F. NIELSEN[b)], B. BATANERO[a)], S. ROLVERING[a)], H.J. SCHÄFER [a)]

a) Organisch-Chemisches Institut der Universität Münster, Corrensstraße 40,
D-48149 Münster, Germany, Fax +49 251 83 39772, E-Mail: schafeh@uni-muenster.de
b) University of Copenhagen, Department of Chemistry, The H.C. Orsted Institute,
Universitetsparken 5, DK-2100 Copenhagen 0, Denmark

ABSTRACT

4-Methylcoumarine (**1**) is enantioselectively reduced in the presence of yohimbine at pH = 3 to afford 4-methyldihydrocoumarine (**2**) in 61 % yield and nearly 70 % *ee*. With 3-methylindenone and strychnine the dihydroproduct is formed with up to 36 % *ee*. The role of the alkaloids is to catalyze the tautomerization of an intermediate enol radical and to enantioselectively protonate a ketoanion.

Enantioselective conversions in organic electrochemistry are rare, examples are the reduction of ketones[1], and 4-methylcoumarine[2,3] in the presence of chiral supporting electrolytes or alkaloids and the oxidation of alcohols with sparteine as chiral inductor[4].

We were interested to increase the yield and enantioselectivity in the reduction of 4-methylcoumarine (**1**), to extend the reaction to 3-methylindenone (**8**) and to get information on the reaction mechanism. 4-Methylcoumarine (**1**) is cathodically reduced in aqueous methanol to the dihydroproduct **2** and the dihydrodimer **3**. In the presence of spartein **2** is formed with 17 % *ee*[2]. By lowering of the pH, the concentration of **1** and by applying a less cathodic reduction potential the yield of **2** could be increased to 61 %, this of the unwanted dihydrodimer **3** lowered to 9 %, and the enantiomeric excess of **2** increased to nearly 70 % *ee*[3] (eq. 1).

(1)

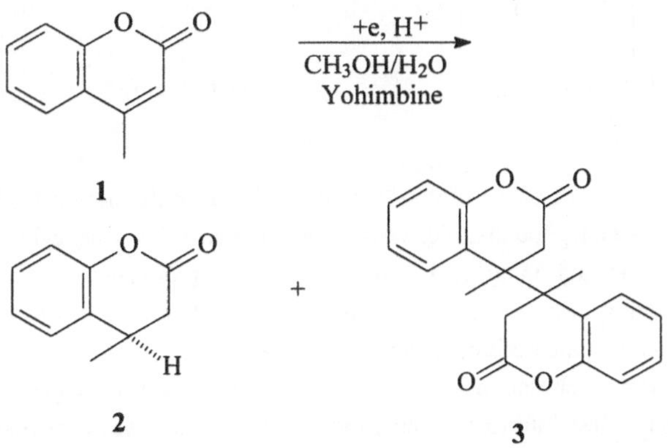

229

The mechanism of the reduction is investigated by cyclovoltammetry[3c] and quantum chemical calculations[3c]. According to these **1-H⁺**, activated by a hydrogen bond to the acid, is reduced to the enol radical **1-H·** (enol) (eq. 2), this undergoes fast dimerization to **3** ($k_2'' > 10^7 M^{-1} s^{-1}$) and slow reduction to **1-H⁻** (enol). The protonated alkaloid yohimbine H⁺ in the electrolyte catalyses the conversion of **1-H·** to the keto radical **1'-H·** (keto) which is easier reduced than **1'-H** (enol) to afford the anion **1-H⁻** (keto), which is subsequently enantioselectively protonated by yohimbine H⁺ to **2**.

(2)

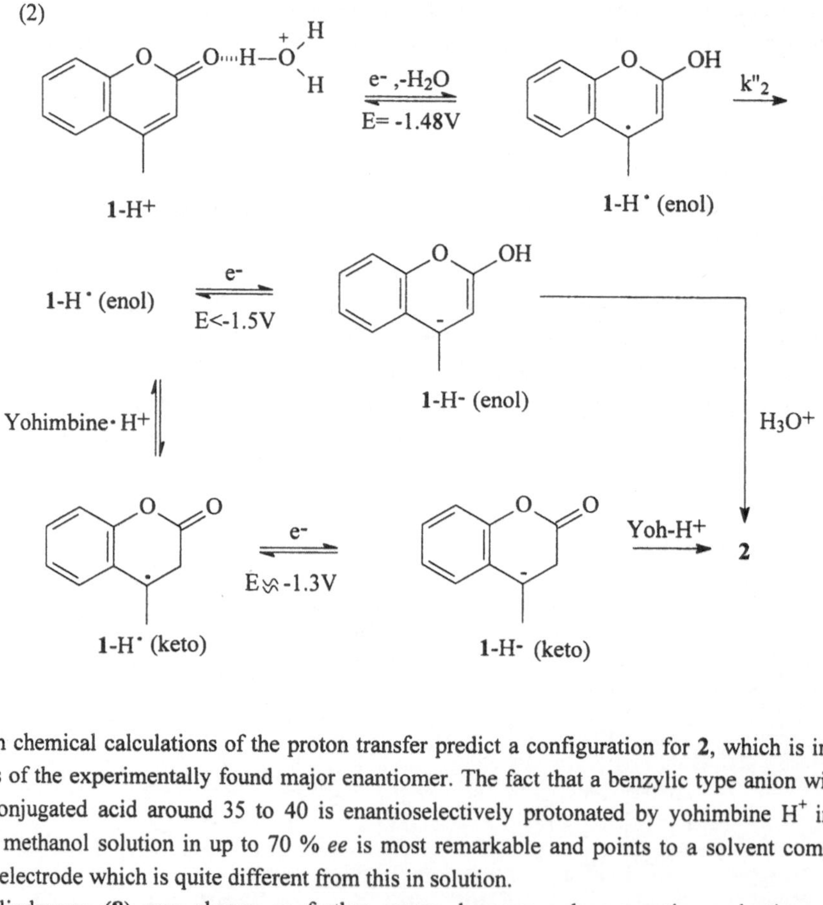

Quantum chemical calculations of the proton transfer predict a configuration for **2**, which is in accord with this of the experimentally found major enantiomer. The fact that a benzylic type anion with a pk_s of the conjugated acid around 35 to 40 is enantioselectively protonated by yohimbine H⁺ in acidic aqueous methanol solution in up to 70 % *ee* is most remarkable and points to a solvent composition near the electrode which is quite different from this in solution.

3-Methylindenone (**8**) was chosen as further enone, because a less negative reduction potential compared to **1** and thus a less interference with hydrogen evolution was expected. Cathodic reduction of **8** in the presence of strychnine affords indanone **9** with up to 36 % *ee* and the hydrodimer **10** (eq. 3).

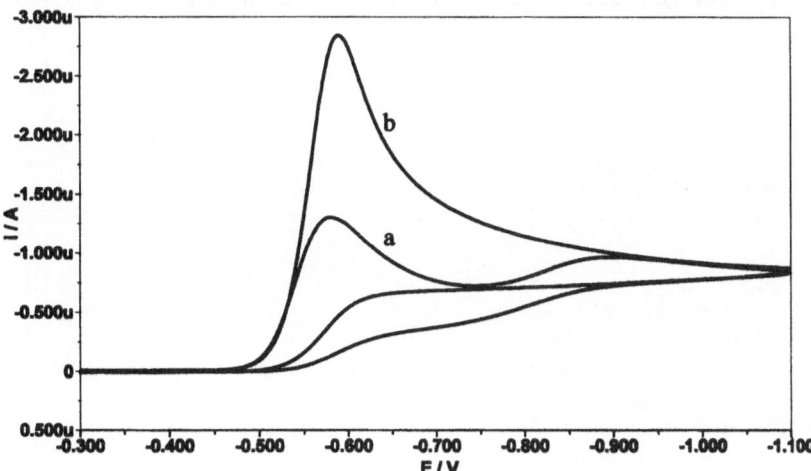

The same mechanism is proposed as for the reduction of **1**. In this case the strychnine adsorption does not interfere with the reduction of **8** in cyclovoltammetry. By comparing the CV of **8** in absence (Fig. 1a) and presence of strychnine (Fig. 1b) one sees in the first case two waves for the reduction of **8** and the intermediate enol radical [corresponding to **1**-H˙ (enol) in the case of **1**] whilst in the presence of strychnine the second peak has disappeared and the first doubled in current. This strongly supports the fast tautomerization of the enol radical to the more easily reduced ketoradical [corresponding to **1**-H˙ (keto) in the case of **1**] catalyzed by strychnine.

Fig. 1a: Cyclovoltammetry of **8** (2.10^{-3} mol/l) in MeOH, 0.1 m LiCl at pH = 3 at mercury, v = 0.2 V/s.
b: as a) with strychnine (1.10^{-3} mol/l)

This view is further supported by the fact that **8** in the absence of strychnine leads to 3 % **9** and 71 % **10** (dimerization of the enol radical), whilst in the presence of strychnine $(2 \cdot 10^{-3} \text{ mol/l})$ 71 % **9** and 3 % **10** are formed.

Acknowledgements: Support of this research by grant ERBCHRXCT 920073 within the EU Human Capital and Mobility Programme and to B.B. by the Ministerio de Educación y Ciencia (Dirección General de Investigación Cientifica y Técnica) of Spain is gratefully acknowledged. M.F.N. also acknowledges the Danish Natural Science Research Council for financial support. S.R. thanks the graduate college „Hochreaktive Mehrfachbindungssysteme" of the Deutsche Forschungsgemeinschaft for support.

References

[1]   L. Horner in *Organic Electrochemistry*, 2. ed. Eds. H. Lund, M.M. Baizer, **1983**, 945; E. Kariv, H.A. Terni, E. Gileadi, *Electrochim. Acta* **1973**, *18*, 433.

[2]   [a] N. Schoo, H.J. Schäfer, *Liebigs Ann. Chem.* **1993**, 601-607; [b] U. Höweler, N. Schoo, H.J. Schäfer, *Liebigs Ann. Chem.* **1993**, 609-614; [c] M.F. Nielsen, B. Batanero, T. Löhl, H.J. Schäfer, E.-U. Würthwein, R. Fröhlich, *Chemistry*, in press.

[3]   R.N. Gourley, J. Grimshaw, P.G. Millar, *J. Chem. Soc. (C)* **1970**, 2318-2323.

[4]   T. Osa, V. Kashiwagi, Y. Yanagisawa, J. Bobbit, *J. Chem. Soc. Chem. Commun.* **1996**, 2535-2537.

ELECTROREDUCTION OF ORGANIC COMPOUNDS WITH TWO REACTIVE CENTERS. II. ISOPHTHAL ALDEHYDE

Petr Zuman, James Bover, and Colin Johnson

Department of Chemistry, Clarkson University, Potsdam, NY 13699-5810, USA

Reduction of all three isomeric diformylbenzenes in aqueous solutions depends on protonation of the aldehydic groups. The reductions of diprotonated forms in acidic media show considerable differences of the behavior for individual isomers. Diprotonated forms of terephthaladehyde and orthophthaladehyde are reduced in a single reversible two-electron step, yielding a stable diradical or a quinomethide. Isophthaladehyde is reduced in acidic media up to pH 5 (where the rate of protonation of both formyl groups is sufficiently fast) in a single two-electron process. The irreversibility is caused by subsequent reactions. Each $CHOH^+$ group is converted into a $CHOH\bullet$ group that can undergo dimerization. The simultaneous transfer of two-electrons in aqueous media is facilitated by protonation which occurs as a surface reaction. This allows an orientation of both $CHOH^-$ groups with equal probability of an electron transfer. This makes the process in a protic medium different from processes in aprotic media, where after the transfer of the first electron the second group acts as a substituent, affecting transfer of the second and subsequent electrons. In the monoprotonated form the protonated - $CHOH^+$ group is reduced first. This reduction occurs in the medium pH - range (pH 5-8) where the uptake of the second electron occurs on radical - $CHOH\bullet$ at more positive potentials than the first electron uptake. First reduction occurs in a single two-electron step, the potential of which is affected by the second CHO group as substituents. The reduction of the unprotonated form at pH > 9 occurs similarly as in aprotic media: First two electrons are transferred in two separated one-electron steps.

Oxidation of isophthalaldehyde occurs in two two-electron anodic processes. In both an addition of hydroxide ions to the formyl group precedes the electron transfer. The resulting geminal diol anion loses two electrons and two protons to yield a carboxylate. Consecutive oxidations of the two formyl groups indicate either a reorientation at the electrode surface after the transfer of first two electrons, or two homogenous successive additions of hydroxide ions - to isophthalaldehyde and 3-formyl benzoate.

The mode of the electron transfer on molecules with two electroactive centers in protic media hence depends on antecedent protonations or additions of hydroxide ions.

SYNTHETIC APPLICATION OF HOMOGENEOUS CHARGE TRANSFER CATALYSIS IN THE ELECTROCARBOXYLATION OF BENZYL HALIDES

Onofrio Scialdone, Giuseppe Filardo, Alessandro Galia, Davide Mantione, and Giuseppe Silvestri
Università di Palermo, Dipartimento di Ingegneria Chimica dei Processi e dei Materiali, Viale delle Scienze, 90128 Palermo, Italy

ABSTRACT

The results of an investigation on the performances of some outer sphere electron transfer homogeneous catalysts in the electrocarboxylation of 1-(p-isobutyl-phenyl)-1-chloroethane to 2-(p-isobutyl-phenyl)-propionic acid (Ibuprofen), using as catalysts the esters of benzoic and o-, m-, and p-phtalic acids, are reported. The performances of the catalysts are evaluated on the basis of the following parameters: faradic yields of the carboxylation and decomposition of the catalyst. The performances of dimethylisophtalate have been examined in greater detail. The rate of decomposition of the catalyst is related to the molar ratio [halide]/[catalyst].

INTRODUCTION

The electrocarboxylation of benzyl halides to the corresponding carboxylic acids (reaction 1) is known since several years[1], and has been object of applicative attention in view of the synthesis of some 2-aryl propionic acids widely used as anti-inflammatory agents[2].

$$Ar\text{-}\underset{X}{CH}\text{-}R + 2\,e\text{-} + CO_2 \longrightarrow Ar\text{-}\underset{COO^-}{CH}\text{-}R + X^-$$

(1)

The development of reaction 1, made with graphite cathodes and magnesium sacrificial anodes in undivided cells[2], evidentiated the insurgence of the electrical passivation of the cathode surface, which was confirmed in our laboratory when magnesium was substituted with aluminum. Assuming that this behavior was due to some reactions of the anion radicals deriving from the reduction of the halide with the cathode surface, leading to the formation of non conductive organic layers, we have attempted to bring the reduction of the halide far from the electrode surface by means of homogeneous charge transfer agents.

Homogeneous catalysts have been widely investigated for the electroreduction of carbon dioxide: good yields are reported by the use of catalysts both inner sphere[3], based on transition metal complexes, and outer sphere, such as nitriles and esters of mono and dicarboxylic aromatic acids[4]. The outer sphere mechanism in the case of esters and nitriles has been recently questioned by Gennaro et al.[4], who showed experimental evidences of a reaction involving the formation of a bond between the anion radical of the catalyst and carbon dioxide. In our case the electron transfer should involve the benzyl halide, and, to our knowledge, the outer sphere mechanism of reduction of such class of compounds by the use of esters and nitriles as electron transfer catalysts is commonly accepted[5].

RESULTS

The results reported here are related to the electrocarboxylation of 1-(p-isobutyl-phenyl)-1-chloroethane to 2-(p-isobutyl-phenyl)-propionic acid (Ibuprofen), using as catalysts the esters of benzoic and o-, m-, and p-phtalic acids.

Cyclovoltammetric data

In Table 1 the $E°$ values of the catalysts are compared to the Ep of the halides and of CO_2.

Cyclovoltammetric data are consistent with a catalytic electron transfer mechanism when esters listed in the table are in the presence of the halide,.

Table 1.   Cyclovoltammetric data of the compounds involved in the electrocarboxylations

Compound	E_{p1}(cath.)	Ip / (2 y Ipd)	$E°_1$
phenylbenzoate	-1,61	1.18	-1,57
dibutylphthalate	-1,54	0.8	-1,45
dimethylisophthalate	-1,48	0.765	-1,41
dimethylterephthalate	-1.2	0.63	-1.25
1-(p-isobutylphenyl)-1-chloroethane	-2,11		
carbon dioxide	-2,55		

Catalyst concentration: 5mM; Substrate concentration: 10mM; CO_2 concentration: ca. 0.2 M (saturated, atm. pressure); SSE: DMF, TBABr 0.1M; reference Ag/AgI/I$^-$ 0.1 M in DMF. Working electrode: Pt 3.14 mm^2 ; Scan rate 0.1 V s^{-1}; Ip: catalytic peak current in presence of substrate; Ipd: peak current without substrate; y = c.sub./c.cat.

Synthetic data

Syntheses were performed in diaphragmless parallel plate cells, with sacrificial aluminum anodes, in N,N-dimethylformamide (DMF) with Bu$_4$NBr 0.1M, at concentrations of the catalysts ranging from 1 to 8mM, and of the halide from 0,05 to 0,3M. Table 3 summarizes the synthetic results related to the use of different catalysts: the main by-product, responsible of the low faradic yields, is the oxalate anion. Other by products are p-isobutylphenylethane and p-isobutylstyrene.

Table 3. Electrocarboxylation of 1-(p-isobutylphenyl)-1-chloroethane with different catalysts

Catalyst	$E°_1$	Product yield, %	Faradic yield, %	Charge passed, C
methylbenzoate	-1,68	82	64.5	2500
phenylbenzoate	-1,57	79	62	2500
dibutylphthalate	-1,45	90	80,4	2200
dimethylisophthalate	-1,41	85	76	2200
dimethylterephthalate	-1,20	48	44	1850

Potentiostatic electrolyses; reference $Ag/AgI/I^-$ 0.1 M in DMF. Cathodic potential was determined by polarisation curves for each catalyst. Cathode: compact graphite; anode 99.99 Al. Tank cells with parallel plate electrode arrangement. Halide concentration: 0.2 M; CO_2 concentration: ca. 0.2 M (saturated, atm. pressure); SSE: DMF, TBABr 0.1M.

Dibutylphthalate allows better results, but its rather high toxicity[6] discouraged its use in the synthesis of pharmaceutical compounds. Dimethylisophthalate, showing lower toxicity, was therefore chosen.

Figures 1 and 2 show the variation of the concentrations of 1-(p-isobutylphenyl)-1-chloroethane, 2-(p-isobutylphenyl)-propionic acid (carboxylate) and dimethylisophthalate with the charge passed. In the first stages of the electrolysis the catalyst is consumed at a very reduced rate (ca. 0.0036 moles per faraday passed), but when the molar ratio [halide]/[catalyst] decreases below a value roughly estimated as 30÷35 the curve of fig. 2 undergoes a dramatic change of slope (point a*), and the catalyst disappears at a much higher rate (ca. 0.033 moles per faraday passed). This behavior has been confirmed at different initial concentrations of halide and catalyst, and for all the catalysts listed in table 1, being the value of the ratio [halide]/[catalyst] at the corresponding point a* higher at more negative values of the E° of the catalyst (e.g. methylbenzoate gave a value of 50 ÷ 55, and dibutylphtalate of 30 ÷ 35).

CONCLUSIONS

The results reported here confirm that homogeneous electron transfer catalysts can be use for synthetic purposes in the electrocarboxylation of benzylic halides. The catalyst is consumed at very low rates, provided that the molar ratio [halide]/[catalyst] is higher than a given threshold value, characteristic of each couple halide - catalyst. In these conditions the mass balance of the syntheses is rather interesting from the applicative point of view: in particular, using dimethylisophthalate, to produce 1 Kg of 2-(p-isobutylphenyl)-propionic acid it is necessary to use 1.12 Kg of 1-(p-isobutyl-phenyl)-1-chloroethane and 52 g of catalyst of wich only 6 g are decomposed.

238

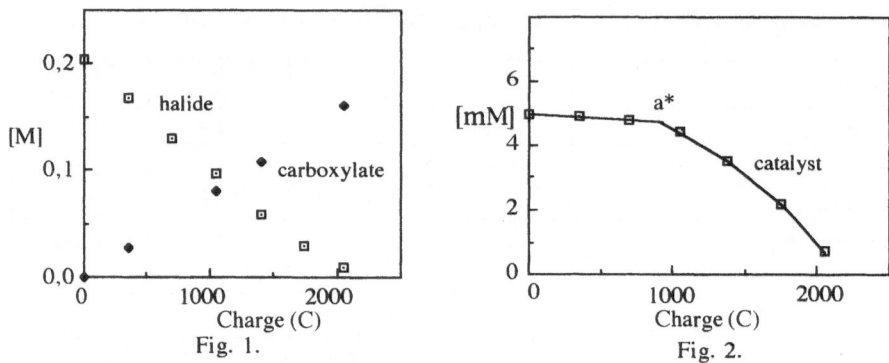

Fig. 1.   Fig. 2.

Fig. 1. Variation of the concentration of 1-(p-isobutyl-phenyl)-1-chloroethane and of 2-(p-isobutylphenyl)-propionate with the charge passed.
Fig. 2. Variation of the concentration of dimethylisophtalate with the charge passed.
Potentiostatic electrolysis; reference Ag/AgI/1⁻ 0.1 M in DMF. Cathodic potential : -1.44 V. Cathode: compact graphite; anode 99.99 Al. Tank cells with parallel plate electrode arrangement. SSE: DMF, TBABr 0.1M.

ACKNOWLEDGEMENTS

This research is supported by the Italian Consiglio Nazionale delle Ricerche, Progetto Strategico.

References
1.   a) M.M. Baizer, J.L. Chruma, J. Org. Chem. 37, 1951 (1972) b) G. Silvestri, S. Gambino, G. Filardo, A. Gulotta, Angew. Chem., 96, 978 (1984) c) O. Sock, M. Troupel, J. Perichon, Tetraehedron Lett., 26, 1509 (1985).
2.   J. Chaussard, in "Electrosynthesis from laboratory, to pilot, to production", J. D. Genders and D. Pletcher Eds., 1990, 165-175.
3.   M. Beley, J.P. Collin, R. Ruppert, J.P. Sauvage, J.Am. Chem. Soc. **108**, 7461 (1986).
4.   A. Gennaro, A.A.Isse, J.M. Savèant, M.G. Severin, E. Vianello, J.Am. Chem. Soc. **118**, 7190 (1996).
5.   C. Andrieux, A. Le Gorande, J.M. Savèant, J.Am. Chem. Soc. **114**, 6892 (1992).
6.   SAX'S Dangerous Properties of Industrial Materials 9° edition, (1996), R. J. Lewis Ed., 1080.

SYNTHESIS OF α,β-UNSATURATED CARBOXYLIC ACIDS BY ELECTROCHEMICAL CARBOXYLATION OF VINYL BROMIDES AND ITS APPLICATION TO THE SYNTHESIS OF ANTI-INFLAMMATORY AGENTS

Masao TOKUDA, Hisato KAMEKAWA, and Hisanori SENBOKU
Division of Molecular Chemistry, Graduate School of Engineering, Hokkaido University, Sapporo 060, Japan

ABSTRACT

Electrochemical carboxylation of various vinyl bromides under an atmospheric pressure of carbon dioxide with a Pt cathode and an Mg anode gave the corresponding α,β-unsaturated carboxylic acids in high yields. In the case of alkyl-substituted vinyl bromides, efficient electrochemical carboxylation was achieved by the addition of 20 mol% of NiBr$_2$•bpy. These facile carboxylations were successfully applied to a synthesis of the precursors of nonsteroidal anti-inflammatory agents.

INTRODUCTION

We have already reported the regioselective synthesis of γ-substituted β,γ-unsaturated acids,[1] allenic acids,[2] and 3-methylene-4-pentenoic acid[3] by electrochemical carboxylation of the corresponding allylic and propargylic halides using a magnesium anode. As part of our continuing studies on the electrochemical synthesis of useful carboxylic acids, we recently carried out electrochemical carboxylation of substituted vinyl halides. In this paper, we report the synthesis of various α,β-unsaturated carboxylic acids by electrochemical carboxylations of vinylic bromides and its application to the synthesis of 2-arylpropenoic acids, the useful precursor of anti-inflammatory agents.

RESULTS AND DISCUSSION

Synthesis of phenyl-substituted α,β-unsaturated acids

Electrochemical carboxylation of phenyl-substituted vinyl bromides (**1**) (R^1, R^2, R^3= H, Ph) under an atmospheric pressure of carbon dioxide with a platinum cathode and a magnesium anode gave the corresponding α,β-unsaturated carboxylic acids (**2**) in isolated yields of 63-92% (Scheme 1).[4] Electrolysis was carried out at a constant current

239

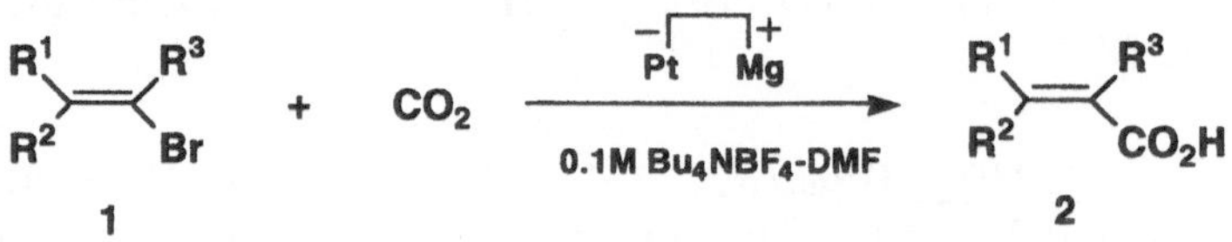

$$R^1 \diagdown R^3 \qquad + \quad CO_2 \xrightarrow[\text{0.1M Bu}_4\text{NBF}_4\text{-DMF}]{\boxed{\text{Pt} \quad \text{Mg}}^+} \qquad R^1 \diagdown R^3$$
$$R^2 \diagup Br \hspace{9cm} R^2 \diagup CO_2H$$

Scheme 1

of 10 mA/cm^2 in a DMF solution containing 0.1M Bu$_4$NBF$_4$ in a one-compartment cell equipped with a platinum cathode and a magnesium rod anode. Electricity of 3 Faradays per mol of **1** is required for efficient carboxylation.

Synthesis of alkyl-substituted α,β-unsaturated acids

Electrochemical carboxylation of alkyl-substituted vinyl bromides (**1**) (R^1, R^2, R^3= H, alkyl) under the same conditions as those of phenyl-substituted vinyl bromides gave the corresponding α,β-unsaturated acids (**2**) in lower yields (14-43 %) (Scheme 1) (Table 1). These electrochemical carboxylations were further studied under a variety of conditions in order to optimize the yields of **2**. We found that the yield of **2** was highly enhanced to 53-82% by the addition of 20 mol% nickel(II) bromide-2,2-bipyridine complex (NiBr$_2$•bpy) to the electrolysis solution. Results are summarized in Table 1.

Application to the synthesis of the precursor of anti-inflammatory agents

The present electrochemical carboxylation can be applied to the synthesis of the precursor of anti-inflammatory agents. 2-Arylpropanoic acids are widely used as non-steroidal anti-inflammatory agents. Various synthetic methods of these acids have been reported; i.e., methylation of arylacetic acids, cyanation or carboxylation of 1-aryl-1-haloethanes, introduction of an aryl group into the α-position of propanoic acids, and rearrangement of propiophenones or their derivatives. However, these synthetic methods are ineffective for an enantioselective synthesis of (*S*)-2-aryl-propanoic acids. For this purpose, a synthesis of α-aryl-substituted α,β-unsaturated acids by the present electrochemical carboxylation followed by asymmetric hydrogenation with a chiral transition metal catalyst[5] would be one of most promising methods.   Synthesis of the

Table 1. Nickel(II)-catalyzed electrochemical carboxylation of alkyl-substituted vinyl bromide [a)]

Substrate	Product	Yield (%)[b)]	
		n=1	73 (18)
		n=2	58 (15)
		n=3	80 (43)
		n=4	82 (30)
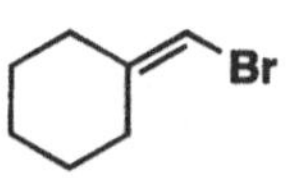	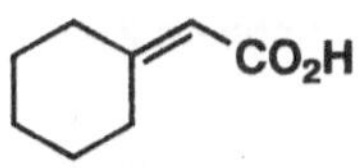	58 (14)	
	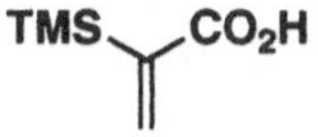	53 (19)	
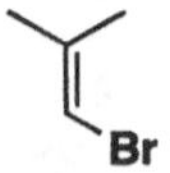	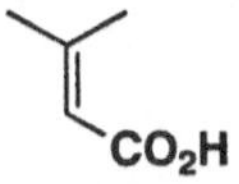	64 (18)	

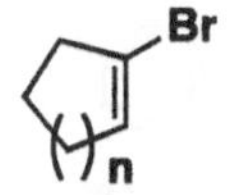

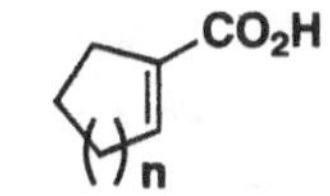

a) Vinyl bromide (3 mmol) in 0.1M Bu_4NBF_4-DMF containing 20 mol% of $NiBr_2 \bullet bpy$ was electrolyzed in the presence of CO_2 (1 atm) with a Pt cathode and an Mg anode.
b) Isolated yields. Yields in parentheses show those in the absence of Ni(II) catalyst.

precursor of *ketoprofen* by means of electrochemical carboxylation of vinyl bromide is shown in Scheme 2. The requisite vinyl bromide **5** was readily prepared in four steps from aryl bromide **3**, with an overall yield of 79%. Electrochemical carboxylation of **5** under the same conditions as those of phenyl-substituted vinyl bromide[4] gave the desired 2-arylpropenoic acid **6**, a precursor of *ketoprofen*, in an 80% yield (Scheme 2).

Similarly, synthesis of the precursors of *ibuprofen*, *naproxen*, *cicloprofen*, and *flurbiprofen* could be achieved in high yields by utilizing the present electrochemical carboxylation of the corresponding aryl-substituted vinyl bromides with a platinum cathode and a magnesium anode as the key step (Chart 1).

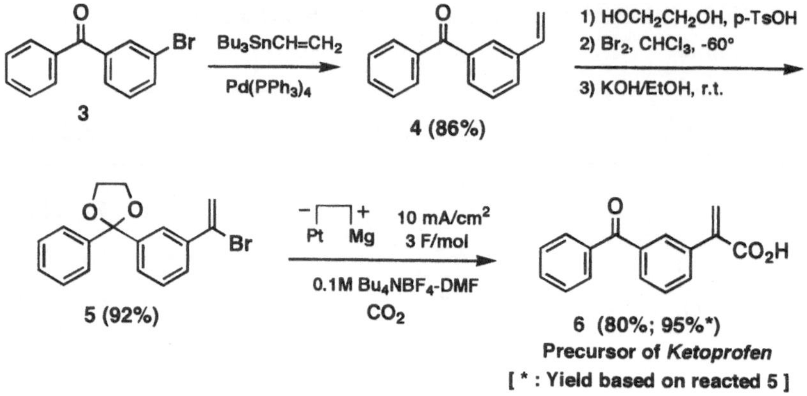

Scheme 2

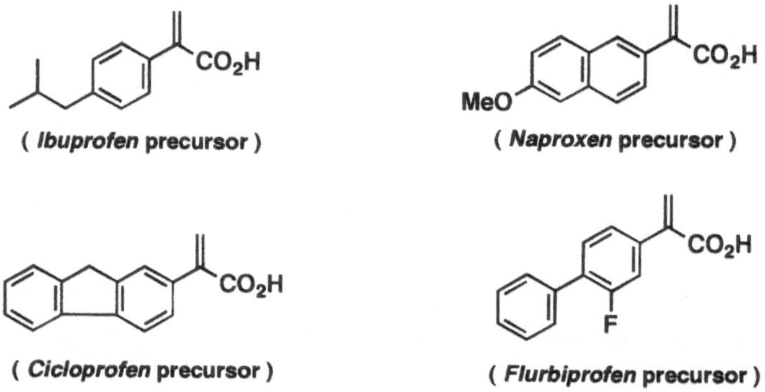

(*Ibuprofen* precursor)

(*Naproxen* precursor)

(*Cicloprofen* precursor)

(*Flurbiprofen* precursor)

Chart 1

References

1) M. Tokuda, T. Kabuki, Y. Katoh, and H. Suginome, *Tetrahedron Lett.*, **1995**, *36*, 3345.
2) M. Tokuda, T. Kabuki, and H. Suginome, *DENKI KAGAKU*, **1994**, *62*, 1144.
3) M. Tokuda, A. Yoshikawa, H. Suginome, and H. Senboku, *SYNTHESIS*, **1997**, in press.
4) H. Kamekawa, H. Senboku, and M. Tokuda, *Electrochimica Acta*, **1997**, *42*, 2117.
5) (a) X. Zhang T. Uemura, K. Katsumura, N. Ayo, H. Kumobayashi, and H. Takaya, *SYNLETT*, **1994**, 501; (b) T. Manimaran, T-C. Wu, W. D. Klobucar, C. H. Kolich, and G. P. Stahly, *Organometallics*, **1993**, *12*, 1467.

NEW AND CONVENIENT SYNTHESIS OF 3-METHYLENE-4-PENTENOIC ACID BY ELECTROCHEMICAL CARBOXYLATION

Hisanori SENBOKU, Yusuke FUJIMURA, Akihiro YOSHIKAWA,
Hiroshi SUGINOME, and Masao TOKUDA
*Division of Molecular Chemistry, Graduate School of Engineering,
Hokkaido University, Sapporo 060, Japan*

Electrochemical carboxylation of 2-bromomethyl-1,4-dibromo-2-butene (**1**)
with atmospheric carbon dioxide in a DMF solution containing Bu_4NI with a Pt cathode
and a manganin alloy or nickel anode gave 3-methylene-4-pentenoic acid (**2**) in 57%
yield. The acid **2** can be used as a diene in an aqueous intermolecular Diels-Alder
reaction with dimethyl fumarate to give the adduct **4** in 67% yield.

Recently, we reported a convenient method for isoprenylation of aldehydes or
ketones by using 2-bromomethyl-1,4-dibromo-2-butene (**1**) (tribromide) in the
presence of electrochemically generated reactive zinc[1] or commercially available zinc
powder.[2] These studies showed that tribromide can work as a synthetic equivalent for
isoprenyl carbanion. Here, we report a convenient method for the synthesis of 3-
methylene-4-pentenoic acid (**2**), a new compound, by electrochemical carboxylation of
2-bromomethyl-1,4-dibromo-2-butene (**1**).

Electrolysis of tribromide **1** in a DMF solution containing a supporting
electrolyte under an atmospheric pressure of carbon dioxide using a one-compartment
cell equipped with a platinum cathode and an appropriate anode gave 3-methylene-4-
pentenoic acid (**2**) in a reasonable yield (Scheme 1). Isolated yields of **2** under various

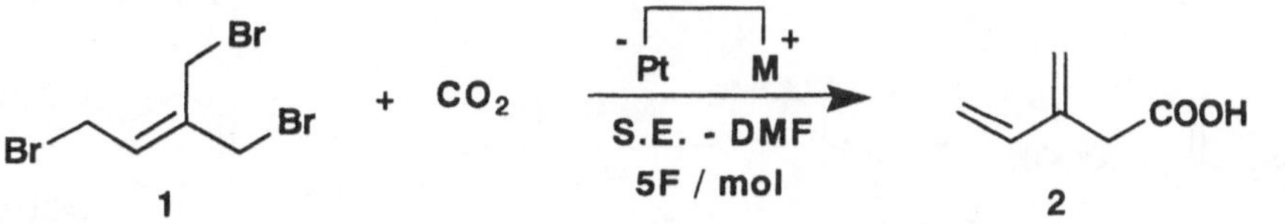

Scheme 1

electrolytic conditions are summarized in Table 1. Nickel and manganin alloy metals
were effective as anodes. The use of tetrabutylammonium iodide (Bu_4NI) as a
supporting electrolyte and the stirring of a DMF solution containing **1** and Bu_4NI

before electrolysis gave 57% yield of **2** (runs 5, 9 and 10). This increase is probably due to a chemical transformation of **3** to give 2-bromomethyl-1,3-butadiene by the iodide-induced 1,4-elimination of bromine. The transformation was proved by the ^{1}H NMR spectrum of **1** in d_7-DMF in the presence of Bu$_4$NI.

Table 1. Electrochemical carboxylation of tribromide **1**.

Run	Supporting Electrolyte	Anode	Current Density (mA / cm^2)	Yield of **2**
1	0.1M Bu$_4$NI	Pt	10	0
2	0.1M Bu$_4$NI	Cu	10	5
3	0.1M Bu$_4$NI	Mg	10	29
4	0.1M Bu$_4$NI	Ni	10	46
5	0.1M Bu$_4$NI	MA[a]	10	47
6	0.1M Bu$_4$NI	MA[a]	2.5	32
7	0.1M Bu$_4$NI	MA[a]	5	38
8	0.1M Bu$_4$NI	MA[a]	20	18
9	0.1M Bu$_4$NClO$_4$	MA[a]	10	42
10	0.1M Bu$_4$NI[b]	MA[a]	10	57

a) MA = manganin (alloy of 12-15% Mn, 2-4% Ni, and 80-86% Cu).
b) Before electrolysis, the solution was stirred for 30 min. at rt.

3-Methylene-4-pentenoic acid (**2**) was useful in the Diels-Alder reactions. In the present study, we showed one successful example in an application to an aqueous intermolecular Diels-Alder reaction.[3] The reaction of sodium salt of **2** with dimethyl fumarate (**3**) in water at 50°C gave a Diels-Alder adduct **4** in 67% yield (Scheme 2).

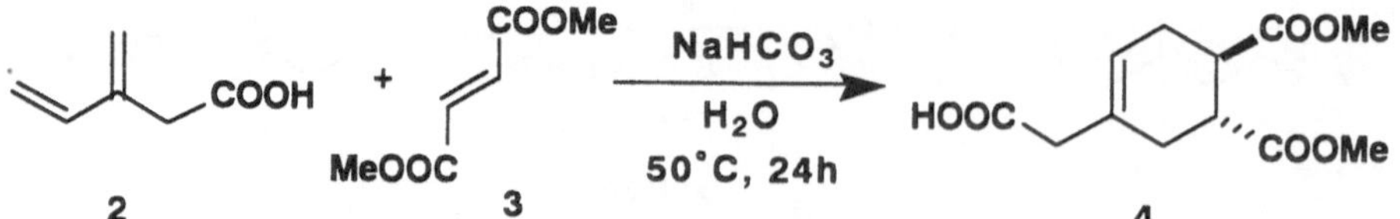

Scheme 2

1) M. Tokuda, N. Mimura, T. Karasawa, H. Fujita, and H. Suginome, *Tetrahedron Lett.*, **1993**, *34*, 7607.
2) M. Tokuda, N. Mimura, K. Yoshioka, T. Karasawa, H. Fujita, and H. Suginome, *Synthesis*, **1993**, 1086.
3) P. A. Grieco, *Aldrichimica Acta*, **1991**, *24*, 59, and the references cited therein.

ELECTROCHEMICAL CARBOXYLATION OF SEVERAL ORGANIC HALIDES IN SUPERCRITICAL CARBON DIOXIDE

Akiyoshi SASAKI, Hiroki KUDOH*, Hisanori SENBOKU*, and Masao TOKUDA*
Hokkaido National Industrial Research Institute, 2-17 Tsukisamu-Higashi, Toyohira-ku Sapporo 062, Japan
Graduate School of Engineering, Hokkaido University, Sapporo 060, Japan

Electrochemical carboxylation of benzyl chloride, cinnamyl chloride, and 2-chloronaphthalene in supercritical carbon dioxide (sc CO_2) with a minimum amount of co-solvent gave the corresponding carboxylic acids in good yields.

Supercritical fluids are unique in that their densities are higher than those in the gas phase. Among these fluids, supercritical carbon dioxide (sc CO_2) can readily be attained under moderate conditions (Tc 31℃, Pc 75.3kg/cm²), and it has significant potential as an environmentally benign solvent to replace hazardous compounds in organic reactions. However, there have only been a few reports on the use of supercritical carbon dioxide in organic reactions.[1] We previously reported the facile electrochemical carboxylation of allylic, propargylic, and vinylic halides using a magnesium anode.[2-4] In this paper, we report the first electrochemical carboxylation of organic halides using supercritical carbon dioxide both as a solvent and as a reagent.

$$R\text{-}X + sc\ CO_2 \xrightarrow[\text{S. E. , Co-solvent}]{\substack{-\boxed{}+ \\ Pt \quad Mg\ or\ Pt}} R\text{-}COOH$$

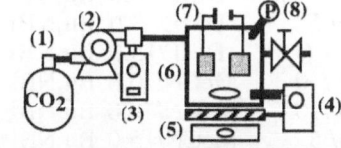

(1) CO_2 cylinder   (2) High pressure pump
(3) Cooling unit   (4) Electric heater system
(5) Magnetic stirrer   (6) Reactor
(7) Power supply   (8) Pressure meter

Fig. 1 Apparatus for the electrochemical carboxylation using sc CO_2

The experimental apparatus is shown in Figure 1. The reactor was a 180ml volume of autoclave. A mixture of organic halide, supporting electrolyte, and co-solvent was added to the reactor. CO_2 was supplied by a high pressure pump, and the temperature of the reactor was controlled by an electric heater system.

Representative results are shown in Table 1. Electrochemical carboxylation of benzyl chloride (BnCl), cinnamyl chloride (Cin-Cl), and 2-chloronaphthalene (2-Cl-

Naph) in the presence of sc CO_2 and 3-15 ml of DMF solution containing 0.1M Bu_4NBr gave the corresponding carboxylic acids in good yields (Entry 1-6). However, in these carboxylations, immersion of electrodes into the DMF solution was necessary for effective carboxylation. On the other hand, the mixture of acetonitrile (MeCN)-Bu_4NBF_4 solution and sc CO_2 became homogeneous at nearly 80 Kg/cm^2 of CO_2 and 40 °C, and carboxylation of BnCl in the mixture gave phenyl acetic acid without immersing the electrodes into the MeCN solution (Entry 8-10).

These results indicate that sc CO_2 can be used as an environmentally benign solvent and reagent by the addition of a small amount of co-solvent in the electrochemical carboxylation.

Table 1 Electrochemical carboxylation of several organic halides in supercritical carbon dioxide[a].

Entry	Halide [mmol]	Co-solvent [ml]	Electrolyte [M]	Electrode [cathode]	[anode]	Press. [kg/cm^2]	Temp. [°C]	Yield [%]
1	BnCl/ 5	DMF/14.4	Bu_4NBr /0.1	Pt	Mg	80	40	81
2	BnCl/ 5	DMF/ 3.0	Bu_4NBr /0.1	Pt	Mg	85	40	54
3	Cin-Cl/ 5	DMF/ 3.0	Bu_4NBr /0.1	Pt	Mg	84	46	64
4	2-Cl-Naph/5	DMF/ 6.0	Bu_4NBr /0.1	Pt	Mg	81	43	61
5	6-Br,2-OH,Naph/5	DMF/ 6.0	Bu_4NBr /0.1	Pt	Mg	79	42	9
6	BnCl/ 5	DMF/ 3.0	Bu_4NBr /0.1	Pt	Pt	79	42	68
7[b]	BnCl / 6	DMF/15.0	Bu_4NBr /0.1	Pt	Mg	1	10	62
8	BnCl /10	MeCN/40.0	Bu_4NBF_4 /0.4	Pt	Mg	69	36	35
9	BnCl /10	MeCN/40.0	Bu_4NBF_4 /0.4	Pt	Pt	50	40	59
10	BnCl/ 5	MeCN/15.0	Bu_4NBF_4 /0.1	Pt	Pt	81	43	49

a) Current density : 37.5 mA/cm^2 / electricity : 3 F/mol of the halide.
b) Electrolysis was carried out in the presence of atmospheric CO_2 using a normal undivided cell (30 ml in volume).

References
1) Y. Ikushima, N. Saito, M. Arai, *J. Phys. Chem.* , **1992**, *96*, 2293
2) M. Tokuda, T. Kabuki, Y. Katoh, and H. Suginome, *Tetrahedron Lett.*, **1995**, *36*, 3345
3) M. Tokuda, T. Kabuki, and H. Suginome, *DENKI KAGAKU*, **1994**, *62*, 1144
4) H. Kamekawa, H. Senboku, and M. Tokuda, *Electrochemica Acta*, **1997**, *42*, 2117

SYNTHESIS OF β-KETO ACIDS BY ELECTROCHEMICAL CARBOXYLATION OF VINYL TRIFLATES

Hisato KAMEKAWA, Hisanori SENBOKU, and Masao TOKUDA

Division of Molecular Chemistry, Graduate School of Engineering, Hokkaido University, Sapporo 060, Japan

Electrochemical reduction of alicyclic vinyl triflates (**1**) under an atmospheric pressure of carbon dioxide with a Pt cathode and a Mg anode resulted in the cleavage of an O-S bond of **1** to give the corresponding β-keto carboxylic acids (**2**) in good yields.

We have already reported regioselective synthesis of γ-substituted β,γ-unsaturated carboxylic acids,[1] allenic acids,[2] α,β-unsaturated carboxylic acids,[3] and 3-methylene-4-pentenoic acid[4] by electrochemical carboxylation using a magnesium anode. In the course of our study on ways to improve the yield in electrochemical carboxylation of alkyl-substituted vinyl bromides, we found a new electrochemical carboxylation of vinyl triflates (**1**) giving β-keto acids (**2**).

Vinyl triflates **1** were readily prepared in 73-93% yields from the corresponding ketones by their reactions with trifluoromethanesulfonic anhydride in the presence of 2,6-di-*t*-butyl-4-methylpyridine. A 4:1 mixture of two isomeric vinyl triflates was obtained in the reaction of β-tetralone. Electrochemical carboxylation of vinyl triflates **1** in a DMF solution containing 0.1M Bu_4NBF_4 under an atmospheric pressure of carbon dioxide gave the corresponding β-keto acids **2** in isolated yields of 28-75% (Scheme 1)(Table 1). Electrolysis was carried out at 5 °C at a constant current of 10 mA/cm^2 in a

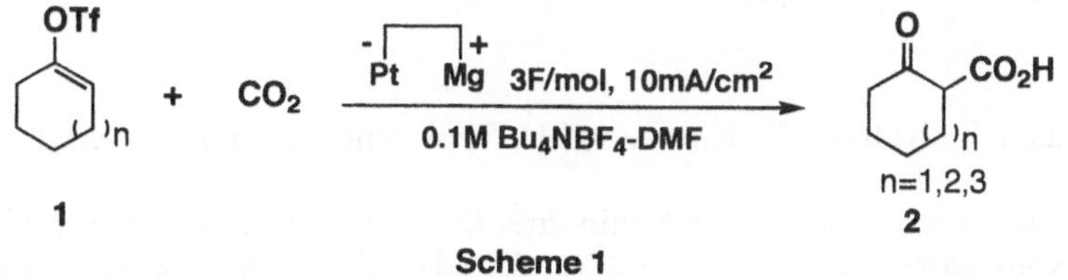

Scheme 1

Table 1. Electrochemical Carboxylation of Vinyl Triflates

Vinyl Triflate	Product	Yield of Products (%)[a]
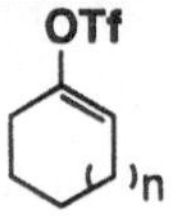 OTf, n=1, n=2, n=3	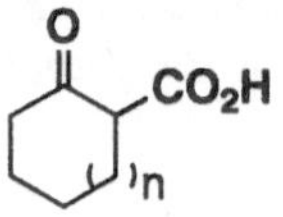 O, CO$_2$H	n=1, 75 (98) n=2, 56 (89) n=3, 28 (59)
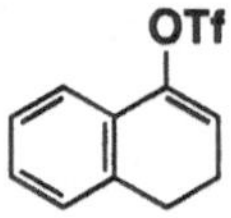 OTf	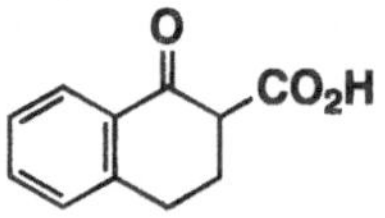 O, CO$_2$H	67 (81)
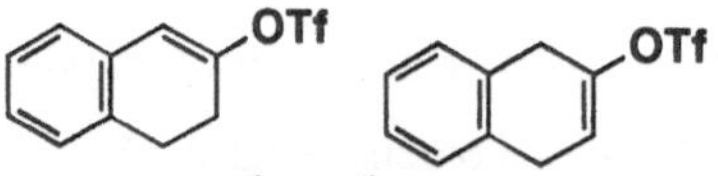 OTf, OTf, 4 : 1	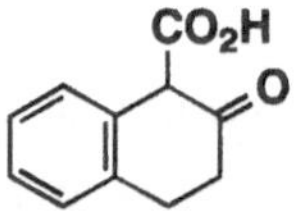 CO$_2$H, O	61 (82)

a) Yields in the parentheses are based on reacted triflates.

one-compartment cell equipped with a platinum cathode and a magnesium rod. Electricity of 3 Faradays per mol of 1 was passed in these carboxylations.

In the present electrochemical carboxylations, a two-electron reduction of vinyl triflates results in the cleavage of an RO-SO$_2$CF$_3$ bond to give enolate anions, which are effectively trapped with carbon dioxide to give β-keto carboxylates. Although a similar reductive cleavage of an RO-SO$_2$CF$_3$ bond was recently reported to occur in the electrochemical reduction of vinyl triflates in the presence of carbon dioxide, the corresponding ketone was only isolated in 46% yield.[5]

References
1) M. Tokuda, T. Kabuki, Y. Katoh, and H. Suginome, *Tetrahedron Lett.*, **1995**, *36*, 3345.
2) M. Tokuda, T. Kabuki, and H. Suginome, *DENKI KAGAKU*, **1994**, *62*, 114.
3) (a) H. Kamekawa, H. Senboku, and M. Tokuda, *Electrochimica Acta*, **1997**, *42*, 2117; (b) H. Kamekawa, H. Senboku, and M. Tokuda, *Chem.Lett.*, **1997**, 917.
4) M. Tokuda, A. Yoshikawa, H.Suginome, and H. Senboku, *SYNTHESIS*, **1997** in press.
5) A. Jutand and S. Négri, *SYNLETT*, **1997**, 719.

USE OF SACRIFICIAL ANODES IN THE TRANSFORMATION OF ORGANIC HALIDES, RESIDUALS FROM INDUSTRY

Marília O. F. GOULART, Márcia B. S. LISBOA, Fabiane C. de ABREU, Márcio H. S. ANDRADE, Josealdo TONHOLO and Marcelo NAVARRO

Departamento de Química, CCEN, Universidade Federal de Alagoas, Maceió, Alagoas, 57072-970, Brazil. E-mail: marilia@fis.ufal.br

ABSTRACT

Cyclic voltammograms (CV) (DMF/TBAP 0.1M, Hg) of 2,2-dichloroethanol (DCOH) and 1,1,2-trichloroethane (TCE) showed irreversible cathodic waves related to the cleavage of C-Cl from the *gem*-dichlorinated group. DCOH worked as a probase, in electrolysis using aluminum sacrificial anode. In the presence of 2-butanone, dimeric alkenes, products from aldolic condensation, were obtained in large amounts. Co-electrolysis of TCE and benzophenone, in aluminum, furnished reduction products from benzophenone, without any evidence of chlorinated alcohol. CV of 2,2-dichloroethylethylfumarate (DCEF) revealed the presence, at low scan rates, of two well defined reduction peaks (Epc_1, Epc_2) and a third one (Epc_3) close to the supporting electrolyte discharge. As the scan rate increases, the first wave turns into a reversible one and a new wave, intermediate between Epc_1 and Epc_2 (Epc_i) appears. Its height (Ipc_i) increases as Ipc_2 decreases. Epc_1 dependence with scan rate suggests an EC process, related to the reduction of the olefin, followed by dimerization. The remaining cathodic waves showed to be irreversible and the last one is related to C-Cl cleavage. Electrohydrodimerization is the main process in electrolyses of DCEF.

INTRODUCTION

DCOH and TCE are lateral products in the MVC manufacturing process. Their transformation into useful products would be highly compensating, not only by the synthetic value *per si*, but also to avoid incineration costs and environmental problems. In order to better understanding the reaction possibilities, in electrolysis, it is necessary to know the reduction potentials of the electroactive groups present in the molecule and establish the relative order of reduction facility and the choice of the better electrodes.

The present article describes the electrochemical studies performed on TCE, DCOH and its derivative, DCEF, prepared in the usual way.

Cyclic voltammograms (CV) were performed in DMF/TBAP, on mercury, platinum and vitreous carbon electrodes with scan rate varying from 0.02 to 20 Vs^{-1}. CV of TCE and DCOH, on mercury, showed, one irreversible wave (Epc_{TCE}=-2.540V, Epc_{DCOH}=-2.850V), followed by an anodic wave (Epa_{TCE}=-0.230V, Epa_{DCOH}=-0.169V), related to the oxidation of Cl^-, generated at Epc. In vitreous carbon, the behavior was similar, without evidence of anodic waves. Coulometry performed in vitreous carbon showed both processes to be monoelectronic. In platinum, there is no evidence of waves. The cleavage of one C-Cl from the *gem*-chlorinated carbon seems the reaction of choice.

DCEF contains two different electroreducible sites Its CV revealed the presence, in low v, of two well defined reduction peaks (Epc_1= -1.335V, Epc_2=-2.340V) and a third one close to the supporting electrolyte discharge. As v increases, the first wave turns into a reversible one. A new wave, intermediate between Epc_1 and Epc_2 (Epc_i) appears (Fig.1 A, B). Its height (Ipc_i) increases as Ipc_2 decreases. Epc_1 and $Ipc_1/v^{1/2}$ dependence with v suggests an EC process ($dEpc_1/dlogv$ = 49mV). The other waves have a irreversible nature.

Electrolyses of DCEF were carried out, at controlled potentials related to Epc_1 and Epc_3, on Hg and vitreous carbon, using divided cells. Electrohydrodimerization (EHD), furnishing the three possible dimers, evidenced by GC/MS, was the main reaction, with applied potentials close to Epc_1. At more cathodic potentials, hydrolysis of the dimers can also occur, generating carboxylic acids. Evidence of cleavage of C-Cl, in electrolysis held on Hg, at Epc_3, was obtained through search of Cl^- by precipitation with $AgNO_3$.

Electrolyses of DCOH were performed, in the presence of 2-butanone, using an undivided cell, fitted with an aluminum sacrificial anode.

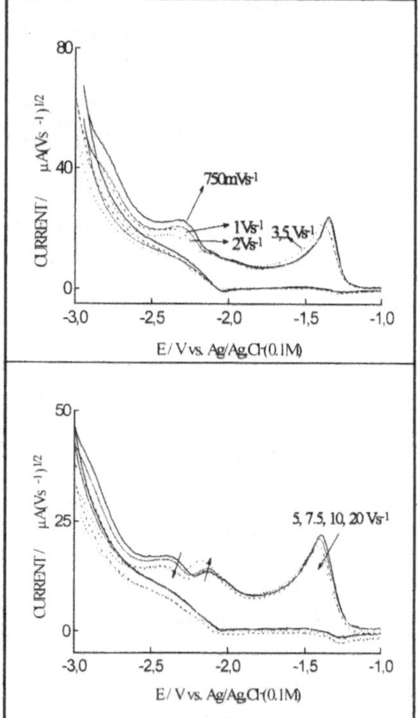

Figure 1: CV of DCEF, E vs. Ag/AgCl, Cl^- 0.1M. [DCEF]10mM. A: $0.75 \leq v \geq 3.5$ Vs^{-1}, B: $5 \leq v \geq 20$Vs^{-1}.

The main products were dimeric alkenes, obtained from aldolic condensation of the enolizable ketone. DCOH was deprotonated in the basic cathodic medium and worked as an electrogenerated base.

Electrolyses of TCE were performed in absence and presence of electrophiles. To avoid enolization, benzophenone was chosen as the electrophile, in spite of its easy reducibility. Its electroreduction alone, in an undivided cell, fitted with an aluminum anode, followed the same pattern of reactivity as in magnesium[1], furnishing benzydrol, benzyl, benzoin, mixed dimers from $(Ph)_2COHCHO$ and $PhCOCPhHOH$ and products from DMF attack. TCE electrolysis, in the presence of benzophenone, was much cleaner, giving benzydrol and hydrocarbons of high molecular weight, without evidences of chlorinated alcohol.

Acknowledgments: To CNPq, RHAE/CNPq, CAPES, TRIKEM, and FAPEAL. To Dr. Heinrich Luftman (University of Münster, Germany) for running the GC/MS.

References: 1) S. Pellegrini; J.C. Folest; J.Y. Nédélec; J. Perichon, *J. Electroanal. Chem.* **1989**, *266*, 349.

ELECTROREDUCTIVE CARBON-HALOGEN BOND FISSION OF ALKYL HALIDES. MICHAEL-TYPE ADDITION OF RADICALS GENERATED BY ONE-ELECTRON REDUCTION

Hideo TANAKA, Yukihiro MISUMI, and Sigeru TORII*

Department of Applied Chemistry, Faculty of Engineering, Okayama University, Tsushima-Naka 3-1-1, Okayama 700, Japan

ABSTRACT

Electroreduction of various alkyl halides in the presence of α,β-unsaturated carboxylate was carried out. Requirements for selective one-electron reduction of alkyl halides leading to radicals will be discussed.

Electroreduction of alkyl halides has intensibly been investigated from mechanistic view points.[1] Recently, Saveant, *et.al.,* reported a precise kinetic study on the reductive cleavage of carbon-halogen bonds of a series of *n-*, *sec-*, *tert-*butyl halides (X = Br and I), showing that upon direct electroreduction, only *sec-* and *tert-*butyl iodides generate radicals through a dissociative one-electron transfer process while other alkyl halides take two electrons at once to give the corresponding carbanions.[2] Studies on the reactions of the radical species generated thus far have, however, scarcely appeared. In order to clarify a new synthetic aspect of the electrogenerated radicals, we investigated electroreduction of alkyl halides in the presence of α,β-unsaturated carboxylate as a radical acceptor. Herein, we describe electroreductive Michael-type addition of *tert-*butyl iodide to 2-phenylethyl acrylate **1** (Scheme 1) and its application to intramolecular carbon-carbon bond formation at *tert-*carbon atom.

Electrolysis was carried out in a divided cell fitted with platinum anode and cathode (1 x 1.5 cm^2). A typical procedure is as follows: Into both compartments of the cell was added a CH$_2$Cl$_2$ solution of Bu$_4$NBF$_4$. To the cathode compartment were added *tert-*butyl iodide and 2-phenylethyl acrylate **1** (3:1) and to the anode compartment was added cyclohexene. A constant current (3.3 mA/cm^2) was supplied under stirring at room temperature. After passage of 2.5 F/mol (based on **1**), work up of the catholytes gave the adduct **3** (65%) and a trace amount of **4** along with recovered **1** (10%).

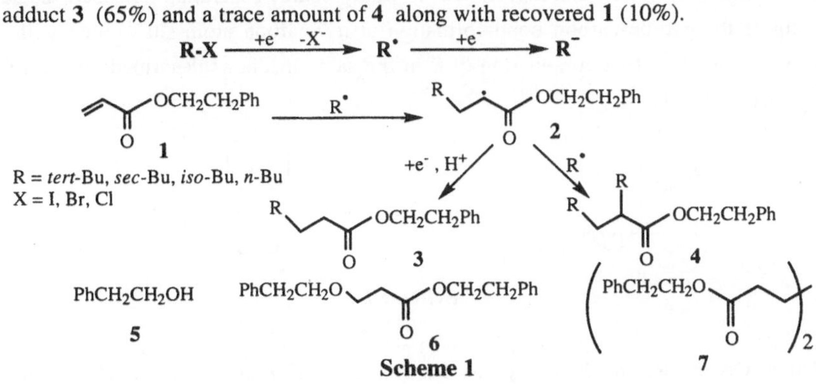

Scheme 1

In a similar manner, electroreduction of a series of butyl halides was attempted. The results are summarized in Table 1. The electroreduction of *sec*-, *iso*-, and *n*-butyl iodides (entries 2-4) afforded only 10-22% yields of adducts **3** together with 2-phenylethanol **5**, its Michael adduct **6** and hydrodimer **7** of the acrylate **1**. *tert*-Butyl bromide and chloride afforded a considerable amount of hydrodimer **7** (35%) and small amounts of **3, 5**, and **6** (less than 13% each). It is interesting to note that in the electroreduction of *tert*-butyl iodide (entry 1), no detectable amounts of **5, 6**, and **7**, which would be formed by the reaction of anionic species, provably, generated from two-electron reduction of butyl halides, and by competitive reductive dimerization of acrylate **1** in the electrolysis media, respectively. This suggests that one electron uptake of *tert*-butyl iodide takes place without further electron transfer and more effectively than that of **1** to afford *tert*-butyl radical which is subsequently allowed to react with **1** to give adduct **2** . The adduct **2** would undergo futher one-electron reduction followed by protonation to give **3**. The formation of a trace amount of **4**, which is reasonably explained by the reaction of **2** with *tert*-butyl radical, may also support the one-electron transfer process.

Table 1. Electroreduction of a Series of Butyl Halides with 2-Phenylethyl Acrylate **1**

Entry	R-X	Products Yield, %[a]					Recovered **1**
		3	**4**	**5**	**6**	**7**	
1	*tert*-Bu-I	65	trace	-	-	-	10
2	*sec*-Bu-I	22	-	15	13	11	13
3	*iso*-Bu-I	10	-	16	3	11	24
4	*n*-Bu-I	19	-	24	11	4	14
5	*tert*-Bu-Br	6	-	13	6	35	8
6	*tert*-Bu-Cl	8	-	10	7	35	10

a) Isolated yields based on **1**

The success in direct one-electron reduction of *tert*-butyl iodide generating *tert*-butyl radical enabled us to investigate the carbon-carbon bond formation at *tert*-carbon atom substituted with iodide as illustrated in Scheme 2. Electroreduction of **8** in the same media as described above afforded the cyclized product **9** (36%) .

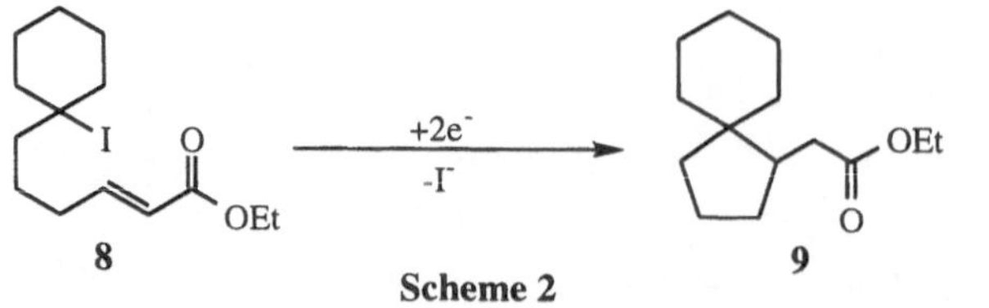

Scheme 2

References

1) D. C. Peters "Organic Electrochemistry", M. M. Baizer and H. Lund, Eds, Marcel Dekker, Inc., New York, **1991**, pp 361-400.

2) C. P. Andrieux, I. G. Gallardo, J.-M. Saveant, and K. Su, *J. Am. Chem. Soc.*, **1986**, 108, 638.

ULTRASONIC EFFECTS ON ELECTROCATALYTIC PROCESSES

Mahito ATOBE, Yoshifumi KADO, and Tsutomu NONAKA
*Department of Electronic Chemistry, Tokyo Institute of Technology, 4259
Nagatsuta, Midori-ku, Yokohama 226-8502, Japan*

ABSTRACT
Clear and significant effects of ultrasonic irradiation on reaction rates could be demonstrated in electrocatalytic processes such as photocatalytic and mediatory reactions, and it was found that the reaction rates were greatly enhanced by ultrasonic waves.

INTRODUCTION
Ultrasonic effects on electrochemical processes have received much attention in recent years.[1, 2] However, ultrasonic waves have not been applied to practical electrochemistry, particularly to electroorganic processes, except for electroplating.[1]

From the above point of view, we have aimed to explore clear effects of ultrasounds on a variety of electroorganic processes. In our previous work, significant ultrasonic effects on the current efficiency and / or product selectivity were also found in the reduction of aldehydes, olefins and carboxylic acids, and the effects could be rationalized as due to the promotion of mass transport of the substrate molecules to the electrode surface from the bulk solution by ultrasonic waves. [3]

Subsequently, in this work, effects of the ultrasonic irradiation on reaction rates could be demonstrated in electrocatalytic processes such as photocatalytic and mediatory reactions.

METHOD
The photocatalytic dehydrogenation of 2-propanol to acetone on a powdered TiO_2 (Anatase, 5 μm) was employed as a model reaction for examining ultrasonic effects (Scheme 1). The TiO_2 (10 mg) was suspended in an aqueous solution (50 cm^{-2}) of the substrate (2 mmol) in a pyrex glass tube. The solution magnetically stirred (1100 rpm) under Ar was irradiated at > 230 nm using an Hg-Xe lamp (500 mW cm^{-2} at 365 nm) with sonication (a 6 mm diameter stepped horn, 20 kHz, 20 W). The reaction products were analyzed by GC.

$$CH_3CH(OH)CH_3 \xrightarrow[- H_2]{TiO_2,\ UV} CH_3COCH_3 \qquad \text{(Scheme 1)}$$

The indirect reduction of benzyl bromide by a radical anion mediator electrogenerated from anthracene in a 0.1 M Bu_4NBF_4 / DMF catholyte (60 cm^3) was employed as a model reaction (Scheme 2). An H-shaped divided cell equipped with platinum cathode (1 x 1 cm) and anode (2 x 3 cm) was used for cyclic voltammetry. A stepped horn (a

1.3 cm diameter titanium rod) connected with a PZT oscillator (20 kHz) was inserted into the cathode chamber.

$$2 \left[C_{14}H_{10} \right]^{\bar{\cdot}} \xrightarrow{\text{Cathode}} 2\, C_{14}H_{10}$$

(Scheme 2)

RESULTS AND DISCUSSION

1. Photocatalytic Reaction

In order to examine an ultrasonic irradiation effect on the formation rate of acetone, the photocatalytic reaction of 2-propanol was carried out using three different reaction modes such as sonication with light, nonsonication with light and sonication without light. As shown in Figure 1, the formation rate under sonication is higher than that under nonsonication. On the other hand, acetone was scarcely formed when the reaction was carried out under sonication without light. From these results, it is clarified that the photocatalytic reaction was enhanced by ultrasound.

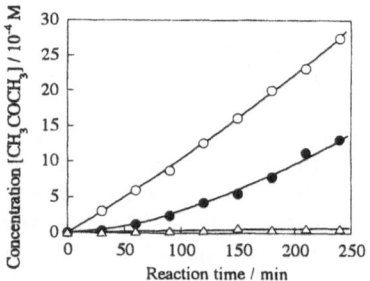

Figure 1. Photocatalytic reaction of 2-propanol to acetone under sonication with light (○), nonsonication with light (●) and sonication without light (△).

2. Redox-mediatory Reaction

Figure 2 shows cyclic voltammograms of anthracene in the presence of benzyl bromide under various stirring modes. Anthracene gives a single irreversible cathodic current peak which indicates the mediatory reduction of benzyl bromide under a still mode. Current for the reduction under ultrasonic irradiation is much larger than those under mechanical stirring and still standing.

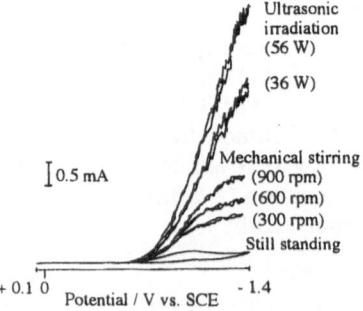

Figure 2. Cyclic voltammograms of anthracene in the presence of benzyl bromide under various stirring and ultrasonic irrdiation modes.

References
1) D. J. Walton and S. S. Phull, "*Advances in Sonochemistry Volume 4* ", T. J. Mason, ed., JAI Press Ltd., Connecticat, **1996**, 205.
2) T. Nonaka and M. Atobe, *Chemical Industry*, **1996**, *47*, 615.
3) M. Atobe, K. Matsuda, and T. Nonaka, *Electroanalysis*, **1996**, *8* , 784.

ELECTROCHEMICAL DARZENS SYNTHESIS

Eugene A.BERDNIKOV, Vakhid A.MAMEDOV and Vitalij V.YANILKIN
Department of Organic Chemistry, Kazan State University,
18 Kremlevskaya Str., Kazan, 420008, Russia
Arbuzov Institute of Organic and Physical Chemistry,
8 Arbuzov Str., Kazan, 420088, Russia

ABSTRACT

Cathodic reduction of mono-, di- and trichloroacetophenones in the presence of aromatic aldehydes in dimethylformamide with 0.1 M $NaClO_4$ as a supporting electrolyte yields epoxyketones as the main products in a Darzens reaction.

INTRODUCTION

A classical version of Darzens' condensation represents the reaction of aldehydes or ketones with esters of α-halogencarbonic acids in the presence of a base to afford α,β-epoxyesters.

$$R - \underset{R^1}{\underset{|}{C}} = O + Hal\underset{R^2}{\underset{|}{C}}H - COOR^3 \xrightarrow{\text{Base}} \underset{R}{\overset{R^1}{\diagdown}} C \overset{O}{\diagup\diagdown} C \underset{COOR^3}{\overset{R^2}{\diagup}}$$

In Darzens' condensation a pivotal intermediate is α-halogencarbanion which is generated from a suitable CH-acid under the action of a base or from dichloromethane derivatives according to the retro-Claisen reaction.

Undoubtedly α-halogencarbanions can be obtained by electrochemical reduction of polyhalogenated compounds [1]. Compared to the generation of α-halogencarbanions chemically the electrochemical method has the advantage that it avoids strong and frequently expensive bases. Earlier [2] it was shown that electroreduction of diethyl(trichloromethyl)phosphonate in the presence of aldehydes or ketones afforded substituted 1,1-dichloroethenes as the result of the Wittig-Horner reaction of electrochemically generated the α,α-dichlorophosphonatecarbanion with carbonyl compounds. Similarly triethyl phosphonodichloroacetate [2b] was reduced in the presence of benzaldehyde or cyclohexanone to yield the α-chloro-α,β-unsaturated esters. An amalgam alkali metal was employed as a reducing agent in the reaction of the methyl α-bromoacetate with the acetone in order to obtain the methyl β,β-dimethylglycidate [3]. Our interest in the Darzens' condensation [4] led us to test the electrochemical Darzens' reaction with some chloroacetophenones and aromatic aldehydes under various conditions.

RESULTS AND DISCUSSION

The preparative electroreduction of mono-, di- and trichloroacetophenones in the presence of aromatic aldehydes has been carried out in the diaphragmal cell in potentiostatic and galvanostatic regimes with the perdeuterodimethylformamide (DMF-d_7) as a solvent and 0.1 M $NaClO_4$ as a supporting electrolyte. The choice of the potential of electrolysis was determined on the basis of polarographic data of chloroacetophenones. The results of the reduction of chloroacetophenones **2**, **3** and **4** in the presence of a number of aromatic aldehydes are presented in the Table 1. The reactions gave a variety of products depending on the type of the used aldehyde and chloroacetophenone. It also depended on the potential of the electrolysis and on the presence of proton donors. The main products were identified as follows:

$PhC(O)CH_3$ (**1**), $PhC(O)CH_2Cl$ (**2**), $PhC(O)CHCl_2$ (**3**), Ar-CH-CH-C(O)Ph (**5**), ArCH(Cl)-C(O)-C(O)Ph (**6**). In addition to these main products we have identified substituted dioxine **7** (10% yield) from the reaction of dichloracetophenone **3** or trichloroacetophenone **4** with p-$CH_3OC_6H_4CHO$, furthermore in the reaction of trichloroacetophenone we have detected (quite unexpectedly) also as by-products benzoic acid **8** and chloroform **9**. The highest yields (52-72%) of epoxides **5** were obtained when concentration of aldehyde exceeded concentration of chloroacetophenone by 1.5 - 2 times, whereas the lowest yields were obtained when the ratio aldehyde/chloro-acetophenone was the opposite. It is suggested that this could be explained by the fact that with the excess of aldehyde side-reactions of α-halogencarbanions with original chloroacetophenones were suppressed. When the electrolysis was conducted in halvanostatic regime the yield of the products of the reduction of chloroacetophenones was increased (**2** or **3** ——-> **1** or **2**). The presence of proton donors (H_2O or PhOH) in the electrolysis solution also increased the yield of **1** or **2**. In the case of trichloroaceto-phenone H_2O changes the direction of reaction dramatically: benzoic acid (with the yield of 48%) and chloroform (with the yield of 50%) are being almost exclusively formed.

The reaction sequence of the electrochemical Darzens' condensation and accompaning reactions is as follows:

$$PhC(O)CH_2Cl \xrightarrow{+2e} PhC(O)\bar{C}H_2 \xrightarrow{+\mathbf{2}} PhC(O)CH_3 + PhC(O)\bar{C}HCl$$
$$(\mathbf{2}) \qquad\qquad (\mathbf{1A}) \qquad\qquad (\mathbf{1}) \qquad (\mathbf{2A})$$

$$PhC(O)CHCl_2 \xrightarrow{+2e} PhC(O)\bar{C}HCl \xrightarrow{+\mathbf{3}} \mathbf{2} + PhC(O)\bar{C}Cl_2$$
$$(\mathbf{3}) \qquad\qquad (\mathbf{2A}) \qquad\qquad (\mathbf{3A})$$

Table 1. Electrochemical Darzens' reactions [a]

N	ArCHO (mmol)	$PhC(O)CH_nCl_{3-n}$ (mmol)	E(V) vs SCE	F/mol	Yeilds, %				
					1	**2**	**3**	**5**	**6**
1	PhCHO (1.9)	$PhC(O)CH_2Cl$ (1.3)	-1.7	1.2	40	—	—	43	—
2	PhCHO (1.9)	$PhC(O)CH_2Cl$ (1.3)	b)	1.2	42	—	—	48	—
3	$mNO_2C_6H_4CHO$ (1.8)	$PhC(O)CH_2Cl$ (1.2)	-1.7	1.1	41	—	—	47	—
4	PhCHO (1.5)	$PhC(O)CHCl_2$ (2.2)	-1.3	1.2	—	15	—	31	18
5	PhCHO (1.5)	$PhC(O)CHCl_2$ (1.5)	-1.3	1.2	—	12	—	42	13
6	PhCHO (1.9)	$PhC(O)CHCl_2$ (1.2)	-1.3	1.3	—	8	—	65	9
7	PhCHO (1.9)	$PhC(O)CHCl_2$ (1.2)	b)	1.1	5	9	—	70	11
8 c)	PhCHO (1.9)	$PhC(O)CHCl_2$ (1.2)	-1.3	1.1	—	31	—	52	—
9	$pMeOC_6H_4CHO$ (2.0)	$PhC(O)CHCl_2$ (1.3)	-1.3	1.2	—	18	—	42	—
10	$mNO_2C_6H_4CHO$ (1.3)	$PhC(O)CHCl_2$ (0.7)	-1.3	1.1	—	7	—	72	9
11	PhCHO (1.9)	$PhC(O)CCl_3$ (1.2)	b)	1.3	tr.	5	17	20	35
12	$mNO_2C_6H_4CHO$ (1.8)	$PhC(O)CCl_3$ (1.1)	-1.0	1.2	—	—	9	10	42
13 c)	$mNO_2C_6H_4CHO$ (1.8)	$PhC(O)CCl_3$ (1.1)	-1.0	0.3	—	—	—	5	<2

a) Mercury cathode, 0.1 mol/l $NaClO_4$, DMF-d_7 as a solvent, 21-23°C

b) In galvanostatic regime: I=20 mA, τ=7200 s (runs 2 and 11) and I=20mA, τ=6000 s (run 7)

c) In the presence of H_2O (2.8 mmol)

$$PhC(O)CCl_3 \xrightarrow{\quad +2e \quad} PhC(O)\overset{-}{C}Cl_2 \xrightarrow[-\overset{-}{D}]{\quad +HD \quad} \underline{\mathbf{3}}$$
$$\quad (\underline{\mathbf{4}}) \qquad\qquad\qquad (\underline{\mathbf{3A}})$$

$$\underline{\mathbf{2A}} + ArCHO \longrightarrow ArCH\!\!-\!\!\underset{\diagdown O \diagup}{CH}\!\!-\!\!C(O)Ph + \overset{-}{C}l$$
$$(\underline{\mathbf{5}})$$

$$\underline{\mathbf{3A}} + ArCHO \longrightarrow ArCH\underset{\underset{O}{|}}{-}CCl_2 - C(O)Ph \longrightarrow ArCH\!\!-\!\!\underset{\diagdown O \diagdown Cl}{\overset{C(O)Ph}{C}}$$

$$\underset{PhC(O) \quad O \quad Ar}{\underset{\overset{\|}{C} \quad \overset{\|}{C}}{\overset{Ar \quad O \quad C(O)Ph}{\overset{\|}{C} \quad \overset{\|}{C}}}} (\underline{\mathbf{7}}) \longleftarrow ArCH - \underset{\overset{\|}{O}}{\overset{}{C}} - \underset{\overset{\|}{O}}{\overset{}{C}} - Ph$$
$$\underset{Cl}{} \qquad (\underline{\mathbf{6}})$$

$$\underline{\mathbf{4}} + \overset{-}{O}H \longrightarrow PhC\overset{-}{O}_2 + CHCl_3$$
$$(\underline{\mathbf{8}}) \quad (\underline{\mathbf{9}})$$

EXPERIMENTAL

Preparative electrolysis was carried out in a glass cell with cathode and anode compartments, separated by a diaphragm (cellulose) at room temperature in argon atmosphere. Hg-pool (s = 1.8 cm^2) and platinum wire were used as working electrode and anode, correspondingly. In a typical experiment the cathode compartment was filled with 3 ml of DMF- d$_7$ containing 1-2 mmol of **2** (or **3**, **4**), 1-2 mmol of an aldehyde and 0.037 g of NaClO$_4$ .In some experiments 0.05 ml of H$_2$O was also added. The anode compartment was filled with 1 ml of DMF-d$_7$, containing 0.012 g of NaClO$_4$ and K$_2$CO$_3$. Dissolved oxygen was removed from the catholyte by argon for 30 minutes prior to the electrolysis. After the completion of electrolysis the catholyte was analysed by NMR ^{1}H (or ^{13}C) spectra to determine the composition and the yields of the products obtained.

References
1) Organic Electrochemistry, Ed. M.M.Baizer, H.Lund. Marcel Dekker, Inc., New York, Basel, **1984**, 123.
2) (a) W.J.M. van Tilborg and C.J.Smit *Rec.*, **1980**, *99*, 202. (b) F.Karrenbrock, H.J.Schafer and I.Langer *Tetrahedron Lett.*, **1979**, 2915.
3) M.Kawamata and S.Fujikake *Chem.Abstr.*, **1973**, *78*, 2147-87.
4) (a) V.A.Mamedov, E.A.Berdnikov, I.A.Litviniv, and L.G.Kuzmina *Izvestiya AN Ser.Chim.*, **1995**, 1294. (b) E.A.Berdnikov and V.A.Mamedov *Izvestiya AN Ser.Chim.*, **1995**, 1544.

ELECTRO-ORGANIC REACTIONS. PART 48: PATHWAYS FOR CARBON-CARBON BOND FORMATION: THE REDOX CHEMISTRY OF QUINODIMETHANES AND OF ALKENYL-SUBSTITUTED HETEROAROMATICS

James H.P. UTLEY[a], Merete Folmer NIELSEN[b], Robert G. JANSSEN[a], Xavier SALVATELLA[a], Sabine SZUNERITS[a], Erada OGUNTOYE[a] and Peter B. WYATT[a]

(a) Department of Chemistry, Queen Mary and Westfield College (University of London), Mile End Road, London E1 4NS, UK; (b) Department of Chemistry, University of Copenhagen, Symbion Science Park, Fruebjergvej 3, DK-2100 Copenhagen Ø, Denmark

ABSTRACT: Carbon-carbon bond-forming reactions involving cathodic reduction of quinonemethides, electrogeneration of quinodimethanes and reduction of some vinylpyridines and vinylquinolines, are compared and contrasted. The quinonemethides undergo "conventional" electrohydrodimerisation *via* radical-anion coupling. Quinodimethanes (QDMs), formed by cathodic elimination from 1,4- and 1,2-di-(halomethyl)arenes, react usefully to give polymers or, for *o*-quinodimethanes, as dienes in Diels-Alder reactions. QDMs are electroactive and have been characterised electrochemically and spectroscopically. The vinylheteroaromatics give *trans*-1,2-di-heteroarylcyclobutanes as major products and these are shown to arise from an overall 0 F reaction (catalytic chain process) but with the key initial step being reaction between radical-anion and the neutral starting material.

INTRODUCTION: Carbon-carbon bond-forming reactions are the most important of organic synthetic conversions and electrochemical methods have much to offer in this regard. The Kolbe reaction (anodic oxidation) and electrohydrodimerisation (EHD, cathodic reduction) are justifiably still the subject of much synthetic and mechanistic effort. Recently the Queen Mary (London) and Copenhagen groups have explored other reactions in which electrogenerated intermediates undergo useful carbon-carbon bond-forming reactions. This article seeks to review that work both from a synthetic and mechanistic viewpoint.

The key electroactive starting materials are quinonemethides (**1**), 1,4-(halomethyl)arenes (e.g. **2**), and vinylheteroaryls (e.g. **3**). Compounds of type (**2**) undergo cathodic elimination to give as reactive intermediates quinodimethanes (e.g. **4**).

(**1**)   (**2**)   (**3**)   (**4**)

RESULTS AND DISCUSSION: *EHD of Quinonemethides*: These reactions are briefly discussed because their redox chemistry led to consideration of the quinodimethanes (**4**) as species which could be studied electrochemically; quinones, quinonemethides and quinodimethanes form a structural series in which the carbonyl oxygen is successively replaced by CH_2. They are all relatively easily reduced to aromatic radical-anions. The radical-anions formed from quinonemethides can undergo dimerisation[1] in good yields but with little stereoselectivity (Table 1). The rates of dimerisation[2] are quite slow and the radical-anions have significant lifetimes when combination is sterically hindered. The general improbability of a radical-anion/substrate route will be discussed later.

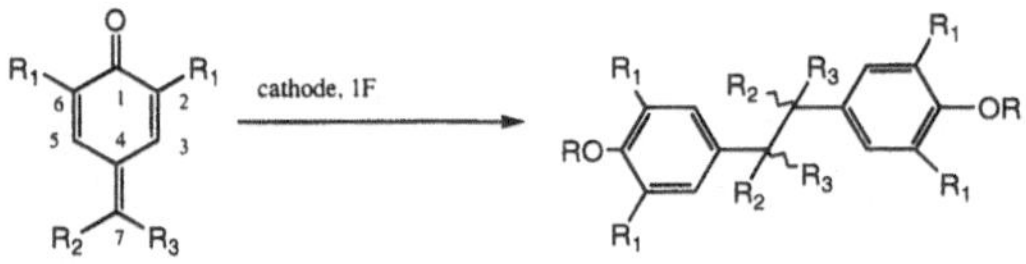

Table 1. Cyclic voltammetric[a], data for quinonemethide reduction

R_1	R_2	R_3	$-E^0$	$-E_{pc}(1)$	$-E_{pc}(2)$	$-E_{pa}(1)$	$10^4 k_2$ $M^{-1} s^{-1}$
t-Bu	H	Ph	1.155[b]	1.205[c]	1.962[c]	1.12[c]	22.5[d]
Me	H	Ph	-	1.073	1.789	-	
t-Bu	C_6H_{11}	Ph	1.397	1.436	-	1.359	*ca.* 5
t-Bu	H	t-Bu	-	1.468[f]	-	-	
t-Bu	t-Bu	Ph	1.387[e]	1.428	1.848	1.348	
t-Bu	Ph	Ph	1.130				

a. Hg/Pt cathode, DMF-Bu$_4$NPF$_6$ (0.1M), potentials *vs.* SCE, substrate 2mM, v = 0.3 V s^{-1}; b. v = 100 V s^{-1}; c. v = 9 V s^{-1}; d. Yield of hydrodimer = 80% (R=Me), *meso* : (±) ratio = 2 (R = Me) and 1 (R = H); e. v = 1 V s^{-1}; f. quasi-reversible at 100 V s^{-1}

The analogous vinylquinonemethides are more reactive but can be electrochemically reduced following generation by *in situ* oxidation of the phenolic precursors. Coupling of the radical-anions is rapid and regioselective; the example given (Scheme 1) is of the preparation of an analogue of the natural product ocimin in which only the required 9-9'-coupled product is formed[3].

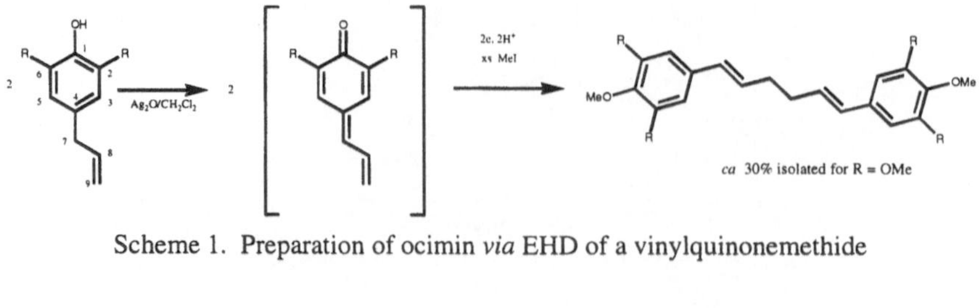

Scheme 1. Preparation of ocimin *via* EHD of a vinylquinonemethide

Electrogenerated quinodimethanes as reactive intermediates: The early work of Gilch[4] and of Covitz[5] established that 1,4-reductive elimination of the bis-(bromomethyl)arenes of type (**2**) gave quinodimethanes which underwent fairly rapid polymerisation. This carbon-carbon bond-forming process has more recently been heavily exploited[6] as a versatile and convenient method for the formation of highly regular poly-*p*-xylylenes (PPXs) and the conjugated analogues, poly-(*p*-phenylene)vinylenes (PPVs) - see Scheme 2.

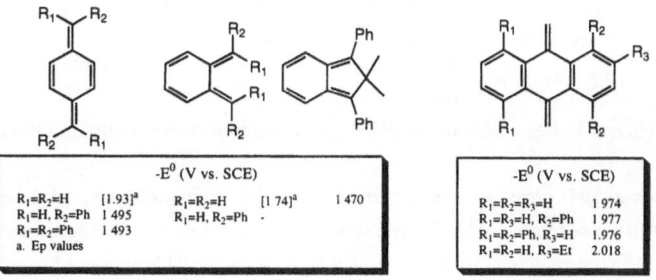

Scheme 2. Polymer formation *via* electrogenerated quinodimethanes.

The work on polymerisation suggested that the intermediate quinodimethanes had radical-like reactivity; the unsymmetrical quinodimethane formed from 2-acetoxy-1,4-(bromomethyl)benzene gave the three possible modes of coupling in statistical proportions[6a]. This has led to attempts at characterising quinodimethanes using electrochemical and spectroscopic methods.

The quinodimethanes formed by reduction of the bis-(bromomethyl)arenes are detectable by cyclic voltammetry in DMF solution and themselves give rise to radical-anions which in most cases react slowly enough to allow the measurement of formal reduction potentials (E^0 values); the parent *o*- and *p*-xylylenes do not exhibit reversible reduction at up to 100 V s-1. In this manner information on the reduction potentials of many quinodimethanes has been obtained[7] and the data are summarised below.

The cyclic voltammogram shown in Figure 1 indicates the clarity with which the intermediates are detected and characterised. The first wave is poorly defined (characteristic of benzyl bromide reductions at Hg) but the second and third waves are well-defined and correspond to the formation of the quinodimethane radical-anion and dianion respectively.

$-E^0$ (V vs. SCE)		
$R_1=R_2=H$ [1.93]^a	$R_1=R_2=H$ [1 74]^a	1 470
$R_1=H, R_2=Ph$ 1 495	$R_1=H, R_2=Ph$ -	
$R_1=R_2=Ph$ 1 493		
a. Ep values		

$-E^0$ (V vs. SCE)	
$R_1=R_2=R_3=H$	1 974
$R_1=R_3=H, R_2=Ph$	1 977
$R_1=R_2=Ph, R_3=H$	1.976
$R_1=R_2=H, R_3=Et$	2.018

The anthracene-derived quinodimethanes, generated in these cases by electroreduction in d_6-DMSO solution of the 9,10-di-(chloromethyl)anthracenes, are sufficiently long-lived to allow characterisation by ^{1}H NMR spectroscopy. Furthermore, kinetic measurements were made[8] by generation of the quinodimethane by rapid electrolysis in a flow cell with following of the subsequent decay of the quinodimethanes by rotating disc voltammetry. The quinodimethanes derived from 9,10-di-(chloromethyl)anthracene and the corresponding 2-ethyl and 2,3-diheptyl compounds decay with second-order kinetics and with rate coefficients in the range 300 - 450 $M^{-1}s^{-1}$.

Electrogenerated QDM from
Ph(Br)CHC$_6$H$_4$CH(Br)Ph; 100 V s^{-1}

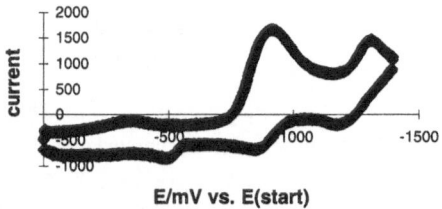

Figure 1. Voltammetry of Ph(Br)CHC$_6$H$_4$CH(Br)Ph at an Hg-Pt cathode in DMF-Bu$_4$NPF$_6$ (0.1M)

The bromides have mostly been used as precursors for quinodimethane generation; the chlorides reduce at more negative potentials at which the quinodimethanes are further reduced. But the chlorides can be used in some cases where redox catalysis operates. Redox mediators such as benzophenone work or where the first-formed quinodimethane gives rise to a persistent radical-anion it will effectively catalyse reduction of the chloromethyl substrate *in situ*.

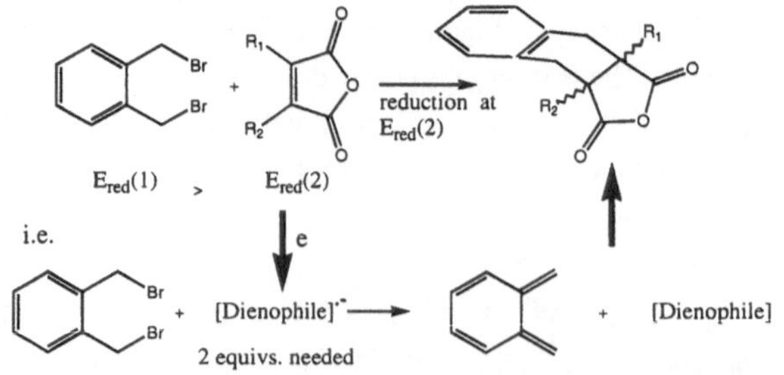

Scheme 3. Redox-catalysed electrogeneration of *ortho*-quinodimethane

Such redox catalysis turns out to be crucial for a preparative application of the electro-generation of *ortho*-quinodimethanes. Early experiments[9] in which 1,2-di-(bromomethyl)benzene was co-electrolysed with maleic anhydride showed that the Diels-Alder adduct from *ortho*-

quinodimethane and the anhydride was formed but at the potential of reduction of the anhydride, which was less negative than that of the bromomethyl substrate. The inference is that the radical-anion from the anhydride is mediating electron transfer to the quinodimethane precursor; the principle is illustrated in Scheme 3.

Compelling evidence for the mediated route is the observation of the characteristic voltammetric behaviour in such cases. In the presence of the less easily reduced substrate the reversibility of reduction of the mediator is lost and a catalytic current is observed which is proportional to the concentration of substrate. The example shown here (Figure 2) is for reduction of 2,3-dimethyl maleic anhydride in the presence of 1,2-di-(bromomethyl)benzene.

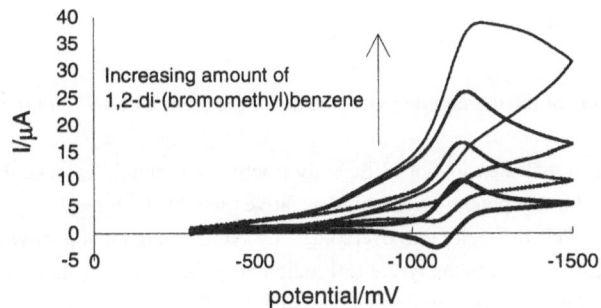

Figure 2. Voltammetric evidence of redox-catalysis in *ortho*-quiinodimethane formation

The electrochemical method for effecting Diels-Alder reactions between *ortho*-quinodimethanes and dienophiles which double as mediators is clean and convenient. Yields as high, or higher, than those obtained by zinc-dust conversion of the 1,2-di-(bromomethyl)arenes are found and the reaction conditions are cleaner and allow easier isolation of the adducts. And the scope of the method is large; a selection of the reactions accomplished[9,10] is given in Scheme 4; in most of the cases shown the yields of adduct were in the 60-80% range. Quinones can be seen also to act as effective mediators and dienophiles for quinodimethane generation.

The voltammetric evidence is overwhelmingly in favour of the redox-catalysis route indicated in Scheme 3. But there is another test which can be applied. It has long been known[11] that the radical-anion formed from diethyl maleate (*cis*) fairly rapidly isomerises to the radical-anion of diethyl fumarate (*trans*). Consequently it can be predicted that in the electrochemical Diels-Alder reaction between maleate esters and an *ortho*-quinodimethane, which will allow isomerisation of the mediator, the adduct will be that of the more stable fumarate ester. In contrast, zinc-dust formation of the quinodimethane in the presence of maleate should give retention of stereochemistry in the adduct.

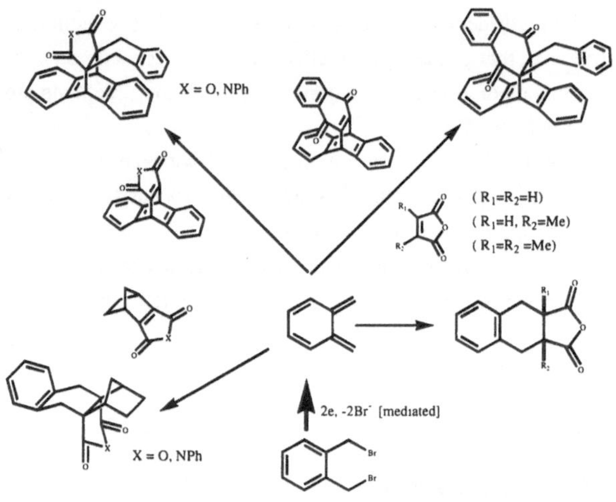

Scheme 4. The scope of the dienophile redox-mediated generation of *ortho*-quinodimethanes

Alkyl fumarates and maleates are not sufficiently reactive as dienophiles to be used in this test and in any event their reduction potentials are more negative than that of 1,2-di-(bromomethyl)benzene. This difficulty was overcome[12] by synthesising the diphenyl esters of maleic and fumaric acids and using 1,4-dimethoxy-2,3-di(bromomethyl)benzene as quinodimethane precursor. The relevant reduction potentials are given in Table 2

Table 2. Reduction potentials[a]

Compound	$-E_{pc}$ (V *vs.* SCE)	$-E_0$ (V *vs.* SCE)
1,2-Di-(bromomethyl)benzene	1.07	-
1,4-Dimethoxy-2,3-di(bromomethyl)benzene	1.27	-
Dimethyl maleate	1.42	-
Diphenyl maleate	1.18	1.09
Diphenyl fumarate	1.17	1.07

a. In DMF-Et$_4$NBr (0.1M); Gold cathode, substrates 3mM; scan rates 0.3 - 10 V s^{-1}

The results of the preparative-scale experiments were strongly indicative but not conclusive. Diels-Alder adducts were obtained using both the electrochemical and zinc-dust methods; the yields from the cathodic route were 85-97% and about 50% by the zinc-dust route. The electrochemical route gave exclusively the *trans*-adduct from diphenyl maleate and from diphenyl fumarate indicating that isomerisation had indeed taken place during electrolysis. The structure of the trans-adduct was confirmed by X-ray crystallography. But the control experiment (zinc-dust) resulted also in predominantly the *trans*-adduct irrespective of starting dienophile isomer. The product mixture contained 5-10% of the *cis*-adduct so it is highly probable that had stereochemistry been retained in the dienophiles used in the cathodic method that at least some *cis*-adduct would have been detected. The difficulties in this approach are probably the result of chemical isomerisation of the *cis*-adduct in the case of zinc-dust initiated reaction, e.g. through one of the products (ZnBr$_2$) acting as a Lewis acid to promote enolisation of the adduct and thereby isomerisation.

Non-EHD carbon-carbon bond formation in the electroreduction of vinylpyridines and vinylquinolines:

There is now much evidence[13,14] to suggest that electrohydrodimerisation (EHD) almost always involves dimerisation of radical-anions as the key step and there are no well-substantiated examples of a much-discussed alternative, the nucleophilic attack of radical-anion upon the neutral substrate. The reported EHD reactions[15] of 2- and 4-vinylpyridine to give the corresponding 1,4-di-pyridylbutanes would appear to be straightforward. However, a re-examination[16] of this reaction, with a view to probing stereoselectivity in related couplings, reveals that upon careful controlled potential electrolysis the major products are *trans*-1,2-di-pyridylcyclobutanes. 2-Vinylquinoline is similarly converted into the corresponding *trans*-1,2-disubstituted cyclobutane. In contrast to the photochemical formation of such products, *cis*-isomers are not formed in detectable amounts. In three cases the cyclobutane products have been characterised through X-ray crystallographic determination of their structures.

Other products in these reductions are the (expected) linear hydrodimer and the product of cathodic hydrogenation, e.g. 2-ethylpyridine from (**3**). Furthermore, co-electrolysis of two different vinylheteroaromatics with substantially separated reduction potentials gives at the less negative potential a significant amount of the cross-coupled cyclobutane product and only a trace of the homo-coupled product arising from the starting material of more negative reduction potential (Scheme 5).

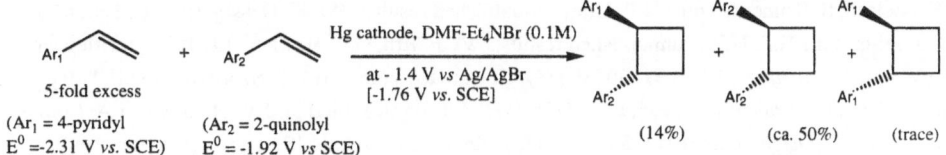

$(Ar_1 = 4\text{-pyridyl}$ $(Ar_2 = 2\text{-quinolyl}$
$E^0 = -2.31$ V *vs.* SCE) $E^0 = -1.92$ V *vs* SCE)

Scheme 5. Cross-coupling in co-electrolysis of vinylheteroaromatics

This is powerful evidence in support of a radical-anion substrate reaction, but the overall mechanism must account for the fact that the cyclobutanes are formally 0 F products!! The charge consumed in these reactions varies with substrate but in early experiments was found to be in the range 0.2 - 0.8 F. A major experimental difficulty is that the cyclobutanes are themselves further reduced, to linear hydrodimer, at potentials only 0.1 - 0.5 V more cathodic than for the starting material. Thus rigorous coulometry was carried out at the lowest practical reduction potential and a reaction profile of product composition versus charge consumption constructed. The result, found to be reproducible in multiple experiments, was conclusive. In experiments on 2-vinylpyridine, consuming 0.7 F, the linear hydrodimer (1 F product) accounted for 0.29 F and 2-ethylpyridine (2 F product) for 0.4 F, giving a total of 0.69 F. Consequently the cyclobutane (50% isolated yield) is indeed formed without overall charge consumption. The conclusion is that these reactions are true radical-anion/substrate reactions, albeit rather unusual. The most likely follow-up steps involve a catalytic chain reaction (Scheme 6), but the exact nature of the dimeric radical-anion (**5**) is uncertain.

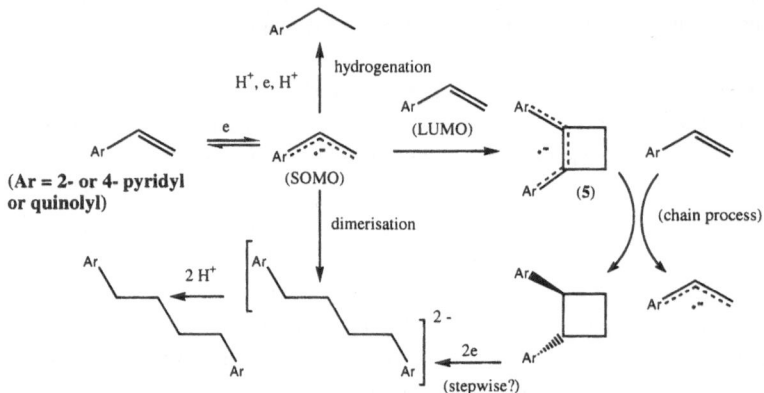

Scheme 6. Cyclobutane formation from vinylheteroaromatics

REFERENCES: *(1)*. M.O.F. Goulart and J.H.P. Utley, *Journal of Organic Chemistry,* 1988, **53**, 2520; R. Ludwig and J.H.P. Utley, unpublished results; *(2)*. By derivative cyclic voltammetry (M.F. Nielsen, Copenhagen); *(3)*. M.O.F. Goulart, C.Z. Smith and J.H.P. Utley, unpublished results; *(4)*. H.G. Gilch, *J. Polymer Science,* 1966, **4**, 1351; *(5)*. F.H. Covitz, *J. Am. Chem. Soc.,* 1967, **89**, 5403; *(6)*. (a) J.H.P. Utley, Y. Gao, J. Gruber and R. Lines, *J. Mater. Chem.,* 1995, **5**, 1297; (b) J.H.P. Utley, Y. Gao, J. Gruber, Y. Zhang and A. Munoz-Escalona, *J. Materials Chem.* , 1995, **5**, 1837; *(7)*. M.F. Nielsen, S. Szunerits and J.H.P. Utley, unpublished results; *(8)*. K. Daasbjerg, S.U. Pedersen, S. Szunerits and J.H.P. Utley, unpublished results; *(9)*. E. Eru, G.E. Hawkes, J.H.P. Utley and P.B. Wyatt, *Tetrahedron,* 1995, **51**, 3033; *(10)*. E. Oguntoye (neé Eru), S. Szunerits, J.H.P. Utley and P.B. Wyatt, *Tetrahedron,* 1996, **52**, 7771; *(11)*. A.J. Bard, J. Puglisi, J.V. Henkel and A. Lomax, *Trans. Faraday Discussions,* 1973, **56**, 353; *(12)*. Xavier Salvatella, MSc Thesis (London, 1997); *(13)*. F. Zhou and A.J. Bard, *J. Am. Chem. Soc.,* 1994, **116**, 393; *(14)*. I. Fussing, M. Güllü, O. Hammerich, A. Hussain, M.F. Nielsen and J.H.P. Utley, *J. Chem. Soc., Perkin Trans. 2,* 1996, 649; O. Hammerich and M.F. Nielsen, *Acta Chem. Scand.,* in press; *(15)*. J.D. Anderson, M.M. Baizer and E.J. Prill, *J. Org. Chem.,* 1965, **30**, 1645; *(16)*. R.G. Janssen, M. Motevalli and J.H.P. Utley, unpublished results

ACKNOWLEDGEMENTS: Partial support came from the EU Human Capital and Mobility Programme # ERBCHRXCT 920073 and #ERBCHBGCT 940590

INTRAMOLECULAR RADICAL CYCLIZATION REACTIONS - SCOPE AND LIMITATIONS FOR ELECTROCHEMICAL PROCESSES

James GRIMSHAW, Jadwiga T. GRIMSHAW, Mandy GIBSON and Marylène DIAS

School of Chemistry, Queen's University, Belfast BT9 5JP, Northern Ireland, UK

Radical-anions formed by one electron attachment to aryl halides undergo cleavage of the carbon-halogen bond in a unimolecular process to give an aryl σ-radical and halide ion. Aryl σ-radicals are highly reactive intermediates and the objective of our work is to find conditions under which they can be trapped in an intramolecular reaction by an adjacent phenyl or alkene substituent. Further steps lead to a stable cyclised product and the process is of interest in synthesis. The corresponding intermolecular reactions between phenyl radicals and either benzene or an alkene have been shown to have bimolecular rate constants[1] in the range 10^5 to 10^8 $M^{-1}s^{-1}$ so that the related intramolecular and unimolecular processes are expected to be very fast. Alternative reactions for the aryl σ-radical intermediates include abstraction of a hydrogen atom from the solvent and further electron transfer at the electrode surface to form a carbanion which undergoes protonation.

All the processes described here were carried out under conditions of constant current density in acetonitrile containing tetraethylammonium tetrafluoroborate in order to simulate a potential industrial process. When we propose the electrochemical method for general synthesis it is absolutely necessary to point out the limitations imposed through the relative kinetics of competing reactions.

Cyclization onto an arene ring

In previous publications[2] we have given many examples of the intramolecular reaction of aryl σ-radicals with an aromatic ring. Reduction of (1) in acetonitrile illustrate this process. Reaction can lead to two products (2) and (3) as a consequence of the competing reactions of the σ-radical intermediates. Reduction of (1, R = H) at a mercury cathode in a divided cell leads to a mixture of (2, R = H) and (3, R = H). This type of reaction can also be carried out at a carbon-steel or a stainless-steel cathode in an undivided cell with a sacrificial magnesium anode.[3] Under these conditions the yield of (2) is essentially quantitative. Using the substrate (1, R = F) we can only detect one product (2, R = F) by ^{19}F-nmr spectroscopy on the crude reaction mixture. This presents a striking demonstration of our previous findings that exploration of different electrode materials can point towards improved conditions for the cyclization process.

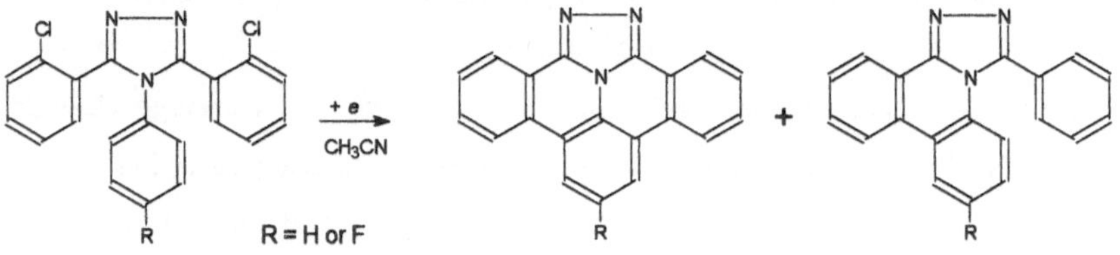

(1)   (2)   (3)

Cyclization onto an alkene

The conversion of (4) to (5) by cyclization of the derived σ-radical is a useful dihydroindole synthesis. Other groups[4,5] have investigated the generation of the required intermediate σ-radical by reaction between the appropriate bromo-compound and tributyltin hydride. This reaction requires a radical initiator to generate the tributyltin radical which will abstract a bromine atom from the substrate. Our objective was to carry out the conversion of chloro-compounds to dihydroindoles in a more environmentally friendly manner. For the process to be useful it must tolerate substituents R_1 and R_2 as both electron withdrawing and electron donating groups, including also hydrogen. The presence of electron donating groups increases the rate of carbon-halogen bond cleavage from the first formed radical-anion so that the σ-radical intermediate is formed closer to the electrode surface and may undergo further electron transfer before cyclization.

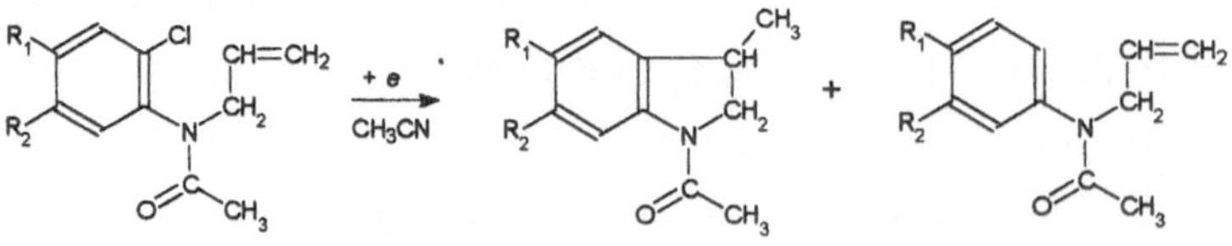

(4)   (5)   (6)

Reduction of two substrates with electron withdrawing nitrile substituents (4, R_1 = CN, R_2 = H or *vice versa*) at a mercury cathode in a divided cell gave the corresponding cyclised product (5) together with some of the acetanilide (6). Samples of (5) were isolated by chromatography and the structures confirmed by ^{1}H-nmr spectroscopy. The CH$_3$CH arrangement is easily recognised. Reduction of N-allyl-2-chloroacetanilide (4, R$_1$ = R$_2$ = H) yielded only N-allylacetanilide and none of the dihydroindole (5, R$_1$ = R$_2$ = H). This illustrates how substituents can influence the course of competing reactions. The half-life of the initially formed radical-anion is critical to successful cyclization. If this half-life (τ_1) is very short then a high concentration of the intermediate σ-radical is formed close to the electrode surface. This intermediate can diffuse back to the electrode and be reduced further when the half-life for cyclization is longer than τ_1.

The technique of indirect electron transfer was used to achieve cyclization of (4, R$_1$ = R$_2$ = H). Reduction was carried out with the addition of *trans*-stilbene at a mercury cathode. Under these conditions, *trans*-stilbene radical-anion is formed by electron transfer at the cathode. This radical-anion migrate from the cathode and transfers an electron to the substrate. Carbon-halogen bond cleavage then leads to a σ-radical which undergoes intramolecular cyclization faster than addition of a second electron ion a bimolecular reaction with more stilbene radical-anion. The product can be isolated by chromatography. Other transfer reagents have also been used.

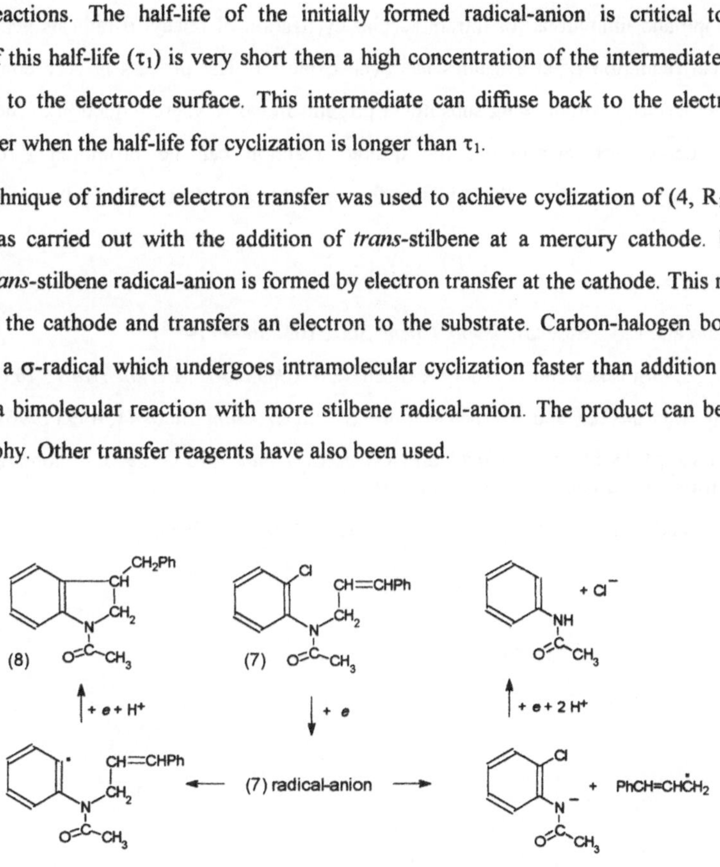

Reduction of 2-chloro-N-cinnamanylacetanilide (7) lead to an unexpected result. Again the reaction was carried out with the addition of *trans*-stilbene at a mercury cathode. The major product was acetanilide with only a low yield of the cyclised product (8). The latter was recognised by its ^{1}H-nmr spectrum. These products arise because the radical-anion of (7) can fragment by two competing reactions. Conversion of the bromo-compound related to (7) into(8) has been effected[5] by reaction with tributyltin hydride where the phenyl σ-radical is formed by abstraction of an atom of bromine and not *via* the radical-anion.

Conclusions

The scope and limitations for intramolecular cyclization of phenyl σ-radicals generated by the electrochemical reduction of aryl halides has been explored. The process is very effective for aryl halides with an electron withdrawing substituent present. In cases where reduction of the intermediate σ-radical competes with cyclization, the desired reaction can be promoted through indirect homogeneous electron transfer to the substrate from a stable and electrochemically generated radical-anion.

References

1) J. C. Scaiano and L. C. Stewart, *J. Am. Chem. Soc.*, **105**, 3609 (1983).

2) S. Donnelly, J. Grimshaw and J. Trocha-Grimshaw, *J. Chem. Soc., Perkin Trans.*, 1557 (1993).

3) S. Donnelly, J. Grimshaw and J. Trocha-Grimshaw, *Electrochim. Acta*, **41**, 489 (1996).

4) K. Jones and J. M. D. Storey, *J. Chem. Soc. Chem. Commun.*, 1766 (1992); K. Jones and J. M. D. Storey, *Tetrahedron*, **49**, 4801 (1993).

5) J. T. Dittami and H. Ramanathan, *Tetrahedron Letters*, **29**, 45 (1988).

ELECTROSYNTHESIS OF HETEROCYCLIC COMPOUNDS AND OTHER ALTERNATIVE ELECTROSYNTHETIC PROCESSES

Fructuoso Barba.

Department of Organic Chemistry. University of Alcalá de Henares. 28871 Alcalá de Henares. Madrid. Spain

ELECTROSYNTHESIS OF 2-BENZHYDRYLIDENE-4,4-DIPHENYL-[1,3]OXATHIOLAN-5-ONE.

The cathodic reduction of 2-bromo-2,2-diphenylacetyl bromide (**1**) in dichloromethane and tetraethylammonium bromide on graphite cathode under argon atmosphere in a divided cell and with a layer of solid sodium thiosulphate on the diaphragm in the anodic side (to prevent the diffusion into the cathodic compartment of the bromine formed in the anodic process) led to an unexpected and surprising heterocyclic compound which, after work up of the catholyte, was isolated and identified as 2-benzhydrylidene-4,4-diphenyl-[1,3]oxathiolan-5-one (**2**).[1]

To determine how a slufur atom has been put in the product, some chemical and electrochemical reactions were carried out[2].

On the other hand, in the literature decomposition of thiosulfuric acid in anhydrous media was described [3]. The anodic compartment always has acidic character, therefore it is logical to consider the formation of thiosulfuric acid, which could supply hydrogen sulfide by decomposition (its smell was detected). Thus, in the anode sodium thiosulfate forms thiosulfuric acid, which decompses to hydrogen sulfide and this one passes through the diaphragm to the catholyte and from there to the cathode surface, where the substrate is reduced to **6**. The latter is a strong base and takes protons from hydrogen sulfide to form a sulfide anion. This anion reacts with starting material, in a redox process with further cross coupling, yielding the thiolate anion **11**, which can undergo either intramolecular cyclization and decarbonylation, giving thiobenzophenone **8**, or fast reaction with diphenylketene **3**, followed by cyclization, leading to the oxathiolanone **2**.

The reduction of **1** in the presence of a saturated solution of H_2S led to a mixture of diphenylketene and products formed by reaction of intermediates with the solvent[4].

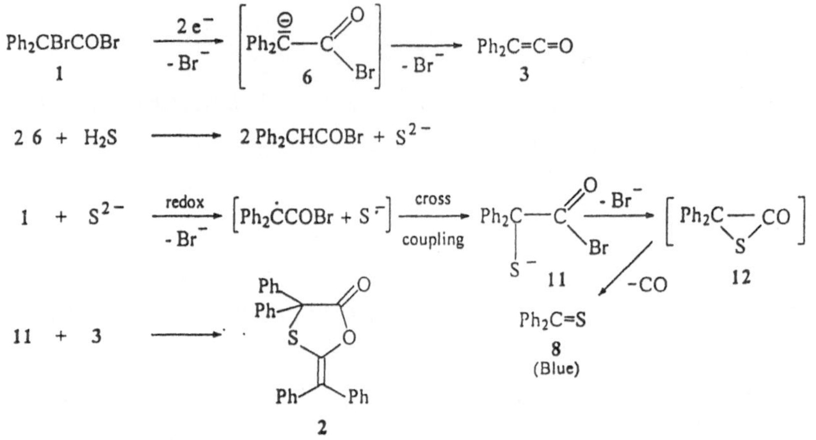

$$Ph_2CBrCOBr \xrightarrow[-Br^-]{2e^-} \left[Ph_2\overset{\ominus}{C}-\overset{O}{\underset{Br}{C}} \right] \xrightarrow{-Br^-} Ph_2C=C=O$$
1   6   3

$$2\ 6 + H_2S \longrightarrow 2\ Ph_2CHCOBr + S^{2-}$$

$$1 + S^{2-} \xrightarrow[-Br^-]{redox} \left[Ph_2\dot{C}COBr + S^{\cdot -} \right] \xrightarrow[coupling]{cross} Ph_2C-\overset{O \ \ -Br^-}{\underset{\underset{S^-}{Br}}{C}} \longrightarrow \left[Ph_2C-CO \atop S \right]$$
11   12

$$Ph_2C=S$$
8
(Blue)

11 + 3 $\longrightarrow$

(structure labeled 2: Ph, Ph on ring carbon; S, O ring; =C with Ph, Ph)

ELECTROSYNTHESIS OF TETRAHYDROFURANOLS

The electrochemical reduction of trans-cinnamaldehyde in protic medium (methanol-NaClO$_4$) on mercury cathode, gives four cyclic hydrodimeric compounds in 80% yield.

Controlled potential coulometry showed the consumption of 1Fmol^{-1}.

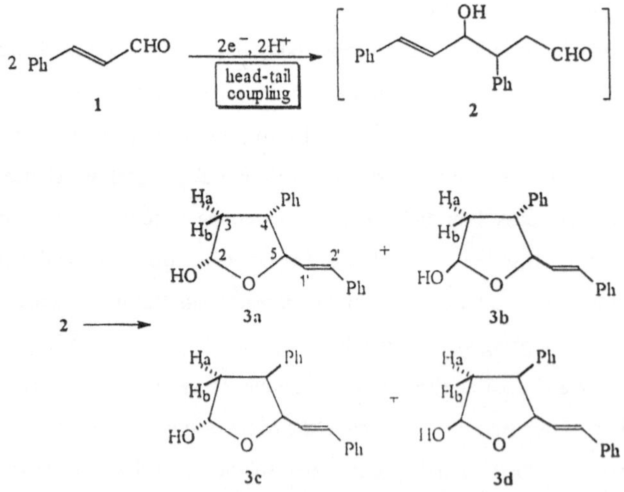

$$2\ Ph\text{—}CHO \xrightarrow[\text{coupling}]{2e^-,\ 2H^+,\ \text{head-tail}} \left[Ph\diagup\overset{OH}{\diagdown}\diagup CHO \atop Ph \right]$$
1   2

2 $\longrightarrow$

(structures 3a, 3b, 3c, 3d: tetrahydrofuran rings with Ha, Hb, Ph, HO, O substituents)

3a   +   3b

3c   +   3d

The first step involves one electron transfer to the substrate with formation of a radical anion. This radical anion may take more than one dimerization path depending on solvation effects and acidity of the medium. In either case, dimerization takes place in a "head to tail" fashion, to yield compound 2 which give the tetrahydrofurans 3a-d by intramolecular cyclization. The anomers in major proportion 3a and 3b (56% yield and in a ratio of 55:45 respectively).

The other pair of isomers 3c and 3d (24% yield and in a ratio 90:10 respectively).

The assignment of the NMR signals to different anomers 3a-d was carried out by the complete analysis of ^{1}H NMR, NOE measurements and 2D-COSY ^{1}H-^{1}H experiments. The assignment of ^{13}C NMR signals was completed with 2D-COSY ^{13}C-^{1}H data NMR

The electrohydrodimerization and subsequent cyclization of trans-cinnamaldehyde in protic medium affords a mixture of products with a ≈ 2:1 trans:cis relative configuration of phenyl and styryl groups in the cyclohydrodimer 3(a,b) : 3(c,d) coupling ratio. This stereochemistry is fixed in the formation of the linear hydrodimers 2 (C-C bond-forming step). Thus, in the transition state for C-C coupling, the staggered arrangement should be of lower energy than the eclipsed arrangement.

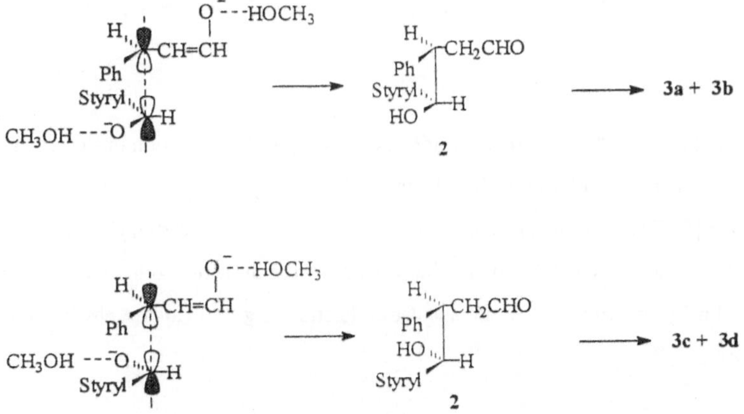

DIASTEREOSELECTIVE ELECTROSYNTHESIS OF (±)-(2R,4S,6R)-6-[(Z)-1'-BROMO-2'-PHENYLETHENYL]-2,4-DIMETHYLTETRAHYDROPYRAN-2,4-DIOL

In the cathodic reduction of α-bromo-cis-cinnamaldehyde in dry acetone and anhydrous LiClO$_4$ on mercury cathode[5] , the major reaction product obtained corresponds to a cyclic structure formed by the addition of two molecules of acetone to a molecule of substrate.

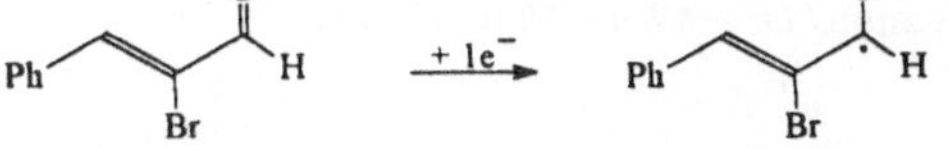

EGB + CH₃-CO-CH₃ ⇌ CH₃-CO-CH₂⁻ + [EBGH]· $\xrightarrow[\text{reduction}]{\text{further}}$ side products

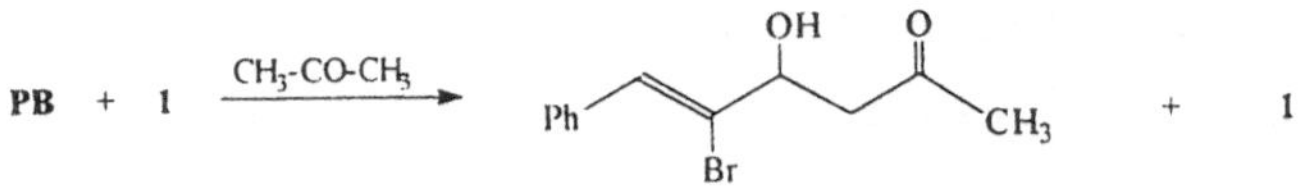

1

PB + 1 $\xrightarrow{\text{CH}_3\text{-CO-CH}_3}$

2

2 + 1 $\xrightarrow{\text{CH}_3\text{-CO-CH}_3}$

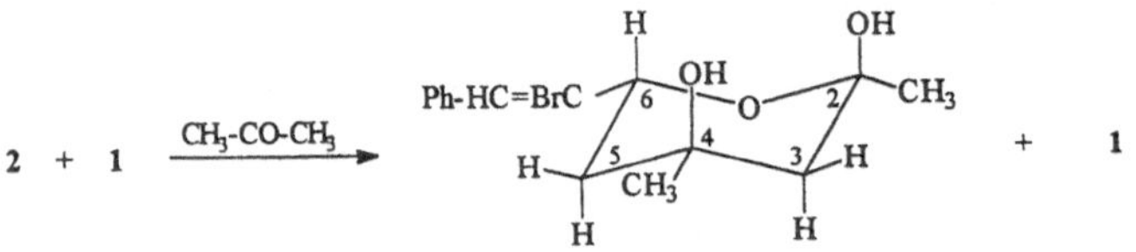

3

The first step involves the transfer of one electron and the formation of the radical anion, EGB. This EGB reacts in solution with a molecule of acetone by abstracting one proton to give the corresponding [EGBH]· and the enolate 1. This enolate 1 adds to the carbonyl group of another molecule of substrate to give 2(12%). The ketone 2 reacts again with another enolate 1 to give a new anion which finally undergoes an intramolecular cyclization to give 3 (the tetrahydropyran 41%) as main product.

References

[1] J. I. Lozano and F. Barba; *Tetrahedron Lett.* **1994**, *35*, 9623

[2] J. I. Lozano and F. Barba; *Tetrahedron* **1996**, *52*, 1259.

[3] A. F. Trotman-Dickerson. *Comprehensive Inorganic Chemistry*. Pergamon Press. **1973**; *2*, 884.

[4] J.I Lozano and F. Barba. *Electrochim. Acta.***1997**. *42*, 2173.

[5] F. Barba, J.L. de la Fuente and M.V. Galakhov. *Tetrahedron*. **1997**, *61*, 8662.

[6] F. Barba, J.L. de la Fuente; *J.Org. Chem.* **1996**, *61*, 8662.

REACTIVITY OF SUPEROXIDE ION WITH ORGANOHALOGENS AND CHALCONES

Ram A. Misra

Department of Chemistry, Faculty of Science,
Banaras Hindu University, Varanasi-221005, INDIA

ABSTRACT

The cathodically *in situ* generated superoxide anion radical a proven bioactive species is a multipotent reagent. As an effective nucleophile it reacts with organohalogens in an addition-elimination process to afford the corresponding phenols. The cleavage of chalcones (enones) to carboxylic acid(s) occurs in a facile step with superoxide mediated (O_2- /O_2) reaction. The reactivity of superoxide anion radical with chalcones in the presence of benzoyl chloride as an affector is different and leads to radical dimer products. The plausible mechanisms have been proposed for the above reactions on the basis of product identification and cyclic voltammetric studies.

INTRODUCTION

During past two decades considerable attention has been focused on the chemistry and reactivity of superoxide anion not only as a biologically active specie[1] but also as a synthetic intermediate[2]. Superoxide anion (O_2-) acts as an effective nucleophile, base, redox specie, a free radical and an electron transfer agent. The generation of superoxide by cathodic reduction of molecular oxygen in aprotic solvent containing tetraalkylammonium salt is most convenient and is a continuous process to give sufficient concentration of O_2- solution in pure form. Efficient regeneration[2e] of phenols from corresponding tosylates has been achieved. Reactivity of superoxide ion as a nucleophile and base for cyclisation and inducing Favorskii rearrangements has also been reported[3]. Formation of phenol from reaction of some reactive aromatic halids with superoxide anion is known[4]. In the present report some other nucleophilic and radical behaviours of superoxide ion are described.

METHOD

Superoxide ion was generated at -1.0V vs SCE in aprotic medium (dry DMF) in presence of tetraalkylammonium salt by reduction of dry oxygen at mercury pool (23.76 cm^2) cathode and platinum foil (3.78 cm^2) anode. The electrochemical cell consists of a double walled glass cell (capacity 150 ml) with a glass frit diaphragm[2c]. After purging with nitrogen gas, the electrolytic solution is first pre-electrolysed at -1.9V vs SCE until the background current falls appreciably (~ 2 mA). Catholyte was saturated with O_2 and then organic substrate was added. Pure and dry oxygen is continued to bubble through the solution during the period of electrolysis. The completion of electrolysis is indicated by appreciable decrease in current and loss of reactant. The catholyte is then worked up to isolate the product.

Reaction of a number of organohalogens was found to be facile with O_2 to give corresponding phenols. In case of aliphatic halides peroxides are isolate finally to afford alcohols. The studies of cyclic voltammetry and product isolation was utilised to propose a plausible mechanism of the reaction. It is assumed that O_2 reacts with haloarenes through a sequence of steps behaving as an effective nucleophile rather than as a radical or electron transfer agent. The displacement possibly proceedes bimolecularly in an addition elimination process to afford the product (Scheme 1).

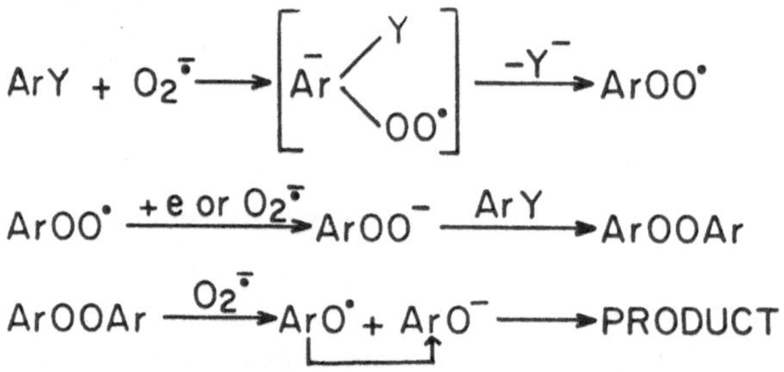

Scheme 1

Thus nitrohaloarenes, ortho-bromotoluene, ortho-iodobenzoic acid, 1-bromo-naphthalene are smoothly transformed to the corresponding phenolic compounds.

The chalcones are biologically important class of compounds possessing enone system. A number of substituted chalcones were reacted with *in situ* electrochemically generated superoxide. These chalcones are observed to be oxidatively cleaved to carboxylic acid(s) in high yield in presence of O_2/O_2 couple in DMF at mercury cathode (- 1.0 vs SCE) :

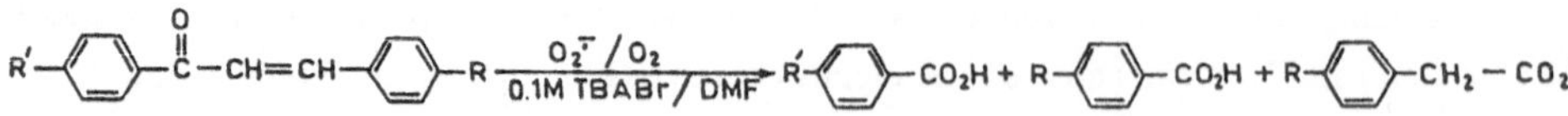

The product isolation and C.V. indicate initial formation of radical anion which then reacts with surrounding molecular oxygen to cleave to carboxylic acid(s). The reactivity of O_2 with chalcones in presence of benzoyl chloride as affector is different and leads to radical dimer products (Scheme 2).

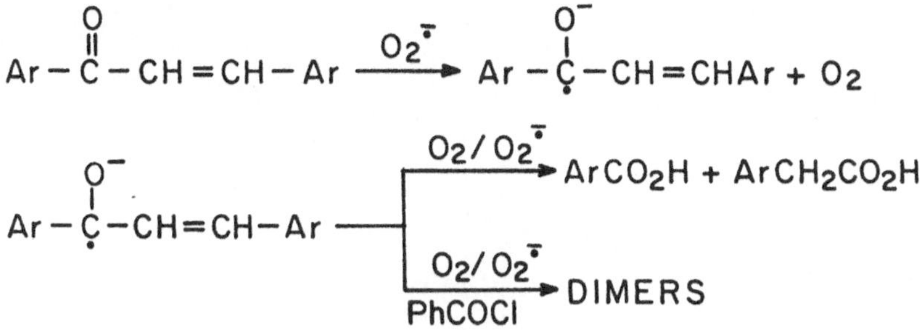

$$Ar - \overset{\overset{O}{\|}}{C} - CH = CH - Ar \xrightarrow{O_2^{\overline{\cdot}}} Ar - \overset{\overset{O^-}{|}}{\underset{\cdot}{C}} - CH = CHAr + O_2$$

$$Ar - \overset{\overset{O^-}{|}}{\underset{\cdot}{C}} - CH = CH - Ar \begin{cases} \xrightarrow{O_2/O_2^{\overline{\cdot}}} ArCO_2H + ArCH_2CO_2H \\ \xrightarrow[PhCOCl]{O_2/O_2^{\overline{\cdot}}} DIMERS \end{cases}$$

Scheme 2

The difference in the reactivity of superoxide anion with chalcone in absence and presence of benzoyl chloride will be discussed.

References :

1) (a) B.A. Omar, S.C. Flores and J.M. McCord, Adv. Pharmacol., (San Diego) **1992**, *23*, 109; (b) D.T. Sawyer and J.S. Valentine, Acc. Chem. Res., **1981**, *14*, 393.

2) (a) E.J. Corey, K.C. Nicolaon and M. Shibasaki, J. Chem. Soc., Chem. Comm., **1975**, 658; (b) M. Singh and R.A. Misra, Synthesis, **1989**, 403; (c) M. Singh, K.N. Singh, S. Dwivedi and R.A. Misra, Synthesis, **1991**, 291; (d) M. Singh, K.N. Singh and R.A. Misra, Bull. Chem. Soc. Jpn., **1991**, *63*, 2599; (e) S. Dwivedi and R.A. Misra, Indian J. Chem., **1992**, *31B*, 282; **1994**, *33B*, 694; (f) M. Singh, K.N. Singh and R.A. Misra, Indian J. Chem., **1994**, *33B*, 173; (g) D. Vasudevan and H. Wendt, J. Electroanalytical Chem., **1995**, *192*, 69.

3) I. Carelli, A. Curruli and A. Inesi, J. Chem. Res.(s), **1989**, 338; **1991**, 10.

4) (a) H. Sagae, M. Fujihara, K. Komazawa, H. Lund and T. Osa, Bull. Chem. Soc. Jpn., **1980**, *53*, 2188; (b) M. Careil, J. Pinson and J.M. Saveant, Nouv. J. Chim., **1981**, *5*, 311.

SOME ELECTROCHEMICAL TRANSFORMATION OF POLYHALOMETHANES.

Vladimir A.PETROSYAN

N.D.Zelinsky Institute of Organic Chemistry, Russian Academy of Sciences, Leninsky prospect 47, Moscow 117913, Russia

ABSTRACT

Some new information on electrosynthesis with participation of carbanions - products of polyhalomethanes reduction was obtained with use of alkylation, cyclopropanation and carboxylation reactions.

INTRODUCTION

Anionic particles - products of polyhalomethane reduction have different reactivity and can be used as precursors of carbenes, nucleophiles and electrogenerated bases. It opens up wide opportunities for use of such anionic particles in organic electrosynthesis. Some new information concerning this problem is given.

RESULTS AND METHODS.

Electrogeneration of halocarbenes usually occurs in aprotic media in the absence of any electrophile, but even under such conditions the reaction of intermediate anion with initial molecule as electrophile can compete with anion fragmentation.

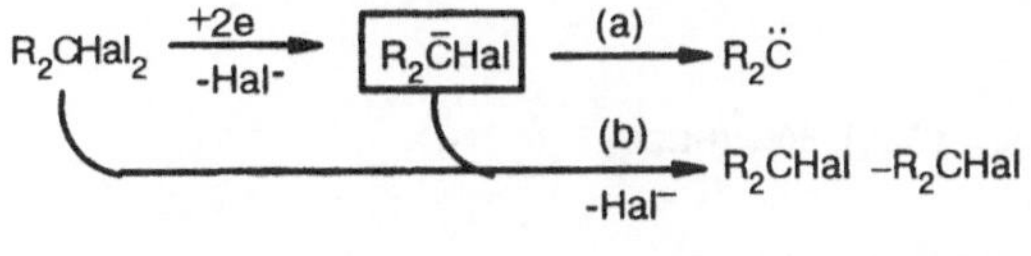

$$\tag{1}$$

Using quantum-chemical calculation data (MNDO, AM-1 and PM3 technique, the solvent effect was introduced within point dipole model) we showed that with increasing of amount of halogen atoms the rates of fragmentation reaction (1a) and nucleophile substitution (1b) change in the opposite direction [1]. The change of medium polarity has the same influence on these reactions. According to obtained results [1] (scheme 2) reduction of tetrahalomethane is preferably occurs along the path leading to dihalocarbene (2a). At the same time selfprotonation is more effective during the reduction of trihalomethane (2b). As a result formation of dihalo- and not monohalocarbene should be expected. The formation of monohalocarbene is possible in the reaction of dihalomethane (2c) where selfprotonation and nucleophile substitution are equal. As it could be expected nucleophilic substitution is realized more effectively during reduction of halomethane.

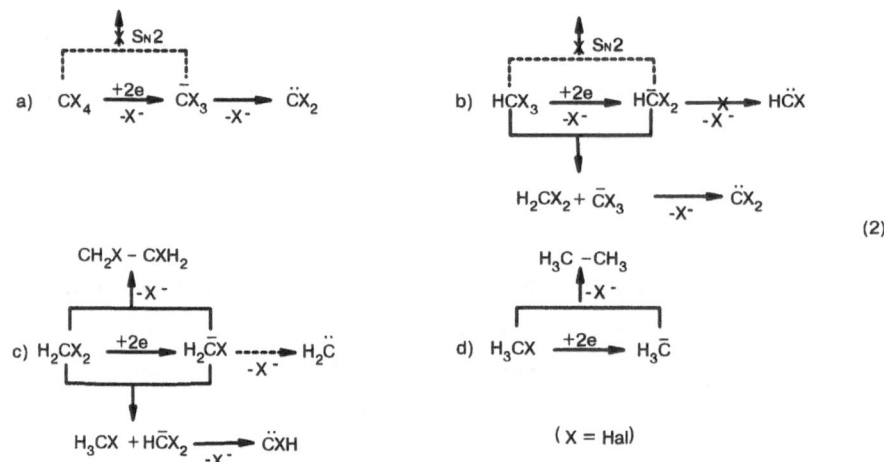

(2)

We showed by many examples [2] the possibility of dichlorocyclopropane electrosynthesis via carbon tetrachloride reduction in presence of nonactivated olefines. (Pt cathode, DMF, divided cell). The same products (scheme 3) were obtained in electroreduction of chloroform though the reaction conditions were not optimized. It conforms with a mechanism given by scheme (2b).

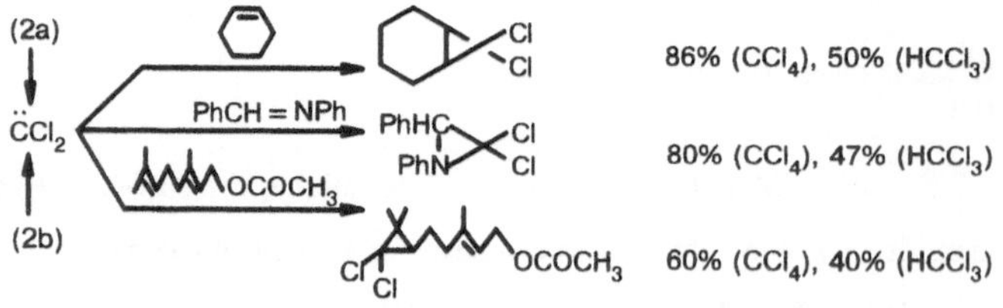

(2a)

$\overset{..}{C}Cl_2$

(2b)

	86% (CCl$_4$), 50% (HCCl$_3$)
PhCH = NPh	80% (CCl$_4$), 47% (HCCl$_3$)
OCOCH$_3$	60% (CCl$_4$), 40% (HCCl$_3$)

(3)

The possibility to realise processes according to scheme (2c) is under investigation.

Electrogenerated anions of polyhalomethanes are in common use as electrophiles. For example, chain reaction in system CCl$_4$/HCCl$_3$/Carbonyl is an effective synthetic method for trichloromethylcarbinol production [3]. It turned out that the process could also be realized in system HCCl$_3$/PhCHO/DMF. The probable mechanism is given by scheme (4), where carbonyl plays a role of co-reagent and probase simultaneously.

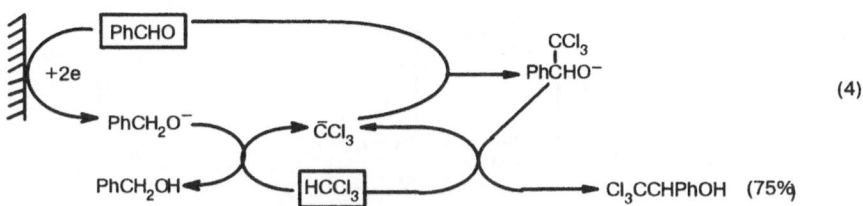

(4)

The yield of carbinol essentially depends on the amount of electricity passed and the ratio of $HCCl_3$ /PhCHO. Under optimum conditions it is equal to 75% and carrent efficiency is 295%.

At the same time we offered convenient approach to synthesis of trichloromethylcarbinol acetate. It was shown that under reduction of CCl_4 (Pt, DMF, divided cell) in the presence of acylales one of their acetic groups was substituted by trichloromethyl anion. The found reaction has a common character and it was spread at acylales of aliphatic, aromatic and α,β-unsaturated aldehydes.

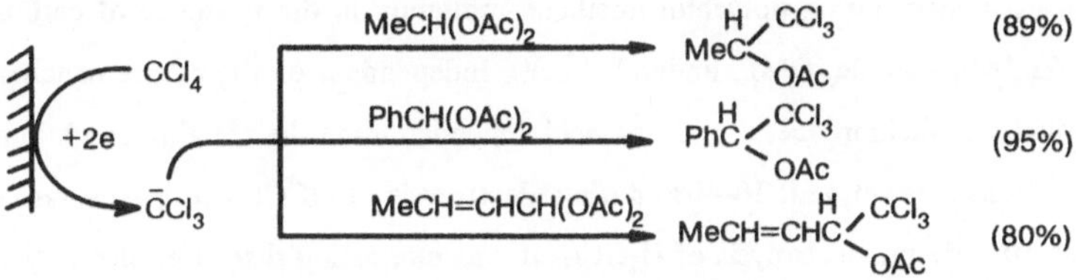

$$
\text{+2e} \quad CCl_4 \longrightarrow \bar{C}Cl_3
\begin{cases}
MeCH(OAc)_2 \longrightarrow MeC{<}^{H\ /CCl_3}_{\ \ OAc} & (89\%) \\
PhCH(OAc)_2 \longrightarrow PhC{<}^{H\ /CCl_3}_{\ \ OAc} & (95\%) \\
MeCH=CHCH(OAc)_2 \longrightarrow MeCH=CHC{<}^{H\ /CCl_3}_{\ \ OAc} & (80\%)
\end{cases}
\quad (5)
$$

In all cases trichloromethylcarbinol acetates were produced with high yield. On the other hand it was shown that reduction of these acetates (Zn-cathode, DMF) is a convenient method of 2-substituted-1,1-dichloroethylene production as useful products in organic synthesis.

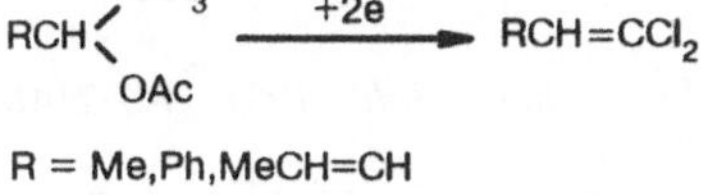

$$
RCH{<}^{CCl_3}_{\ \ OAc} \xrightarrow{\text{+2e}} RCH=CCl_2 \qquad (\geq 70\%) \qquad (6)
$$

R = Me,Ph,MeCH=CH

Few examples of electrosynthesis with dihalomethanes and practical lack of information about regularities of these processes stimulated such investigation. It was shown [4] that amperostatic electrolysis of different types of organic AH-acids (aminoderivatives, thiols, alcohols, substituted malonates, phosphonates) in the presence of dihalomethanes led to corresponding alkylated products with 40-90% yield. Proposed [4] mechanism of these reactions (scheme 7) had a partly formal character.

$$AH \xrightarrow{+e} \bar{A} \xrightarrow[Hal^-]{CH_2Hal_2} ACH_2Hal \longrightarrow \begin{array}{l} \xrightarrow[-Cl^-]{+2e} A-CH_3 \\ \qquad\qquad \Big\uparrow V \\ \qquad +2e \Big| -A^- \\ \xrightarrow[-Hal^-]{+A^-} A-CH_2-A \end{array} \qquad (7)$$

$$\text{I} \qquad\quad \text{II} \qquad\qquad\quad \text{III} \qquad\qquad\qquad\qquad \text{IV}$$

Depending on the nature of II passing 1F of electricity leads to III or IV as final products (the acetales form in the case of thiols and aromatic alcohols reduction). More prolonged electrolysis gives V. At the same time regularities of I →III transformation need refinements. So co-electrolysis of alkyl or aryl malonates in the absence of dihalomethane really gives II with quantitative yield. However salt II does not react with methylene chloride neither chemically, nor under electrolysis conditions. Hence it follows that formation of III during co-electrolysis of methylene chloride and malonate is a result of more complicated transformation initiated by previous formation of methylene chloride anion-radical.

Interesting results were obtained during polychloromethane reduction in the presence of carbon dioxide (Pt, Fe - cathode, Zn, Mg - anode, DMF, undivided cell). Independent on the nature of initial halomethane the main product was dichloroacetic acid. Its yield depended upon the reaction conditions and was 15-50% when CCl_4 was electrolyzed; 10-40% during electrolysis of $HCCl_3$ and it was hardly surprising that its yield was 10-30% on electrolysis of H_2CCl_2. It was interesting that after electrolysis of CCl_4 only traces of trichloroacetic acid were found among reaction products ($\leq$ 2%). However monochloroacetic acid presented practically in every experiment and its yield was between 1 and 15%. Besides that small amounts of oxalate and formic acid were noted. This information has a preliminary character and we shall continue the work for optimization of these process.

References.
1) I.A.Matchkarovskaya, K.Ya. Burshtein, and V.A.Petrosyan *Russ. Chem. Bull.*, **1995**, *44*, 2148 (Engl. transl.).
2) V.A.Petrosyan, M.E. Niyazymbetov, and T.K.Baryshnikova, *Izv. Akad. Nauk SSSR, Ser. Khim.*, **1988**, *37*, 91.
3) T.Shono, N.Kise, M.Masuda, and T.Suzumoto, *J. Org. Chem.*, **1985**, *50*, 2527.
4) V.A.Petrosyan, *Elektrokhimiya*, **1996**, *32*, 53.

INVESTIGATION OF THE COUPLING REACTION BETWEEN AROMATIC RADICAL ANIONS AND ALKYL RADICALS.

Steen Uttrup Pedersen[*], Torben Lund[$], Kim Daasbjerg[*], Mihaela Pop[§], Ingrid Fussing[*] and Henning Lund[*]

[*]Department of Organic Chemistry, University of Aarhus, 8000 Aarhus C, Denmark
[$]Institute of Life Science and Chemistry, University of Roskilde, 4000 Roskilde, Denmark
[§]Romanian Academy, Institute of Organic Chemistry, Ro-71141 Bucharest, Romania

Radical-radical coupling reactions are generally fast when steric hindrance is moderate and charge repulsion is small. Even small radical anions of activated olefins prefer radical-radical coupling at rates often approaching the diffusion-limit. Only few studies have been made[1-2] on the coupling of alkyl radicals (R·) with delocalised radicals like the aromatic radical anions (A·⁻) although this reaction (2) is relevant in several applications.

Alkyl radicals are produced by indirect reduction of alkyl halides (RX) by aromatic mediators (A/A·⁻) and the following mechanism has been described in numerous papers[3]:

Scheme 1
$$A + e \quad\quad A^{-\cdot} \quad\quad (0)$$
$$A^{-\cdot} + RX \xrightarrow{k_1} A + R\cdot + X^- \quad (1)$$
$$A^{-\cdot} + R\cdot \xrightarrow{k_2} A\text{-}R^- \quad\quad (2)$$
$$A^{-\cdot} + R\cdot \xrightarrow{k_3} A + R^- \quad\quad (3)$$
$$R^-, AR^- + H^+ \xrightarrow{k_4} RH, ARH \quad (4)$$

where (1) is the rate determining dissociative ET resulting in the direct generation of an alkyl radical, R·. This radical can then react with another A·⁻ either by coupling (2) or by reduction (3). If E^o_{A/A^-} is more positive than the reduction potential of R·, $E_{R\cdot}$, then the coupling reaction will be favoured compared to ET-reaction (3); however, if the potential of A·⁻ becomes less than E_R then the down-hill ET reaction (3) from A·⁻ to R· will take over. The competition between (2) and (3) determines the product distribution and the consumption of electrons. Because both (2) and (3) occur after the rate determining step (1) no direct measurements of the absolute rate constants, k_2 and k_3, can be made.

A useful strategy is to introduce an alternative reaction to distract some R· from reacting with A·⁻. In previous studies radical cyclisation probes were used as R· and the alternative reaction was a first order cyclisation reaction[1-2]. In this study two different types of competition reactions have been introduced to distract R·.[4] The first one investigated is the addition of R· to an activated olefin and the second one is the hydrogen atom transfer to an alkyl radical.

The recently obtained reduction potentials for alkyl radicals can be used very efficiently to reduce otherwise complicated reaction schemes. In fig 1 is shown in schematic form the reduction potentials for some typical radicals.

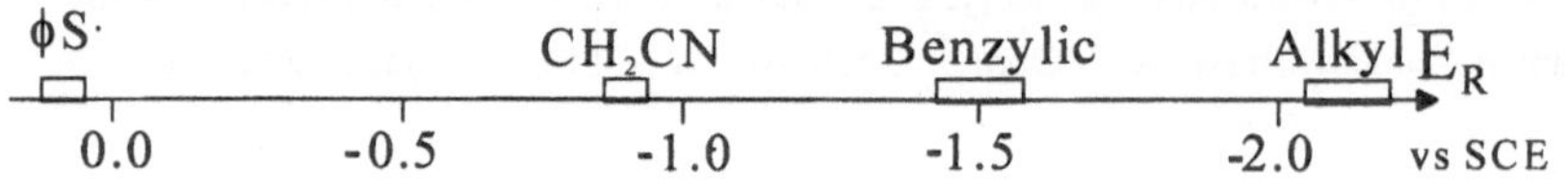

$\phi S^{\cdot}$ CH_2CN Benzylic   Alkyl E_R

| 0.0 | -0.5 | -1.0 | -1.5 | -2.0 | vs SCE | Fig. 1. |

The indirect reduction of alkyl halides (RX) in the presence of an activated olefin like styrene or ethyl cinnamate are outlined in the following reaction scheme 2.

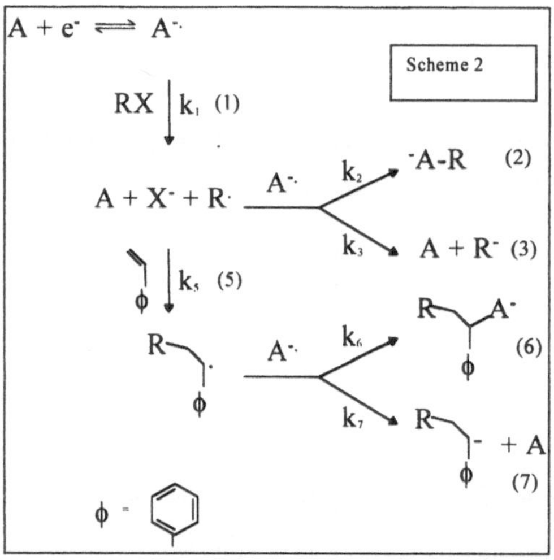

Scheme 2

If only mediators with E^o_{A/A^-} higher than -2.1 V vs. SCE are used the balance will be entirely shifted from (3) to (2). It is well known from the literature that an alkyl radical will add to styrene to form a benzylic radical and if then only mediators with E^o_{A/A^-} lower the -1.45 V vs SCE are applied this benzylic radical will not couple with A^- (6), but will be reduced to the benzylic anion (7). The competition for the alkyl radical is then between the coupling reaction (2) and the addition reaction (5) and the competition

parameter, $\xi, = \dfrac{k_5}{k_2} \cdot \dfrac{[O]}{[A^-]}$ (8), where O is the activated olefin. The first ratio is between the rate

constants k_5 and k_2. Unless severe steric hindrance is present in both $R^{\cdot}$ and $A^{\cdot -}$ the rate constant for the coupling, k_2, will be very high. Previous results for primary alkyl radicals indicate that typical values for k_2 are $1 \cdot 10^9$ $M^{-1}s^{-1}$. The rate constant for the addition of alkyl radicals to activated olefins is, even in the most favourable situation, less than $2 \cdot 10^6$ $M^{-1}s^{-1}$ and the ratio k_5/k_2 will therefore be smaller than 10^{-3}. The second ratio, $[O]/[A^-]$, can be controlled experimentally. In a direct LSV experiment [A] is 1-2 mM and $[A^-]$ will be of the same order in the reaction layer during the reduction, and high concentration of the olefin (> 0.1M) is therefore necessary for the competition to be effective in LSV, $\xi \approx 1$. Such high concentrations of activated olefins are problematic because they can be indirectly reduced and form oligomers or polymers which can change the electron consumption and the form of the voltammograms. The competition between (2) and (5) can, however, also be accentuated by decreasing the concentration of $A^{-\cdot}$ in the reaction layer.

When the rate determining reaction (1) is slow, the reaction layer is extended to the bulk. Steady-state conditions will develop for $A^{-\cdot}$ when the electrochemical rate of formation of $A^{-\cdot}$ is equal to the consumption by the initial ET-reaction (1). The electrochemical rate of formation is controlled by the current, i, in a galvanostatic electrolysis. The steady-state condition is expressed by eqn. (9) and the steady-state concentration of A^{-} , $[A^{-}]_{ss}$, obtained can be inserted in eqn. (8) to give eqn. (10).

Visible spectroscopy with a dip-in probe was used to monitor $[A^{-}]_{ss}$ during the electrolysis and concentrations in the range from 16 to 242 μM were typically found; experimentally controlled through the adjustment of the galvanostatic current, i, and the concentration of alkyl halide, [RX].

$$(\partial[A^{-\cdot}]/\partial t)_{formation} = \frac{-i}{FV} = 2 \cdot k_1[A^{-\cdot}][RX] = (\partial[A^{-\cdot}]/\partial t)_{consumption} \qquad (9)$$

$$\xi = \frac{[ORH]}{[ARH]} = \frac{2FVk_1k_5[O][RX]}{-ik_2} \qquad (10)$$

Preparative electrolysis under galvanostatic control where [O] and [RX] are kept constant during the experiment give a mixture of addition product (ORH) and coupling products (ARH) which can be quantified by normal GC-analysis and the rate constant k_2 can therefore be calculated from eqn. (10). Two different systems (A/O) have been used 1: anthracene/styrene + n- ,sec- or tert-butyl chloride and 2: p-diacetylbenzen/ethyl cinnamate + n- ,sec- or tert-butyl bromide. Typical data are shown in Table 1 for anthracene/styrene.

Alkyl chloride	-i mA	[A] mM	[RX] mM	[O] mM	V ml	n	$[A^{-\cdot}]_{ss}$ /μM	$\frac{ORH}{ARH}$	$10^4 \cdot k_2/k_5$	$10^{-9} k_2/$ $M^{-1}s^{-1}$
tert-Butyl	10.0	3.0	60.5	60.1	38.0	1.88	30	0.44	0.45	0.62
tert-Butyl	5.3	4.1	71.7	80.5	38.5	1.92	13	0.90	0.67	0.92
sec-Butyl	5.0	8.0	80.9	80.3	35.0	2.01	22	0.12	3.0	1.3
sec-Butyl	10.0	2.8	52.4	60.1	36.0	2.02	65	0.045	2.0	0.84
Butyl	5.2	8.0	79.8	81.4	36.0	1.99	59	0.064	2.2	0.68
Butyl	4.1	12.5	39.9	80.0	36.0	2.00	125	0.020	3.1	0.96

The second method used in this study consisted of adding a good hydrogen atom donor like thiophenol to the solution containing the aromatic compound and the alkyl halide. If only mediators with E^o_{A/A^-} higher than -2.1 V vs. SCE are used then the reduction of R can be neglected. The competition for R will then be between the coupling reaction (2) and the hydrogen atom transfer reaction (11). The formed thiophenoxy radical in (11) will be further reduced if E^o_{A/A^-} is lower than +0.12 V vs. SCE which is the common situation. The focused reaction scheme is presented below.

$$A + e^- \rightleftharpoons A^{\cdot -}$$

$$RX \Big\downarrow k_1 \quad (1)$$

$$A + X^- + R^{\cdot} \xrightarrow[k_2]{A^{\cdot -}} {}^-A\text{-}R \quad (2)$$

$$SH \Big\downarrow k_{11} \quad (11)$$

$$RH + S^{\cdot} \xrightarrow[k_{12}]{A^{\cdot -}} S^- + A \quad (12)$$

The competition parameter ξ_{SH} is defined as $\dfrac{k_{11}}{k_2}\dfrac{[SH]}{[A^-]}$.

The ratio between the rate constants is about 0.1 when the SH is thiophenol and it is therefore possible with even a small excess of thiophenol, [SH]/[A], to shift the mechanism from coupling to hydrogen atom transfer. The change in mechanism is accompanied with a change in electron consumption from overall 2 electrons per A to 2 electron per RX. In LSV an increase in the peak current is noticed when the thiophenol is added. Comparison with simulated voltammograms permitted the determination of k_{11}/k_2 and thus k_2 since literature values are available.

Table 2

Alkyl radical	Aromatic radical anion	$10^{-9} \cdot k_2/M^{-1}s^{-1}$	method
tert-Butyl	anthracene	0.8	Addition to olefin
tert-Butyl	*p*-diacetylbenzene	1.7	Addition to olefin
tert-Butyl	*p*-dicyanobenzene	1.0	Hydrogen atom trans
sec-Butyl	anthracene	1.0	Addition to olefin
sec-Butyl	*p*-diacetylbenzene	1.3	Addition to olefin
Isopropyl	*p*-dicyanobenzene	2.0	Hydrogen atom trans
Butyl	anthracene	0.9	Addition to olefin
Butyl	*p*-diacetylbenzene	2.0	Addition to olefin
Butyl	*p*-dicyanobenzene	2.6	Hydrogen atom trans
Butyl	benzophenone	1.6	Hydrogen atom trans

The results obtained so far are collected in Table 2 and they show that the coupling reaction (2) is fast in all cases. The measured rate constants k_2 range from $0.5 \cdot 10^9$ to $3.5 \cdot 10^9$ $M^{-1}s^{-1}$ which are approximately one order of magnitude lower than the diffusion controlled limit.

References:
1. Garst, J.F. and Smith, C.D., J. Am. Chem. Soc., **1976**, *98*, 1520.
2. Pedersen, S.U. and Lund, T., Acta Chem Scand., **1991**, *45*, 397.
3. Lund, H., Daasbjerg, K., Occhialini, D., and Pedersen, S.U., Russian Journal of Electrochemistry, **1995**, *31*, 865.
4. Pedersen, S.U, Lund, T, Daasbjerg, K., Pop, M., Fussing, I. And Lund, H., Acta Chem. Scand. **1998**, *52*, in press.

ELECTROCHEMICAL GENERATION OF FLUORINATED ACTIVE SPECIES - A NEW APPROACH TO THE SYNTHESIS OF ORGANOFLUORINE COMPOUNDS

Nikolai IGNAT'EV, Sergii DATSENKO and Elena SMERTENKO

Institute of Organic Chemistry Ukrainian National Academy of Sciences, Laboratory
Electrochemistry of Organoelement Compounds, Murmanskaya 5, 253660 Kiev-94, Ukraine

ABSTRACT

Cyclic voltammetry is a powerful method for the estimation of relative reactivity of various nucleophilic agents towards perfluoroalkyl halides. This method can be applied also for monitoring the reactivity of aromatic compounds towards trifluoromethyl radicals, generated by the electrochemical oxidation of the trifluoromethanesulfinate anion.

INTRODUCTION

Perfluoroalkyl halides as a source of perfluoroalkyl radicals play an important part in organofluorine chemistry. Heat, UV light or a radical source (peroxides, azo initiators) are classically used to initiate radical reactions of perfluoroalkyl halides [1]:

$$R_F{-}I \xrightarrow{\text{thermolysis or UV irradiation}} R_F$$

The first example of electrochemical reduction of a perfluoroalkyl iodide on a mercury cathode was reported by A.Commeyras at al. [2]. In fact this process leads to the formation of $R_F HgI$.

RESULTS AND DISCUSSION

Direct electrochemical reduction of perfluoroalkyl halides on solid electrodes (glassy carbon, Pt and others) was studied by means of the cyclic voltammetry method at several laboratories [3a-c] at the beginning of 90th. Process of electrochemical reduction of primary perfluoroalkyl iodides and bromides is irreversible and leads to the generation of perfluoroalkyl radicals.

$$R_F X + e \longrightarrow \left[R_F X \right]^{\overset{\bullet}{-}} \longrightarrow R_F + X^-$$
$$X = I,\ Br$$

Perfluoroalkyliodonium salts can be applied for the electrochemical generation of perfluoroalkyl radicals at low cathodic potentials [4].

$$C_6H_5 - \overset{+}{\underset{X^-}{I}} - R_F \quad + \quad e \quad \longrightarrow \quad C_6H_5I \quad + \quad R_{\overset{\cdot}{F}} \quad + \quad X^-$$

$$X^- = BF_4^-, \ CF_3SO_3^-, \ CF_3COO^-$$

Recently we have shown that electrochemical oxidation of perfluoroalkyl sulfinates results in the formation of perfluoroalkyl radicals as well [5a].

$$CF_3SO_2^- \quad - \quad e \quad \longrightarrow \quad [CF_3SO_2]^{\cdot} \quad \xrightarrow{-SO_2} \quad CF_3^{\cdot}$$

A combination of these methods gives the possibility of carrying out electrochemical perfluoroalkylation of different organic compounds. In our previous publications [3a,6a-c] we have shown that electrochemical initiation of the interaction of perfluoroalkyl halides with carbon-, oxygen-, and sulfur-containig nucleophiles is a useful method for the electrosynthesis of organofluorine compounds. The process can be described by the following equations:

$$R_FX \quad + \quad e \quad \longrightarrow \quad [R_FX]^{\cdot -} \quad \longrightarrow \quad R_{\overset{\cdot}{F}} \quad + \quad X^- \quad (1$$

$$R_{\overset{\cdot}{F}} \quad + \quad Nu^- \quad \longrightarrow \quad [R_FNu]^{\cdot -} \qquad\qquad\uparrow \qquad (2)$$

$$[R_FNu]^{\cdot -} \quad + \quad R_FX \quad \rightleftharpoons \quad R_FNu \quad + \quad [R_FX]^{\cdot -} \quad (3)$$

To estimate the reactivity of perfluoroalkyl halides towards various nucleophiles, a simple method based on cyclic voltammetry measurements has been developed.

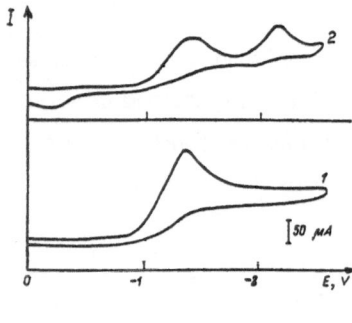

Fig 1.

Fig.1 presents a cyclic voltammogram, which was recorded on glassy carbon electrode for the acetonitrile solution of heptafluoropropyl iodide itself (curve 1) and in the presence of tetrabutylammonium p-chlorothiophenolate (curve 2). We have found that the dependence between a decrease in the reduction peak current of the perfluoroalkyl halide and the nuclephilic agent concentration is of an exponential type:

$$\frac{I}{I_0} = \exp \beta \frac{[Nu]}{C} \qquad \gamma = \frac{[Nu]^*}{[Nu]} \quad \text{at the:} \quad \frac{I}{I_0} = \frac{1}{2}$$

where C is the concentration of the perfluoroalkyl halide, I_0 is the reduction peak current of the perfluoroalkyl halide in the absence of a nucleophile and I is the current at the same reduction peak in the presence of the nucleophile. The coefficient β can be used to characterize the reactivity of different perfluoroalkyl halides towards nucleophilic agents under electrochemical initiation conditions. Table 1 presents the results which were obtained by that method [7].

Table 1

Reactivity of Perfluoroalkyl Halides towards Sodium 8-Mercaptoquinolinate

Perfluoroalkyl halide	$E_{1/2}^{red}$, V (vs. SCE)	$-\beta$	γ
n- $C_6F_{13}I$	- 1.10	1.03	1.00
n- C_3F_7I	- 1.18	0.99	0.98
$CF_3OCF_2CF_2I$	- 1.24	0.56	0.80
i - C_3F_7I	- 0.94	0.37	0.52
n - $C_6F_{13}Br$	- 1.54	0.17	0.25

Electrode: glassy carbon ; Electrolyte: 0.1 M $(C_4H_9)N^+ BF_4^-$ in DMF

This simple and rapid method has the disadvantage related to the unpredictable influence of the side processes (4) on the results of the measurements.

$$R_F^{\cdot} + e \longrightarrow R_F^- \xrightarrow{SolH} R_FH \qquad (4)$$

The effect of the process (4) probably can be neglected by using indirect electrochemical reduction of perfluoroalkyl halides mediated by some stable radical-anions.

$$Q + e \rightleftharpoons Q^{\cdot -} \qquad (5)$$

$$Q^{\cdot -} + R_FX \longrightarrow Q + \left[R_FX\right]^{\cdot -} \qquad (6)$$

$$\left[R_FX\right]^{\cdot -} \longrightarrow R_F^{\cdot} + X^- \qquad (7)$$

The theory of this method was developed by J.-M. Saveant [8]. We simplify the data processing and apply this method for the determination of relative reactivity of perfluoroalkyl halides towards various nucleophilic agents.

Fig.2 presents cyclic voltammogramms which reflect the process of the electrochemical reduction of 4-nitrobenzaldehyde (mediator Q) itself (curve 1), 4-nitrobenzaldehyde in the presence of C_3F_7I (curve 2) and 4-nitrobenzaldehyde in the presence of C_3F_7I and 4-methylthiophenol (curve 3). An increase in the current of the mediator (curve 2) is proportional to the amount of heptafluoropropyl iodide added (Fig 3) and reflects the process which can be described by equations (5,6).

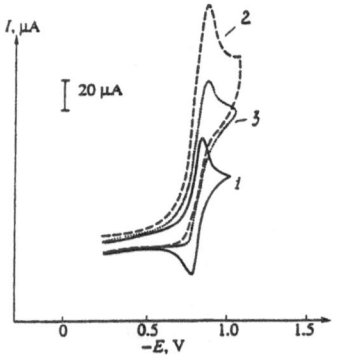

Fig.2. Homogeneous electrocatalytic reduction of C_3F_7I by 4-nitrobenzaldehyde radical-anion.

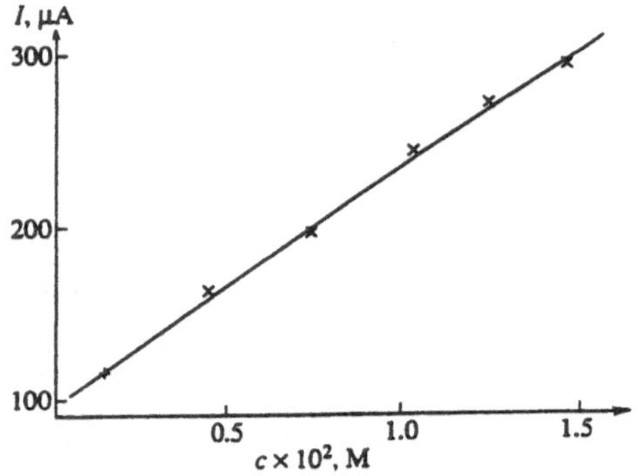

Fig.3. Peak current for 4-nitrobenzaldehyde against the concentration of C_3F_7I.

The decrease in the peak current of 4-nitrobenzaldehyde (curve 3) after addition of nucleophilic agent (4-methylthiophenol) is caused by fast chain processes (reactions 7,2,3) and depends on the concentration of nucleophilic agent (Fig 4). This dependence can be described by the exponential equation. Coefficient β' reflects the reactivity of nucleophilic agents towards perfluoroalkyl halides [7]. Table 2 presents the results which were obtained by that method.

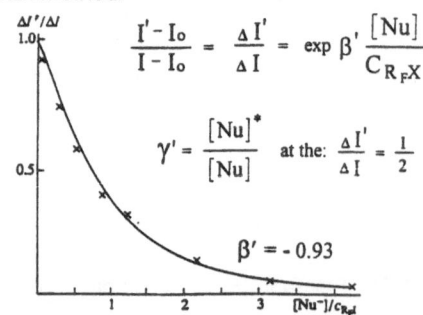

$$\frac{I'-I_o}{I-I_o} = \frac{\Delta I'}{\Delta I} = \exp \beta' \frac{[Nu]}{C_{R_FX}}$$

$$\gamma' = \frac{[Nu]^*}{[Nu]} \quad \text{at the:} \quad \frac{\Delta I'}{\Delta I} = \frac{1}{2}$$

$$\beta' = -0.93$$

I_o - peak current of 4-nitrobenzaldehyde.

I - peak current of 4-nitrobenzaldehyde in the presence of perfluoroalkyl halide.

I' - peak current of 4-nitrobenzaldehyde in the presence of perfluoroalkyl halide and nucleophilic agent.

Fig.4. Decrease of the peak current on cyclic voltammogram of 4-nitrobenzdehyde in the presence of C_3F_7I against the amount of sodium 8-mercaptoquinolinate added.

Table 2

Reactivity of Heptafluoropropyl Iodide Towards Various Nucleophiles

Nucleophile	$E_{1/2}^{ox}$, V (vs. SCE)	$-\beta'$	γ'
Tetramethylammonium 2-nitropropanide	0.05	1.4	1.00
Tetramethylammonium nonanethiolate	- 0.12	1.3	0.98
Tetramethylammonium 4-methylthiophenolate	- 0.11	1.2	0.89
Tetramethylammonium 4-chlorothiophenolate	- 0.06	1.0	0.73
Sodium 8-mercapto-quinolinate	0.00	0.93	0.59
Tetramethylammonium diethylphosphite	0.81	0.76	0.55
4-Methylthiophenol	1.17	0.53	0.43
Tetramethylammonium triazolate	0.86	0.51	0.41
Tetramethylammonium succinimidate	0.23	0.23	0.17
Triethylbenzylammonium 4-chlorophenyl sulfinate	1,16	0.12	0.10
Triethyl phosphite	0.83	0.01	0.01
Aniline	1.04	0.02	0.02

The data which are listed in the Table 2 show that „soft" thiolate anions and carbanions are the most reactive nucleophiles towards perfluoroalkyl halides (high absolute value of β').

The similar approach can be applied for the monitoring of the reactivity of trifluoromethyl radicals, generated by anodic oxidation of trifluoromethanesulfinate anions, towards aromatic compounds. We have used the ratio between oxidation peak current of corresponding aromatic compounds in the absence and presence of sodium trifluoromethanesulfinate as the key measurement. This gives the possibility of deriving a simple equation which can be used to estimate the relative reactivity of different aromatic compounds towards trifluoromethyl radicals, generated by anodic oxidation of trifluoromethanesulfinate anions [5b].

$$\frac{I}{I_o} = \exp \alpha \frac{[CF_3SO_2Na]}{[ArH]}$$

Table 3

Relative Reactivity of Aromatic Compounds towards Trifluoromethyl Radicals

Aromatic Compound	E_p^{ox}, V (vs.SCE)	$-\alpha$
Naphthalene	1.80	0.1
Durene	1.94	0.07
p- Xylene	2.14	0.02
Toluene	2.24	0.003
Benzene	2.60	0.0002

Electrode: glassy carbon ; Electrolyte: 0.1 M $(C_4H_9)N^+ BF_4^-$ in Acetonitrile.

The data (Table 3) show that more easy oxidizable aromatic compounds are more reactive towards trifluoromethyl radicals (high absolute value of α).

References

1) C.Wakselman and A.Lantz, in *Organofluorine Chemistry,* ed. R.E. Banks, B.E. Smart and J.C. Tatlow, Plenum Press, **1994**, ch.8 and references cited therein.
2) P.Calas, P.Moreau, A.Commeyras, J. Electroanal. Chem. **1977**, *78,* 271.
3) (a) N.Ignat'ev, S.Datsenko, S.Pazenok, L.Yagupolskii, Zh. Org. Khim.(Russ), **1990**, *26,* 1740; (b) C.P.Andrieux, L.Gelis, M.Medebielle, J.Pinson and J.-M.Saveant, J. Am. Chem. Soc. **1990**, *112*, 3509. (c) P.Calas, C.Amatore, L.Gomez, A.Commeyras, *J. Fluor. Chem*, **1990**, *49*, 247.
4) N.Ignat'ev, L.Nechitailo, A.Mironova, I.Maletina, V.Orda, *Electrochimia (Russ)*, **1992**, *28*, 502.
5) (a) S.Datsenko, E.Smertenko, N.Ignat'ev, *Communications of 11th.European Symposium on Fluorine Chemistry*, Bled, Slovenia, **1995**, 206; (b) S.Datsenko, E.Smertenko, N.Ignat'ev, *Russian J. of Electrochemistry*, **1997**, in press.
6) (a) N.V. Ignat'ev, S.D. Datsenko and L.M. Yagupol'skii, *Zh. Org. Khim*, **1991**, *27*, 905; (b) S.D. Datsenko, N.V. Ignat'ev and L.M. Yagupol'skii, *Soviet Electrochemistry*, **1991**, *27*, 1016; (c) S.D. Datsenko, N.V. Ignat'ev and L.M. Yagupol'skii, *Soviet Electrochemistry,* **1991**, *27*, 1484.
7) N.Ignat'ev, S. Datsenko, *Russian J. of Electrochemistry*,.**1995**, *31*, 1235.
8) J.-M. Saveant, *Acc. Chem. Res.*, **1980**, *13*, 323.

ELECTROREDUCTION OF DIHALOPERFLUOROALKANES AND PERFLUOROALKYLHALIDES IN NEAR-CRITICAL AND SUPERCRITICAL CO2 -METHANOL MEDIUM

Vladimir M. MAZIN , Sergey R. STERLIN, and Vitali A.GRINBERG

A.N.Frumkin Institute of Electrochemistry, Russian Academy of Sciences,

31 Leninsky prospekt, 117071, Moscow, Russia

ABSTRACT

The electrocatalytical carboxylation of some perfluoroalkylhalides and α,ω-dihaloperfluoroalkanes on copper and stainless steel electrodes in near- and supercritical CO_2 - methanol mixtures have been investigated. The main products and intermediates were determinated. It was shown that the difference in the composition of products in these two cases indicates at quite different catalytic activity of copper and stainless steel electrodes as to regards the electroreduction reaction of the intermediates and quite different adsorption of these intermediates on the surface of the electrodes.

INTRODUCTION

It was reported earlier [1] that perfluoroalkylhalides electrocarboxylation occurred with the participation of perfluorocarbanions. Under the conditions of the normal chemical reactions the interaction of the carbanions in protic media with electrophilic particles other than protons seems to be hardly probable. In the course of the electrolysis the reduction of hydrogen ions takes place together with the carbanions generation. This is equivalent to the decrease of the protons concentration at the surface of the electrode and one could expect that more or less favorable conditions for the carbanion - CO_2 interaction could be maintained.

RESULTS AND DISCUSSION

The electrocatalytical carboxylation of trifluoromethyliodide, perfluorobutyliodide and 1,4-dibromoperfluorobutane on copper and stainless steel electrodes in near- and supercritical CO_2 - methanol mixtures was investigated. It was determined that electroreduction of 1,4-dibromoperfluorobutane on Cu electrode in supercritical CO_2-methanol mixture (the current efficiency 85-90%) led to the formation of 1,4-dihydroperfluorobutane (main product), ω-hydroperfluorocarboxylic acids and olefins. The main products and intermediates of electrolysis on Cu electrode could be presented as following:

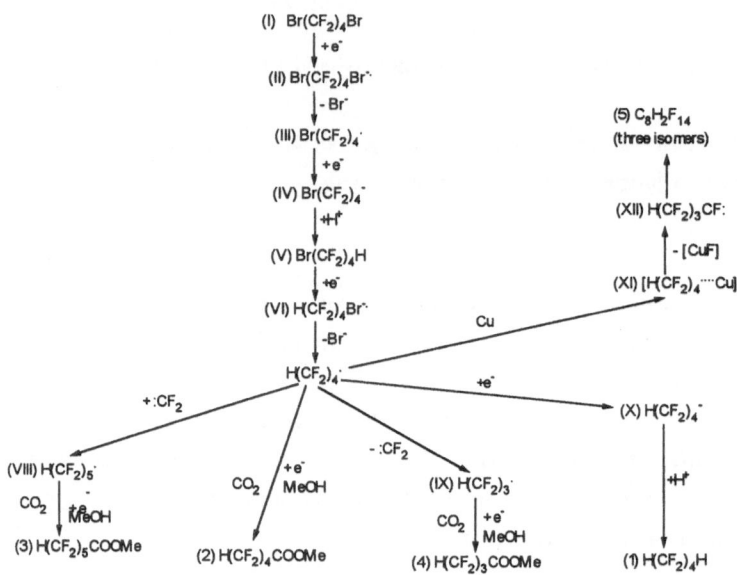

The products composition of the electrolysis performed on the stainless steel electrode (current efficiency 60%) dramatically differs both qualitatively and quantitatively from the above discussed situation. $H(CF_2)_4Br$ (V) was the main product, GCMS showed presence of $Br(CF_2)_4COOCH_3$ (6); $Br(CF_2)_3COOCH_3$ (7) and $Br(CF_2)_8Br$ (8). The products (1), (2), (3), (4) are found in amount less than 0.01% and olefin (5) is not found at all. The difference in the composition of products indicated quite different catalytic activity of copper and stainless steel electrodes as regards to the electroreduction reaction of the intermediates and quite different adsorption of these intermediates on the surface of the electrodes.

The electroreduction of trifluoromethyliodide on Cu electrode (the current efficiency 57%) led to the formation of CF_3H (main product) and CF_3COOCH_3.

In the case of electroreduction of C_4F_9I on Cu and stainless steel there was no products of carboxylation. Only C_4F_9H (the current efficiency 57%) was detected.

It was shown that in spite of high CO_2 concentration on the surface of the electrode in near- and supercritical CO_2 - methanol mixtures the reaction of electrophilic CO_2 addition to the intermediately generated fluorinated anions occurs with insufficient rate. The main process appears to be the protonation of the intermediate particles.

References

1) P. Calas , A. Commeyras , *J. Electroanal.Chem.* **1978,** 89, 363.

Electroreductive Defluorination of Trifluoromethyl Imines

Tsuyoshi Kato and Kenji Uneyama

Department of Applied Chemistry, Faculty of Engineering, Okayama University, 3-1-1, Tsushima-naka, Okayama 700, Japan

Difluoromethylene - containing compounds are receiving considerable attention because of their potential biological activities. In this study, we propose a new approach to obtain N-silylated β,β-difluoroenamines as useful precursors of difluoromethylene - containing compounds by electroreductive defluorination of trifluoromethyl imines.

Cathodic reduction of trifluoromethyl imines (**1**) in the presence of TMSCl resulted in defluorination of one of the three fluorines in CF_3 group, affording N-silylated β,β-difluoroenamines (**2**) (eq 1).

$$\text{(1)}$$

Reaction conditions were surveyed with **1a** and the results were summarized in Table 1. The best

Table 1 Electroreducion of Ethyl 3,3,3-trifluoro-2-(N-p-anisyl)iminopropanoate (**1a**)

Entry	Temp (°C)	TMSCl (mmol)	Cathode material	Supporting electrolyte	Current density (mA/cm^2)	Yield of 2a $(\%)^a$
1	-12	3.0	Pb	LiClO$_4$	20	58
2	0	3.0	Pb	LiClO$_4$	20	86
3	rt	3.0	Pb	LiClO$_4$	20	52
4	0	1.0	Pb	LiClO$_4$	20	26
5	0	2.0	Pb	LiClO$_4$	20	46
6	0	5.0	Pb	LiClO$_4$	20	74
7	0	3.0	Pt	LiClO$_4$	20	65
8	0	3.0	Zn	LiClO$_4$	20	50
9	0	3.0	Ni	LiClO$_4$	20	55
10	0	3.0	Pb	Et$_4$NBr	20	62
11	0	3.0	Pb	Bu$_4$NBr	20	67
12	0	3.0	Pb	Et$_4$NClO$_4$	20	57
13	0	3.0	Pb	LiClO$_4$	6.7	82
14	0	3.0	Pb	LiClO$_4$	33	75
15	0	3.0	Pb	LiClO$_4$	67	58

a) Yields were determinded by ^{19}F NMR.

result was obtained when lead as a cathode and LiClO$_4$ as a supporting electrolyte were employed. Current density of 20 mA/cm^2 or less was favored. More than 3 mmol of TMSCl was required for the satisfactory yields of **2**. Trapping of fluoride anion eliminated from the imine were essential to avoid desilylation of N-silylated β,β-difluoroenamine. Because water in these reaction systems hydrolyzed TMSCl and N-silylated β,β-difluoroenamines, careful removal of water was also required.

Table 2 Electroreductive Defluorination of Trifluoromethyl Imines (1)

Entry	—R		Yield of 2 (%)	Entry	—R		Yield of 2 (%)
1[c]	—CO$_2$Et	(1a)	78[a]	5	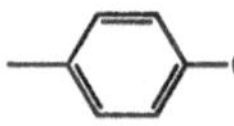 ⟨aryl⟩—Cl	(1e)	74[a]
2[c,d]	—CO$_2$Bn	(1b)	50[a]	6	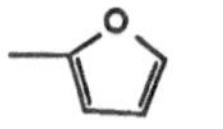 ⟨furan⟩	(1f)	57[a]
3	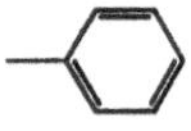 ⟨phenyl⟩	(1c)	75[a]	7[e]	—Et	(1g)	50[b]
4[d]	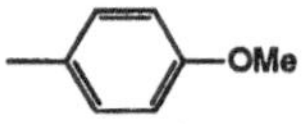 ⟨aryl⟩—OMe	(1d)	58[a]	8	—H	(1h)	47[a]

a) Isolated yield b) Yield was determined by ^{19}F NMR analysis. c) 4.0 mmol of TMSCl and 4.2 mmol of Et$_3$N were used. d) CH$_3$CN as a solvent and Bu$_4$NBr as a supporting electrolyte were used. e) Reaction was conducted at rt.

As shown in Table 2, imines having either an alkyl group or an aryl group gave the corresponding N-silylated β,β-difluoroenamines in good to moderate yields. Reduction of imines bearing aryl group gave better results. All these electroreductive defluorinations proceeded effectively to complete at 2.0 F/mol. As for temperature, the best result was obtained when reduction of **1a** was conducted at 0 ℃ and other imines gave good results even at rt. Fortunately reduction of trifluoromethyl imine bearing ester group (**1a**, **1b**) gave N-silylated amino β,β-difluoroacrylate in good yields. The acrylates were expected to be useful precursors of β,β-difluoro-α-amino acids. Particularly, radical reaction with 2-iodopropane to the acrylate (**2a**) gave ethyl 2-(N-anisyl)imino-3,3-difluoro-4-methyl pentanoate (**3a**), which would be readily converted to β,β-difluoroleucine (eq 2).

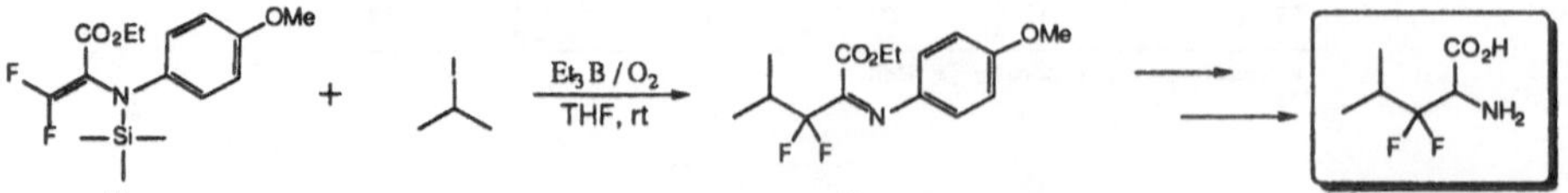

$$\text{(2)}$$

CATHODIC CLEAVAGE OF 1-TRIFLUOROMETHYL 1-ALKENYL SULFONE AND SULFOXIDE

Akira KUNUGI, Md. Abdul JABBAR, and Hidemitsu UNO†

Department of Chemical Science and Technology, Faculty of Engineering,
The University of Tokushima ,Minamijosanjima-cho, Tokushima 770, Japan
† Advanced Instrumentation Center for Chemical Analysis,
Ehime University, Bunkyo-cho, Matsuyama 790, Japan

The electrolytic desulfurization of organosulfur compounds has received considerable attention because most of them are valuable synthetic precursors. The previous paper showed that α,β-unsaturated α-fluoro sulfones such as 2-aryl-1-fluorovinyl phenyl sulfones were subjected to cleavage of carbon-sulfur and/or carbon-fluorine bonds to give 2-aryl-1-fluoro-ethylenes and arylethylenes with a molar ratio of about 1:1, in the presence of efficient proton donors such as phenol, acetic acid, and benzoic acid [1]. On the other hand, the electrolytic reduction of α,β-unsaturated α-fluoro sulfoxide such as 2-(4-biphenylyl)-1-fluorovinyl phenyl sulfoxide involves desulfinylation, defluorination followed by reduction of the sulfinyl group, and reduction to the corresponding sulfide, resulting in the formation of 2-(4-biphenylyl)-1-fluoroethylene, 2-(4-biphenylyl)vinyl phenyl sulfide, and 2-(4-biphenylyl)-1-fluorovinyl phenyl sulfide, respectively [2].

The present work deals with cathodic cleavage of α,β-unsaturated sulfone and sulfoxide having a trifluoromethyl group in N,N-dimethylformamide (DMF). Such an electroreductive method may possess high potentiality in the formation of trifluoromethyl substituted olefins *via* the title compounds. We chose 2-(4-biphenylyl)-1-trifluoromethylvinyl phenyl sulfone (**1**) and sulfoxide (**2**) as the representative compounds because all products would be detected even if some degradation or extensive reduction would occur.

A series of controlled potential macroelectrolyses of **1** and **2** at each plateau potential of the first reduction wave were carried out using the different cathodes such as Hg, Pt, and glassy carbon (*gc*) and the anodes such as Pt and Mg in DMF containing tetrabutylammonium tetrafluoroborate in the presence of benzoic acid, carbon dioxide, and acetic anhydride as an additive at room temperature. The results are listed in Table. The coulometric n-values (electrons per molecule) were obtained from the amount of the substrate added and the charge passed until the termination of the electrolysis. The controlled potential macroelectrolysis of **1** using a Hg pool cathode and a Pt anode in the absence of any additive afforded a trifluoromethyl substituted olefin, 2-(4-biphenylyl)-1-trifluoromethylethylene (**3**) in a low yield together with an unexpected product (**4**) (run 1). As shown in run 2, the use of a sacrificial Mg anode caused a large increase in the yields of products **4** and **5**. On the other hand, the addition of excess benzoic acid gave high yields of **3** and the formation of **4** and **5** was not observed (runs 3 and 4). The cathode material had no influence on the product yields in the presence of the proton donor (runs 5 and 6). We thought that dimerization of the anion radical species generated from single electron transfer to **1** could be suppressed by other electrophiles than proton. As can be seen from runs 7-9, however, any of carbonylated products was not obtained in the presence of acetic anhydride or carbon dioxide, and the simple reduction to **3** was dominant. In these cases, the cyclopentene **4** was also produced when the sacrificial Mg anode was employed.

In the controlled potential macroelectrolysis of the sulfoxide **2** at the Hg pool cathode, **3** was produced as the major product accompanied with small amounts of **6** and **7**, which would be generated by reduction of the sulfoxide moiety itself. The use of the sacrificial Mg anode in place of a platinum anode caused the yield of **3** to increase largely from 50 to 91%. Similarly, the controlled potential macroelectrolysis of **2** in CO_2-saturated DMF using the Hg pool cathode and the Mg anode did not yield any carboxylated product but **3** in 57% yield with a small amount of **6** and **7** [3].

Table. Controlled potential electrolysis of **1** and **2** at the first reduction potential in DMF containing 0.1 mol dm^{-3} Bu$_4$NBF$_4$

Run	Sub-strate	Cathode	Anode	Additive	Charge (F/mol)	Yield/%[a]				
						3	**4**	**5**	**6**	**7**
1	1	Hg	Pt	None	1.8	4	5	—		
2		Hg	Mg	None	2.2	5	37	26		
3		Hg	Pt	PhCO$_2$H	1.9	78	—	—		
4		Hg	Mg	PhCO$_2$H	2.0	91	—	—		
5		Pt	Mg	PhCO$_2$H	2.0	92	—	—		
6		gc	Mg	PhCO$_2$H	1.9	92	—	—		
7		Hg	Mg	Ac$_2$O	3.2	83	6	—		
8		Hg	Pt	CO$_2$	1.8	42	—	—		
9		Hg	Mg	CO$_2$	1.9	31	51	—		
10	2	Hg	Mg	None	2.9	11			6	5
11		Hg	Pt	PhCO$_2$H	3.5	50			—	13
12		Hg	Mg	PhCO$_2$H	3.8	91			5	1
13		Hg	Mg	CO$_2$	2.9	57			2	10

a) Isolated yield by the preparative TLC.

Ar = 4-PhC$_6$H$_4$

References and notes

1) A. Kunugi, K. Yamane, M. Yasuzawa, H. Matsui, H. Uno, and K. Sakamoto, *Electrochim Acta*, **38**, 1037 (1993).
2) A. Kunugi, S. Mori, S. Komatsu, H. Matsui, H. Uno, and K. Sakamoto, *Electrochim. Acta*, **40**, 829 (1995).
3) In the case of 2-(4-biphenylyl)-1-fluorovinyl phenyl sulfoxide, the carboxylated product, 3-(4-biphenylyl)-2-fluoropropenoic acid was produced. See ref 2.

ELECTROCHEMICAL ROUTE TO FLUORINATED KETHOESTERS

B.I. Martynov[1], A.A. Stepanov[2]
[1]State Research Institute of Organic Chemistry & Technology,
23, Shosse Entuziastov, Moscow, 111024 Russia
[2]A.N.Nesmeyanov Institute of Organoelement Compounds Russian Academy of
Sciences, 28, Vavilov str., Moscow, 117813 Russia

ABSTRACT
Electrochemical reduction of organic fluorohalogen compounds in the presence of trimethylchlorosilane enables to obtain the fluorinated silanes in synthetic yields. Trifluoroacetic acid esters at the same conditions forms condensation product - pentafluoroacetoacetic ester.

INTRODUCTION
Silanes, containing fluorinated substituent, are convenient synthetic reagents [1]. Traditional methods of their preparation are laborious enough. At the same time the electrochemical silylation of organohalogen compounds is described [2]. Our purpose was to investigate the electrochemical reduction of various fluororganics in the presence of trimethylchlorosilane.

METHOD
Preparative electrolyses were carried out in acetonitrile or DMF at the stainless steel or Ni cathode sacrificial Zn or Al anode in the presence of trimethylchlorosilane excess at galvanostatic conditions. The reduction potentials values adduced vs. SCE.

Table 1. Electrosilylation of fluoroorganic compounds.

Compound	$-E^{red}$, V	Product	Current efficiency, %
C_6F_6	2,11 2,33 2,61	$C_6F_5SiMe_3$	20
C_6F_5Cl	2,04 2,37 2,68	$C_6F_5 SiMe_3$	80
$C_6F_5CF_3$	2,01	$4-Me_3SiC_6F_4CF_3$	33
C_5F_5N	2,11 2,26 2,38 2,62	$4-Me_3SiC_5F_4N$	24
$CFCl_3$	1,82	$CFCl_2SiMe_3$	57
CF_3CCl_3	1,61	$CF_3CCl_2SiMe_3$	18
CF_2Br_2	1,51	-	-
CF_3I	1,72	CF_3SiMe_3	32
CF_3Br	1,92	CF_3SiMe_3	[3]
$CFCl=CFCl$	2,63	$CFCl=CFSiMe_3$	41
$CF_2=CFBr$	1,90	$CF_2=CFSiMe_3$	28
$CF_3CF=CF_2$	2,14	$CF_3CF=CFSiMe_3$	<10

The electrosilylation is a two-electron process, including the electrochemical generation of fluoroorganic carbanion and its silylation:

$$R_f X \xrightarrow{+2e} R_f^- + X^- \xrightarrow{Me_3SiCl} R_f SiMe_3 + Cl^- \qquad (1)$$

The silylation stage is not fast enough and some other processes may take place. In the case of bromodifluoromethanephosphonic acid it is protonation and for trifluoroacetic acid esters such process is a Kleisen-type condensation. The intermediate carbanion preferably reacts with starting ester (scheme 2).

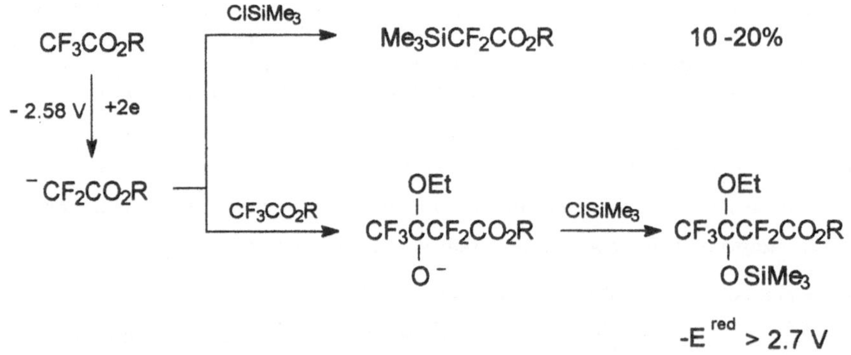

$$
\begin{array}{l}
CF_3CO_2R \xrightarrow{ClSiMe_3} Me_3SiCF_2CO_2R \qquad 10\text{-}20\% \\[2mm]
\downarrow -2.58\,V \;\; +2e \\[2mm]
{}^- CF_2CO_2R \xrightarrow[CF_3CO_2R]{}
\begin{array}{c} OEt \\ | \\ CF_3CCF_2CO_2R \\ | \\ O^- \end{array}
\xrightarrow{ClSiMe_3}
\begin{array}{c} OEt \\ | \\ CF_3CCF_2CO_2R \\ | \\ O\,SiMe_3 \end{array}
\end{array}
\qquad (2)
$$

$$-E^{red} > 2.7\,V$$

The yield of pentafluoroacetoacetic ester silylacetal is 53% (R = Et) and 37% (R = Me). The presence of TMCS in electrolyte is essential because the final products are not electroactive at working conditions, while perfluoroacetoacetic ester (E^{red} -1.96 V), forming in the absence of TMCS undergoes further reduction. Pentafluoroactoacetic ester is available by hydrolysis of silylacetal in mild conditions.

ACKNOWLEDGMENTS

The authors thank the International Science & Technology Center (ITSC project № 136-94) for financial support of this research.

REFERENCES
1) G.K.S. Prakash, R. Krishnamuri, G.A.Olah, *J. Am. Chem. Soc.*, **1989**, *111*, 2449.
2) P. Pons, C. Biran, M. Bordeau, J. Dunogues, *J. Organomet. Chem.*, **1988**, *358*, 31.
3) G.K.S. Prakash, D. Deffieux, A.K. Yudin, G.A.Olah, *Synth. Lett.*, **1994**, 1057.

ELECTROCHEMICAL SYNTHESES OF ORGANOFLUORINE COMPOUNDS

Kenji Uneyama, Kazushige Maeda, Tsuyoshi Kato and Toshimasa Katagiri

Department of Applied Chemistry, Faculty of Engineering, Okayama University, Tsushima-naka 3-1-1, Okayama 700, Japan

Difluoromethylene compounds are one of the current synthetic targets because of their unique biological activity. Among many synthetic methods of difluoromethylene compounds, selective defluorination of trifluoromethyl compounds is promising due to easy availability of trifluoromethylated compounds. Question is how to create the difluoromethylene unit from trifluoromethyl moiety. Deprotonation from difluoromethyl group and dehalogenation from halodifluoromethyl group have been well established. However, there is very few successful demonofluorination from trifluoromethyl group. Bond breaking of C-F bond is not easy due to the large bond energy, however, the bond breaking does easily occur when CF_3 group is attached to π-electron system as shown **1**, where the added electron would push out a fluoride ion. One of the problems in the conversion of **1** to **2**, however, is a further reduction of the initial product **2** and contamination of monofluoro-compounds because **2** mostly has a similar reduction potential to **1**. Whereas, demonofluorination would exclusively predominate in the conversion of **1** to **3** due to the large difference of reduction potentials between **1** and **3**. In fact, LUMO of difluoromethyl phenyl ketone is close to that of trifluoromethyl phenyl ketone which is much lower than that of silylenolether of difluoromethyl phenyl ketone **1** (Scheme 1). Therefore, silylenolethers are desirable products which would not be reduced under the electrolysis conditions and are synthetic synthons of the enolate of difluoroketones, and should be employed as useful building blocks for difluoro methylene compounds. Here, we describe an electrochemical transformation of trifluoromethyl ketones and imines to the β,β-difluoroenolsilyl ethers and enamines, respectively.

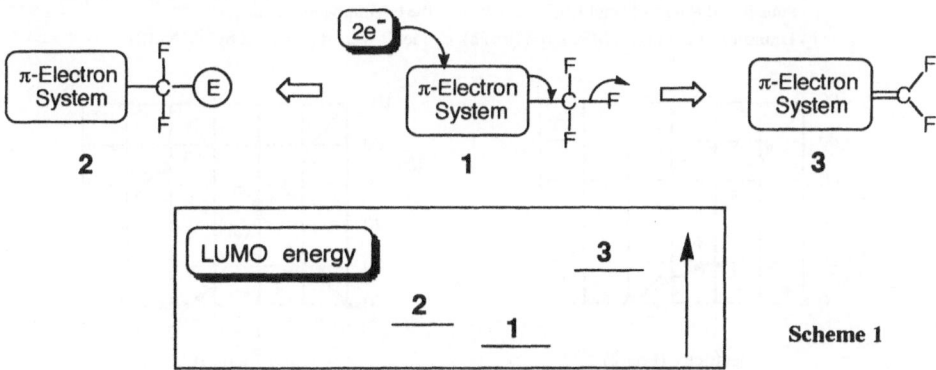

Scheme 1

The electroreductive defluorination of trifluoromethyl ketone (1) (1 mmol) was carried out using a Pb cathode (1 × 2 cm^2) and a carbon anode in dry acetonitrile (7 + 7 ml) containing n-Bu$_4$NBr (4 mmol) and TMSCl (3 mmol) in an H-type divided cell. Electricity was passed under a constant current of 30 mA at 0 ˚C until **1** was consumed (2 F/mol). After the electrolysis, Et$_3$N (3 mmol) was added, and the yield of difluorosilylenolether **5** was determined by ^{19}F NMR. At first, effects of solvents and supporting electrolytes were examined. n-Bu$_4$NBr was a better supporting electrolyte than LiClO$_4$. An n-Bu$_4$NBr-CH$_3$CN system gave a better result (80 %) than those of a LiClO$_4$-DMF (43 %) and a LiClO$_4$-CH$_3$CN (39 %) system. The results obtained under several reaction conditions are summarized in Table 1. Temperature did not affect the yield of **5** (entry 1-3) in the temperature range so far as examined. The lower current density at around 10 mA/cm^2 was found to be effective (entries 4-6). Cathode of Pb gave a good result so far as examined [Pb (80 %), carbon (77 %), and Pt (65 %)].

Table 1. Effects of Temperature, Current Density, and Amount of TMSCl$^{a)}$

Entry	Temp. (˚C)	Current density (mA/cm^2)	Equivalent of$^{b)}$ TMSCl (eq.)	Yield (%) $^{c)}$
1	- 20	15	3	81
2	0	15	3	80
3	30	15	3	80
4	0	4	3	78
5	0	15	3	80
6	0	50	3	71
7	0	15	1	0
8	0	15	2	53
9	0	15	3	80
10	0	15	5	63

a) Substituent R was Ph, and Et$_3$N was added at the beginning of electloysis.
b) Equimolar amount of TMSCl to **4** (R=Ph). c) Yields were determined by ^{19}F NMR.

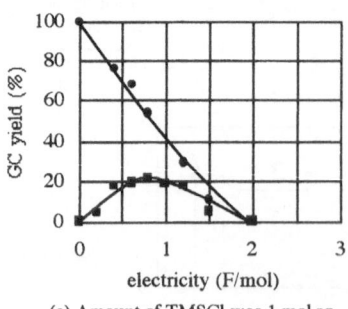

(a) Amount of TMSCl was 1 mol eq.

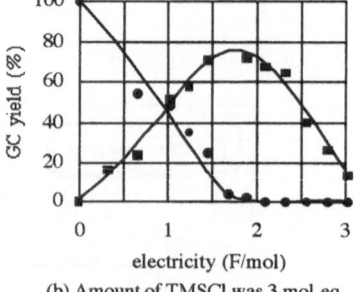

(b) Amount of TMSCl was 3 mol eq.

Figure 1. GC yields of 4 and 5 vs electricity: ● and ■ show 4 and 5, respectively.

One of the critical factors for the yield of **5** was the amount of TMSCl (entries 7-10).   When three equivalents of TMSCl for **4** were used, the yield of **5** was 80 % (entry 9). However, when only one equivalent of TMSCl for **4** was used (entry 7), **5** was not obtained.   To study the effect of the amount of TMSCl, the time-dependent distributions of **4** and **5** were analyzed (Figure 1).   According to Figure 1(b), the best yield of **5** was obtained at about 2 F/mol in the presence of 3 equivalents of TMSCl. However, when one equivalent of TMSCl was used, the yield of **5** reached the maximum on passing 1 F/mol of electricity and then gradually decreased, and **5** finally disappeared.   As the reaction proceeds, a fluoride ion is produced and it reacts with both TMSCl and product **5**, thus losing **5** (Figure 1(a)). These experimental facts suggest that trapping of the fluoride ion produced *in situ* by TMSCl is essential.

This reaction can be applied to a variety of substituted trifluoromethyl ketones (**4**) (Table 2). Both aromatic and aliphatic trifluoromethyl ketones provided **5** in reasonable yields.   Reactions of active methylene compounds (entry 6) and a heteroaromatic compound worked well.   Both aromatic and aliphatic ketones provided **5** in reasonable yields as shown in Table 2.

Scheme 2

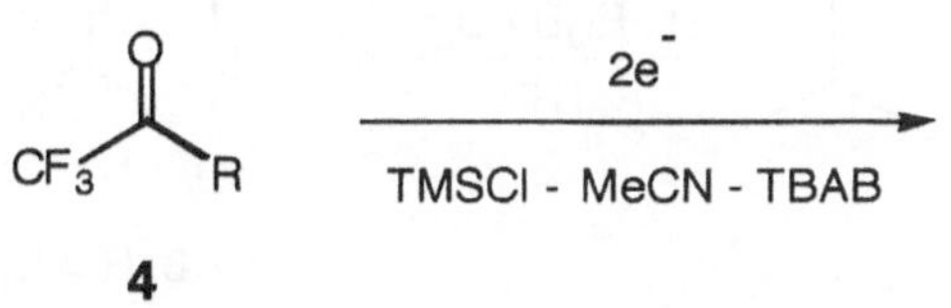

Table 2   Yield of β, β-difluoroenolsilyl ethers **5**

R	Ph	CH$_2$Ph	(CH$_2$)$_2$Ph	n-Hex	c-Hex	CH$_2$CO$_2$Me	2-Furyl
Yield (%)	80	44	74	71	66	50	55

The electrochemical defluorination can be applicable for imines **6** which are transformed to β,β-difluoroenamines **7**.   A system of LiClO$_4$-DMF-(C)-(Pb) was found to be useful in the electrolysis. Both aromatic and aliphatic imines gave **7** in resonable yields.   The enamines are promising precursors for β,β-difluoro-α-aminoacids when R is carboalkoxyl group.   The corresponding imino ester **7** (R=CO$_2$Et) is obtained in 86% yield (Table 3. by GC-analysis).

Scheme 3

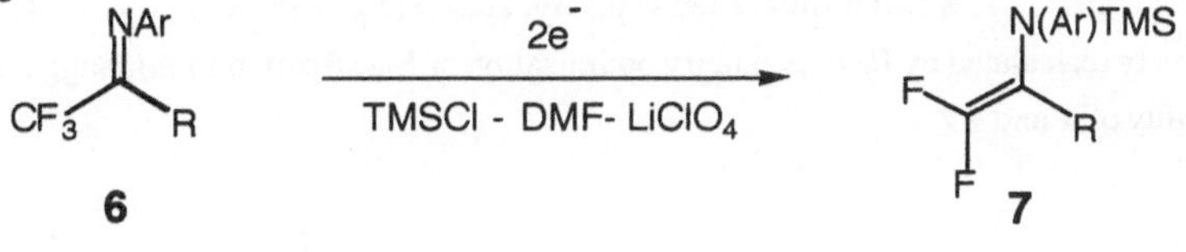

Table 3　Yield of β, β-difluoroenamines **7**

R	Ph	p-MeO-C$_6$H$_4$	p-Cl-C$_6$H$_4$	2-Furyl	CO$_2$Et	Et
Yield (%)	72	60	75	41	86	50

Alkylation of the flurorinated β-carbon atom of **7a** proceeded by both radical addition and the fluoride ion-promoted reaction. Thus, an isopropyl radical added to the β,β-difluoroacrylates **7a** affording 3,3-difluoro-2-imino-4-methylpentanoate **8** respectively, precursors of β,β-difluoro leucine[12] (Scheme 4).

Scheme 4

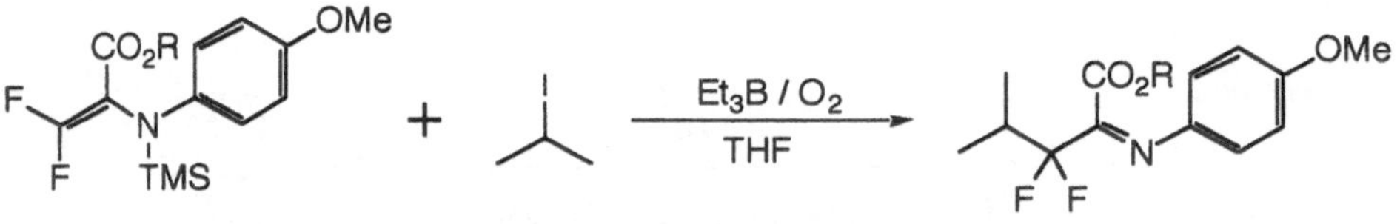

7a: R = Et　　　　　　　　　　　　　　　　**8**: R = Et (33 %)

Treatment of *N*-trimethylsilylated β,β-difluoroenamine **9** with a fluoride ion promoted the generation of 2,2-difluoro-2-imino-1-phenylethyl carbanion and its reaction with benzaldehyde leading to the formation of the adduct **11** in 76 % overall yield (Scheme 5).

Scheme 5

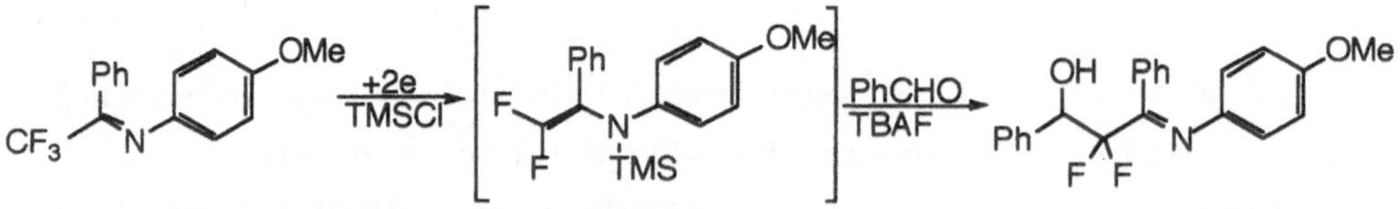

9　　　　　　　　　　　**10**　　　　　　　　　　**11** (72 % overall yield)

Note

1) LUMO energies of **4**, **5** and difluoromethyl phenyl ketone (**12**) were-0.97 eV, -0.41 and -0.87 eV, respectively (calculated by PM3 geometry optimization of Mac Spartan Plus), suggesting similar reducibility of **4** and **12**.

ELECTROCHEMICAL INTRODUCTION OF FLUORINATED SUBSTITUENTS INTO ORGANIC MOLECULES

Nikolai IGNAT'EV, Sergii DATSENKO and Elena SMERTENKO

Institute of Organic Chemistry Ukrainian National Academy of Sciences, Laboratory
Electrochemistry of Organoelement Compounds, Murmanskaya 5, 253660 Kiev-94, Ukraine

ABSTRACT

Electrochemical transformations of organofluorine compounds lead to the generation of active fluoro-containing species (radicals, radical-anions, anions etc.), which are able to react with different substrates. This approach gives the possibility to introduce fluorinated substituents into organic molecules by electrochemical methods.

INTRODUCTION

Traditionally electrochemistry is used for the preparation of fluorine and its compounds. Production of fluorine and electrochemical fluorination in anhydrous hydrogen fluoride (Simons process) are examples of industrial application of this method [1]. Nowadays different kind of fluoro organic compounds are produced at industrial scale. Electrochemical transformation of these substances is a promising way for the synthesis of new types of fluoro organic compounds (building-block approach) [2].

RESULTS AND DISCUSSION

In contrast to the alkyl halides, perfluoroalkyl halides show a high resistance against the displacement of halogen atom via S_N2 or S_N1 mechanism. But owing to their reduction potential values [3], perfluoroalkyl iodides and bromides can be converted to radical intermediates via single electron transfer processes [4]. For example, we have found that the electrochemical reduction of perfluoro- and polyfluoroalkyl halides in the presence of thiolate anions leads to the formation of substituted products [5a,b].

$$R_FX \quad + \quad RS^- \quad \xrightarrow{\ e\ } \quad R_FSR \quad \text{Yield}: 60\text{-}80\%$$

$$R_F = CF_3, C_3F_7, C_4F_9, C_4F_8H, C_6F_{13}; \quad X = I, Br$$

$$R = C_8H_{17}, C_6H_5CH_2, p\text{-}CH_3C_6H_4, p\text{-}ClC_6H_4$$

Low consumption of electricity, less than 0.3 F/mol testifies to a chain radical-ion mechanism.

Polyfluoroalkyl iodides of general formula $H(CF_2CF_2)_nCH_2I$ where n = 1-3 are reduced at more negative potentials than the perfluoroalkyl iodides. Due to this fact, a higher cathodic potential is required for the initiation of their reaction with thiolate anions[5b].

$$H(CF_2CF_2)_nCH_2I \quad + \quad p\text{-}ClC_6H_4S^- \quad \xrightarrow[DMF]{e} \quad H(CF_2CF_2)_nCH_2SC_6H_4Cl$$

$$n = 1 - 3$$

Yield: 63 - 72 %

Current yield: 0.3 - 0.5 F/mol

The methyl sulfite anion, $CH_3OSO_2^-$ is attacked by electrophilic perfluoroalkyl radicals exclusively on the sulfur atom.

$$CH_3OS\overset{O}{\underset{O}{\lessdot}}{}^- \ Na^+ \quad + \quad CF_3X \quad + \quad 2e \quad \longrightarrow \quad CF_3SO_2^- \quad + \quad CH_3O^- \quad + \quad X^-$$

$$X = I, \ Br \qquad \qquad Yield \ (NMR): 95 \%$$

Perfluoroalkylation of phenolates on the oxygen position can be carried out by the electrochemical reduction of perfluoroalkyl halides in the presence of ring substituted phenolate-anions [6], for example:

$$2,4,6\text{-}(CH_3)_3C_6H_2O^- \quad + \quad C_3F_7I \quad \xrightarrow{e} \quad 2,4,6\text{-}(CH_3)_3C_6H_2O\,C_3F_7$$

Yield : 44 %

It was shown that oxygen acts as a mediator in this process.

$$O_2 \quad + \quad e \quad \rightleftharpoons \quad O_2^{\cdot-}$$

$$R_FX \quad + \quad O_2^{\cdot-} \quad \longrightarrow \quad [R_FX]^{\cdot-} \quad + \quad O_2$$

$$[R_FX]^{\cdot-} \quad \longrightarrow \quad R_F^{\cdot} \quad + \quad X^-$$

Perfluoroalkyl radicals can be generated not only by electrochemical reduction of perfluoroalkyl halides, but also by electroreduction of perfluoroalkyliodonium salts. In this case the process proceeds at much lower cathodic potentials[7].

$$C_6H_5 - \overset{+}{I} - R_F \quad \xrightarrow{\;e\;} \quad C_6H_5I \quad + \quad R_F^{\cdot}$$

$$X^-$$

$$R_F = C_3F_7, \; i\text{-}C_3F_7, \; C_6F_{13}$$

$$X = BF_4^-, \; CF_3SO_3^-, \; CF_3COO^-$$

$$C_3F_7^{\cdot} \quad + \quad p\text{-}ClC_6H_4SO_2^- \quad \xrightarrow{\;-e\;} \quad p\text{-}ClC_6H_4SO_2C_3F_7$$

Perfluoroacyl halides are convenient source for the *in situ* electrochemical generation of α-diketones [7].

$$R_F - C\!\!\overset{\displaystyle O}{\underset{\displaystyle X}{<}} \quad \xrightarrow{\;e\;} \quad \left[R_F - C\!\!\overset{\displaystyle O}{\underset{\displaystyle X}{<}} \right]^{\cdot -} \quad \xrightarrow{\;-X^-\;} \quad R_F - C\!\!\overset{\displaystyle O}{\cdot}$$

$$\downarrow \text{dimerization}$$

$$R_F = C_4F_9 \qquad X = I, \; Br \qquad\qquad R_F - \underset{\displaystyle O}{\overset{\displaystyle ||}{C}} - \underset{\displaystyle O}{\overset{\displaystyle ||}{C}} - R_F \quad \text{Yield: } 15\text{ - }20\,\%$$

Electrochemical oxidation of sodium trifluoromethanesulfinate leads to the formation of trifluoromethyl radical, that gives the possibility to introduce of CF_3 - group into organic molecules via the electrooxidation process[8].

$$CF_3SO_2^- \quad \xrightarrow{\;-e\;} \quad \left[CF_3SO_2 \right]^{\cdot} \quad \xrightarrow[-SO_2]{} \quad CF_3^{\cdot}$$

$$CF_3^{\cdot} \quad + \quad ArH \quad \longrightarrow \quad \left[CF_3ArH \right]^{\cdot}$$

$$\left[CF_3ArH \right]^{\cdot} \quad \xrightarrow{\;e, -H^+\;} \quad Ar\,CF_3 \quad \text{Yield up to 50\%}$$

Ar = benzene, toluene, durene, naphthalene, xylene

Preparative scale electrochemical oxidation of the trifluoromethanesulfinate-anion in the presence of benzene and oxygen results in the formation of trifluoromethoxybenzene.

$$C_6H_6 \quad + \quad O_2 \quad + \quad CF_3SO_2^- \quad \xrightarrow{\;-e\;} \quad C_6H_5OCF_3 \quad \text{Yield : 30\%}$$

Perfluoroalkanesulfonamides [9] are a convenient source for the electrochemical generation of $(CF_3)_2N^-$ anion and $(CF_3)_2N^{\cdot}$ radical and can be applied to the electrochemical introduction of $(CF_3)_2N$ - group into organic molecules.

Electrochemical reduction of $CF_3SO_2N(CF_3)_2$ in the presence of $BrCH_2COOC_2H_5$ leads to bromine substitution:

$$CF_3SO_2N(CF_3)_2 \quad + \quad BrCH_2COOEt \quad \xrightarrow{\ e\ } \quad (CF_3)_2NCH_2COOEt$$

main product

The reaction proceeds via a electrocatalytic way with low consumption of electricity. The possible mechanism of this process is the following:

$$BrCH_2COOEt \quad \xrightarrow{\ e\ } \quad Br^- \quad + \quad {}^-CH_2COOEt$$
$$E_p \; = \; -1.56 \text{ V (vs. SCE)}$$

$$CF_3SO_2N(CF_3)_2 \quad + \quad Br^- \quad \longrightarrow \quad {}^-N(CF_3)_2 \quad + \quad CF_3SO_2Br$$

$$BrCH_2COOEt \quad + \quad {}^-N(CF_3)_2 \quad \longrightarrow \quad Br^- \quad + \quad (CF_3)_2NCH_2COOEt$$

The electrochemical oxidation of the $(CF_3)_2N^-$ anion results in the formation of the $(CF_3)_2N^{\bullet}$ radical which can be trapped in the reaction with aromatic compounds.

$$(CF_3)_2N^- \quad - \quad e \quad \longrightarrow \quad (CF_3)_2N^{\bullet}$$
$$E_p \; = \; 1.72 \text{ V (vs. SCE)}$$

$$(CF_3)_2N^{\bullet} \quad + \quad ArH \quad \xrightarrow{-e,-H^+} \quad Ar-N(CF_3)_2$$

Ar = benzene, naphthalene

References

1) *Organofluorine Chemistry*, ed. R.E. Banks, B.E. Smart and J.C. Tatlow, Plenum Press, N.Y., **1994**.
2) T. Fuchigami, Electrochemical Reactions of Fluoro Organic Compounds, *Topics in Current Chemistry*, Vol. 170, **1994**.
3) N.Ignat'ev, S.Datsenko, S.Pazenok, L.Yagupolskii, Zh. Org. Khim.(Russ), **1990**, *26*, 1740.
4) C.Wakselman and A.Lantz, in *Organofluorine Chemistry,* ed. R.E. Banks, B.E. Smart and J.C. Tatlow, Plenum Press, N.Y., **1994**, ch.8, 178.
5) (a) N.V. Ignat'ev, S.D. Datsenko and L.M. Yagupol'skii, *Zh. Org. Khim*, **1991**, 27, 905; (b) S.D. Datsenko, N.V. Ignat'ev and L.M. Yagupol'skii, *Soviet Electrochemistry*, **1991**, 27, 1484.
6) S.D. Datsenko, N.V. Ignat'ev and L.M.Yagupol'skii, *Soviet Electrochemistry,* **1991**, 27, 1016
7) E.A. Smertenko, S.D. Datsenko, N.V. Ignat'ev, L.E. Deev, I.K. Bil'dinov and P.V. Podsevalov, *Russian J. of Electrochemistry*, **1994**, 30, 1172.
8) S.Datsenko, E.Smertenko, N.Ignat'ev, *Russian J. of Electrochemistry*, **1997**, in press.
9) P. Sartori , N. Ignat'ev and S. Datsenko , *J. Fluorine Chem.*, **1995**, 75, 157.

INVESTIGATION OF THE MECHANISMS OF ELECTROGENERATION OF ANIONS

Stephen E. TREIMER and Dennis H. EVANS

Department of Chemistry and Biochemistry, University of Delaware, Newark, Delaware 19716, USA

ABSTRACT

Recent studies of the mechanism of reduction of weak acids resulting in the formation of active anions are reviewed. The acids that were investigated included CH, NH and OH acids with pK_a (DMSO) ranging from 1.6 to 19.8. Acids with pK_a up to about 6 were reduced by a CE scheme involving prior dissociation of the acid giving the anion (conjugate base) and the solvated proton, the latter being discharged at the platinum cathode to produce dihydrogen. Acids with pK_a greater than 6 cannot dissociate rapidly enough to produce the diffusion-controlled reduction currents that are seen. These acids are reduced by a different mechanism, direct discharge with simultaneous electron transfer and element-hydrogen bond breaking producing the anion and an adsorbed hydrogen atom. The CH acids in this group are thought to undergo tautomerization to OH or NH forms prior to reduction. In the case of ethyl nitroacetate and 2,4-pentanedione tautomerization is slow enough to become the rate-limiting step in the overall reaction scheme.

INTRODUCTION

There are several ways to generate anions by electrochemistry including the use of electrogenerated bases and the direct reduction of an element-hydrogen bond, *e.g.*, $ROH + e^- \rightarrow RO^- + \frac{1}{2} H_2$. The anions so produced can be made the basis of useful electroorganic synthetic methods.[1] Investigation of the mechanism of formation of active anions may lead to improvements in the efficiency and selectivity of these synthetic methods. Here we will review our recent studies[2,3] of the electrochemical reduction of weak acids which have led to a categorization of mechanisms. In this research, the mechanisms of the reduction of a series of OH, NH and CH acids were investigated. The reactions were studied at platinum electrodes in dimethyl sulfoxide to take advantage of the well developed absolute acidity scale that exists in this solvent.

309

DISCUSSION

The acids have been categorized according to their thermodynamic and kinetic properties. Category 1 acids are relatively strong acids that undergo facile proton transfer reactions and which do not have low-lying LUMOs which would allow reduction to the radical anion prior to their reduction as acids. These acids are reduced by a CE mechanism in which dissociation of the acid occurs forming the solvated proton and the conjugate base of the acid. The solvated proton is then discharged at the platinum electrode leading to the formation of H_2. Five acids between pK_a 1.6 and 5.1 were shown to be reduced *via* this CE mechanism.

When studied at a platinum electrode, the reaction occurs quite reversibly. The cyclic voltammetric peaks are close to the formal potential for the $2\,HA + 2\,e^- = 2\,A^- + H_2$ couple (HA denotes the weak acid and A^- its conjugate base). The use of a catalytic surface such as platinum is essential for obtaining a reversible voltammetric response. Glassy carbon or mercury are much less active leading to high overpotentials for discharge of the solvated proton.

Category 2 acids are weaker acids which cannot dissociate rapidly enough to produce the currents that are seen. (The rate constant for dissociation, k_{dissoc}, cannot be larger than $K_a k_{assoc}$, where k_{assoc} is the rate constant for recombination of the solvated proton with the conjugate base. Of course, k_{assoc} cannot exceed the diffusion-controlled limit). In fact, however, it was observed that these acids are reduced in a diffusion-controlled electrode reaction that has been interpreted as being due to direct discharge of the acid without prior dissociation. For example, for dichloroacetic acid (pK_a (DMSO) = 6.4) the proposed reaction is:

$$CHCl_2COOH + Pt + e^- \longrightarrow CHCl_2CO_2^- + H-Pt$$

where H-Pt indicates an adsorbed hydrogen atom. Subsequent reactions of H-Pt eventually yield H_2 either by surface combination of hydrogen atoms or by discharge of another molecule of acid at H-Pt. These reactions are analogous to the well-studied discharge of water from neutral or basic aqueous solutions. A large number of acids between pK_a 6.4 and 19.8 were shown to be reduced as Category 2 acids under the conditions used in the studies.

Category 3 acids are those which have a low-lying LUMO so that they are able to accept an electron to form the radical anion at less negative potentials than are required for their reduction as acids. 4-Nitrobenzoic acid (pK_a = 9.0) is an example of such an acid. Many examples of category 3 acids are known. The mechanisms often feature a parent-child reaction in which the initially formed radical anion is protonated by the weak acid (HA) giving a neutral radical and A⁻. The final products arise from the subsequent reactions of the neutral radicals. Dihydrogen is normally not a product of the reduction of category 3 acids.

Among the Category 2 acids (those that are reduced by direct discharge of the acid at the platinum cathode) were examples of OH, NH and CH acids. Though the mechanism for reduction of category 2 acids has changed from that prevailing for category 1 acids, the midpoint between cathodic and anodic peak potentials obtained by slow scan rate cyclic voltammetry was found to be very close to the voltammetric half-wave potential computed from the formal potential for the H^+/H_2 couple and the pK_a. However, the separation between the voltammetric peaks becomes quite large for category 2 acids and it increases with increasing pK_a reaching 0.9 V for pyrazole ($pK_a = 19.8$) at 0.10 V/s. Thus the position of the set of peaks along the potential axis is governed by the thermodynamics of the overall reaction but the kinetics is influenced by the acidity, with the weaker acids being kinetically less facile.

Of twenty acids investigated in the pK_a range of 6.4 to 19.8, only three deviated significantly from the trends discussed above. These acids were 2-naphthol, phenol and 2,6-di-*tert*-butylphenol and each showed a separation of peak potentials that was significantly larger than other acids of comparable acidity (mainly NH heterocycles). The direct discharge mechanism can be considered to be a type of dissociative electron transfer for which the bond-dissociation energy is a major contributor to the activation barrier. Thus, the less reversible behavior of the arenols may be traced to the fact that OH bond-dissociation energies are significantly larger than those for NH bonds.

A major objective of the studies was to determine what differences, if any, exist for CH acids compared to OH and NH acids of comparable acidity. The basis for this interest is the well-known fact that the rates of acidic dissociation (heterolysis) for CH acids are abnormally small. As pointed out above, the direct discharge mechanism for a CH acid would involve simultaneous electron transfer and breaking of the CH bond so it was anticipated that the reduction of CH acids might occur with greater overpotential than OH or NH acids. A major complicating factor in such a study is the fact that CH acids that fall in the range of pK_a-values studied are caused to be fairly acidic by the presence of strong electron-withdrawing groups (nitro, cyano, acetyl, benzoyl, carbomethoxy, etc.). Thus, they are not typical CH acids. Furthermore, many of them exist partially or totally in tautomeric forms which are either OH or NH acids so they are not truly CH acids at all. For example, triacetylmethane is completely enolized in DMSO and its electrochemical behavior is very similar to that of NH and OH acids of comparable acidity. 1,3-Diphenyl-1,3-propanedione is also completely enolized and it appears to be reduced as an OH acid but the issue is complicated by interference by its reduction as a category 3 acid, i.e., insertion of an electron into its LUMO forming an intermediate radical anion.

Two examples of relatively strong CH acids that exist as such are 2,2,5-trimethyl-1,3-dioxane-4,6-dione (pK$_a$ (DMSO) = 7.4) and malononitrile (pK$_a$(DMSO) = 11.0). The behavior of each of these was found to be very similar to that of OH and NH acids of comparable acidity including diffusion-controlled reduction at scan rates up to 40 V/s. Thus, these CH acids are either reduced by direct discharge or, as discussed below, they are capable of very fast tautomerization to form OH and NH forms, respectively, that can be reduced by direct discharge.

Ethyl nitroacetate exists as a CH acid in DMSO and, at low scan rates, its behavior is very similar to that seen for OH and NH acids whose pK$_a$-values are similar (pK$_a$ (DMSO) = 9.2 for ethyl nitroacetate). However, as the scan rate is increased from 0.1 to 40 V/s, the main reduction peak diminishes relative to a new peak that grows in at more negative potentials. This second peak occurs very close to the position seen for reduction of ethyl nitroacetate at a non-catalytic surface such as glassy carbon suggesting that it is due to reduction as a category 3 acid, i.e., formation of the radical anion. Thus the first peak appears to be under kinetic control. The interpretation that was suggested is that the chemical reaction that limits the current seen at the first peak is the tautomerizaton of the CH form of ethyl nitroacetate to a hypothetical OH form (either a nitronic acid or an enol) which is the species that can be reduced by direct discharge at the potential of the first peak.

Indications of this same type of prior tautomerization were found for 2,4-pentanedione and ethyl acetoacetate. Thus, it was not possible to reach general conclusions concerning the reduction of CH acids though it does appear that in some cases direct discharge of the CH bond is kinetically hindered and prior tautomerization is necessary in order for reduction to occur.

REFERENCES
1) For a review see M. E. Niyazymbetov and D. H. Evans, *Tetrahedron*, **1993**, *49*, 9627.
2) S. E. Treimer and D. H. Evans, *J. Electroanal. Chem.*, (in press).
3) S. E. Treimer and D. H. Evans, submitted to *J. Electroanal. Chem.*

ELECTROCHEMICAL FORMATION OF FAVORSKII REARRANGEMENT TYPE INTERMEDIATES AND THEIR REACTIVITIES

Toshiro CHIBA, Isao SAITHO, and Mitsuhiro OKIMOTO
*Department of Applied Chemistry, Kitami Institute of Technology,
Kitami, Japan 090*

ABSTRACT

The indirect electrooxidation of enamines of alicyclic ketones (**1**) in NaOMe-MeOH using iodide ion as a mediator provided ring fused cyclopropane aminoethers (**2**). The resulting products were then subjected to acid-catalyzed hydrolysis, alcoholysis, the reductive elimination of the methoxy group with $LiAlH_4$, and the replacement of the methoxy group by a cyano group with Me_3SiCN in the presence of a Lewis acid.

INTRODUCTION

While the reaction of enamines with various kinds of electrophiles has been studied, little attention has been given to their oxidation, especially electrochemical oxidation. In our continuing studies of the electrochemical oxidation of nitrogen containing compounds,[1] we found that the electrolytic oxidation of enamines **1** in methanol containing sodium methoxide and a catalytic amount of KI induced their intramolecular cyclization and a simultanious methoxylation to give ring fused cyclopropane aminonitriles, such as **2**.

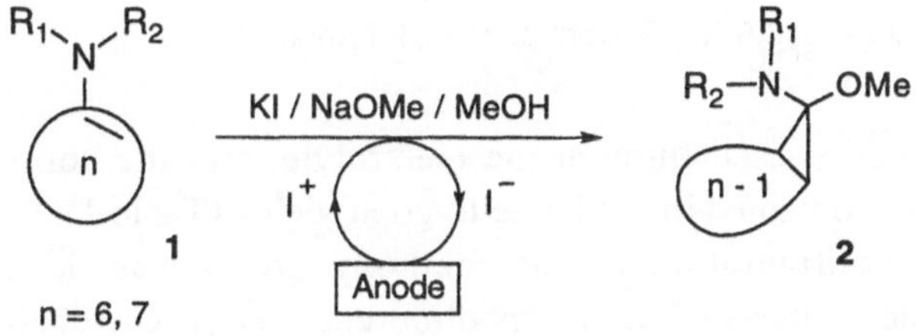

The obtained products **2** have the ketal or aminal structures of cyclopropanone, and they can be regarded as one of the intermediates in the Favorskii rearrangement.[2] Although the Favorskii rearrangement has been extensively studied by various investigators, such a cyclopropane intermediate has scarcely been isolated. Thus, we explored the present electrolysis using various enamines and examined the chemical properties of these unique products.

RESULTS AND DISCUSSION

Preparative electrolyses were carried out in a divided cell equipped with a platinum gauze anode, a platinum coil cathode, and a porous ceramic diaphragm. A solution of an enamine (30 mmol) in 1 M NaOMe-MeOH (80 mL) containing KI (30 mmol) was electrolyzed under a constant current of 0.5 A. In most cases, the starting enamines were almost totally consumed by the time approximately 2 F/mol

Table 1. Electrooxidation of Enamines in KI/NaOMe/MeOH^a

	Starting enamine **1**			Cyclopropaneaminoether **2**
	R_1	R_2	n	(yield, %)
1a	Et	Et	6	**2a** (67)
1b	Me	Ph	6	**2b** (69)
1c	-(CH$_2$)$_4$-		6	**2c** (63)
1d	-(CH$_2$)$_2$-O-(CH$_2$)$_2$-		6	**2d** (74)
1e	-(CH$_2$)$_4$-		7	**2e** (63)
1f	-(CH$_2$)$_2$-O-(CH$_2$)$_2$-		7	**2f** (56)

^{a}Enamine (20 mmol), NaOMe (30 mmol), KI (30 mmol) in MeOH (80 mL).
Constant current, 0.5 A. Current passed, 2 F/mol.

of electricity had passed through the electrolyte, and the corresponding bicyclic compounds were obtained in moderate to good yields (Table 1).

The *exo* configuration of the methoxy group was determined by X-ray crystallographic analysis of the morpholino enamine of cycloheptanone (**2f**) which crystallized at room temperature. In our experiments, the *endo*-isomers could not be detected.

The resulting products **2** were highly susceptible toward acidic aqueous conditions, and a different reactivity between the bicyclohexane and bicycloheptane aminoethers was obsurved for the acid-catalyzed hydrolyses and alcoholyses. For example, when the bicyclohexane aminoether (**2c**) was shaken with dil. H$_2$SO$_4$ in ether, α-hydroxycyclohexanone (**8**) immediately formed. In contrast to this, the hydrolysis of the bicycloheptane aminoether (**2e**) under similar conditions resulted in the formation of the hemiaminal (**9**), instead of the α-hydroxyketone. The resulting hemiaminal **9** could be selectively transformed into 7-*endo*-norcaranol by treating with NaBH₄, as has been already reported.[2b)]

On the other hand, the treatment of the bicyclohexane aminoether (**2d**) with excess methanol afforded the methoxyenamine (**6**). Likewise, the reaction with ethanol gave the ethoxyenamine (**7**). However, during the ethanolysis of the bicycloheptane aminoether (**2f**) under similar conditions, replacement of the methoxy by an ethoxy group took place without causing carbon skeleton rearrangement leading to the ethoxynorcarane (**10**).

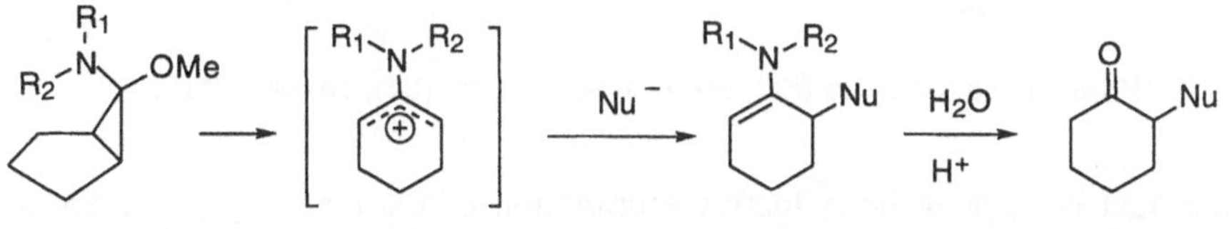

2a-d   **5**   **6, 7**   **8**

8, R₁, R₂ = -(CH₂)₄-, Nu = OH (90%)
6, R₁, R₂ = -(CH₂)₂-O-(CH₂)₂-, Nu = OMe (78%)
7, R₁, R₂ = -(CH₂)₂-O-(CH₂)₂-, Nu = OEt (76%)

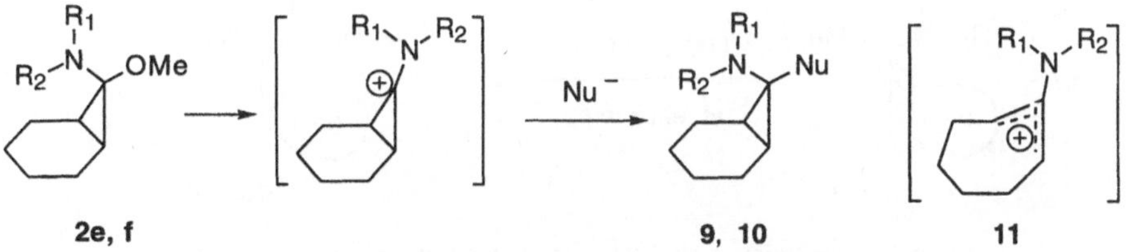

2e, f   **9, 10**   **11**

9, R₁, R₂ = -(CH₂)₄-, Nu = OH (80%)
10, R₁, R₂ = -(CH₂)₂-O-(CH₂)₂-, Nu =OEt (78%)

Some of these reactions appear to proceed through the iminium cation. The different reactivity between the bicyclohexane and bicycloheptane aminoethers may be due to the strain in the bicyclic system. The cyclopropane ring fused to a five-membered ring seems to favor the ring opening reaction to give the monocyclic system (**5**), due to relieving strain, whereas the cyclopropane ring fused to a six-membered ring does not rearrange into the strained cycloheptenyl cation (**11**).

It has been known that the action of Lewis acids on α-methoxyamides brings about the generation of iminium cations, which can be trapped *in situ* with various nucleophiles.[3]   An attempt to replace the methoxy group by a cyano group using Me_3SiCN as a nucleophile was successfully performed when BF_3 etherate was used as the Lewis acid at - 78 °C.   The cyanation occurs stereoselectively, and the corresponding *exo*-nitriles (3) were exclusively obtained in good yields.

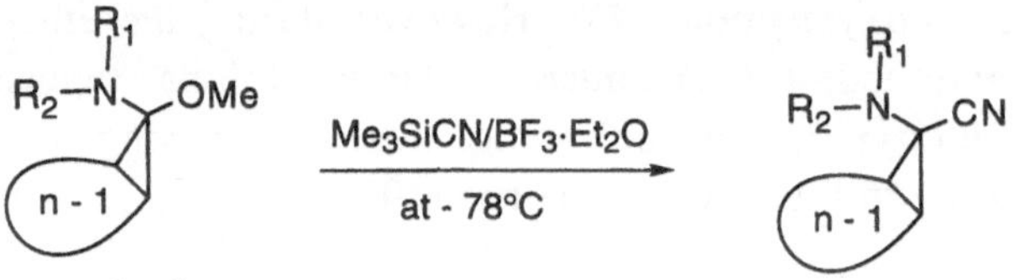

Product, yield (%): **3a** (84), **3b** (86), **3c** (63), **3d** (54), **3e** (90), **3f** (73)

We next attempted the reductive elimination of the methoxy group from the electrolysis products **2**, since ketals have been found to undergo hydrogenolysis with appropriate metal hydrides.[4]   When aminoether **2** was refluxed with a 1.5 fold excess of $LiAlH_4$ in Et_2O, the corresponding *endo*-amines (4) could be obtained in good yields, except for **2b**.

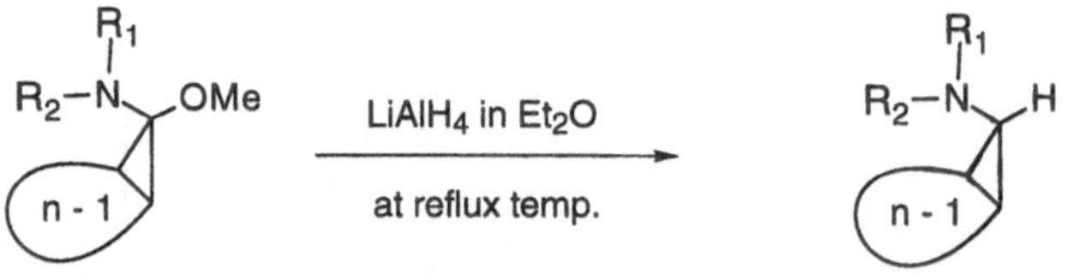

Product, yield (%): **4a** (84), **4c** (85), **4d** (88), **4e** (87), **4f** (74)

References
1) T. Chiba, H. Sakagami, M. Murata, M. Okimoto, *J. Org. Chem.*, **1995**, *60*, 6764.
2) (a) A. S. Kende, *Org. Reactions*, **1960**, *11*, 261; see also (b) J. Szmuszkovicz, E. Cerda, M. F. Grostic, J. F. Zieserl, Jr., *Tetrahedron Lett.*, **1967**, 3969.
3) H. E. Zaugg, W. B. Martin, *Org. Reactions*, **1965**, *14*, 52.
4) E. L. Eliel, R. A. Daignault, *J. Org. Chem.*, **1965**, *30*, 2450.

STUDIES OF CATHODICALLY-PROMOTED ADDITION REACTIONS OF NITROMETHANE WITH VARIOUS ALDEHYDES

Zukhra NIAZIMBETOVA and Dennis H. EVANS

Department of Chemistry and Biochemistry, University of Delaware, Newark, Delaware 19716, USA

ABSTRACT

Addition of nitromethane to five heterocyclic aldehydes, anthracene-9-carboxaldehyde and ferrocenecarboxaldehyde has been studied. In all cases, the corresponding nitroalcohols are formed initially and, for the less basic aldehydes, this product can be isolated in good to excellent yield. For the more basic heterocyclic aldehydes, the initially formed nitroalcohol dehydrates under the reaction conditions forming the nitroolefin. This is particularly evident for the reaction of N-methylpyrrole-2-carboxaldehyde where the nitroolefin was found to add a second molecule of nitromethane forming 2-(N-methyl-2-pyrrolyl)-1,3-dinitropropane. Similar results were obtained with ferrocenecarbox-aldehyde where it was possible to obtain the 1,3-dinitro compound as the sole product.

INTRODUCTION

This paper provides a review of our recent work on the use of nitronate anions in electroorganic synthesis.[1,2] When studying the electrogeneration of the anion of ethyl nitroacetate,[3] it was discovered that efficient generation required the use of an electrogenerated base, superoxide anion. Attempts to generate the anion by direct reduction were quite inefficient due to side reactions of the initially formed radical anion. When formed *via* electrogenerated base, the anions of ethyl nitroacetate were found to undergo efficient alkylation and Michael addition reactions. The method was extended to include a variety of nitro compounds particularly with respect to Michael addition reactions[4] wherein it was observed that reaction of the anion of nitromethane with levoglucosenone resulted in a product derived from one molecule of nitromethane and two of levoglucosenone. Multiple additions were also observed for nitromethane and methyl vinyl ketone where a cyclic product derived from one nitromethane and three molecules of methyl vinyl ketone was obtained.[5]

Extension of these studies to the addition of nitromethane to carbonyl compounds led to the discovery that the reaction was very slow when conducted under standard conditions with acetonitrile as solvent.[1] However, when nitromethane itself was used as solvent, very efficient addition to benzaldehyde was achieved. A 95% yield of the nitroalcohol ((1-hydroxy-2-nitroethyl)benzene) was obtained and the reaction was catalytic, requiring only 0.06 Faraday charge per mole of aldehyde. Using this procedure, good to excellent yields of nitroalcohols from pentanal as well as mono- and bis-nitroadducts of 1,4- and 1,3-benzenedicarboxaldehyde were obtained.[1]

Treatment of the nitroalcohols with hot 5% aqueous H_3PO_4 resulted in dehydration with formation of the corresponding nitroolefins (Scheme 1). The addition of nitromethane to the nitro-

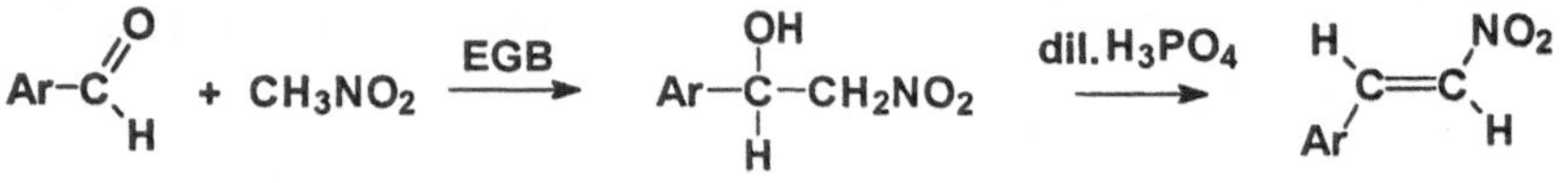

$$Ar-\overset{O}{\underset{H}{C}} \;+\; CH_3NO_2 \;\;\xrightarrow{EGB}\;\; Ar-\overset{OH}{\underset{H}{C}}-CH_2NO_2 \;\;\xrightarrow{dil.\,H_3PO_4}\;\; \overset{H}{\underset{Ar}{}}C=C\overset{NO_2}{\underset{H}{}}$$

Scheme 1

olefin derived from benzaldehyde (Ar = C_6H_5) was achieved in good yield forming 1,3-dinitro-2-phenylpropane.

RESULTS AND DISCUSSION

To investigate the scope of the synthetic procedure, a number of other aldehydes were studied (Table 1). The reaction conditions were as previously described.[1] A very good yield of nitroalcohol was obtained from thiophene-2-carboxaldehyde but with furan-2-carboxaldehyde (furfural) a significant amount of dehydration occurred in the electrolysis mixture. In this case, a "solventless" electrolysis was carried out, as described previously,[1] in which a 1.2/1.0 molar ratio of nitromethane to aldehyde was used without added solvent. The overall yields were significantly lower than obtained under standard conditions. By contrast, 5-nitrofuran-2-carboxaldehyde produced only nitroalcohol, in moderate yield, with no evidence for dehydration occurring in the electrolysis cell. In fact, this nitroalcohol was somewhat resistant to dehydration by dilute phosphoric acid though dehydration could still be accomplished.

Treatment of pyrrole-2-carboxaldehyde produced only small amounts of the desired products. It is suspected that air-oxidation was taking place. By contrast, a reasonable yield of the 1,3-dinitro compound derived from addition of two molecules of nitromethane to N-methylpyrrole-2-carboxaldehyde was obtained. Here, the initially formed nitroalcohol undergoes rapid dehydration in the cell leading to the nitroolefin which also readily adds nitromethane.

Anthracene-9-carboxaldehyde produced a good yield of nitroalcohol when the electrolysis was conducted in a divided cell, the other principal product being nitroolefin (Table 2). The nitroalcohol is

Table 1. Addition of nitromethane to heterocyclic aldehydes.[a]

Compound	Product	Charge (F/mol)	Yield	Conditions
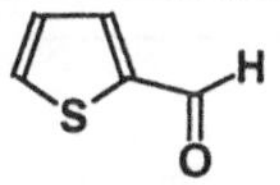	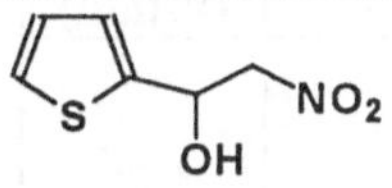	0.13	78%	Undivided
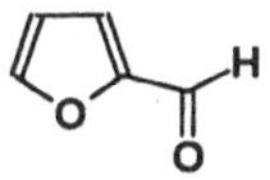	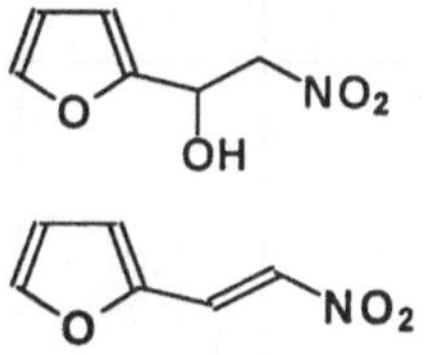	0.10 (0.04)	74% (45%) 25% (17%)	Undivided (Solventless)
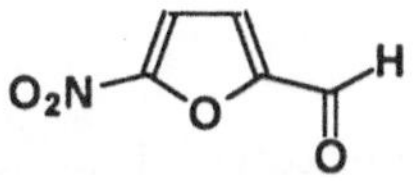	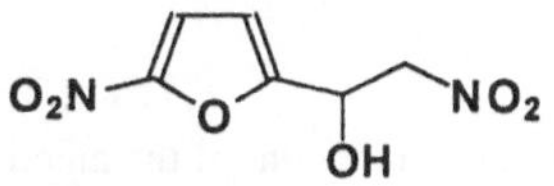	0.036 0.13	63% 61%	Undivided Divided
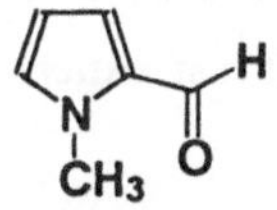	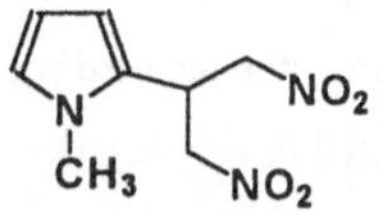	0.08 0.26	25%[b] 50%[c]	Undivided Divided

[a] Electrolyses conducted in nitromethane solvent with 0.10 M Bu$_4$NPF$_6$ with a platinum cathode and magnesium anode. Yields are isolated yields. Current densities: 0.8-2.5 mA/cm^2.
[b] By NMR.
[c] By NMR; isolated yield: 40%.

difficult to purify by chromatography because dehydration occurs on the column. The yields reported in Table 2 are by NMR of the crude product mixture.

Finally, ferrocenecarboxaldehyde was found to give nitroalcohol in moderate yield under the standard conditions. Attempts to increase the yield by increasing the amount of charge passed were not successful as product decomposition and conversion to the 1,3-dinitro derivative were observed. For this reason, a pre-electrolysis[1] was attempted in which 0.50 F/mol charge was passed in the

Table 2. Addition of nitromethane to anthracene- and ferrocenecarboxaldehyde.[a]

Compound	Product	Charge (F/mol)	Yield	Conditions
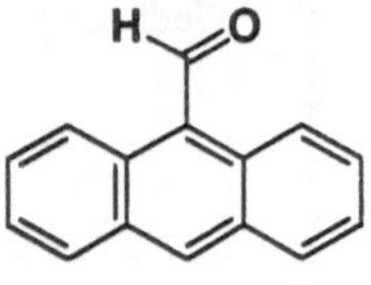 H—C=O (anthracene-9-carboxaldehyde)	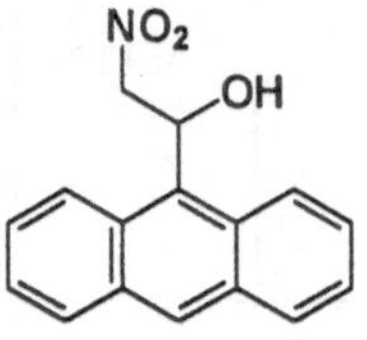 NO₂ / OH (1-(anthracen-9-yl)-2-nitroethanol)	0.05 / 0.22	73% / 84%	Divided / Undivided
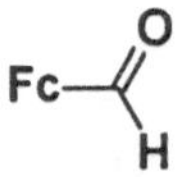 Fc—C(=O)H	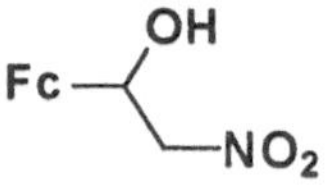 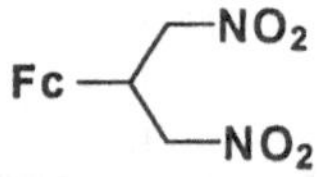 OH / Fc—CH(OH)—CH₂—NO₂ and Fc—CH(—NO₂)—CH₂—NO₂	0.27 (0.76) ((0.50))	57% (13%) ((0%)) / 5% (59%) ((95%))	Undivided (Divided) ((Pre-electrolysis))

[a] Experimental conditions as in Table 1.

absence of aldehyde in order to produce a substantial amount of the anion of nitromethane. Then the aldehyde was added and the course of the reaction was monitored by thin layer chromatography. Nitroalcohol formed in the first few minutes followed by the appearance of the dinitro derivative, 1,3-dinitro-2-ferrocenylpropane. The amount of nitroolefin in the mixture was never significant. After standing overnight, the 1,3-dinitro derivative was isolated in very high yield.

REFERENCES

1) C. Suba, M. E. Niyazymbetov and D. H. Evans, *Electrochim. Acta*, **1997**, *42*, 2247.

2) Z. Niazimbetova, S. E. Treimer and D. H. Evans, submitted.

3) M. E. Niyazymbetov and D. H. Evans, *J. Org. Chem.*, **1993**, *58*, 779.

4) A. L. Laikhter, M. E. Niyazymbetov, D. H. Evans, A. V. Samet and V. V. Semenov, *Tetrahedron Lett.*, **1993**, *34*, 4465.

5) M. E. Niyazymbetov and D. H. Evans, *Denki Kagaku*, **1994**, *62*, 1139.

NEW ELECTROORGANIC RESULTS IN THE REDUCTION OF PHENACYL AZIDE DERIVATIVES

Belén Batanero

Department of Organic Chemistry. University of Alcalá. 28871 Alcalá de Henares(Madrid).Spain

The electrochemical reduction of phenacyl bromide semicarbazones in aprotic medium afforded the dimeric semicarbazones[1] or 3,7-diaryl-*2H*-imidazo[2,1-b][1,3,4]oxadiazines[2] depending on the experimental conditions.

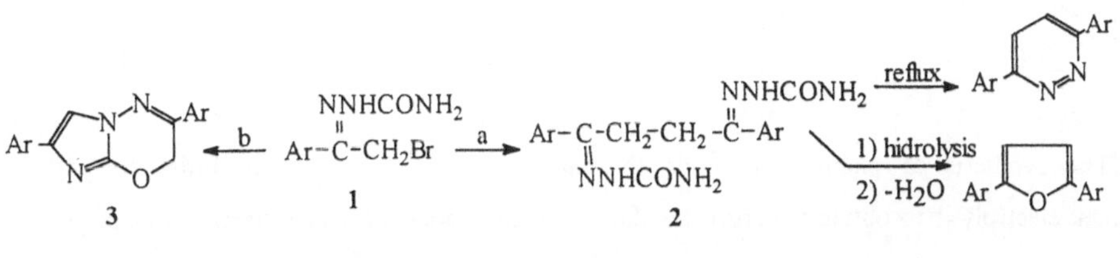

$$(1)$$

Experimental conditions: DMF/LiClO$_4$ Mercury cathode. Potentiostatic conditions E= -1V(*vs SCE*). a) Initial concentration of substrate in the cell : 2.10^{-3} mol/40 ml. b) 2.10^{-3} mol of **1** solved in 20ml DMF were added during 8h into the cathodic compartment with 30ml SSE.

Proposed mechanism to form **3**

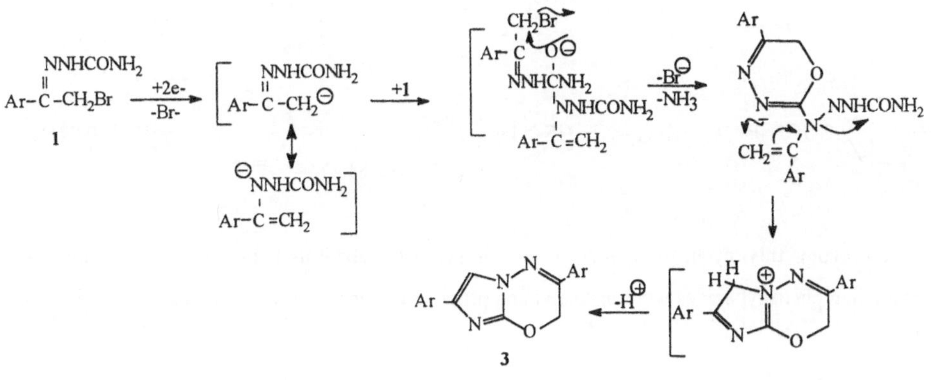

$$(2)$$

The electrogenerated imidazo-oxadiazines allowed to get new series of heterocyclic compounds:pyrrolo[2,1-b][1,3,4]oxadiazines[3], mesoionic imidazol derivatives[4] or N-substituted

imidazolones[5]

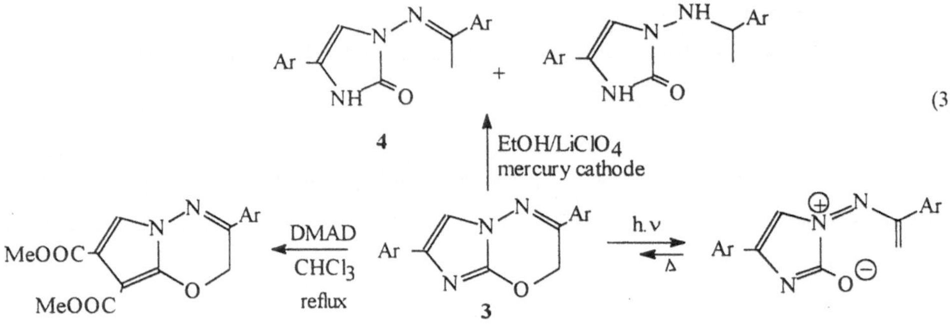

(3)

The synthesis of phenacyl bromide thiosemicarbazones was tried in order to perform subsequent electrolysis to obtain the corresponding imidazo-thiadiazines. The reaction was performed using the same procedure as used to prepare phenacyl bromide semicarbazones, and was unsuccessful, getting in this case undesirable 1,3,4-thiadiazines[6a] and 1,3-thiazols[6b]:

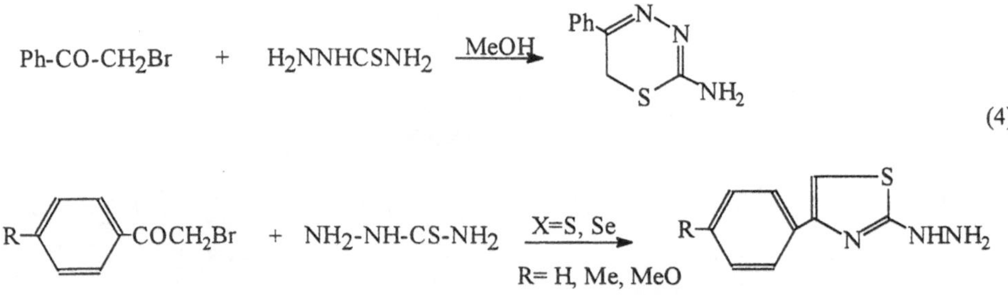

$$Ph\text{-}CO\text{-}CH_2Br \quad + \quad H_2NNHCSNH_2 \xrightarrow{MeOH}$$

(4)

$$R\text{-}\bigcirc\text{-}COCH_2Br \quad + \quad NH_2\text{-}NH\text{-}CS\text{-}NH_2 \xrightarrow[R=H, Me, MeO]{X=S, Se}$$

To avoid this cyclization the bromo atom was substituted by an azide group. The corresponding phenacyl azides were prepared and protected under the thiosemicarbazone form in very good yields:

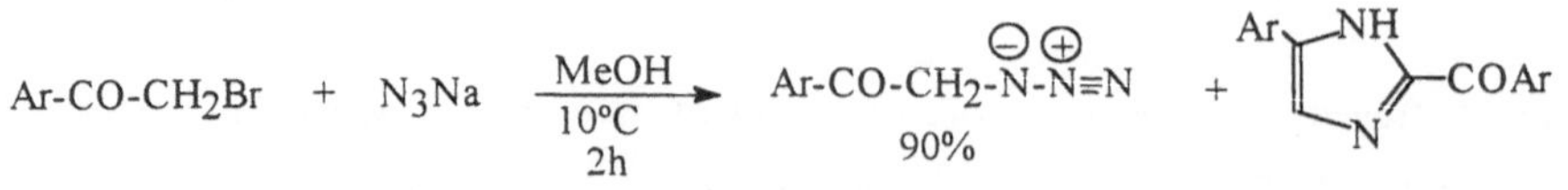

$$Ar\text{-}CO\text{-}CH_2Br \quad + \quad N_3Na \xrightarrow[\substack{10°C \\ 2h}]{MeOH} \quad Ar\text{-}CO\text{-}CH_2\text{-}\overset{\ominus}{N}\text{-}\overset{\oplus}{N}{\equiv}N \quad + \quad \underset{90\%}{}$$

(5)

$$\text{Ar-CO-CH}_2\text{-N-N}{\equiv}\overset{\oplus}{\text{N}} \quad + \quad 2\ \text{NH}_2\text{-NH-CS-NH}_2 \quad \xrightarrow[\substack{10°C \\ 24h}]{\text{MeOH/H}_2\text{O}(1{:}1)/\text{H}^+} \quad \begin{array}{c} \text{Ar-C-CH}_2\text{N}_3 \\ \| \\ \text{NNHCSNH}_2 \end{array} \qquad (6)$$

There are few literature references about the reduction of azides. Only *H. Lund.*[7] studied their electrochemical behaviour:

$$\overset{\ominus}{\text{R}}\text{-N-N}{\equiv}\overset{\oplus}{\text{N}} \quad \xrightarrow[\text{R= Ph, Bz}]{+2\overset{\ominus}{e} + 2\overset{\oplus}{H}} \quad \text{R-NH}_2 \ + \ \text{N}_2 \qquad \text{R-CH}_2\text{-N-N}{\equiv}\overset{\oplus}{\text{N}} \quad \xrightarrow[\text{R= Ph-CO}]{+2\overset{\ominus}{e} + 2\overset{\oplus}{H}} \quad \text{R-CH}_3 \ + \ \overset{\ominus}{\text{N}}_3 \qquad (7)$$

CYCLIC VOLTAMMETRY

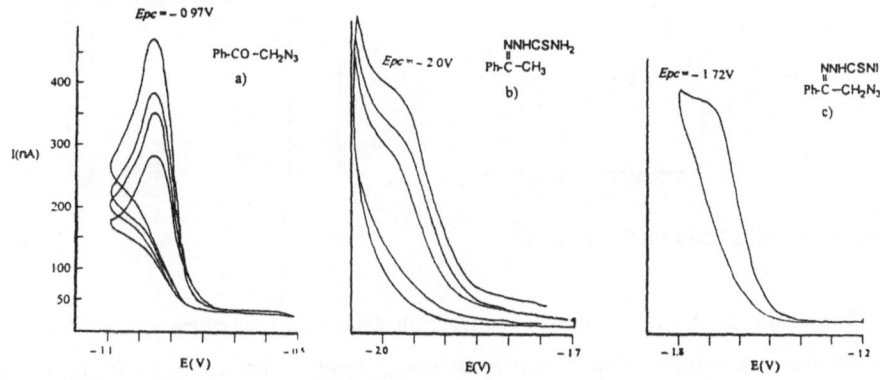

Voltammograms of: a) phenacyl azide b) acetophenone thiosemicarbazone c) phenacyl azide thiosemicarbazone. DMF/LiClO$_4$. HMDE/Ag/Ag$^+$. 0.5V/s. c = 2.10^{-3}M.

CATHODIC REDUCTION OF PHENACYL AZIDE THIOSEMICARBAZONES.

$$2\ \begin{array}{c}\text{NNHCSNH}_2 \\ \| \\ \text{Ar-C} \ \text{-CH}_2\text{N}_3 \\ \mathbf{1'} \end{array} \quad \xrightarrow[\substack{-\text{NH}_3 \\ -(\text{NH}_2\text{-CS-NH}_2)}]{\substack{+2e\ \ominus \\ -\text{N}_3\ \ominus}} \quad \left[\text{structure } \mathbf{3'} \right] \quad \xrightarrow[+2\text{H}\ \oplus]{+2e\ \ominus} \quad \text{structure } \mathbf{4'} \qquad (8)$$

Ar = a: Ph, b: 4-MeO-C$_6$H$_4$, c: 4-Cl-C$_6$H$_4$

Experimental conditions: DMF/LiClO$_4$ Mercury cathode. Potentiostatic conditions: E= -1.7V(*vs SCE*). Initial concentration of substrate in the cell: 2.10^{-3} moles/40 ml.

The reaction takes place through the corresponding imidazo-thiadiazine which is immediately reduced to the opened 4-aryl-1-(1-aryl-ethylidenamino)-1,3-dihydro-2-imidazothiolones (**4'**).

^{1}HNMR, ^{13}CNMR, MS spectrometry and other spectroscopic techniques supported the nature of the electrogenerated structure.

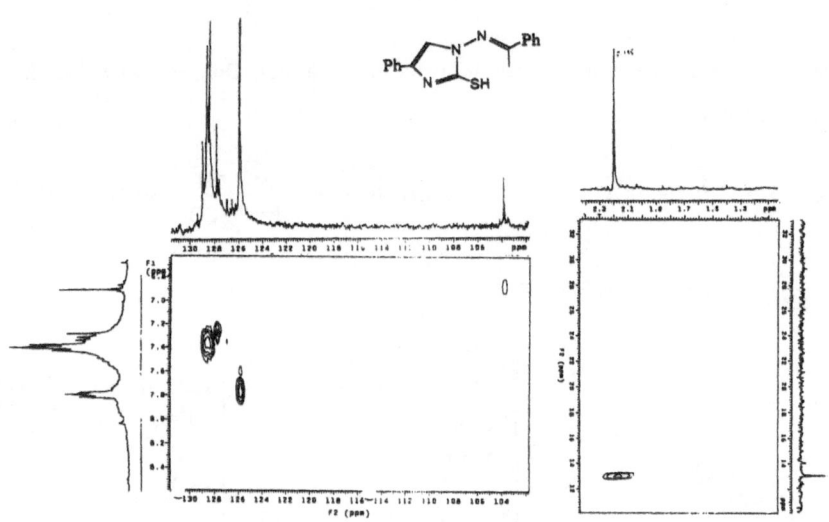

Bidimensional ^{1}H-^{13}C NMR of **4'a** in CDCl$_3$.

On the other hand, when phenacyl azide semicarbazone is electrochemically reduced under the same conditions, the corresponding N-ethylidenamino-1,3-dihydro-2-imidazolone **4** is obtained in a low selective reaction compared to the direct reduction in EtOH of the imidazo-oxadiazine **1**.

References

[1] F. Barba, M.D. Velasco and A. Guirado.*Synthesis*. **1984**, *593*. F. Barba, M.D. Velasco and A. Guirado. *Synth. Comm.* **1985**, *15*, 939.
[2] F. Barba and B. Batanero. *J. Org. Chem.* **1993**, *59*, 6889.
[3] F. Barba and B. Batanero. *Synthesis*. **1994**, 555.
[4] F. Barba and B. Batanero. *Tetrahedron Lett.* **1994**, *35*, 6355. F. Barba and B. Batanero. *Tetrahedron.* **1994**, *51*, 2023.
[5] F. Barba and B. Batanero. *Electrochim. Acta.* **1995**, *40*, 2779.
[6] a) Postovskii Y. *Khim Tekhnol. Org. Soedin.* 14th, **1975**, 207. Jones Winton, Miller Ger. Offen. 3,031,703 US Appl 71, 952, 04 Sept, 179. b) Nalepa Karel. *Acta Univ. Palacki Olomuc. Fac. Rerum. Nat.* **1978**, *57* (Chem. 17), 191.
[7] H. Lund. *Oesterreich. Chem. Z.* **1967**, *68*, 43.

CHEMICAL AND ELECTROCHEMICAL REDUCTIVE CYCLIZATIONS OF O-NITROARENES TOWARD NITROGEN CONTAINING HETEROCYCLES

Byeong Hyo KIM*, Yoon Seok LEE, Yong Rack CHOI, Sun Kyong KIM,
Young Moo JUN, and Woonphil BAIK[†]

Department of Chemistry, Kwangwoon University, Seoul, 139-701, Korea

[†]*Department of Chemistry, Myong Ji University, Kyung Ki Do, 449-728, Korea*

ABSTRACT

A series of o-nitroarenes such as 2-nitrobenzaldehydes, 2'-nitroacetophenone, N-(2-nitrobenzylidene)anilines, or 2-nitrophenylazobenzenes was examined for reductive cyclizations by applying 2-bromo-2-nitropropane/Zn, SmI$_2$, or cathodic electrolysis. Chemical or electrochemical reactions of o-nitroarenes involving electron transfer processes were found to give heterocycles such as 2,1-benzisoxazoles, or benzotriazoles in excellent yields. Electrolysis of 2-nitrophenylazobenzenes produced benzotriazole N-oxides or benzotriazoles exclusively depending upon the reaction condition.

INTRODUCTION

The N-containing 5-membered heterocyclic compounds have been employed for medicinal and industrial purposes. For example, 2,1-benzisoxazoles have a modest number of patented usages, *i.e.* antiinflammatory, antituberculotic, lipodemia, and analogs of psilocene and muscomal. On the other hands, benzotriazoles are widely used as ultraviolet absorbers for the protection of commercially important plastics against sunlight. Those useful heterocycles could be obtained by reductive cyclizations of o-nitroarenes.

Recently, we have reported the selective reduction of aromatic nitro or nitroso compounds containing o-, m-, or p-electron withdrawing groups to their corresponding anilines by Bakers' Yeast-NaOH[1] and Turner[2] reported regioselective reduction of aromatic dinitro compounds to their nitroanilines with Bakers' Yeast without NaOH. In addition to well known metal reduction, SmI$_2$ was employed by several groups for the reduction of nitro compounds to give either the corresponding hydroxylamine or primary amines.[3] Besides those kinds of chemical methods, electrochemical approach could be a powerful methodology for the reductive reactions of nitroarenes. Herein we describe an efficient and mild reductive cyclization of ortho nitro substituted arenes to the corresponding nitrogen containing heterocycles by applying 2-bromo-2-nitropropane/Zn, SmI$_2$, or cathodic electrolysis.

RESULTS AND DISCUSSION

In the course of our study on reductive cyclization reaction of 2-nitroarenes,[4] we found an efficient synthetic method of 2,1-benzisoxazoles by using 2-bromo-2-nitropropane (BNP) and Zn dust.[5] The reaction of 2-nitrobenzaldehydes, 2'-nitroacetophenone, or N-(2-nitrobenzylidene)anilines,

1 with BNP (1.2 equiv.) and Zn (5 equiv.) in methanol at 50 °C (2-nitrobenzaldehydes, 2'-nitroacetophenone) or at room temperature [for *N*-(2-nitrobenzylidene)anilines] produced 2,1-benzisoxazoles **2** in 73 - 98% yields. The role of BNP is likely to be an electron acceptor due to its low lying antibonding π-orbital which has been employed in $S_{RN}1$ process, and the utility of BNP has been observed in elsewhere.[6]

$$R_1 = \text{-H, -Cl, -OMe, -NO}_2\text{, -OCH}_2\text{O-}$$
$$R_2 = \text{-H, -CH}_3; \quad X = \text{O, N-Ph}$$

(1)

As a part of our continuing study on *o*-nitroarenes, we have tried cathodic electrochemical cyclization reactions of *o*-nitroacylbenzenes under the controlled potential electrolysis conditions [Pb or Pt cathode, 0.4 M $LiClO_4$/MeOH, -0.50 ~ -1.00 V vs. Ag/AgCl] and 2,1-benisoxazoles were obtained in 52 - 100% yields.

In extensions of the reductive cyclization reactions, we have studied the reductive cyclizations of *o*-nitrophenylazobenzenes (**3**) using BNP/Zn, SmI_2, or electrolysis. Similar to 2-nitrobenzaldehyde, **3** with BNP (1.2 equiv.)/Zn (12 equiv.) in $MeOH/CH_2Cl_2$ (5:1 v/v) at room temperature was cyclized to the corresponding benzotriazoles (**4**) in reasonable yields (71 - 88%) (eq. 2).

We tried more powerful electron transferring reagent, SmI_2 instead of BNP/Zn and it worked marvelously in THF.[7] To a solution of SmI_2 (7 equiv.) in THF, **3** were added dropwise and the reaction mixture was stirred for 40 min. - 160 min. at room temperature. After the general work-up process, **4** were obtained in 84 - 97% yields (eq. 2). Because of powerful electron donating ability of SmI_2, benzotriazole *N*-oxide intermediates (**5**) were not detectable.

Little mechanistic information is currently available for the reactions of nitrogen compounds with SmI_2. Evans reported that 2 equiv. of $Sm(C_5Me_5)_2$ deoxygenated pyridine *N*-oxide or 1,2-epoxybutane and was transformed into the complex $(C_5Me_5)_2Sm\text{-}O\text{-}(C_5Me_5)_2$ whose structure was

(2)

	BNP/Zn/ MeOH/CH₂Cl₂/rt	SmI₂ THF, rt
3a ; R₁ = H, R₂ = H, R₃ = *t*-Bu, R₄ = *t*-Bu	**4a** (72%)	**4a** (94%)
3b ; R₁ = H, R₂ = Cl, R₃ = *t*-Bu, R₄ = *t*-Bu	**4b** (88%)	**4b** (84%)
3c ; R₁ = H, R₂ = H, R₃ = *t*-pentyl, R₄ = *t*-pentyl	**4c** (71%)	**4c** (85%)
3d ; R₁ = H, R₂ = H, R₃ = Me, R₄ = H	**4d** (83%)	**4d** (97%)
3e ; R₁ = Cl, R₂ = H, R₃ = *t*-Bu, R₄ = *t*-Bu	**4e** (79%)	**4e** (91%)

established by X-ray crystallography.[8] Zhang and Lin also demonstrated the deoxygenation of pyridine N-oxides by SmI_2.[9] By analogy, when SmI_2 is either used to cyclize o-nitrophenylazobenzene derivatives (3) by transferring electrons to nitro group or used to remove an oxygen atom from N-oxide, the formation of a complex I_2Sm-O-SmI_2 can be postulated. We also observed the exclusive transformation of isolated benzotriazole N-oxide (5a) to 4a by the SmI_2 reduction. Based on Evans' and our results, a possible reaction mechanism is shown in Scheme.

Scheme

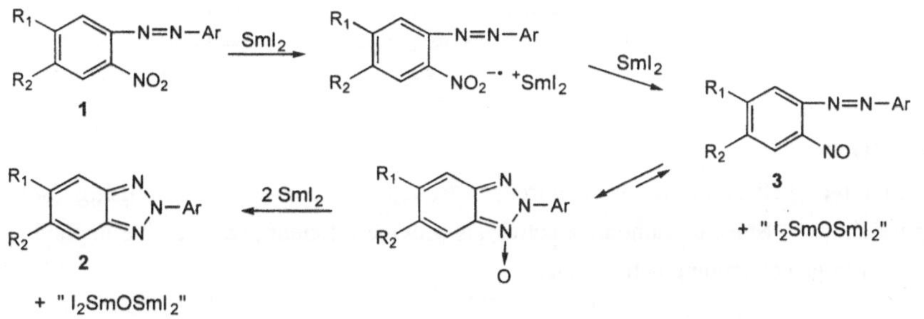

For mechanistic purposes and comparison with chemical reactions, electrolysis reactions were carried out also. Based on cyclic voltametric behavior, 3 were reduced by the cathodic electrolysis under the controlled potential condition [Pt cathode, 0.4 M $LiClO_4$/(MeOH : CH_2Cl_2 = 1 : 1, v/v), -0.70 ~ -0.95 V vs. Ag/AgCl]. As we expected, benzotriazole N-oxides (5) were obtained exclusively in 85 ~ 98% yields without any formation of 4 (eq. 3). Furthermore, we could reduce isolated 5 to 4 by electrolysis reaction in the presence of a base in THF/ H_2O (1 : 1, v/v). [5b, Pt cathode, -1.2 V vs. Ag/AgCl, KOH (0.8 M), $LiClO_4$ (0.4 M)→ 4b, 97%] similar to Lund's result.[10]

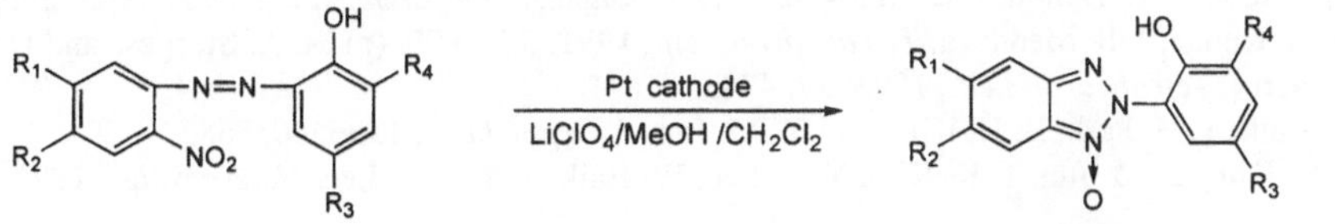

(3)

3a ; R_1 = H, R_2 = H, R_3 = t-Bu, R_4 = t-Bu   -0.75 V   5a (93%)
3b ; R_1 = H, R_2 = Cl, R_3 = t-Bu, R_4 = t-Bu   -0.70 V   5b (98%)
3c ; R_1 = H, R_2 = H, R_3 = t-pentyl, R_4 = t-pentyl   -0.95 V   5c (90%)
3d ; R_1 = H, R_2 = H, R_3 = Me, R_4 = H   -0.80 V   5d (88%)
3e ; R_1 = Cl, R_2 = H, R_3 = t-Bu, R_4 = t-Bu   -0.80 V   5e (85%)

After the various basic electrolysis conditions examined, we obtained the optimum electrolysis condition of direct transformation of 3 to 4. By applying the higher reduction potential in the presence of NaOH, 3 were reduced to 4 [Pt cathode, 0.2 M NaOH/(THF : H_2O = 1 : 1, v/v), -1.3 ~

-1.5 V vs. Ag/AgCl] in excellent yields (eq. 4). It is apparent that electron transfer ability controls the reductive cyclization reactions.

$$\begin{array}{c} \text{Pt cathode} \\ \hline \text{NaOH/THF /H}_2\text{O} \end{array} \quad (4)$$

3a ;	$R_1 = H$,	$R_2 = H$,	$R_3 = t\text{-Bu}$,	$R_4 = t\text{-Bu}$	-1.4 V
3b ;	$R_1 = H$,	$R_2 = Cl$,	$R_3 = t\text{-Bu}$,	$R_4 = t\text{-Bu}$	-1.3 V
3c ;	$R_1 = H$,	$R_2 = H$,	$R_3 = t\text{-pentyl}$,	$R_4 = t\text{-pentyl}$	-1.5 V
3d ;	$R_1 = H$,	$R_2 = H$,	$R_3 = Me$,	$R_4 = H$	-1.5 V
3e ;	$R_1 = Cl$,	$R_2 = H$,	$R_3 = t\text{-Bu}$,	$R_4 = t\text{-Bu}$	-1.4 V

4a (89%)
4b (92%)
4c (77%)
4d (89%)
4e (98%)

CONCLUSIONS

The reductive cyclizations of ortho nitro substituted aromatic compounds using 2-bromo-2-nitropropane/Zn dust, SmI$_2$, or cathodic electrolysis provide efficient and selective methods for the syntheses of nitrogen containing heterocycles.

ACKNOWLEDGMENTS: The financial supports from Korea Science and Engineering Foundation (95-0501-06-01-3 and RRC in part), and BSRI program, Korea Ministry of Education are greatly acknowledged.

REFERENCES

1) (a) W. Baik, J. L. Han, N. H. Lee, B. H. Kim, and J. T. Hahn, *Tetrahedron Lett.*, **1994**, *35*, 396. (b) W. Baik, J. U. Rhee, S. H. Lee, N. H. Lee, B. H. Kim, and K. S. Kim, *Tetrahedron Lett.*, **1995**, *36*, 2793.
2) C. L. Davey, L. W. Powell, N. J. Turner, and A. Wells, *Tetrahedron Lett.*, **1994**, *35*, 7867.
3) (a) J. Souppe, L. Danon, J. L. Namy, and H. B. Kagan, *J. Organometal. Chem.*, **1983**, *250*, 227. (b) A. Kende, J. S. Mendoza, *Tetrahedron Lett.*, **1991**, *32*, 1699. (c) M. A. Sturgess and D. J. Yarberry, *Tetrahedron Lett.*, **1993**, *34*, 4743.
4) W. Baik, T. H. Park, B. H. Kim, and Y. M. Jun, *J. Org. Chem.*, **1995**, *60*, 5683.
5) B. H. Kim, Y. M. Jun, T. K. Kim, Y. S. Lee, W. Baik, and B. M. Lee, *Heterocycles*, **1997**, *45*, 235.
6) a) G. A. Russell, M. Jawdosiuk and M. Makosza, *J. Am. Chem. Soc.*, **1979**, *101*, 2355. b) G. A. Russell and A. R. Metcalfe, *J. Am. Chem. Soc.*, **1979**, *101*, 2359. c) G. A. Russell and B. Mydryk, *J. Org. Chem.*, **1982**, *47*, 1879. d) R. A. Russell and W. Baik, *J. Chem. Soc. Chem. Comm.*, **1988**, 196.
7) B. H. Kim, S. K. Kim, Y. S. Lee, Y. M. Jun, W. Baik, and B. M. Lee, *Tetrahedron Lett.*, **1997**, *38*, in press.
8) W. J. Evans, J. W. Grate, I. Bloom, W. E. Hunter, and J. L. Atwood, *J. Am. Chem. Soc.*, **1985**, *107*, 405.
9) Y. Zhang and R. Lin, *Synth. Commun.*, **1987**, *17*, 329.
10) H. Lund and S. Kwee, *Acta Chem. Scand.*, **1968**, *22*, 2879.

THE SYNTHESIS OF SUBSTITUTED AMINOINDOLES BY CHEMICAL AND ELECTROCHEMICAL REDUCTION OF NITROINDOLES

Ian MARCOTTE, Alexandre LEMIRE and Jean LESSARD

Centre de Recherche en Électrochimie et Électrocatalyse,, Département de Chimie,
Université de Sherbrooke, Sherbrooke, Québec, Canada J1K 2R1

ABSTRACT

The electrochemical reduction of 4-, 5-, 6- and 7-nitroindoles in basic and acidic aqueous methanol gave substituted aminoindoles and/or the corresponding aminoindole in a ratio depending on the nitroindole, the pH and the strength of the nucleophile present in the medium. In the case of 5-nitroindole and 3-hydroxypropyl-5-nitroindole, in basic medium and in the presence of EtS⁻, only the corresponding 4-ethanethio-5-aminoindole was formed in very good yields (regiospecific and highly selective reaction).

INTRODUCTION

A patent issued to Bristol-Myers-Squibb Company (1) reported that the zinc reduction of 3-(3-hydroxypropyl)-5-nitroindole (**1**) in DMF containing NH_4Cl as proton source and EtSH in excess (10 eq.) gave the desired 4-ethylthio-5-aminoindole **2** in a low 20% isolated yield and a low 27% selectivity (selectivity = 100×[%**2**/(%**2**+%**3**)]), the major product (55%) being 3-(3-hydroxypropyl)-5-aminoindole (**3**) (Scheme 1). This led us to study the electroreduction of 4-, 5-, 6- and 7-nitroindoles (**4**) to assess the efficiency of the electrochemical method to synthesize monosubstituted aminoindoles **5** (Scheme 1).

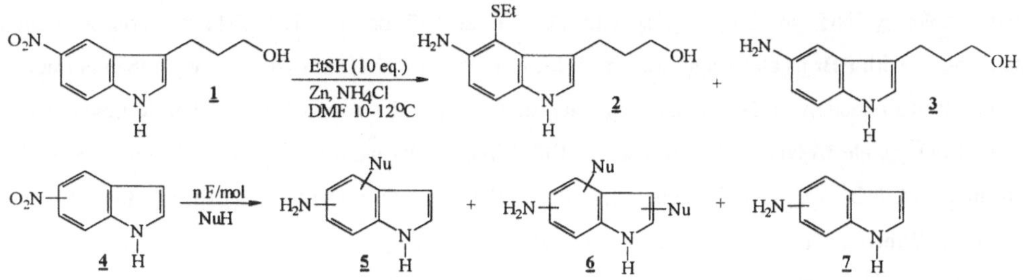

SCHEME 1

RESULTS AND DISCUSSION

The electroreductions were carried out at Hg under controlled potential conditions in MeOH-H_2O (95:5, v/v) acidic (HX 0.15 M, pH = 0.3) and basic (KOH 0.15 M, pH > 13) solutions in the absence or in the presence of an excess of RSH (R = Et, Ph). Some of the results are presented in Table 1. They show that the ratio of monosubstituted aminoindole **5** to aminoindole **7** varies (from >99:1 to <1:99) with the nitroindole **4**, the pH and the strength of the nucleophile (the stronger the nucleophile, the larger the **5**/**7** ratio)

The best yields of monosubstituted aminoindole were obtained with 5-nitroindole (**4**, NO_2 at C-5) in the presence of a good nucleophile (Br^- in acidic medium, RS^- in basic medium), in which cases, the reaction was 100% selective (with Br^- and PhS^- (result not shown)) or highly selective (83% with EtS^-, result not shown) (% selectivity = 100[%**5**/(%**5** +%**6** + %**7**)]) and regiospecific (substituent at C-4 only): 71-86% of **5** (NH_2 at C-5, Br at C-4); 70% of **5** (NH_2 at C-5, EtS at C-4); 59% of **5** (NH_2 at C-5, SPh at C-4, result not shown). The reaction was also regiospecific and 100% selective in H_2SO_4 containing an excess of PhSH but the yield of 4-substituted 5-aminoindole **5** (NH_2 at C-5, Nu at C-4) was low (35%) and a mixture of 4-methoxy (Nu = OMe, 20%) and 4-phenylthio (Nu = PhS, 15%) was formed. It is noteworthy that the synthesis of the Bristol-Meyers-Squibb compound **2** (see Scheme 1) by electroreduction of 3-hydroxypropyl-5-nitroindole (**1**) in basic medium in the presence of ethanethiol (8 eq.) proved distinctly superior (100% selectivity, 55-72% yield of **2**) to that described in the patent (27% selectivity, 20% yield of **2**) (1). The hydroxypropyl side chain at position 3 appears to have no influence on the yield, selectivity and regioselectivity.

The electroreduction of 6-nitroindole (**4**, NO_2 at C-6) in acidic medium gave also monosubstituted 6-aminoindoles **5** (NH_2 at C-6) regiospecifically (Nu at C-7 only). The yield of monosubstituted aminoindole with HBr as electrolyte was much lower (31% of **6**, NH_2 at C-6, Br at C-7) than in the case of the electroreduction of 5-nitroindole (**4**, NO_2 at C-5) and the selectivity was slightly lower (91%). With H_2SO_4 as electrolyte in the presence of EtSH, the results were similar to those obtained with 5-nitroindole (**4**, NO_2 at C-4) in the same electrolyte in the presence of PhSH: regiospecific formation of a mixture of 7-methoxy (Nu = OMe, 21%) and 7-ethylthio- (Nu = EtS, 14%) 6-aminoindole (**5**, NH_2 at C-6). In basic medium in the presence of EtSH, the electroreduction of 6-nitroindole (**4**, NO_2 at C-6) did not lead to the regiospecific formation of monosubstituted 6-aminoindole **5** (NH_2 at C-6), giving a mixture of 7-ethylthio- (Nu = SEt at C-7, 24%) and 2-ethylthio (Nu = SEt at C-2, 31%) derivatives.

In the case of the electroreduction of 4-nitroindole (**4**, NO_2 at C-4) in the presence of PhSH, the reaction was quite selective giving mainly or exclusively two isomeric monosubstituted 4-aminoindoles:

Table 1. Preparative electrolyses of 4-, 5-, 6- and 7-nitroindoles, at Hg, in acidic (pH = 0.3) and basic (pH > 13) media: the influence of nucleophiles

Substrate (4)	Electrolyte[a]	E[b] (V)	Charge (F/mol)	Products (Scheme 1)						
				Monosubstituted aminoindole (5)			Disubstituted aminoindole (6)			Amine (7)
				Nu	Position	Yield (%)	Nu	Position	Yield(%)	Yield (%)
5-Nitroindole	KOH	1.31	5.6	OMe	C-4	43-47				30-34
	KOH/EtSH	-1.31	4.7	SEt	C-4	70				15
	H_2SO_4	-0.77	5	OMe	C-4	14				12
	H_2SO_4/PhSH	-0.77	4.3	SPh	C-4	20				
				SPh	C-3	15				
	HBr	-0.61	4.3	Br	C-4	71-86				
6-Nitroindole	KOH	-1.24	6							53-65
	KOH/EtSH	-1.48	4.4	SEt	C-2	31	SEt	C-2/C-7	2	
				SEt	C-7	24				
	H_2SO_4		4.6	OMe	C-7	15				
	H_2SO_4/EtSH		4.8	OMe	C-7	21				23
				SEt	C-7	14				
	HBr		4.8	Br	C-7	31				3
4-Nitroindole	KOH[c]	-1.40	5.1	SPh	C-7	56-69	SPh	?	3-5	37-44
				SPh	C-3	2-12	SPh	?	3-6	
	KOH/PhSH	-1.40	4.3	SPh	C-7	35-40				
				SPh	C-3	15				
	H_2SO_4/PhSH	-0.62	4.0	Br	C-7	37-47				
	HBr	-0.62	4.3	OMe	C-6	71				traces
7-Nitroindole	H_2SO_4	-0.56	4.2	Br	C-6(C-4)	44-49	Br	?	traces	traces
	HBr	-0.56	4.4	Br	C-4(C-6)	traces				traces
	KOH	-1.05	5.2-6.1	OMe	C-6	15-20				30-32

a: Solvent: MeOH-H_2O (95:5 v/v); electrolytes: KOH 0.15 M, H_2SO_4 0.15 M, HBr 0.15 M; added nucleophile: EtSH 48 mM, PhSH 12 mM; substrate concentration: 6 mM.
b: Reference electrode: saturated calomel electrode (SCE).
c: An azo dimer has been isolated in a 10-18% yield.

the major isomer being 7-phenylthio-4-aminoindole ($\underline{5}$, NH$_2$ at C-4, PhS at C-7) (56-69% yield in basic medium, 35-40% in acidic medium) and the minor isomer being 3-phenylthio-4-aminoindole ($\underline{5}$: NH$_2$ at C-4, PhS at C-3) (2-12% yield in basic medium, 15% yield in acidic medium). With HBr as electrolyte, the electroreduction gave exclusively 7-bromo-4-aminoindole ($\underline{5}$: NH$_2$ at C-4, Br at C-7) in 37-47% yield.

The electroreduction of 7-nitroindole ($\underline{4}$: NO$_2$ at C-7) was highly selective and regioselective in acidic medium with the almost exclusive formation of 6-monosubstituted 7-aminoindoles ($\underline{5}$, NH$_2$ at C-7, Nu at C-6). Surprisingly, the yield of $\underline{5}$ (NH$_2$ at C-7) was higher with H$_2$SO$_4$ (OMe at C-6, 71%) than with HBr as electrolyte (Br at C-6, 44-49%) contrary to what has been observed in the case of 5-nitroindole ($\underline{4}$: NO$_2$ at C-5) and 6-nitroindole ($\underline{4}$: NO$_2$ at C-6) and in disagreement with the fact that Br$^-$ should be a better nucleophile than MeOH. With KOH as electrolyte, 6-methoxy-7-aminoindole ($\underline{5}$: NH$_2$ at C-7, OMe at C-6) was the minor product (15-20% yield, regiospecific) and the major product was 7-aminoindole ($\underline{7}$: NH$_2$ at C-7; 30-32% yield).

A disubstituted aminoindole $\underline{6}$ (Nu = RS) has been isolated in low yields (2 to 11%) in the electroreduction of 4- and 6-nitroindoles ($\underline{4}$: NO$_2$ at C-4 and C-6 respectively) in basic medium in the presence of RSH, and a dibromo-7-aminoindole ($\underline{6}$: NH$_2$ at C-7, Nu = Br) has been detected by GC/MS in the products of electroreduction of 7-nitroindole ($\underline{4}$: NO$_2$ at C-7).

The formation of monosubstituted and disubstituted aminoindoles $\underline{5}$ and $\underline{6}$ in the electroreduction of nitroindoles $\underline{4}$ can be explained by a mechanism involving the formation of intermediates (protonated in acidic medium, unprotonated in basic medium) such as *quinonemethaneimines* $\underline{8}$ and $\underline{9}$ (Y = OH and/or H) in the case of 4- and 6-nitroindoles ($\underline{4}$: NO$_2$ at C-4 or C-6 respectively) and *diiminoquinones* $\underline{10}$ and $\underline{11}$ (Y = OH and/or H) in the reduction of 5-and 7-nitroindoles ($\underline{4}$: NO$_2$ at C-5 or C-7 respectively) as discussed in another paper in this issue (2). A protonated diiminoquinone $\underline{10H}^+$ (Y = OH and/or H) has been trapped with cyclopentadiene in a pericyclic reaction (2).

REFERENCES

1) J. A. Jonas, E. H. Ruediger, C. Crola and P. Dextrase, *European Patent* 0 645 385 A1, **1995**.

2) I. Marcotte, J..M. Chapuzet, Y. Dory and J. Lessard, *This issue.*

TECHNOLOGICAL ASPECTS OF ELECTROSYNTHESIS OF N, N - DIMETHYL - p PHENYLENDIAMINE

Igor CRETESCU and Matei MACOVEANU

Environmental Department, Faculty of Industrial Chemistry, Technical University "Gh. Asachi" Iasi, Bd D. Mangeron, Nr. 71 A, 6600 Iasi, ROMANIA

ABSTRACT

In this paper are presented some technological aspects of electrosynthesis of N,N - dimethyl - p - phenylendiamine approached by reduction of N,N - dimethyl - p -nitrosoaniline, both directly and in the presence of a mediatory system.

INTRODUCTION

The aim of every technological electrosynthesis process is to obtain maximal yield of product and good reliability of electrolysis equipment,   at   minimal energetical costs. In this respect we have investigated two technological modalities for the industrial preparing of N,N - dimethyl - p - phenylendiamine approached by the reduction of N,N - dimethyl - p - nitrosoaniline both directly [1] and in the presence of a mediatory system[2].

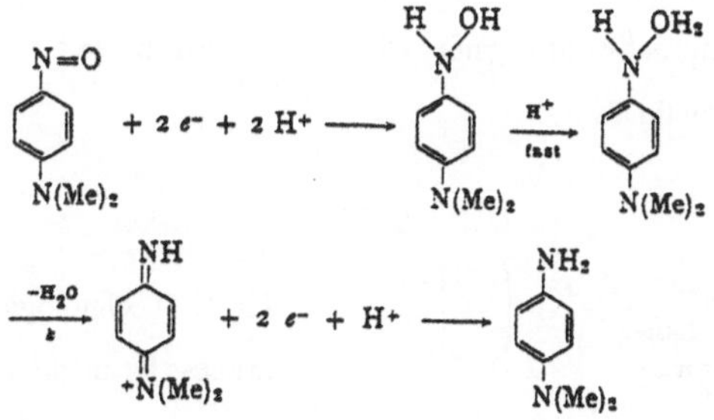

$$ArNO + 4Ti^{3+} + 4H^{+} \rightarrow ArNH_{2} + H_{2}O + 4Ti^{4+}$$

$$4Ti^{4+} + 4e^{-} \rightarrow 4Ti^{3+}$$

The choice of the best technological alternative is realised according to the simultaneous fulfilling of as much requirements above mentioned as possible. To these are to be added certain technological requirements such as: product separation, recovering and regeneration of the solvent or reagents etc. Generally this choice is a difficult task, but in the case of the present system it is very difficult because of the oxidation sensibility both of the reagents and the products and because of the possibility of a reaction between them.

After the selection of the appropriate alternative, an essential role is played by the performance of the electrochemical reactor [3]. This depends on its construction and nature of the electrodic materials and also on the knowledge of the complex kinetic relationships resulting from the combination of the reaction mechanisms and the macroscopic condition of electrochemical reactors.

In the following it will be presented the two technological alternatives with their advantages and disadvantages. It will be also presented the variation of the composition of the reacting amount and there will be determined certain performance indicators for the electrochemical reactor: fractional conversion, mean current efficiency and integral space time yield.

METHOD

a. Reduction approched by Ti^{3+} mediatory system.

This technological alternative of organic reduction process, bassed on the high reduction power of $TiCl_3$ is presented in the fig.1.

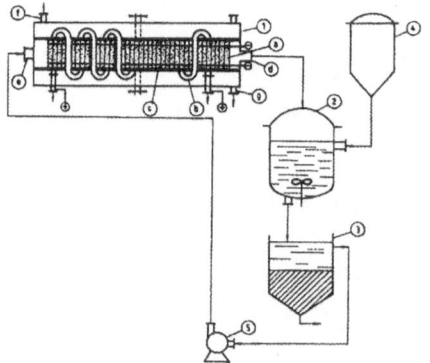

Fig.1. Technological diagram of reduction process by mediatory system.

1. electrochemical reactor

2. chemical reactor

3. phases separator

4. measurement vessel for the organic compound

5. pump

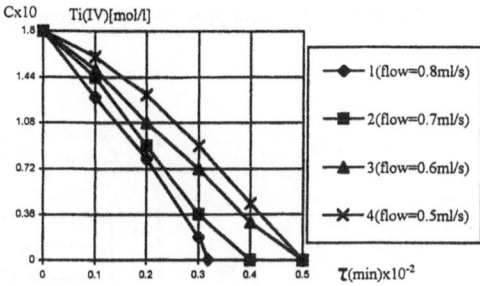

cathod-carbon felt electrode (a)
anod-lead spiral (b)
cation exchange membrane MK40 (c)
cathodic volume: $20 \times 75 = 1500$ cm^3
electrolyte: HCl - 1 mol/l
t - 25 $^\circ$C
i - 30-50 A/m^2
current efficiency - 80%

Fig.2. Electrochemical regeneration of mediator agent.

b. Directly electrochemical reduction

This technological alternative of organic reduction process, bassed on the electronic injection in LUMO molecular orbitals of organic compound have been carried out using an electrochemical small cell under galvanostatic conditions and inert atmosfer, which can be assimilated to the non-stationary stirred tank reactor, fig.3.

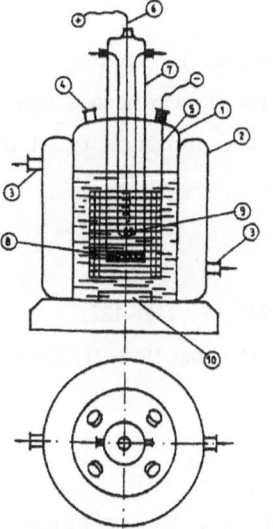

Fig. 3. Electrochemical batch reactor (stirred tank reactor nonstationary).

1. Electrochemical batch reactor
2. Thermostatic cover
3. Fittings for input-output of the thermal agent
4. Reactor fittings
5. Cathode
6. Anode
7. Anodic compartiment
8. Membrane
9. Thermostatic coil
10. Magnetic stirred

Evolution of the reaction mixture during the electrochemical process is presented in fig. 4.

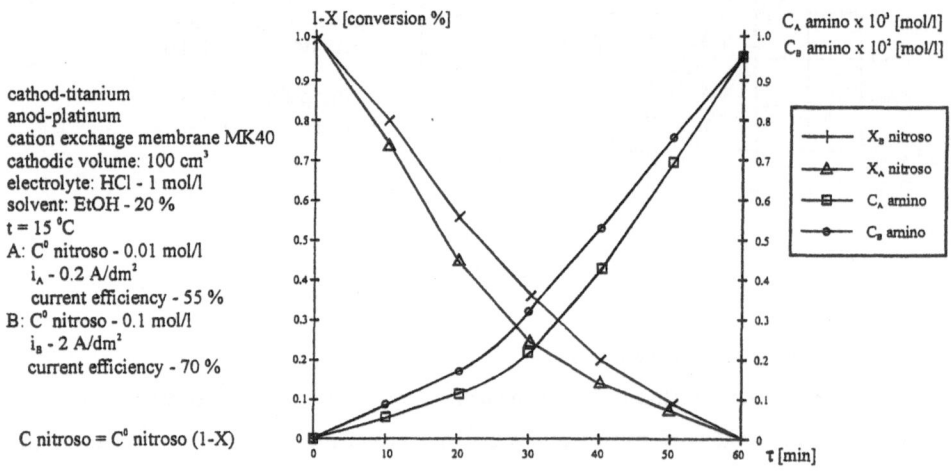

cathod-titanium
anod-platinum
cation exchange membrane MK40
cathodic volume: 100 cm³
electrolyte: HCl - 1 mol/l
solvent: EtOH - 20 %
t = 15 °C
A: C° nitroso - 0.01 mol/l
   i_A - 0.2 A/dm²
   current efficiency - 55 %
B: C° nitroso - 0.1 mol/l
   i_B - 2 A/dm²
   current efficiency - 70 %

C nitroso = C° nitroso (1-X)

Fig. 4. Variation of fractional conversion and concentration during the directly electrochemical reduction of N,N - dimethyl - p -nitrosoaniline.

The analytical control of electrosynthesis was achieved by polarographic and chromatographic methods. On the basis of the evolution of the reaction mixture during the electrochemical process, we have found the macrokinetic and mathematical model of the electrochemical reactor.

References

1. a) Donald W. Leedy, Ralph N. Adams, J. Electroanal. Chem., **1967,** 14 , 119 - 122,.

   b) N.C. Sarada, I. Ajit Kumar Reddy, K. Mastan Rao, J. Indian Chem.Soc., **1994,** 71, 729-731,

2. a) Jacques Simonet, L'actualité Chimique, **1982,** 11,19-28.

   b) C. Moinet, E. Raoult, Bull. Soc. Chim. France, **1991,** 214.

3. a).J. Chaussard , L'actualité Chimique, **1982,** 11, 29-32.

   b)Ewald Heitz, Gerhard Kreysa, PRINCIPLES OF ELECTROCHEMICAL ENGINEERING, VCH Publishers, **1986**

ELECTROREDUCTION OF ORGANIC COMPOUNDS WITH TWO REACTIVE CENTERS. I. 1,2,4-TRIAZINES

Petr Zuman[a], František Riedl[b], František Liška[c], and Jiří Ludvík[b]

[a]Department of Chemistry, Clarkson University, Potsdam, NY 13699-5810, USA
[b]J. Heyrovský Inst. of Physical Chemistry, Dolejškova 3, 18223 Prague 8, Czech Republic
[c]Department of Organic Chemistry, Inst. of Chemical Tech., Technická 5, 16628, Prague 6, Czech Republic

In most organic molecules a conjugation of two adjacent double bonds facilitates the electroreduction of the more easily reducible group. An exception of this rule has been observed for some 1,2,4-triazines. In two model compounds metamitron(I) (4-amino-3-methyl-6-Phenyl-1,2,4-triazin-5(4H)-one) and metribuzin(II) (4-amino-6-tert-butyl-3-methylthio-1,2,4-triazin-5(4H)-one), which cannot undergo tautomeric changes, two azomethine bonds are present. Their electrochemical reduction on mercury electrodes occurs in two two-electron steps. The 1,6-bond C=N is reduced in protic solvents in the protonated form at potentials by about 0.5 V more positive than the protonated C=N bond in position 2, 3. Potentials of both reduction steps are shifted with increasing pH to more negative values, limiting currents of both steps decrease with increasing pH in a potential range where the protonotion becomes insufficiently fast. Limiting currents of both steps are affected by covalent hydration.

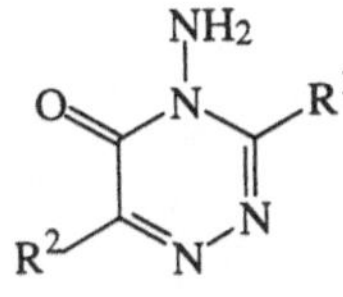

$R^1 = CH_3, R^2 = C_6H_5$
METAMITRON

$R^1 = SCH_3, R^2 = t - C_4H_9$
METRIBUZIN

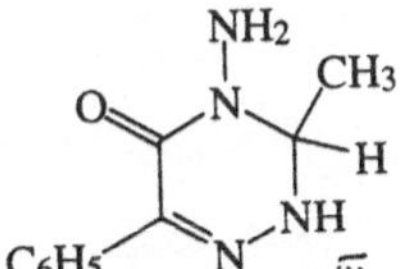

The first reduction step of metamitron(I) corresponding to the reduction of the 1,6-bond was compared with the reduction of the azomethine bond in 2,3-dihydrometamitron(II) prepared by chemical reduction using sodium borohydride. The reduction of the 1,6-bond in both compounds occurs over the entire pH range at the same potential (± 0.003 V) and the shapes of $E_{1/2} = f(pH)$ and $i = f(pH)$ plots are identical.

337

Thus, whereas for two adjacent double bonds linked by two carbon atoms (e.g., C=C–C=C, O=C–C=C, O=C–C=O, N=C–C=C, N=C–C=O, N=C–C=N) conjugation occurs due to an overlap of orbitals on C, in C=N–N=C no interaction between two azomethine bonds occurs. The presence of the second C=N group does not affect the electroactivity of the first C=N group. Electron density maps indicate absence of overlap of p-orbitals on the two nitrogen atoms.

SINGLE ELECTRON-TRANSFER REACTIONS OF ARYLIMINES

David W.BROWN, Adrian FISHER, Aleyamma NINAN, Malcolm SAINSBURY and Virginia WOOD

School of Chemistry, University of Bath, BA2 7AY, UK

We are interested in the design and synthesis of chain-breaking antioxidants. Early examples are the tetrahydroindenoindoles (**1**), which form stable radical cations and which are recycled *in vivo* by ascorbate. Other compounds of interest are the benzoxazine (**2**) and the Astra compound (**3**).

Compounds **2** and **3** were synthesised from the parent imines by the addition of cyclohexyl lithium (generated from ʹbutyllithium and cyclohexyl chloride). In such reactions conjugate addition of ʹbutyl radicals was often a major side reaction.

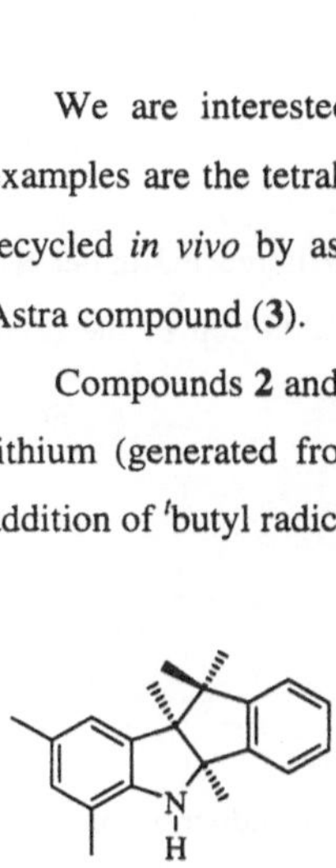

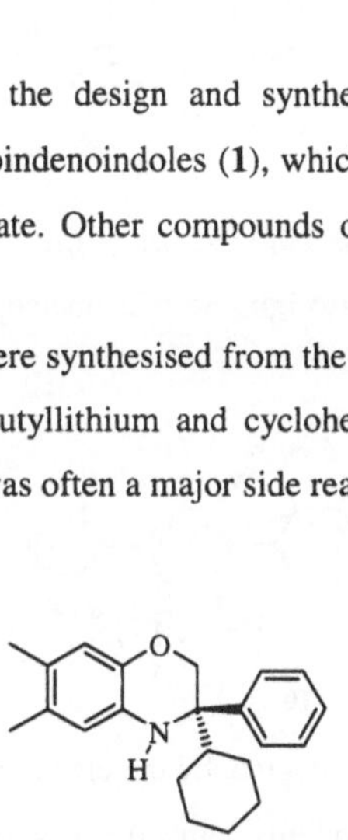

It is probable that BOTH nucleophilic addition and conjugative coupling proceed by a common first step, which involves a single electron-transfer (SET) between the alkyllithium and the imine double bond, leading to a caged unit containing the alkyl radical, the radical anion of the imine and a lithium cation. The radical and the radical anion may now react to give the anion of the corresponding alkylarylamine. Conjugation within the radical anion or the arylimine now permits coupling of the alkyl radical at centres "remote" from the original imine bond. For example, when the 2*H*-1,4-benzoxazine **4** is reacted with cyclohexyllithium the amine **2** was formed in 40% yield, plus by-products in which the ʹbutyl unit is bonded to the phenyl side chain.

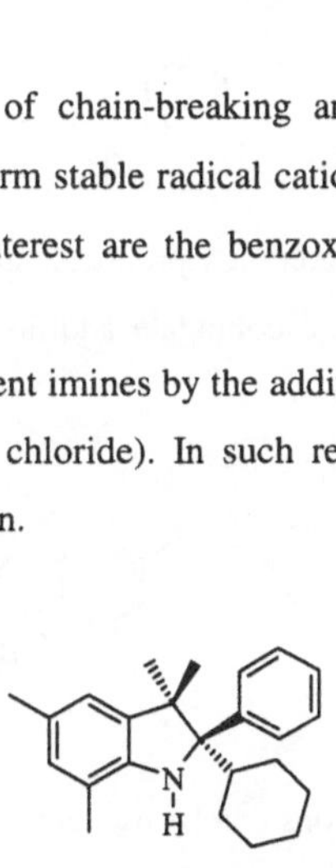

Similarly, when **4** is reacted *directly* with excess ʼbutyllithium and tetrahydrofuran it gives **5** (27%), **6** (4%),   **7** (2%) and the diastereomeric dimers **8** (1.5%).

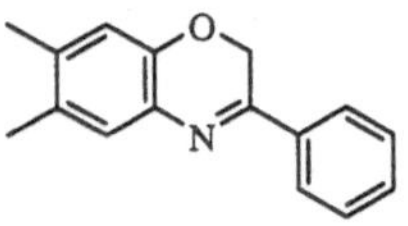

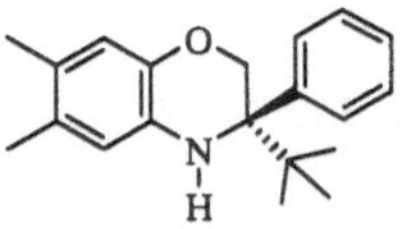

Even more complex products, such as **10,** arise when the imine **9** is reacted under these conditions and here conjugate addition outweighs normal nucleophilic addition to the double bond of the parent.

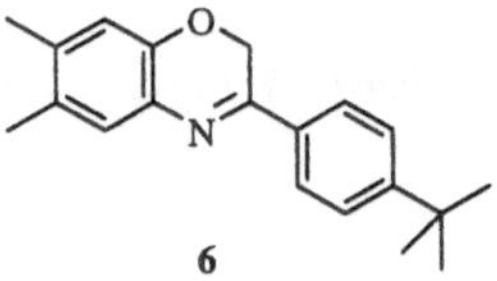

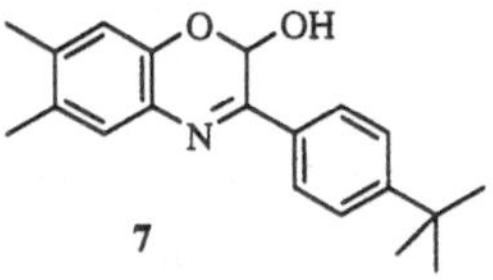

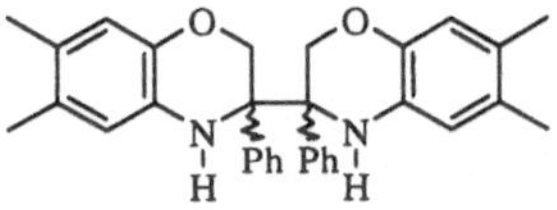

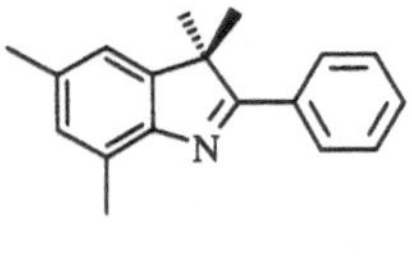

These reactions can be replicated, to a degree, by the electrochemical reduction of suitable alkyl halides in the presence of arylimines and this points the way to the synthesis of heterocycles through the intramolecular cyclisation of suitable substrates containing both types of functional groups, for example **11** → **12**. This work extends the pioneering studies of Degrand *et al.* (Tetrahedron Letts., **1978**, *33*, 3023).

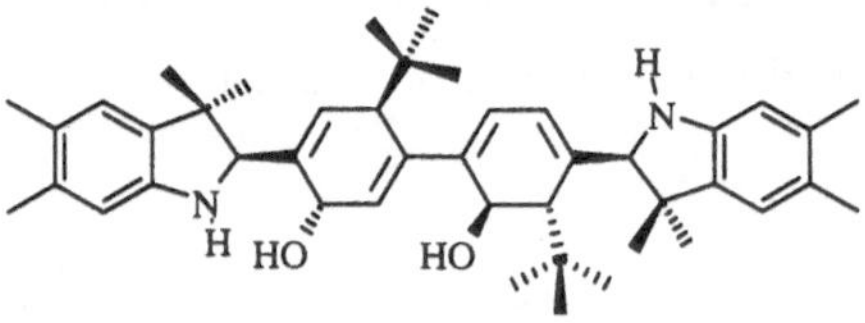

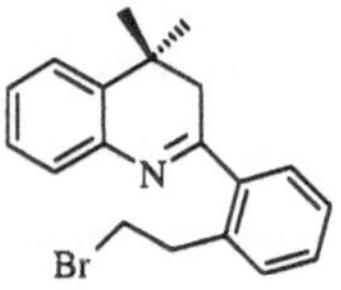

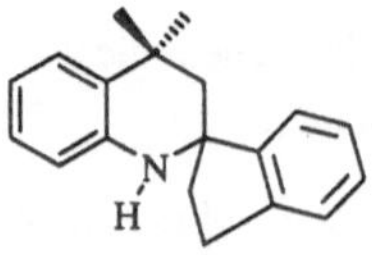

Acknowledgements This programme was generously funded by AB Astra; details of this work are to be published shortly in J. Chem. Soc., Perkin Trans. 1, 1997.

ALKYL TRANSFER REACTIONS OF TRIALKYLBORANE TO THE CARBONYL GROUPS BY ELECTROCHEMICAL METHOD

Junghoon Choi[*], Jongsung Yoem, Jongha Lee, Taekee Hong[**], Jinsoon Cha[***]
Department of Chemistry, Hanyang University, Seoul, 133-791, Korea;
[**]Department of Chemistry, Hanseo University, Chung-Nam 352-820,
Korea; [***]Department of Chemistry, Yeungnam University, Kyongsan
712-749, Korea

ABSTRACT

Alkyl groups in trialkylboranes were successfully transferred to carbonyl compounds upon electrolysis with sacrifial anodes to produce various substituted alcohols in good to fair yields. This new, mild electrochemical alkyl transfer reaction is described. Our new electrochemical method for the synthesis of alcohols can be especially useful for the systems where organometallic addition reactions cannot be employed.

INTRODUCTION

Electrochemistry has been used as a convenient tool in generating many active species which may not be accessible by other conventional chemical methods, with an additional advantage avoiding handling of hazardous, toxic reagents and byproducts. In his pioneering study, Suzuki has first reported the formation of coupling products from alkyl organoboranes upon electrolysis in DMF. Suzuki has also showed later on that the use of sacrifial anode in electrochemical process can greatly reduce possible oxidative destruction of the carbanions in anode when conducted in undivided cells[1,2,3]. Sacrificial anode, made of an electropositive metal, is oxidized first before oxidation of the carbanions occurs. This electrochemically generated carbanions have since been used for alkylation of electrophiles, but not of carbonyl compounds (in 1,2-addition fashion). In conjuction with our on going research program involving regioselective electrochemical reduction of carbonyl compounds, we were interested in the use of sacrificial anode for the above reaction. Under the condition described in the text, the alkyl groups (primary, secondary) were successfully transferred from trialkylboranes to carbonyl compounds for the first time to produce various substituted alcohols.

METHODS

The electrochemical reactions were carried out with EG & G PARK Model 173 potentiostat and BAS 100 β potentiostat under the constant cathode potential at -3.2 or -3.6 V *vs* SCE(saturated calomel electrode).

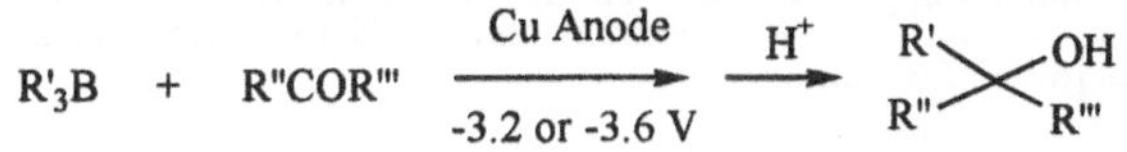

$$R'_3B \ + \ R''COR''' \xrightarrow[\text{-3.2 or -3.6 V}]{\text{Cu Anode}} \xrightarrow{H^+} \underset{R''}{\overset{R'}{\diagdown}}\underset{R'''}{\overset{OH}{\diagup}} \qquad \text{(equation 1)}$$

RESULTS AND DISCUSSION

As shown in Table 1, various types of carbonyl compounds underwent facile alkyl transfer reactions with both triethylborane and tri *sec*-butylborane to produce the corresponding alcohols in good to fair yields. For more sterically demanding *sec*-butyl transfer reactions compared to triethylborane, slightly higher voltage (-3.6 V) was required. In contrast to similar Grignard or other organometallic reactions, no noticeable differences were observed in both reaction rate and chemical yields between 1) tri *sec*-butylborane and triethylborane, 2) aliphatic and aromatic carbonyl compounds, 3) aldehydes and ketones. We do not understand the reason for the independence of our method to the steric hindrance, which might imply the exact molecular mechanism of such electrochemical reaction. Apart from the mechanistic issue, our method could be used for the synthesis of the alcohols from the carbonyl compounds carrying functional groups sensitive to the conventional organometallic reagents. Additional advantage over the lithium and magnesium based organometallic reactions would be that structurally versatile trialkylboranes can be readily synthesized from the corresponding alkenes through the well known hydroboration. In conclusion, our results represent the first successful electrochemical alkyl transfer reactions of carbonyl compounds with trialkylboranes. This new method is expected to be widely used in synthetic organic chemistry due to the mild reaction condition.

Table 1. Electrochemical Reaction with Carbonyl Compounds Using Cu Sacrificial Anode[a,b,c].

R_3B	Carbonyl compounds	Chemical yields(%)[d]	Current yields(%)[e]	voltage(V)
R=ethyl	Aliphatic aldehydes	70-75	40-45	
	Aromatic aldehydes	65-75	30-40	-3.2
	Aliphatic ketones	76-78	38-41	
	Aromatic ketones	62-77	30-41	
	Alicyclic ketones	73-75	33-35	
R=sec-butyl	Aliphatic aldehydes	62-77	33-36	
	Aromatic aldehydes	61-64	34-39	-3.6
	Aliphatic ketones	61-62	33-34	
	Aromatic ketones	52-75	35-36	

[a]One equivalent of trialkylborane was used for each carbonyl compound. [b]The reactions were carried out at concentration of 0.1 M and room temperature in DMF. [c]Tetrabutylammonium iodide (0.05 M) was used as supporting electrolyte. [d]Yields were determined by GC. [e]Current yields were determined by the ratio of the actual amount consumed to theoretical amount of electricity required to effect a given reaction.

REFERENCES

1) Takahashi, Tokuda, Itoh, Suzuki, Chem.Lett., **1975**, 573
2) Takahashi, Yuasa, Tokuda, Itoh, Suzuki, Bull.Chem.Soc.Jpn., **1978**, *51*, 339
3) Takahashi, Tokuda, Itoh, Suzuki, Chem.Lett., **1980**, 461

HYDROBORATION OF ALKENES OR ALKYNES BY ELECTROLYSIS OF SODIUM BOROHYDRIDE

Ikuzo NISHIGUCHI, Kotaro ITOH, Hirofumi MAEKAWA, Yoshiharu MATSUBARA+, and Ryushi SHUNDOU+

Department of Chemistry, Nagaoka University of Technology, Nagaoka, Niigata940-21, Japan

+Kinki University, Higashi-Osaka577, Japan

ABSTRACT

We have already reported that electrochemical oxidation of the solution containing sodium borohydride led to the generation of diboranes (BH_3). Application of this method to reductive reactions of aliphatic carboxylic acids was successful to form the corresponding primary aliphatic alcohols selectively.[1] These reactions was quite familiar to us as one of the typical transformations by use of diboranes.

RESULTS AND DISCUSSIONS

In this study, facile and efficient generation of diborane by anodic oxidation of sodium borohydride in diglyme tempted us the application to electrochemical hydroboration of alkenes and alkynes. The electrolysis was carried out according to the similar procedure as before except treatment of the resulting mixture with alkali-hydrogen peroxide instead of acid-catalyzed hydrolysis after the reaction.

$$R\diagdown\diagup \xrightarrow[\text{NaBH}_4 \,/\, \text{NaI}]{\text{Electrolysis}} \xrightarrow{\text{NaOH/H}_2\text{O}_2} R\diagdown\diagup\diagdown\text{OH} \qquad (1)$$

Table 1. Electrochemical hydroboration of β-pinene (1)

Supporting Electrolyte	Mole ratio ($NaBH_4/1$)	Yield of 2 (%)	Supporting Electrolyte	Mole ratio ($NaBH_4/1$)	Yield of 2 (%)
NaI	1.25	82	-	2.0	71
NaI	0.5	81	-	1.25	0
KI	1.25	71	$NaBF_4$	2.0	44
LiI	1.25	82	$LiClO_4$	1.25	12
LiI	0.5	0	Et_4NOTs	1.25	0

Yield of the obtained alcohol was found to be influenced by nature and amount of a supporting electrolyte, as shown in Table 1, in which (-)-2(10)-pinene (1) was transformed to (-)-cis-myrtanol (2). It was interesting that just half equivalent mole of sodium borohydride based on 1 was enough when sodium iodide was used as a supporting electrolyte, while more than uniequivalent mole of sodium borohydride was required for the efficient transformation of 1 to 2 for an other supporting electrolyte than sodium iodide.

Table 2 shows some of the results of the present electrochemical hydroboration of a variety of alkenes under the optimum conditions. It was found that the products were formed regio- and stereoselectively in good yields, as observed in the conventional methods.

As one of unsaturated systems, electrochemical hydroboration of acetylenes under the similar reaction conditions was also carried out to afford predominantly the corresponding alcohol.

Table 2. Electrochemical hydroboration of various olefins

Olefins	Products	Yield (%)	Olefins	Products	Yield (%)
		81			69
		78			82
		72			72
		70			75

$$R\diagup\!\!\!\equiv\!\!\!\diagdown H \xrightarrow[\text{NaBD}_4/\text{NaI/Diglyme}]{\text{Electrolysis}} \xrightarrow{\text{NaOH/H}_2\text{O}_2} \begin{array}{l} R\text{---}CD_2\text{---}CDH\text{---}OH \\ R\text{---}CD_2\text{---}CD_2\text{---}OH \end{array} \quad (2)$$

$$\begin{array}{c} Ph\diagup\!\!\!\equiv\!\!\!\diagdown R \end{array} \xrightarrow[\text{NaBH}_4/\text{NaI/Diglyme}]{\text{Electrolysis}} \xrightarrow{\text{NaOH/H}_2\text{O}_2} \left(Ph\diagdown\!\!\diagup R \right)_2 \begin{array}{l} R = H \quad 64\,\% \\ R = Me \quad 49\,\% \end{array} \quad (3)$$

It may be also noteworthy that the present electrochemical hydroboration of alkynes under the similar conditions afforded predominantly the corresponding alcohols with formation of only a trace amount of the corresponding carbonyl compounds. Use of NaBD4 instead of NaBH4 in this reaction brought about introduction of 3 or 4 deuterium atoms to α- and β-carbon atoms of the obtained the hydroxyl group, suggesting possibility of reduction of the intermediate vinyl borane or NaBH4.

On the other hand, phenylacetylene as a kind of acetylenes was electrochemically transformed to diphenylbutane, a reductive dimer of phenylacetylene. This dimerization may proceed through generation of tristyrenyl borane intermediates. Detailed reaction mechanism for the formation of this product and reactivity of other acetylene derivatives are now under investigation.

Reference
1) R. Shundou, Y. Matsubara, I. Nishiguchi, and T. Hirashima, Bull. Chem. Soc. Jpn., **1992**, *65*, 530.

BACK-BITING REACTIONS OF σ-CONJUGATED POLYMERS IN ELECTROCHEMICAL SYSTEMS

Mitsutoshi OKANO, Takeshi TORIUMI, Hiroyuki OHTSUKA, and Hiroshi HAMANO
Department of Photo-Optical Engineering, Faculty of Engineering,
Tokyo Institute of Polytechnics, Kanagawa 243-02, Japan

INTRODUCTION

σ-Conjugated polymers (polysilanes and polygermanes) have been the subject of considerable interest in recent years because of their unique physical and chemical properties. Electrochemical reduction of dihalogeno-silanes and -germanes is one of the best pathways to the polymers. However, only relatively small molecular weight polymers are obtained by electrochemical method compared to those obtained by Wurtz-type coupling reaction.

In order to find out causes for the relatively small molecular weight, back-biting reaction (see Scheme 1) was examined if it is taking place in electrochemical systems, because back-biting reactions are often discussed in chemical preparation of σ-conjugated polymers as a cause which decreases molecular weights of them.

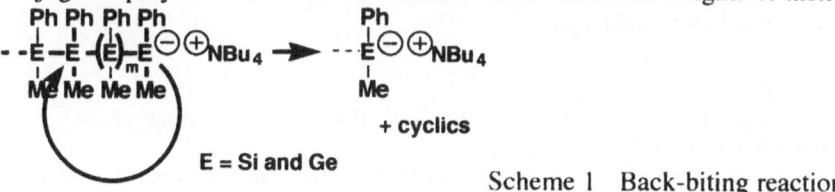

Scheme 1 Back-biting reaction

EXPERIMENTAL

All polymers were synthesized electrochemically by the method reported earlier.[1]

Generation and accumulation of the anion (key species in back-biting reaction) were carried out at -20°C. The exsistence of the anion species were confirmed by ^{1}H NMR using the techniques reported ealier.[2]

The time courses of back-biting reactions were examined by UV spectra and GPC.

RESULTS AND DISCUSSION

Firstly, in order to obtain experimental support for the generation of the anions (see Scheme 1) in electrochemical systems, poly(methylphenylgermane) was reduced at -20°C in a two-compartment cell. The catholyte solution was examined by ^{1}H NMR after electrolysis. Figure 1 shows the spectrum. Signals attributable to phenyl substituted germyl anions were observed between δ = 6.6 to 7.0. Thus, generation of anions in the electrochemical system was proved.

Secondly, the above anion solution was examined if it causes back-biting reaction. A portion of the anion solution was added to a solution of poly(phenylmethylgermane) and decomposition of the polymer was observed. Figure 2 shows the GPC profiles. Decomposition of high molecular weight polymers and formation of low molecular weight oligomers (cyclics) without formation of medium molecular weight polymers. This result strongly suggests the back-biting reaction.

Therefore, it was concluded that back-biting reactions may occur in electrochemical systems. The above experiments can be carried out only when the electrochemically prepared anions are stable (require limited compounds and somewhat low temperature). Simply reducing polymers in a two-compartment cell was found to be an alternative method to examine back-biting reactions. With the alternative method, it was found that the rates of back-biting reactions are suppressed at low temperatures.

References
1) M. Okano, T. Toriumi, K. Takeda, Y. Kurimoto, T. Ohyama, and H. Hamano, *Denki Kagaku* , **62**,1163(1994).
2) M. Okano, T. Kugita, D. Ohtani, and H. Hamano, *Bull. Chem. Soc. Jpn.*, **68**, 759(1995).

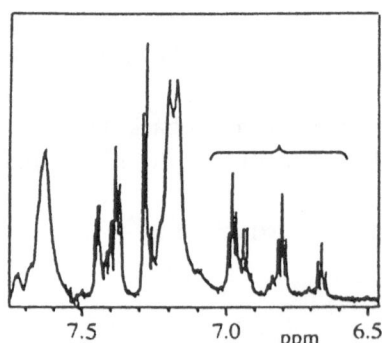

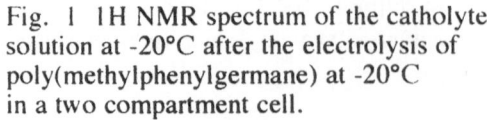

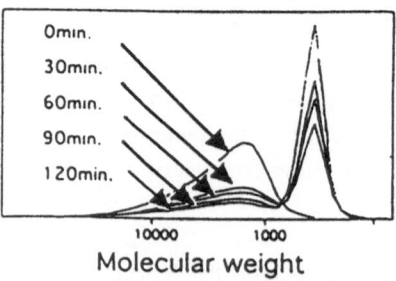

Fig. 1 1H NMR spectrum of the catholyte solution at -20°C after the electrolysis of poly(methylphenylgermane) at -20°C in a two compartment cell.

Fig. 2 GPC profiles of the poly-(methylphenylgermane) before and after the reaction with the electro-chemically prepared anion solution.

FORMATION OF Sn-Sn BOND AND SYNTHESIS OF POLYSTANNANE BY ELECTROLYSIS

Junzo NOKAMI, Kenji YOSHIZANE, Satoshi GODA, and Hiroshi SETO

Okayama Unversity of Science, Faculty of Engineering,

Department of Applied Chemistry, Ridai, Okayama 700, Japan

ABSTRACT

Hexaorganoditin was synthesized by electrolysis of R_3SnX (R=alykyl, phenyl; X=OCHO, SPh, H) in high yield, and polystannane was alos prepared from $Ph_2Sn(SPh)_2$ in good yield.

INTRODUCTION

The Wurtz-coupling of organotinhalide by alkali metals (Na, Li) is well known as a usual method for Sn-Sn bond formation. On the other hand, for polystannane synthesis, transition metal-catalyzed dehydropolymerization of diorganotindihydride (eq.-2)[1] was also useful as well as the Wurts-coupling reaction of diorganotindichloride (eq.-1).[2] Moreover, reaction of organotinhydride with organotin amide was reported as a useful method for Sn-Sn bond formation and was employed for synthesis of polystannane oligomers *via* stepwise elongation of Sn-Sn linkage (eq.-3).[3]

$$R_2SnCl_2 \quad + \quad 2\,Na \quad \xrightarrow{\text{15-crown-5 (cat.)}} \quad R_2ClSn(R_2Sn)_nSnR_2Cl \qquad \text{eq.} -1$$

$$R_2SnH_2 \quad \xrightarrow{\text{Me}_2C(\eta^5\text{-C}_5H_4)_2Zr[Si(SiMe_3)_3]Me\,(cat.)} \quad R_2HSn(R_2Sn)_nSnR_2H \qquad \text{eq.} -2$$

$$R_3Sn(R_2Sn)_mSnR_2H \quad + \quad R'_2NSnR_3 \longrightarrow \quad R_3Sn(R_2Sn)_{m+1}SnR_3 \qquad \text{eq.} -3$$

METHOD

We have found out that triorganotin derivatives **1** was conveniently electrolized in an undivided cell to give hexaorganoditins **2** in high yield (eq.-4).[4]

$$\underset{\textbf{1}}{R_3SnX} \quad \xrightarrow[\text{high yield}]{\text{electrolysis}} \quad \underset{\textbf{2}}{R_3Sn\!-\!SnR_3} \qquad \text{eq.} -4$$

R = n-Bu, CH_2Ph, Ph
X = OCHO, SPh, H

In this case, triorganotin formate (X = OCHO in **1**) seemed to be most useful because it was easily available and conveniently electrolyzed (in good current efficiency without using supporting electrolyte) to give hexaorganoditins in good yield. However, it is noteworthy that phenylthiotriorganostannane (X = SPh in **1**) is also useful for this pourpose, though it requires catalytic amount of supporting electrolyte (0.01 M in DMA, 2.6 mol% to **1**) for electrolysis.

347

Based on these results, we investigated an electrosynthesis of polystannane **4** by using diorganotin derivative (R_2SnX_2) **3** and found out that bis(phenylthio)diorganostannane (X = SPh in **3**) was useful for this purpose (eq.-5), but diorganotindiformate (X = OCHO in **3**) was not available.

$$\text{Ph}_2\text{Sn(SPh)}_2 \xrightarrow{\text{electrolysis}} \text{(Ph}_2\text{Sn)}_n \quad \text{PhS(Ph}_2\text{Sn)}_2\text{SPh} \qquad \text{eq.}-5$$

3   **4**   **5**

Table 1   **Electrolysis of $Ph_2Sn(SPh)_2$ to $(Ph_2Sn)_n$**[a]

Substrate (g)	Solvent[b] (ml)	Electrolyte[c] (mg)	Current mA/cm^2	Voltage (V)	Time (h)	Electricity F/mol	Yield (%) 4	5
3 (0.5)	**DMA** (3.5)	5	50→7	29	13		97	
3 (0.5)	**DMA** (3.5)	5	50	18→30		12	81	
3 (1.0)	**DMA** (4.0)	40	50→41	10	5			73
3 (1.0)	**DME** (3.0)	150	20	13→20		6		91
3 (1.0)	**THF** (3.0)	50	20	25→29		6		90
5 (0.5)	**THF** (8.0)	300	30	20→27		6	87	

a. Carried out using an undivided cell and Pt electrodes at 7-8 °C.
b. DMA = N,N-dimethylacetamide; DME = 1,2-dimethoxyethene.
c. Bu_4NClO_4 was used as a supporting electrolyte.

A white precipitate was deposited during the electrolysis under the reaction conditions (Table 1). After filtration of the reaction mixture, the filtrate was washed with methanol and chloroform to give a white crystal which proved to be Ph_2Sn by elemental analysis, though it was insolved in solvent. The crystalline product should be polystannane which might be decomposed to stable oligomer. Actually, $(Ph_2Sn)_4$, slightly soluble in benzene or chloroform, was obtained by thermolysis of the product in refluxing benzene for several hours (until the crystal was disappeared). Furthermore, we found out that electrolysis of **3** by ca. 6 F/mol of electricity (half of the synthesis of **4**) selectively gave **5** which was converted to **4** by further electrolysis.

As described above, polystannane was easily synthesized by the convenient electrolysis of bis(phenylthio)diphenylsatannane, $Ph_2Sn(SPh)_2$, using catalytic amount of supporting electrolyte (without using any alkali metals or metal hydrides) in an undivided cell under a usual reaction conditions.

References
1) T. Imori, V. Lu, H. Cai, T. D. Tilley, *J. Am. Chem. Soc.*, *117*, 9931 (**1995**).
2) N. Devylder, M. Hill, K. C. Molloy, G. J. Price, *J.C.S., Chem. Commun.*, **1996**, 711.
3) L. R. Sita, *Organometal. 11*, 1442 (**1992**).
4) J. Nokami, H. Nose, R. Okawara, *J. Organometal. Chem.*, *212* (**1981**) 325.

A CONSIDERATION ON MECHANISM OF TIN-MEDIATED ELECTRO-ALLYLATION OF CARBONYL COMPOUNDS

Anny JUTAND, Hiroyuki KAWAFUCHI,[*] Christian AMATORE,
Manabu KUROBOSHI,[**] and Sigeru TORII[**]

Department de Chimie, URA CNRS 1679, Ecole Normal Superieure, 24 Rue Lhomond, 75231 Paris Cedex 05, FRANCE, [*]Department of Industrial Chemistry, Toyama National College of Technology, Hongo 13, Toyama 939, JAPAN, [**]Department of Applied Chemistry, Faculty of Engineering, Okayama University, Tsushima-Naka 3-1-1, Okayama 700, JAPAN

ABSTRACT

Allylation of aldehydes and ketones with electrogenerated allyltin reagents proceeds in a MeOH/AcOH(4/1)-(Pt) system to give the corresponding homoallyl alcohol in 72-91% yields. Mechanistic assumption on tin-mediated electroallylation of carbonyl compounds is proposed by use of chemical, electrochemical, and spectroscopic methods.

INTRODUCTION

Although allylation of carbonyl group is a well-studied reaction, there are only a few reports dealing with re-generation of allylation reagents. As the result of our continuing efforts, we have already reported that electroallylation of aldehydes and ketones proceeds in a MeOH/AcOH(4/1)-(Pt) system to give the corresponding homoallyl alcohol in 72-91% yields using a catalytic amount of Sn(0) in the presence of allyl bromide.[1] However, the reaction mechanism is still unclear. We attempted to clarify the generation and reactivities of allyltin species under electrolysis conditions.

METHOD

The allylation is considered to be the following three processes. In the first process, diallyltin dibromide (Allyl)$_2$SnBr$_2$ is generated from the reaction of Sn(0) with allyl bromide. Next, (Allyl)$_2$SnBr$_2$ reacts with carbonyl compounds to give the corresponding homoallyl alcohols. Finally, low-valent tin reagents, Sn(II) and/or Sn(0), are re-generated from the electroreduction of Sn(IV) species. In order to know reactivities of tin reagents, the chemical allylation of benzaldehyde with metallic, divalent, and tetravalent tin derivatives in the presence (or absence) of allyl bromide has been investigated. As the result, SnBr$_2$ did not react with allyl bromide in a MeOH/AcOH(4/1) system in the presence of benzaldehyde, since the homoallyl alcohol was not formed. The allylation reaction of benzaldehyde with allyl bromide in the presence of SnBr$_4$ failed. The reliable way to get the homoallyl alcohol was to react benzaldehyde with allyltin derivatives, (Allyl)$_x$SnBr$_{(4-x)}$ (x = 1,2,3 or 4), or to start from Sn(0) and allyl bromide. The reduction potentials of tin derivatives were measured by cyclic voltammetry in methanol and in the mixture of methanol and acetic acid (4/1) (n-Bu$_4$NBF$_4$-(Pt)-(Pt)-(SCE), 200 mV/S). Sn(0) slowly reacts with allyl bromide in the mixture of methanol and acetic acid (4/1) and a reduction peak is observed at -0.75 V vs. SCE and, with time, a shoulder appears at -0.60 V. The reduction peak of SnBr$_2$ in methanol is observed at -0.51 V. On the reverse scans, an adsorption peak is present characteristic of a deposit of Sn(0) on the electrode surface. The cyclic voltammetry of SnBr4 performed in methanol exhibited two reduction peaks at -0.34 and -0.72 V. The first one (-0.34 V) is quasi-reversible and involves two electrons. In the presence of allyl bromide, the first peak (-0.34 V) becomes irreversible, suggesting that the tin derivative generated by the electrochemical reduction of SnBr4 is highly reactive with allyl bromide. The two reduction peaks of (Allyl)2SnBr2 in methanol are observed at -0.58 and -1.20 V. The first one (-0.58 V) is quasi-reversible and involves two electrons. The reaction of (Allyl)2SnBr2 with benzaldehyde was monitored by cyclic voltammetry. After 3 h, two reduction peaks are observed at -0.39 and -0.84 V. The reduction potential of the first peak (-0.39 V) of them looks like that of the first peak (-0.34 V) of SnBr4.

In conclusion, Sn(0) first reacts with two equivalents of allyl bromide to give (Allyl)2SnBr2, which reacts with benzaldehyde to give PhCH(OH)CH2CH=CH2 and Sn(IV)Br2(OY)2. These Sn(IV) species can be reduced to Sn(II) species [Sn(II)Br2(OY)2]$^{2-}$, which readily reacts with allyl bromide to give [allylSnBr3(OY)2]$^{2-}$. This tetravalent allyltin species could react with benzaldehyde to give the homoallyl alcohol, PhCH(OH)CH2CH=CH2, and non-allylated Sn(IV) species, such as Sn(IV)Br2(OY)2.

From the above result, the plausible reaction system is as Scheme 1.

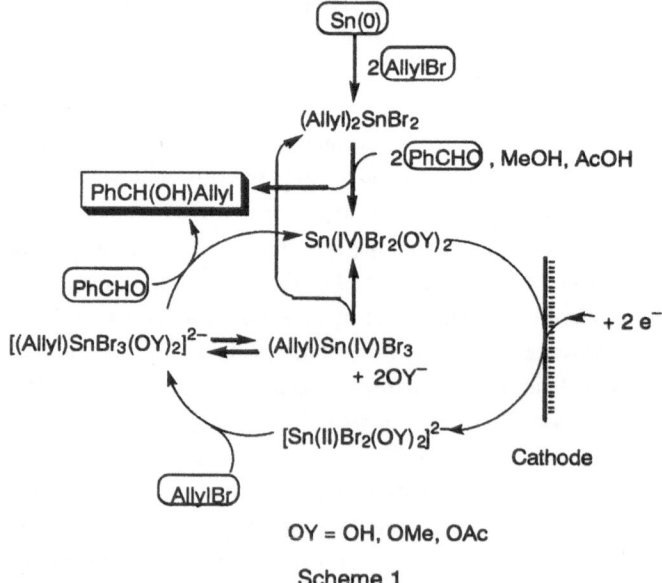

Scheme 1

Reference

1) K. Uneyama, H. Matsuda and S. Torii, *Tetrahedron Lett.*, **25**, 6017 (1984).

ASYMMETRIC CARBON-CARBON AND CARBON-SILICON BOND FORMATION OF ACTIVATED OLEFINS PROMOTED BY Mg-METAL AND ELECTROREDUCTION

Ikuzo NISHIGUCHI

Nagaoka University of Technology

1603-1, Kamitomioka-cho, Nagaoka, Niigata 940-21, JAPAN

ABSTRACT

It has been found in this study that Mg-promoted electron transfer reaction in DMF / trimethylsilyl chloride (TMSCl), and electroreduction using reactive metal anodes of aromatic α, β-unsaturated systems in the presence of aldehydes, TMSCl, and acid chlorides (or acid anhydrides) brought about efficient and selective reductive cross-coupling reactions to give the corresponding γ-butyrolactones, β-silylated and β-acylated products in good yields, respectively. Furthermore, some diastereo-selectivity (40-50%) was observed in the β-silylation and β-acylation using chiral α, β-unsaturated amides as the substrates.

INTRODUCTION

Electron transfer on heterogenous interface between solutions and electrodes or metals may not only generate "specific active chemical species" possessing unique reactivities and characters, unexpected from full knowledges of conventional organic chemistry, but also involve "dipole inversion (Umpolung)" of nucleophiles or electrophiles through donation-acceptance of electrons. For example, donation of one electron to an electrophile (A) leads to transformation of it to the corresponding anion radical, a nucleo-philie, which may react another electrophile (B) to give the cross-coupling products selectively.

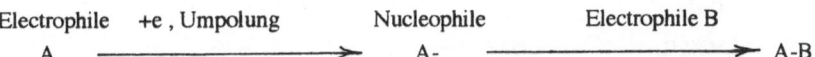

Electrophile　　+e , Umpolung　　　Nucleophile　　　Electrophile B

A ────────────────────➤ A-　────────────────────➤ A-B

Thus, we have found in this study that a facile new method for efficient and regioselective C-C and C-Si bond formation of α,β-unsaturated compounds with trimethylsilyl chloride (TMSCl), aliphatic aldehydes and acid chlorides (or anhydrides) was also successfully developed through electroreductiion[1,2] in a DMF/Et₄NOTs system using Mg as a reactive metal anode, and Mg-promoted reduction at room temperature in DMF containing TMSCl to give the corresponding β-silyl carbonyl compounds[3], γ-butyrolactones[4] and 1,4-dicarbonyl compounds[5] in good yields, respectively.

RESULTS AND DISCUSSION

(1) Electroreductive and Mg-Promoted Reductive β-Silylation of Activated Olefins[1,3]

We now wish to report two new methods for regioselective introduction of trimethylsilyl chloride(TMSCl) to a β-position of activated olefins, such as β-arylacrylates, nitriles and ketones, by electroreduction using a reactive Mg anode in an undivided cell, and Mg-promoted Michael-type addition at room temperature in N,N-dimethylformamide(DMF) to give the corresponding β-silyl-propionates, nitriles and ketones, respectively.

Among a variety of metals in the reaction of ethyl cinnamate, use of magnesium led to the best result in both of reductive coupling (yield of 3-trimethylsilyl-3-phenyl propionate : 79% by electroreduction , 92% by Mg-promoted reaction), while use of Zn or Al instead of Mg resulted in no conversion of the activated olefin or formation of the complex product mixture.   DMF or N,N-dimethylacetamide was found to be the most suitable for this reductive coupling while use of other solvents, such as tetrahydrofuran, acetonitrile or diethyl ether, resulted in almost quantitative recovery of ethyl cinnamate. As shown in a Table, a variety of activated olefins was treated under the similar conditions of the two methods to give the corresponding β-silylated coupling products in good yields.   It may be noteworthy that the presence of an electron-withdrawing group on the phenyl ring of ethyl cinnamate led to exclusive formation of simply reduced product, ethyl 3-phenyl propionate.   These facts may indicate much similarity of reaction mechanism of the present two reductive silylation.   The reaction may proceed through electrophilic attack of TMSCl to a radical anion species generated by initial electron transfer from Mg-metal to activated olefins , as shown in a Scheme

R^1	Z	Isolated Yield(%)	
		Electroreduction[a]	Mg-Promoted[b]
C$_6$H$_5$	CO$_2$Et	79	92
p-MeC$_6$H$_4$	CO$_2$Et	76	71
p-MeOC$_6$H$_4$	CO$_2$Et	79	80
p-ClC$_6$H$_4$	CO$_2$Et	70	—
p-NCC$_6$H$_4$	CO$_2$Et	0	0
p-NO$_2$C$_6$H$_4$	CO$_2$Et	0	0
(furyl)	CO$_2$Et	72	77
(thienyl)	CO$_2$Et	59	70
C$_6$H$_5$	CN	70	71
C$_6$H$_5$	COMe	68	80

On the other hand, some of diastereoselective asymmetric introduction (d.e.=31-52%)   was found in either of electroreductiive   or   Mg-promoted β–silylation   of   tha chial aromatic α,β-unsaturated amides , prepared from amidation of aromatic α,β-unsaturated acid chloride and (S)-(-)-4-benzyl-2-oxazolidione, to give the corresponding chiral β–phenyl-β-trimethylsilylpropionamide.   It may be noteworthy that the reaction at lower temperature led to some increase in the distereoselectivity.

Ar	Electroredction		Mg-Promoted	
	Isolated Yield(%)	de(%)	Isolated Yield(%)	de(%)
a C_6H_5	63(48)	37(44)	69(51)	39(52)
b (furyl)	70	37	73	38
c p-MeC$_6$H$_4$	82	31	85	39

The numbers in the parenthesis shows the de values of the reaction at -40°C

(2) Electroreductive and Mg-Promoted β–Acylation of Activated Olefins[2,4]

1,4-addition of an acyl anion or its chemical equivalents to α, β–unsaturated carbonyl compounds may be of importance as one of attractive methods. It was reported that novel acylation of some activated olefins such as α, β-unsaturated esters or nitriles was successfully accomplished by electroreduction using a divided cell equipped with carbon rod electrodes to afford the corresponding γ-ketoesters or nitriles.[5] These products can be regarded as those formed from 1,4-addition of an acyl anion to these activated olefins, although this electrochemical reaction may not involve the acyl anion itself as the active species.   In the electroreduction system, use of a reactive metal anode (sacrificial anode) in an undivided cell has some remarkable advantages over that of electrochemical non-sacrificed anode like carbon or platinum in a divided cell.   These facts prompted us to examine the feasibility of electroreductive acylation of α, β-unsaturated esters, ketones and nitriles using a reactive metal anode. The results shows the results obtained for several other α, β-unsaturated esters, ketones and nitriles with acetic anhydride. The reaction of aromatic ring-substituted ethyl cinnamates (4-Me, 4-Cl, 4-MeO, 4-MeOOC) with acetic anhydride gave the corresponding acetylated compounds smoothly.

X=CO$_2$Et, COMe, CN

Furthermore, the similar regioselective acylation on the the β–carbon atom of the activated olefins was found to proceed quite smoothly by Mg-promoted reduction at room temperature.   This method is clearly superior to the reported electroreductive method because more readily available acid chlorides and cyclic anhydrides can be used as acylating agents in addition to better yield of acylated products, much more simplicity of instrument, and facile procedure in the present Mg-romoted coupling, giving high potentiality for large-scale production and wide application in organic synthesis.

At the next stage , diastereoselective asymmetric introduction was investigated in these selective

$$\text{Ph}\diagdown\!\!\diagdown_{\text{Z}} \quad\xrightarrow[\text{DMF}]{\substack{\text{TMSCl /Acid Chloride / Mg} \\ \text{TMSCl /Acid Anhydride / Mg}}}\quad \text{Ph}\diagdown\!\!\underset{\underset{\text{R}}{\overset{\text{O}}{\parallel}}}{\diagup}\!\!\diagdown_{\text{Z}}$$

Mg-Promoted Acylation of Active Olefins

Olefins (Z)	Acid Anhydride (RCO)$_2$O	Yield (%)[a]	Olefins (Z)	Acid Chloride (RCOCl)	Yield (%)[a]
CO$_2$Et	Ac$_2$O	62	CO$_2$Et	AcCl	60
CO$_2$Et	(CH$_3$CH$_2$CO)$_2$O	84	CO$_2$Et	CH$_3$CH$_2$COCl	73
			CO$_2$Et	CH$_3$CH$_2$CH$_2$CH$_2$CH$_2$COCl	54
CO$_2$Et	O=⟨⟩=O (succinic anhydride)	85	CO$_2$Et	EtO$_2$CCH$_2$CH$_2$COCl	100
CO$_2$Et	O=⟨⟩=O (glutaric anhydride)	61	CO$_2$Et	PhCH$_2$COCl	76
			CO$_2$Et	ClCH$_2$CH$_2$CH$_2$COCl	93
COCH$_3$	Ac$_2$O	67	COCH$_3$	AcCl	43
CN	Ac$_2$O	76	CN	AcCl	44

a : Olefin (5mmol), Mg 3eq, TMSCl 3eq, Acid Chloride 15eq, DMF 50ml

carbon-acyl and carbon-silicon bond formation through electrochemical and Mg-promoted reduction of aromatic α,β-unsaturated carbonyl componds.   Only low diastereoselectivities($d.e.$=4-19%) were observed in both of the electroreductive and the Mg-promoted acylation of chiral cinnamate esters(**1**), prepared from the reaction of cinnamoyl chloride and chiral alcohols such as (-)-menthol, (-)-8-phenylmenthol, and (-)-isopinocamphenol.   However, electroreduction of the chiral cinnamyl amide, prepared from the amidation of cinnamoyl chloride with (S)-(-)-4-benzyl-2-oxazolidione, in the presence of acetic anhydride brought about considerably high diastereoselectivitiy($d.e.$=55%) to give the chiral β−phenyl-β-acetylpropionamide(**3**) while Mg-promoted acetylation of the same starting chiral amide resulted in low diastereoselectivitiy ($d.e.$=27%).

$$\underset{\mathbf{2a}}{\text{Ar}\diagdown\!\!\diagdown\!\!\underset{\overset{\text{O}}{\parallel}}{\text{C}}\!\!-\!\!N\!\!\underset{\overset{\text{O}}{\parallel}}{\diagup\!\!\diagdown}\!\!O\,/\,^{\text{PhCH}_2}}\quad +\ (\text{CH}_3\text{CO})_2\ \xrightarrow[\substack{\text{DMF}\\ \text{r.t.}}]{+2e}\quad \underset{\mathbf{3}}{\text{Ar}\!\!\overset{*}{\diagdown}\!\!\diagdown\!\!\underset{\text{CH}_3\text{OC}}{}\!\!N\!\!\underset{}{}O\,/\,^{\text{PhCH}_2}}$$

Electroreduction :	Y= 41%	de= 55%
Mg/Me$_3$SiCl :	Y= 50%	de= 27%

References
1)T.Ohno,H.Nakahiro,K.Sanemitu,T.Hirashima, I.Nishiguchi*, *Tetrahedron Lett.*, **3 3**,3515(1992).
2)T.Ohno,H.Aramaki,H.Nakahiro,I.Nishiguchi*, *Tetrahedron*, **5 2**, 1943(1996).
3)I.Nishiguchi*, T.Ohno, M.Sawada, H.Maekawa, *to be published*.
4)T.Ohno,Y.Ishino,Y.Tsumagari,I.Nishiguchi*, *J. Org. Chem.*, **60**, 458(1995).
5)T.Ohno,Y.Ishino,H.Maekawa, I.Nishiguchi*, *to be published*.

RECENT PROGRESS IN THE ELECTROCHEMICAL SYNTHESIS OF POLYSILANES

Shigenori Kashimura

Department of Metallurgy, Faculty of Science and Engineering,
Kin-dai University, 3-4-1, Kowakae Higashiosaka 577, Japan

Polymers containing silicon in the main chain have attracted considerable attention due to their usefulness as the precursors for thermally stable ceramics or materials for microlithography, and also due to their potentiality in the preparation of new types of material showing conducting, photoconducting, or nonlinear optical property. One of the most important innovation required in this field is the introduction of suitable functional groups into the polymer since the property of silicon containing polymers must remarkably be modified by such functional groups. We have recently found that the electroreduction of chlorosilanes with Mg electrodes is highly useful for the practical method for the synthesis of polysilane (PS) high polymers (Scheme 1).[1,2]

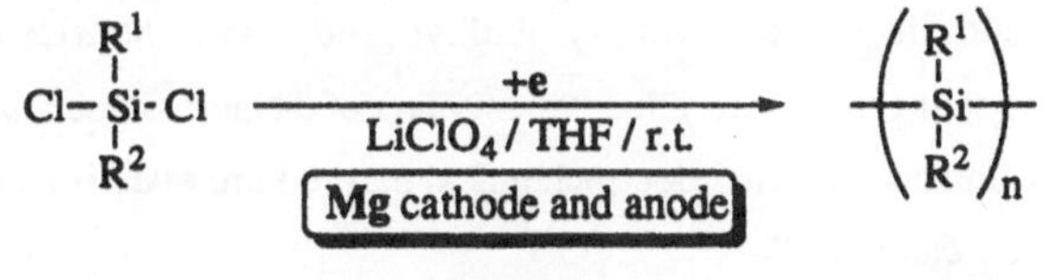

Scheme 1

The mildness of the reaction conditions of the electroreductive method is remarkably favorable for the synthesis of the polysilanes having a variety of hydroxyl-related functional groups. As illustrated in Scheme 2, the electroreduction of a mixture of **1a** and a dichlorosilane having a protected hydroxyphenyl group (**1b-e**) with Mg electrode afforded the corresponding copolymer 2 (**2b-e**), and the deprotection of the resulting copolymer gave polysilane having a hydroxyl group.[3,4] Typical results are shown in table 1.

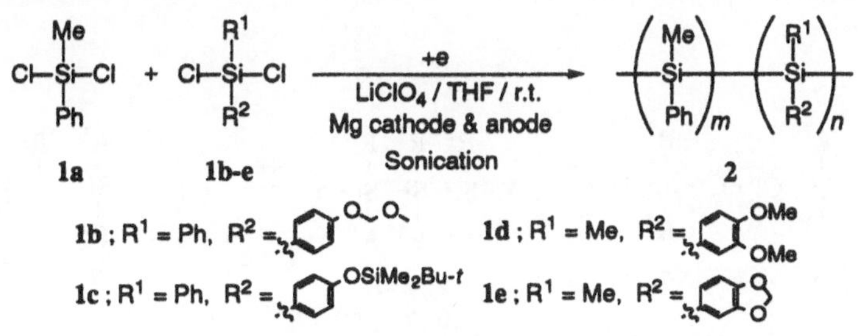

Scheme 2

355

Table 1. Electroreductive Synthesis of Functionalized Polysilanes[a]

run	charged mol% of 1b-d[b]	yield of 2, %[c, d]		$\overline{M}n$	$\overline{M}w/\overline{M}n$	run	charged mol% of 1b-d[b]	yield of 2, %[c, d]		$\overline{M}n$	$\overline{M}w/\overline{M}n$
1	7 (1b)	79	(7)	9900	1.9	6	10 (1d)	50	(6)	4500	1.3
2	10 (1b)	57	(12)	6900	1.7	7	50 (1d)	22	(46)	4600	1.3
3	100 (1b)	28	(100)	1100	1.2	8	100 (1d)	—[e]	(100)	1700	1.3
4	10 (1c)	36	(11)	6100	1.5	9	10 (1e)	56	(17)	4600	1.3
5	100 (1c)	—[e]	(100)	1100	1.2	10	100 (1e)	57	(100)	4000	1.1

a) The electroreduction was carried out by using Mg electrodes under sonication (47 kHz) ,and anode and cathode were alternated with the interval of 15 sec. Total momomer concentration, 0.67 mol/L;Supplied electricity, 4 F/mol. b) 1b-e /(1b-e+1a) x 100. c) Purified by reprecipitation from benzene-EtOH. d) The values in parentheses indicate the mol% of 1b-d units in the resulting copolymers determined by ^{1}H NMR. e) Polymer was not obtained by reprecipitation.

Because the nature of PS is greatly effected by the alkyl groups on the polymer side chain we have also studied the introduction of a variety of alkyl groups into the side chain of PS. It has been found in this study that the electroreductive coupling of dichlorosilanes with hydrosilane, and polymer reaction of PS prepared by the electrochemical method are effective for the introduction of alkyl groups into the side chain of PS.

Electroreductive coupling of dichlorosilane (3) with hydrosilane (4) (Schemes 3) and that of dichlorodisilane (8) with 4 (Schemes 4) have been studied and found that these reactions gave the corresponding trisilane (5) and tetrasilane (9) in reasonable yields, respectively. The products 5 and 9 are useful intermediates for the synthesis of PS having unique structures. For example, the electroreduction of dichlorotrisilane (6) prepared from 5 with Mg electrodes gave polysilane (7). Under similar reaction conditions, the electroreduction of dichlorotetrasilane (10) prepared from 9 gave polysilane (11).

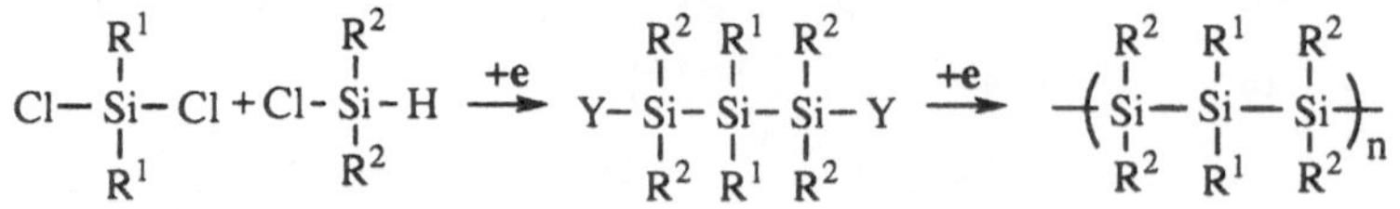

$$\underset{3}{\overset{R^1}{\underset{R^1}{Cl-\underset{|}{\overset{|}{Si}}-Cl}}} + \underset{4}{\overset{R^2}{\underset{R^2}{Cl-\underset{|}{\overset{|}{Si}}-H}}} \xrightarrow{+e} \underset{5;\ Y=H\ \ 6;\ Y=Cl}{\overset{R^2\ R^1\ R^2}{\underset{R^2\ R^1\ R^2}{Y-\underset{|}{\overset{|}{Si}}-\underset{|}{\overset{|}{Si}}-\underset{|}{\overset{|}{Si}}-Y}}} \xrightarrow{+e} \underset{7}{\left(\overset{R^2\ R^1\ R^2}{\underset{R^2\ R^1\ R^2}{\underset{|}{\overset{|}{Si}}-\underset{|}{\overset{|}{Si}}-\underset{|}{\overset{|}{Si}}}}\right)_n}$$

Scheme 3

$$4 + \text{Cl}-\overset{\overset{R^1}{|}}{\underset{\underset{R^1}{|}}{\text{Si}}}-\overset{\overset{R^3}{|}}{\underset{\underset{R^3}{|}}{\text{Si}}}-\text{Cl} \xrightarrow{+e} \text{Y}-\overset{\overset{R^2}{|}}{\underset{\underset{R^2}{|}}{\text{Si}}}-\overset{\overset{R^1}{|}}{\underset{\underset{R^1}{|}}{\text{Si}}}-\overset{\overset{R^3}{|}}{\underset{\underset{R^3}{|}}{\text{Si}}}-\overset{\overset{R^2}{|}}{\underset{\underset{R^2}{|}}{\text{Si}}}-\text{Y} \xrightarrow{+e} \left(\overset{\overset{R^2}{|}}{\underset{\underset{R^2}{|}}{\text{Si}}}-\overset{\overset{R^1}{|}}{\underset{\underset{R^1}{|}}{\text{Si}}}-\overset{\overset{R^3}{|}}{\underset{\underset{R^3}{|}}{\text{Si}}}-\overset{\overset{R^2}{|}}{\underset{\underset{R^2}{|}}{\text{Si}}}\right)_n$$

8   **9**; Y= H   **11**

10; Y= Cl

Scheme 4

Polymer reaction of PS is also found to be useful for the transformation of the structure of PS. As shown in scheme 5, the transformation of phenyl substituent of phemylmethyl polysilane to chloro group (**12**) followed by the reaction with the Grignard reagents gave alkyl-substituted PS. Typical examples of the reaction of **12** with some nucleophiles are shown in table 2.

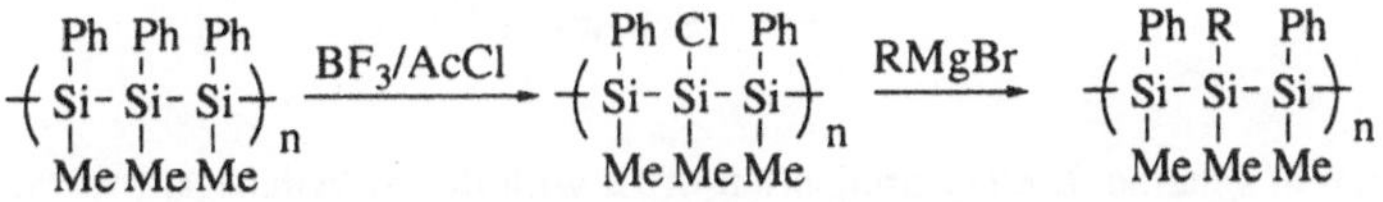

$$\left(\overset{\overset{Ph}{|}}{\underset{\underset{Me}{|}}{\text{Si}}}-\overset{\overset{Ph}{|}}{\underset{\underset{Me}{|}}{\text{Si}}}-\overset{\overset{Ph}{|}}{\underset{\underset{Me}{|}}{\text{Si}}}\right)_n \xrightarrow{BF_3/AcCl} \left(\overset{\overset{Ph}{|}}{\underset{\underset{Me}{|}}{\text{Si}}}-\overset{\overset{Cl}{|}}{\underset{\underset{Me}{|}}{\text{Si}}}-\overset{\overset{Ph}{|}}{\underset{\underset{Me}{|}}{\text{Si}}}\right)_n \xrightarrow{RMgBr} \left(\overset{\overset{Ph}{|}}{\underset{\underset{Me}{|}}{\text{Si}}}-\overset{\overset{R}{|}}{\underset{\underset{Me}{|}}{\text{Si}}}-\overset{\overset{Ph}{|}}{\underset{\underset{Me}{|}}{\text{Si}}}\right)_n$$

12

Scheme 5

Table 2 Reaction of PS with some nucleophiles.

Run	Ph →Cl (%)	RMgBr	Other Nucleophile	Mn	Mw/Mn
1	10	Et		7600	1.4
2	10	Pr		6700	1.3
3	10	Bu		8200	1.4
4	10	Oct		6900	1.3
5	10	 ∿		6400	1.3
6	10	 ∧		6700	1.4
7	10		NH_3	7500	1.5
8	10		BuLi	6800	1.3
9	20	Et		6300	1.5
10	20	Pr		6700	1.2
11	20	Bu		8300	1.5

The reaction of **12** with the di-Grignard reagent gave linked PS and resulted in the increase of the molecular weight of PS (Scheme 6).

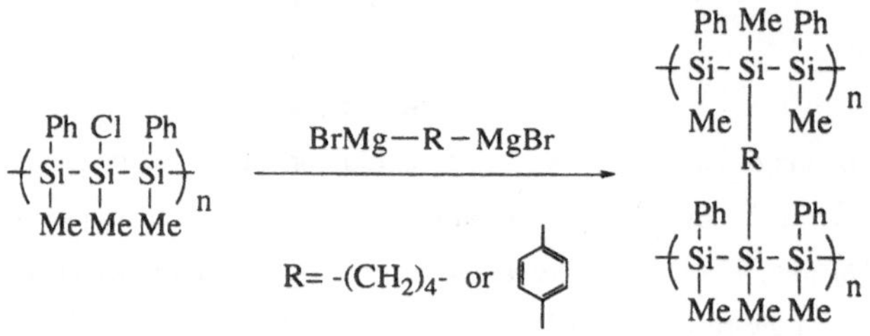

Scheme 6

As shown in scheme 7, electroreduction of **12** with dichlorosilane led to the formation of PS having silicon linkage.

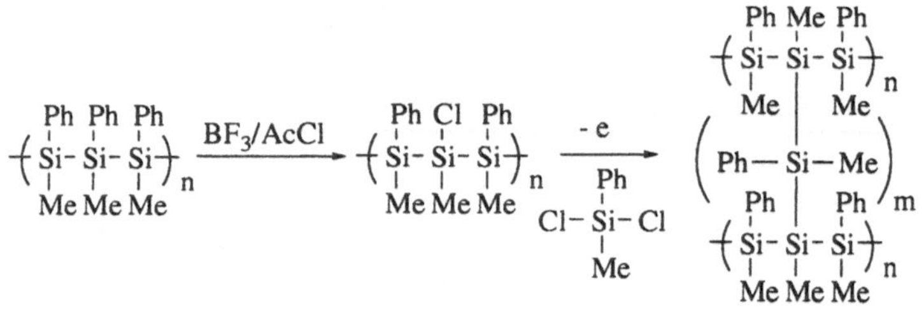

Scheme 7

References

1) T.Shono, S. Kashimura, M. Ishifune, and R. Nishida. *J. Chem. Soc., Chem. Commun.*, 1160 (1990).
2) T.Shono, S. Kashimura, and H. Murase. *J. Chem. Soc., Chem. Commun.*, 896 (1992).
3) Kashimura, M. Ishifune, H. Murase, R. Nishida, and T. Shono, *J. Org. Chem.*, 62, in press (1997).
4) Kashimura, M. Ishifune, H. Murase, R. Nishida, S.Kawasaki, and T. Shono, *Tetrahedron Lett.,38,* 4607 (1997).

THE ELECTROCHEMISTRY OF BIOLOGICALLY ACTIVE 2-HYDROXY NAPHTHOQUINONES: THE ROLE OF HYDROGEN BONDING. REDUCTIVE SILYLATION AND ACETYLATION, USEFUL ELECTROCHEMICAL TRANSFORMATIONS

Marília O. F. GOULART[a], Fabiane C. DE ABREU[a],
Patrícia A. L. FERRAZ[a], Josealdo TONHOLO[a] and Victor GLEZER[b]

[a]Depto. de Química, CCEN, UFAL, Maceió, Alagoas, 57072-970, Brazil
[b]Lab. of Environ. Chemistry, The Hebrew University of Jerusalem, Israel

ABSTRACT

2-Hydroxy-3-alkenylnaphthoquinones showed typical voltammograms in aprotic medium (DMF/TBAP 0.1M). They are represented by two pair of peaks with reversible or *quasi*-reversible nature, depending on the position of the double bond, and one or more intermediate shoulders. The first pair of peaks, in lapachol, corresponds to a *quasi*-reversible monoelectronic transfer, forming a semiquinone strongly stabilized by intramolecular hydrogen bonding, contrary to the reported proton reduction of one of the possible tautomeric forms of the hydroxyquinones. High yield reductive acetylation was obtained by co-eletrolysis of lapachol with acetic anhydride, at the potential of the first wave, on mercury or platinum electrodes. The reduction process depends on the nature and concentration of proton sources. Silyl ethers, key intermediates for modification of heterocyclic quinones, were prepared through electrolysis at Ep_{c2}, specially in THF. Triacetyldihydrolapachol showed a significant trypanocidal activity.

INTRODUCTION

Numerous quinones play vital roles in the biochemistry of living cells and exert important biological activities, such as antitumoral, anti-protozoan and antibiotic. The mechanism of action requires their bioreduction that depends on their redox properties. The understanding of how structural features of the quinones are related to these properties is an important step to comprehend their mechanism of action and predict modifications to improve their biological activity. Quinones **1-4** are biologically active[1]. A relationship between the reduction'ease and the *in vitro* activity on trypomastigotes of *T. cruzi,* ethiological agent of Chagas'disease was shown[1]. Among the assayed quinones, the active 2-hydroxynaphthoquinones showed some of the least negative potentials. For comparison between biological activities and electrochemical parameters, a similar electrodic mechanism would be essential. The present study aimed to clarify the mechanism and to demonstrate the possibilities of electrochemical modifications, as an auxiliary in the mechanism elucidation, as well as for synthesis and biological assays.

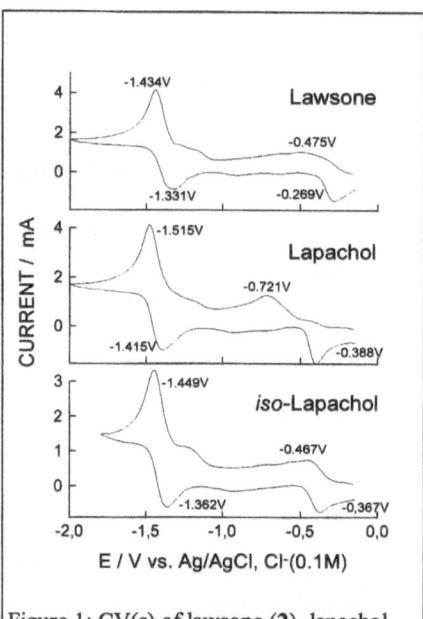

1 $R_1 = H, R_2 = CH_2CH=CMe_2$
2 $R_1 = H; R_2 = H$
3 $R_1 = H; R_2 = CH=CHCHMe_2$
4 $R_1 = H; R_2 = CH=CHEt$
5 $R_1 = Ac; R_2 = CH_2CH=Cme_2$

The electrochemical studies of **1-5** (cyclic voltammetry, coulometry and electrolysis) were performed, on Hg and Pt, in aprotic medium (DMF/TBAP 0.1M) (E *vs* Ag/AgCl 0.1M), with several added reagents. Lapachol (**1**) was chosen for a complete study. Electrolyses were performed with several electrophiles (acetic anhydride, silyl chlorides and alkyl halides), in order to avoid the usual recovering of the quinone, by air reoxydation of the electrogenerated dianion. For silylation, ACN and THF were also used as solvents.

Typical cyclic voltammograms (CV) for **2** and for non conjugated (**1**) and conjugated 3-alkenyl side chain (**3**) quinones, at 0.1 V.s^{-1}, are shown in Figure 1. All the following experiments are related to **1**. The first waves and shoulders correspond to irreversible reduction processes. The poor definition of the shoulders preclude accurate potential and wave height measurements. The latter couple of waves can be considered as a *quasi*-reversible system. In Pt, the behavior was the same, except for a more positive Ep_{a1} (0.112 V).

Earlier studies on the electrochemistry of lawsone (**2**) and lapachol (**1**)[2,3], on Pt electrode, in DMSO/TEAP, suggested that the first and second irreversible steps of

Figure 1: CV(s) of lawsone (**2**), lapachol (**1**) and isolapachol (**3**); DMF/TBAP, Hg electrode; 100 mVs^{-1}.

reduction could be related to proton reduction of *ortho-* and *para*-quinones' isomeric forms. However, no comments were made about a possible influence of intramolecular hydrogen bonding[4], neither the reason by which the proton reduction is so easy, compared to other acidic protons in the same conditions. So, additional experiments were run for **1**: electrolysis at the first wave for evidence of H_2 or quinone modification; addition of base and acids and analysis of their effects on CV. In spite of different supporting electrolytes, the electrochemical behavior of **1**, on Pt and Hg, is similar, allowing comparison.

Electrolyses of lapachol were held in different conditions. After consumption of *ca* of 2 F/mol, at -1.7 V, lapachol was recovered unchanged [non electrochemical acetylation, by work up with acetic anhydride, leads to the formation of **5** (80%)]. However, high yield bielectronic reductive acetylation, giving triacetyldihydrolapachol **6** (90%) was achieved, by co-electrolysis with acetic anhydride, in Pt or Hg, at potentials related to Ep_{c1} and Ep_{c2}, without any evidence of H_2 evolution. This result disclaims the suggested mechanism[2,3] and is compatible to the reduction, at Ep_{c1}, of the quinone to the semiquinone, that suffers disproportionation reactions, furnishing a modified catechol system, stabilized by reaction with electrophiles. Additional proof came from voltammetric studies with added reagents. The behavior of **1** in the presence of TBAH, associated to the quinone anion, is similar to the reported one[2]. When the concentration of TBAH is twice that of **1**, an unique and reversible pair of waves, at potential slightly less negative than the original one, with peak heights twice its original size, remained. It is evident that the effect on the quinone' reduction of removing the α-carbonyl hydroxyl proton is considerable: a negative shift of 0.81 V (Fig.2A), along with the disappearance of the shoulders and first pair of waves, definitely related to its presence. Coulometric experiments established the bielectronic nature for this process, that corresponds to the full reduction of the anionic form of the quinone, contrary to the reported monoelectronic reduction of the quinone anion.

Addition of different proton sources, reported for the first time, led to different results. As expected, benzoic acid provoked a classical change (Fig.2B): the complete merge into one bielectronic wave. The potential of the first anodic wave remains, with a significant current increase. Phenol addition caused no change on the first pair of waves, but produce a gradual positive shifting of the second main wave, turning it irreversible (Fig.2C). This effect is usually interpreted as a step of protonation followed by a second electron transfer[4]. Phenol is not acid enough to protonate the electrogenerated semiquinone. The higher current should be due to reduction in hydrogen bonded homo- and heteroassociated dimers[4]. The combined results are definitely compatible with the usual reduction scheme for quinone' reduction. The intramolecular hydrogen bonding explains the

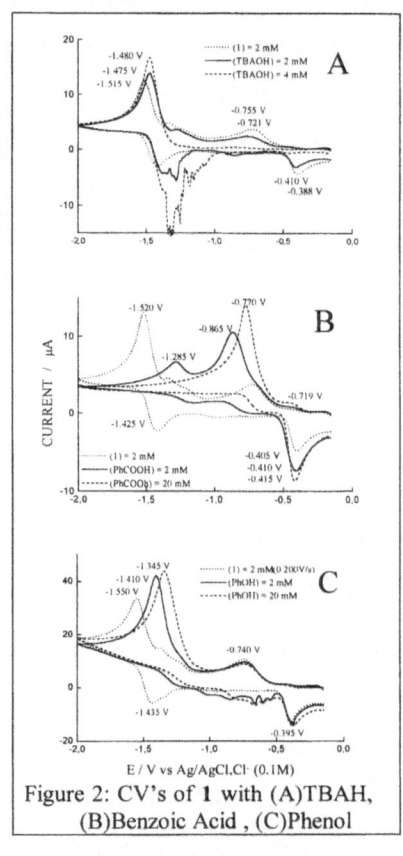

Figure 2: CV's of **1** with (A)TBAH, (B)Benzoic Acid , (C)Phenol

easier reduction of 2-hydroxy naphthoquinones[1]. Protected hydroquinones are important intermediates for structural modification. Silylated catechols were synthesized, in spite of their instability through hydrolysis. After electrolysis with TMSCl, TBDMSCl, in DMF or THF, they should be extracted with pentane or kept in THF. The triacetoxy derivative **6** was assayed and showed a relevant trypanocidal activity.

Acknowledgments: To CNPq, RHAE, CAPES, PADCT, FAPEAL.
References: 1) M.O.F. Goulart, C. L. Zani; J. Tonholo, L. R. Freitas; F. C. de Abreu, D. S. Raslan; S. Starling; A. B. Oliveira; E. Chiari, *Bioorg. Med. Chem. Lett.* **1997**, in press. 2) M. E. Bodini, and V. Aranciaba, *Polyedron* **1989**, *8*, 1407. 3) M. E. Bodini, *Polyedron* **1994**, *13*, 497. 4) I. Piljac, and R.W. Murray, *J. Electrochem. Soc.* **1971**, *118*, 1758.

ELECTROCHEMISTRY OF ORGANOSILICON COMPOUNDS: FROM HALOSILANES TO ETHYNYLSILANES, SILOLES, AND ORGANOSILICON POLYMERS

Atsutaka KUNAI, Osamu OHNISHI, Masataka MORISHITA,
Joji OHSHITA, and Mitsuo ISHIKAWA*
Department of Applied Chemistry, Faculty of Engineering, Hiroshima University, Higashi-Hiroshima 739, Japan, and * Department of Chemical Technology, Kurashiki University of Science and the Arts, Kurashiki 712, Japan

ABSTRACT

Electrolysis of various halosilanes such as iodo-, chloro-, and fluorosilanes in the presence of phenylacetylene on a Pt electrode leads to the formation of Si–sp-C bonds to give phenylethynylated products. Silole derivatives were synthesized from the electrolysis of bis(phenylethynyl)silanes thus obtained. The co-electrolysis of p-bis(chlorosilyl)benzenes and p-diethynylbenzenes afforded polymers composed of an alternating arrangement of p-bis(silanylene)phenylene and p-diethynylenephenylene units.

INTRODUCTION

It is well known[1] that electrolytic reduction of allyl, benzyl, and aryl halides in the presence of chlorosilanes leads to the formation of silicon–carbon bonds. Electrochemical silylation of unsaturated bonds in activated olefins, dienes, and hydroaromatics with chlorosilanes has also been reported. More recently, it has been demonstrated that compounds with the Si–Si bonds are produced in good yields by the electrolytic reduction of chlorosilanes in an undivided cell using a sacrificed metal anode or a hydrogen electrode.

The utilization of chlorosilanes in the field of electrochemistry is thus well developed to date but is limited to the formation of the silicon–sp^3- and sp^2-carbon bonds as well as the silicon–silicon bonds. In this paper, we report the formation of the silicon–sp-carbon bonds by the electrolysis of various halosilanes in the presence of phenylacetylene,[1] and the formation of a silole ring starting from diethynylsilanes thus obtained, and moreover, application of the present method to the synthesis of polymers composed of a silanylene unit and π-electron system.

RESULTS AND DISCUSSION

In a typical run, 4–10 mmol of halosilane was electrolyzed in 25 cm^3 of a solvent in the presence of 2 equiv of phenylacetylene in an undivided cell by using Pt (6 cm^2) as the cathode, Al (12 cm^2) or Pt (6 cm^2) as the anode, and Bu4NBPh4 (0.5 g) as the supporting electrolyte until halosilane disappeared. Products were isolated by usual work-up. Results are summarized in Table 1.

When a mixture of chlorodimethylphenylsilane and phenylacetylene was electrolyzed by using Al as the sacrificed anode in pivalonitrile (PN), an ethynylated product, dimethylphenyl(phenylethynyl)silane was obtained in 13% yield, besides a 51% yield of tetramethyl-1,2-diphenyldisilane (run 1). In contrast to this, when iodosilane was used instead of chlorosilane, the ethynylated product was obtained as the main product (runs 2 and 3).

We later found, however, that the electrochemical ethynylation proceeds quite smoothly without the use of the sacrificed anode. Various phenylethynylsilanes were obtained by the use of the Pt-Pt electrode system in high yields, regardless of starting halosilanes (see runs 4–9).

$$R^1R^2R^3SiX + PhC\equiv CH \xrightarrow{+e^-} R^1R^2R^3SiC\equiv CPh$$

$$R^1R^2SiX_2 + PhC\equiv CH \xrightarrow{+e^-} R^1R^2Si(C\equiv CPh)_2$$

Table 1. Electrolysis of Halosilanes in the Presence of Phenylacetylene

Run	Halosilane	Electrode	Electrolyte	$F\cdot mol^{-1}$	Product	Yield/%
1	$Me_2PhSiCl$	Al/Pt	BPh_4^-/PN	1.9	$(Me_2PhSi)_2$	51
					+ $Me_2PhSiC\equiv CPh$	13
2	Me_2PhSiI	Al/Pt	BPh_4^-/PN	1.6	$Me_2PhSiC\equiv CPh$	88
3	Et_3SiI	Al/Pt	BPh_4^-/PN	1.8	$Et_3SiC\equiv CPh$	52
4	$Me_2PhSiCl$	Pt/Pt	BPh_4^-/PN	1.8	$Me_2PhSiC\equiv CPh$	84
5	Et_3SiCl	Pt/Pt	BPh_4^-/PN	2.0	$Et_3SiC\equiv CPh$	89
6	Me_3SiCl	Pt/Pt	BPh_4^-/PN	2.6	$Me_3SiC\equiv CPh$	91
7	$MePhSiCl_2$	Pt/Pt	BPh_4^-/PN	5.4	$MePhSi(C\equiv CPh)_2$	56
8	Me_2PhSiI	Pt/Pt	BPh_4^-/PN	1.9	$Me_2PhSiC\equiv CPh$	82
9	Et_3SiI	Pt/Pt	BPh_4^-/PN	1.3	$Et_3SiC\equiv CPh$	80
10	$MePh_2SiF$	Pt/Pt	BF_4^-/DME	3.4	$MePh_2SiC\equiv CPh$	42 (GC)
11	Me_2PhSiF	Pt/Pt	BPh_4^-/DME	1.5	$Me_2PhSiC\equiv CPh$	50 (GC)

In the present reaction, phenylacetylene may be reduced on the platinum cathode to give phenylethynyl carbanion as the reactive intermediate. In order to confirm this, we carried out the reaction of fluorosilanes, which are known to be highly resistive towards electrolytic reduction. Thus, the electrolysis of fluoromethyldiphenylsilane and fluorodimethylphenylsilane in dimethoxyethane (DME) in the presence of phenylacetylene afforded the ethynylation products, methyldiphenyl(phenylethynyl)silane and dimethylphenyl(phenylethynyl)silane in 42% and 50% yields, respectively (runs 10 and 11).

$$PhC\equiv CH + e^- \longrightarrow PhC\equiv C^- + 1/2\ H_2$$

$$MePhRSiF + PhC\equiv C^- \longrightarrow MePhRSiC\equiv CPh + F^-$$

(R=Ph or Me)   (R=Ph or Me)

Silole derivatives can be synthesized from bis(phenylethynyl)silanes thus obtained (Scheme 1). Namely, electroreductive cyclization of dimethylbis(phenylethynyl)silane and methylphenylbis(phenylethynyl)silane afforded 1,1-dimethyl-3,4-diphenylsilole (47%) and 1-methyl-1,3,4-triphenylsilole (45%), respectively.

Scheme 1

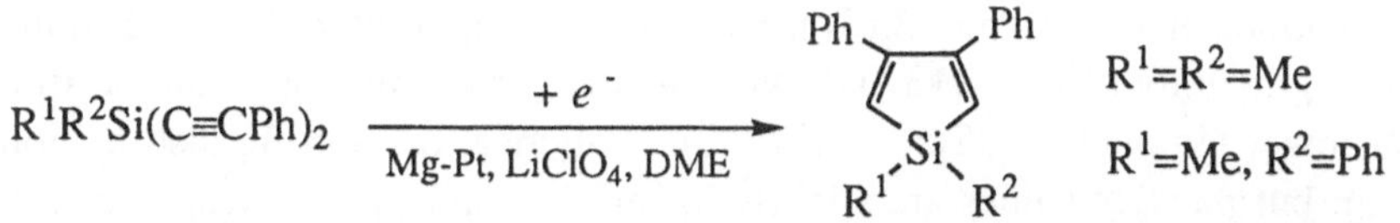

$$R^1R^2Si(C\equiv CPh)_2 \xrightarrow[\text{Mg-Pt, LiClO}_4\text{, DME}]{+e^-}$$

$R^1=R^2=Me$

$R^1=Me,\ R^2=Ph$

The present ethynylation reaction of chlorosilanes can be applied to the synthesis of polymers composed of a silanylene unit and π-electron system (Scheme 2). Thus, a 1:1 mixture of 1,4-bis(chlorosilyl)benzene (**1a–c**) and *p*-diethynylbenzene (**2**) was electrolyzed on the Pt-Pt electrodes in DME. From the reaction of **1a**, a 13% yield of an alternating polymer **3a** was obtained as yellow solids, whose molecular weight was determined to be Mw = 11,000 (Mw/Mn = 3.6) by GPC (Table 2). On the other hand, polymers were obtained in much higher yields with respect to the electrolysis of chloro-dialkylsilyl derivatives. Thus, when a mixture of **1b** and **2** was electrolyzed, a 60% yield of the polymer **3b** was obtained after reprecipitation from ethanol, while similar co-electrolysis of **1c** and **2** followed by reprecipitation afforded polymer **3c** in 48% yield. In all cases, small amounts of insoluble polymer films were deposited on the cathode surface during the electrolysis.

Scheme 2

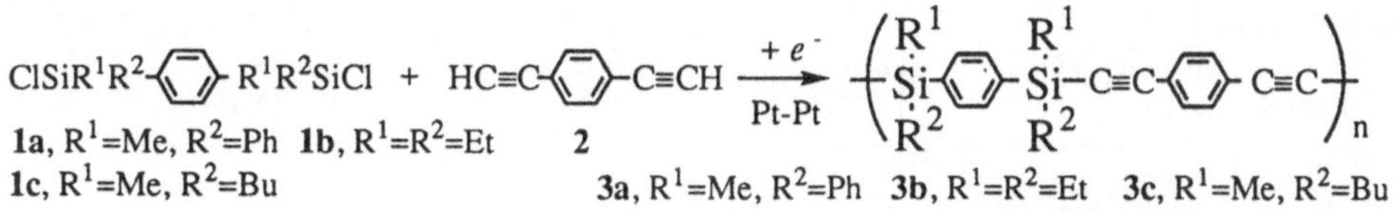

$$ClSiR^1R^2\text{—}\langle\bigcirc\rangle\text{—}R^1R^2SiCl + HC\equiv C\text{—}\langle\bigcirc\rangle\text{—}C\equiv CH \xrightarrow[\text{Pt-Pt}]{+e^-} \left(\begin{matrix}R^1 & & R^1\\ Si\text{—}\langle\bigcirc\rangle\text{—}Si\text{—}C\equiv C\text{—}\langle\bigcirc\rangle\text{—}C\equiv C\\ R^2 & & R^2\end{matrix}\right)_n$$

1a, R^1=Me, R^2=Ph   **1b**, R^1=R^2=Et   **2**

1c, R^1=Me, R^2=Bu   **3a**, R^1=Me, R^2=Ph   **3b**, R^1=R^2=Et   **3c**, R^1=Me, R^2=Bu

Polymers **3a–c** thus obtained are solids and soluble in common organic solvents such as benzene, THF, and halocarbons. The structures for **3a–c** were verified by spectroscopic and elemental analyses. For example, ^{1}H NMR spectrum for **3c** reveals signals at δ 0.48 (s, 6H, SiMe), 0.88 and 1.36 (br m, 18H, SiBu), and 7.44–7.69 (m, 8H, aromatic ring protons), while the ^{13}C NMR spectrum exhibits those at δ 1.0 (MeSi), 13.8, 15.2, 25.2, 26.0 (BuSi), 93.7, and 106.6 (C≡C), in addition to four signals due to phenylene ring carbons in a lower field, suggesting that **3c** has the regular alternating structure shown in Scheme 2. In ^{29}Si NMR spectrum, two resonances are observed at δ 19.47 and 19.67, which may arise from diastereomeric bis(silyl)phenylene units in the polymer backbone. For comparison, we synthesized polymer **3c** by a chemical method. All of the NMR and IR spectra for this polymer were identical with those for the polymer prepared electrochemically.

Thermal properties of **3a–c** were examined by thermogravimetric analysis (TGA) under a nitrogen atmosphere. Polymer **3a** is thermally stable up to 400 °C, and then the weight decreases gradually (Td_5 = 413 °C), whereas the corresponding insoluble film decomposes at lower temperature (Td_5 = 287 °C). Polymers **3b** and **3c** also exhibit similar behavior, but the Td_5 values are slightly lower (378 and 376 °C, respectively), compared to that of **3a**. The weight loss values for **3a–c** at 1,000 °C were found to be 39%, 41%, and 49%, respectively.

Polymers **3a–c** are insulators but become conducting on doping. Thus, when thin films of **3a–c** casted on glass plates by spin-coating were exposed to $FeCl_3$ vapor in vacuo (1 mmHg), conducting films were obtained. The conductivities were found to be 4.5×10^{-4}, 1.6×10^{-3}, and 6.7×10^{-3} S/cm for **3a–c**, respectively. Similarly, when films of insoluble polymers were doped with $FeCl_3$, they became conducting. However, the conductivities were not well reproducible in these cases.

Table 2. Electrochemical Polymerization of **1** and **2** on Pt/Pt Electrode in DME containing Bu_4NBPh_4 and Properties of Polymers **3a–c**

Monomer	**1a** and **2**, 1:1	**1b** and **2**, 1:1	**1c** and **2**, 1:1
F/mol	1.6	1.8	2.3
Polymer	**3a**, 13%	**3b**, 60%	**3c**, 48%
Mw (Mw/Mn)	11,000 (3.6)	9,400 (3.6)	16,500 (2.7)
Td_5 / °C	413	378	376
Weight loss at 1000°C /%	38.8	40.6	49.0
Doping time with $FeCl_3$	12 h	6 h	7.5 h
Conductivity /S cm^{-1}	4.5×10^{-4}	1.6×10^{-3}	6.7×10^{-3}

Reference
1) For detailed discussion, see previous report: A. Kunai, O. Ohnishi, T. Sakurai, and M. Ishikawa, *Chem. Lett.* **1995**, 1051; and references cited therein.

THE GENERATION OF HYDROXYMETHYL RADICALS: PHOTOINDUCED ELECTRON TRANSFER AS OPPOSED TO ELECTROCHEMICAL ELECTRON TRANSFER

Guido Gutenberger, Eric Meggers and Eberhard Steckhan*

Kekulé-Institut für Organische Chemie und Biochemie der Universität Bonn,
Gerhard-Domagk-Str. 1, D-53121 Bonn, Germany

The intermolecular addition of hydroxy(alkoxy)methyl radicals to electron-poor double bonds is of strategic importance for organic synthesis. The reason is that these functions can thus be introduced in the form of the umpoled hydroxy(alkoxy)methyl anion synthons.

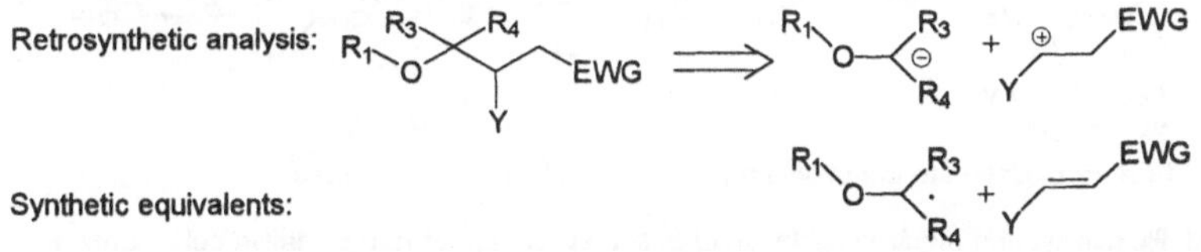

Scheme 1: Addition of hydroxy(alkoxy)methyl radicals to electron-poor double bonds as synthetic equivalent for hydroxy(alkoxy)methyl anion synthons.

In principle, hydroxy(alkoxy)methyl radicals could be generated *via* single electron oxidation followed by deprotonation. However, the thus formed radicals are much easier oxidized than the starting materials and therefore can not be trapped by alkenes. Additionally, the oxidation potentials are very positive especially in the case of ethers. These problems can be circumvented, if the heteroatoms are in α-position substituted by a silyl group as electrophilic leaving group thus lowering the oxidation potentials by about 0.5 V.[1] To prevent further oxidation of the thus formed radicals, the oxidation has to take place in a **non-oxidizing** environment. This is not possible by direct electrolysis but by phtoinduced electron transfer (PET) oxidation (Scheme 2). In analogy to the generation of aminomethyl radicals,[2] we found that by redox photosensitization using 9,10-dicyanoanthracene (DCA) as photoexcited acceptor and biphenyl (BP) as primary donor these reactions can be performed successfully.[2] (Scheme 3, Table 1).

The mechanism of the reaction can be explained in the following way (Scheme 4): BP (E_{pox} = 1.98 V vs. SCE) as the primary electron donor is oxidized by the excited DCA* (E_{red}(S1) = 2.0 V vs. SCE) to its radical cation which subsequently oxidizes the α-silylether (E_{ox} = 1.41 - 1.66 V vs. Ag/AgNO$_3$). The thus formed α-silylether radical cation immediately looses the silyl group as an electrofuge giving the desired alkoxymethyl radical which is trapped by elektron-poor double bonds. The newly formed adduct radical is further reduced presumably by the DCA radical anion (back electron transfer). The obtained carbanion is protonated by the solvent. This was proved by working in CH$_3$OD giving the deuterated product in more than 95 % yield. Additionally, it was shown that on the product stage, no H/D exchange occurs.

HOW TO GENERATE A HYDROXY(ALKOXY)METHYL RADICAL

1. Possibility: one electron oxidation of protected hydroxymethyl groups:

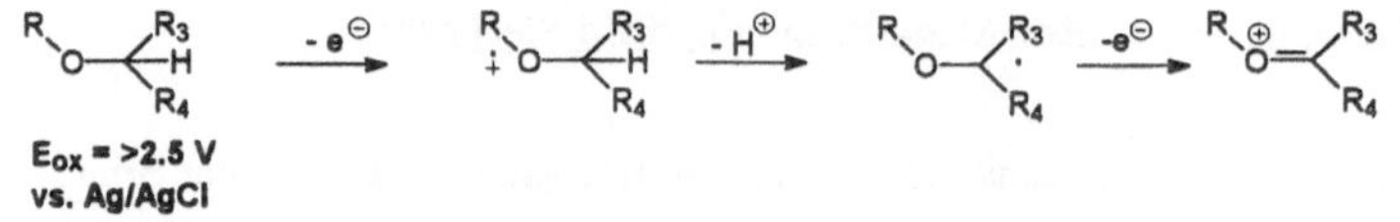

$E_{ox} = >2.5$ V
vs. Ag/AgCl

Oxidation at very high potentials. Further oxidation to the oxonium ion can usually not be prevented (second oxidation step by more than 1.0 V easier).

2. Possibilty: one electron oxidation of α-silyl ethers: α-silyl goups lower the oxidation potentials by 800-900 mV. α-Silyl groups are good electrofugs.

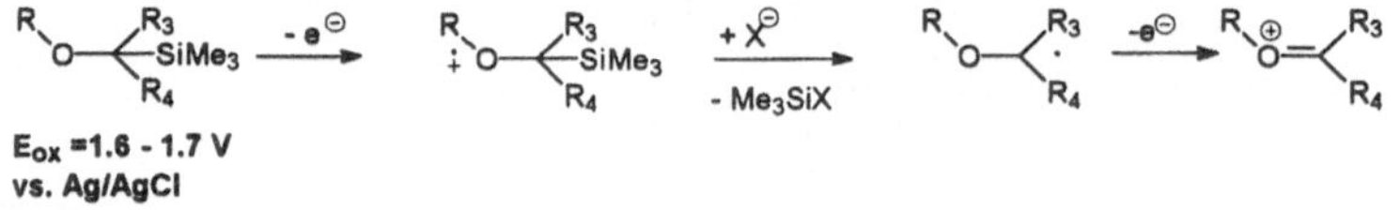

$E_{ox} = 1.6 - 1.7$ V
vs. Ag/AgCl

Further oxidation to the oxonium ion can usually not be prevented.

3. Possibilty: one electron oxidation of α-silyl ethers under non-oxidative conditions: Application of photo-electron transfer (PET).

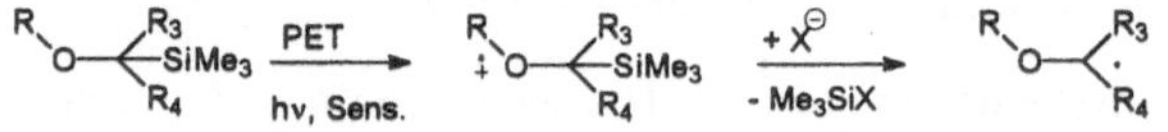

Scheme 2: Principal pathways to hydroxy(alkoxy)methyl radicals.

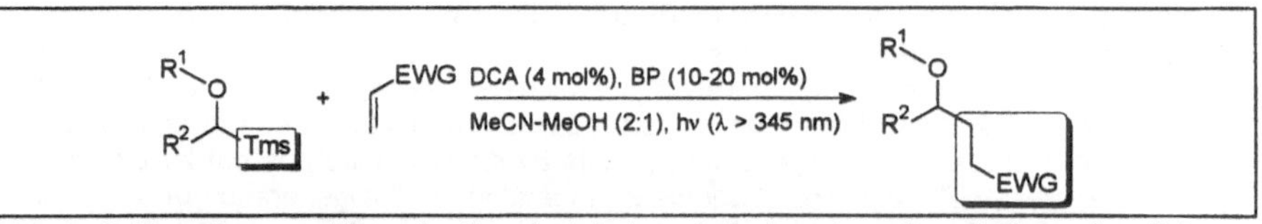

Scheme 3: Reaction conditions for the photoinduced electron transfer addition of α-silyl-ethers to acceptor substituted alkenes (BP = biphenyl; DCA = 9,10-dicyano anthracene;450 W xenon arc lamp; Wavelength filter λ> 345 nm; ratio of silyl ether to alkene = 1 : 1.5 - 2.

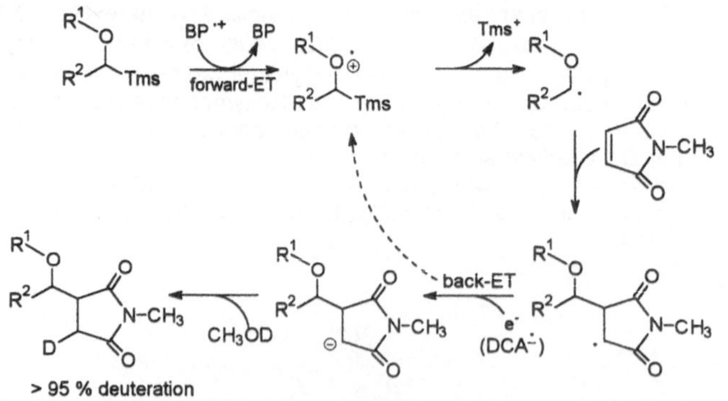

Scheme 4: Mechanism of the PET-catalyzed addition of alkoxymethylradicals to acceptor substituted double bonds.

Table1: Results of the PET-catalyzed addition of α-silyl-ethers to acceptor-substituted alkenes.

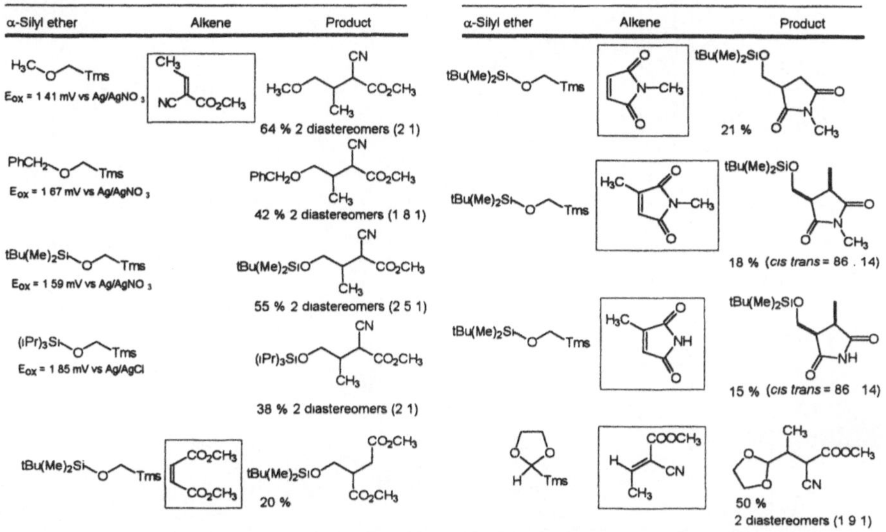

α-Silyl ether	Alkene	Product	α-Silyl ether	Alkene	Product
H$_3$C–O–Tms E_{ox} = 1 41 mV vs Ag/AgNO$_3$	CH$_3$ NC CO$_2$CH$_3$	H$_3$CO CN CO$_2$CH$_3$ CH$_3$ 64 % 2 diastereomers (2 1)	tBu(Me)$_2$Si–O–Tms	O N–CH$_3$ O	tBu(Me)$_2$SiO O N–CH$_3$ O 21 %
PhCH$_2$–O–Tms E_{ox} = 1 67 mV vs Ag/AgNO$_3$		PhCH$_2$O CN CO$_2$CH$_3$ CH$_3$ 42 % 2 diastereomers (1 8 1)	tBu(Me)$_2$Si–O–Tms	H$_3$C O N–CH$_3$ O	tBu(Me)$_2$SiO O N–CH$_3$ O 18 % (cis trans = 86 : 14)
tBu(Me)$_2$Si–O–Tms E_{ox} = 1 59 mV vs Ag/AgNO$_3$		tBu(Me)$_2$SiO CN CO$_2$CH$_3$ CH$_3$ 55 % 2 diastereomers (2 5 1)	tBu(Me)$_2$Si–O–Tms	H$_3$C O NH O	tBu(Me)$_2$SiO O NH O 15 % (cis trans = 86 14)
(iPr)$_3$Si–O–Tms E_{ox} = 1 85 mV vs Ag/AgCl		(iPr)$_3$SiO CN CO$_2$CH$_3$ CH$_3$ 38 % 2 diastereomers (2 1)			
tBu(Me)$_2$Si–O–Tms	CO$_2$CH$_3$ CO$_2$CH$_3$	tBu(Me)$_2$SiO CO$_2$CH$_3$ CO$_2$CH$_3$ 20 %	O O H Tms	COOCH$_3$ H CN CH$_3$	CH$_3$ O O COOCH$_3$ CN 50 % 2 diastereomers (1 9 1)

The *cis*-diastereoselectivity in the case of the methyl substituted maleimides can easily be explained by a kinetically controlled protonation of the intermediate carbanion from the less hindered side according to Scheme 5. In the presence of sodium methoxide, the *cis*-isomer is transformed into the thermodynamically favored *trans*-compound.

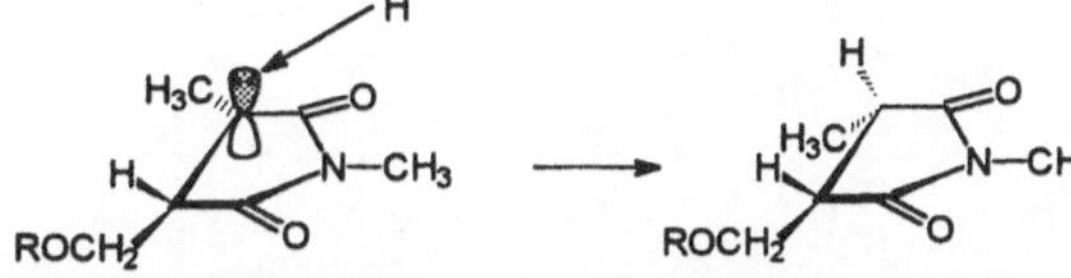

Scheme 5: Kinetically controlled protonation of the intermediate carbanion from the less hindered side leading to the *cis*-product.

In conclusion, it is demonstrated that hydroxy(alkoxy)methyl radicals as synthetic equivalents for hydroxymethyl anions can be generated by photinduced electron transfer but not by direct electrochemical oxidation. An electrochemical alternative, however, might be an indirect electrochemical process using a redox mediator which is oxidized at low current density thus leading to very low homogeneous stationary concentrations of the oxidizing agent. Such studies are currently under way.

Acknowledgements. Financial support by the Volkswagen-Stiftung (I/71 748), the Fonds der Chemischen Industrie and the BASF Aktiengesellschaft is gratefully acknowledged.

References
[1] J. Yoshida, *Top. Curr. Chem.* **1994**, *170*, 39.
[2] E. Meggers, E. Steckhan, S. Blechert, *Angew.Chem.* **1995**, *107*, 2317; *Angew.Chem.Int.Ed.Engl.* **1995**, *34*, 2137.

ELECTROREDUCTIVE SYNTHESIS OF SEQUENCE-ORDERED POLYSILANES USING Mg ELECTRODES

Manabu ISHIFUNE, Hang-Bom BU, Shigenori KASHIMURA, Natsuki YAMASHITA, and Tatsuya SHONO

Department of Applied Chemistry, Faculty of Science and Technology, Kinki University, 3-4-1 Kowakae, Higashi-Osaka 577, Japan

ABSTRACT

The stepwise elongation of Si-Si chain was achieved by the electroreductive cross-coupling reaction of chlorohydrosilanes with dichlorooligosilanes. The electroreductive polymerization of the resulting dichlorooligosilanes using Mg electrodes is highly promising for the synthesis of sequence-ordered polysilanes. Dichlorotrisilanes were found to be good monomers for the electroreductive synthesis of the polysilanes, units of which are ordered in three sequences.

INTRODUCTION

Polysilanes are attracting considerable attention due to their potentiality in the preparation of new types of material showing conducting, photoconducting, or nonlinear optical property. They have been usually prepared by the Wurts type coupling reaction with Na metal, but this method has a disadvantage in controlling the unit-structure. Several novel methods have opened new approaches to the synthesis of structure-ordered polysilanes, which is essential to the further development in the polysilane chemistry. The anionic polymerization of masked disilanes has provided a useful method to prepare some poly(disilanylene)s whose units are ordered in two sequences.[1] The anionic ring-opening polymerization of cyclic oligosilanes is another resolution to give structure-controlled polysilanes.[2]

RESULTS AND DISCUSSION

We have previously reported that use of Mg electrodes is highly effective for the electroreductive formation of Si-Si bonds under mild reaction conditions.[3-5] The mildness of the reaction conditions of this electroreductive method is remarkably favorable for the synthesis of the polysilanes having Si-H bonds which are known to be reactive under radical or anionic condition. The electroreductive cross-coupling reaction of chlorodimethylsilane **1** with dichlorodiphenylsilane **2**, for instance, gave the corresponding trisilane **3** in good yield (Scheme 1), which was readily transformed to the corresponding dichlorotrisilane **4**. In the same way, dichlorotetrasilane **7** can be prepared by the reaction between **1** and dichlorodisilane **5**. Thus, this electroreductive cross-coupling reaction of chlorohydrosilanes with dichlorooligosilanes provides a new powerful method for the stepwise elongation of Si-Si bonds and synthesis of sequence-controlled oligosilanes.

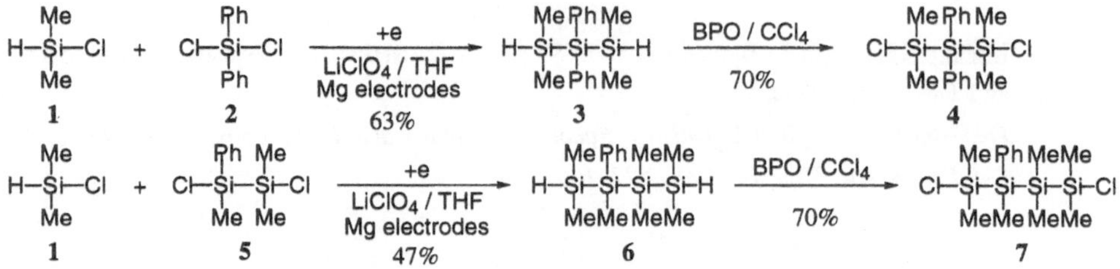

Scheme 1

Me
|
H—Si—Cl + Cl—Si—Cl →(+e, LiClO₄ / THF, Mg electrodes, 63%)→ H—Si—Si—Si—H →(BPO / CCl₄, 70%)→ Cl—Si—Si—Si—Cl
|
Me

$$\text{Me} \quad \text{Ph} \quad \overset{+e}{\underset{\substack{\text{LiClO}_4 / \text{THF} \\ \text{Mg electrodes} \\ 63\%}}{\longrightarrow}} \quad \text{Me Ph Me} \quad \overset{\text{BPO / CCl}_4}{\underset{70\%}{\longrightarrow}} \quad \text{Me Ph Me}$$

1 + **2** → **3** → **4**

1 + **5** →(+e, LiClO₄ / THF, Mg electrodes, 47%)→ **6** →(BPO / CCl₄, 70%)→ **7**

Scheme 1

The electroreductive polymerization of the dichlorooligosilanes is highly promising for the synthesis of sequence-ordered polysilanes. Dichlorooligosilanes, such as dichlorotrisilane **4** and tetrasilane **7**, were found to be good monomers for the electroreductive synthesis of the polysilanes, units of which are ordered in longer sequences (Schemes 2, 3). As summarized in Table, the reaction temperature control is very important in the electroreductive polymerization of **4**, that is, at higher temperature the backbiting reaction of the propagating polymer proceeded forming cyclohexasilane as a by-product. This side reaction was successfully suppressed below 0°C, and polysilanes having relatively high molecular weight were obtained (runs 3, 4). In the optimized reaction condition, the electroreduction of dichlorotetrasilane **7** gave the corresponding polysilane **9**, units of which were ordered in four sequences in satisfactory yield (Scheme 3).

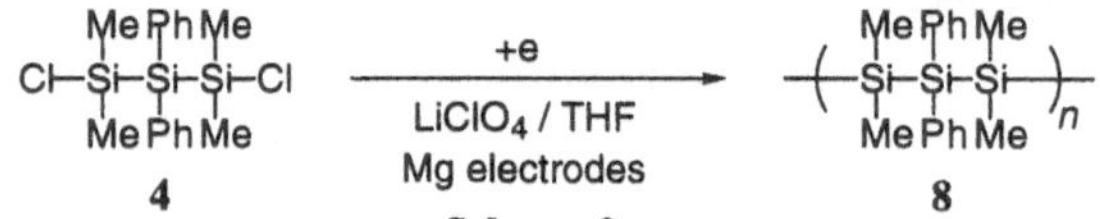

Scheme 2

4 →(+e, LiClO₄ / THF, Mg electrodes)→ **8**

Table. Electroreductive Polymerizartion of Dichlorotrisilane [a]

run	Reaction Temperature	$\overline{M}$n	$\overline{M}$w / $\overline{M}$n	Yield of **8**, %[b]
1	18	3800	1.44	(42)[c]
2	0	4700	1.87	(50)[c]
3	-10	5500	1.54	35
4	-15	5400	1.56	55

a) The electroreduction was carried out by using Mg electrodes, and ultrasound (47 kHz) was applied during the reaction. Anode and cathode were alternated with the interval of 15 sec. [Dichlorotrisilane] = 0.27 mol/L; [LiClO₄] = 0.35 mol/L; Supplied electricity, 4 F/mol. b) Purified by reprecipitation from benzene-EtOH. c) Contaminated by 1,1,2,2,4,4,5,5-octamethyl-3,3,6,6-tetraphenylcyclohexasilane.

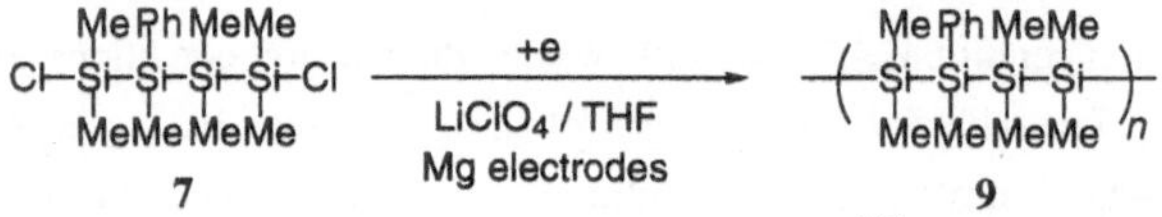

Scheme 3

7 →(+e, LiClO₄ / THF, Mg electrodes)→ **9**

Yield, 40% $\overline{M}$n = 3200

$\overline{M}$w / $\overline{M}$n = 1.30

References

1) M. Yoshida, T. Seki, F. Nakanishi, K. Sakamoto, and H. Sakurai Nishida *J. Chem. Soc., Chem. Commun.* **1996**, 1381.
2) M. Suzuki, J. Kotani, S. Gyobu, T. Kaneko, T. Saegusa *Macromolecules* **1994**, *27*, 2360.
3) T. Shono, S. Kashimura, M. Ishifune, and R. Nishida *J. Chem. Soc., Chem. Commun.* **1990**, 1160.
4) T. Shono, S. Kashimura, H. Murase *J. Chem. Soc., Chem. Commun.* **1992**, 896.
5) S. Kashimura, M. Ishifune, H.-B. Bu, M. Takebayashi, S. Kitajima, D. Yoshihara, R. Nishida, S. Kawasaki, H. Murase, and T. Shono *Tetrahedron. Lett.* **1997**, *38*, 4607.

CATALYTIC REDUCTION OF HALOGENATED ORGANIC COMPOUNDS WITH ELECTROGENERATED METAL(I) SALEN COMPLEXES

Dennis G. PETERS, Kent S. ALLEMAN, and Michael J. SAMIDE

Department of Chemistry, Indiana University, Bloomington, Indiana 47405, USA

ABSTRACT

Cobalt(I) salen, electrogenerated at a carbon cathode in dimethylformamide or acetonitrile containing a tetraalkylammonium salt, can be employed to effect the catalytic reductions of iodoethane and 3-chloro-2,4-pentanedione. To elucidate the mechanistic features of these processes, we have used cyclic voltammetry and controlled-potential electrolysis, along with product identification and quantitation.

INTRODUCTION

Reversible one-electron reduction of cobalt(II) salen is the first step in the catalytic process:

$$Co(II)\ salen\ +\ e^-\ \rightleftharpoons\ [Co(I)\ salen]^-$$

As quickly as it is formed, cobalt(I) salen reacts with the organohalogen compound (RX) to form an organocobalt(III) salen intermediate

$$[Co(I)\ salen]^-\ +\ RX\ \longrightarrow\ RCo(III)\ salen\ +\ X^-$$

which can undergo further one-electron reduction:

$$RCo(III)\ salen\ +\ e^-\ \rightleftharpoons\ [RCo(II)\ salen]^-$$

A central issue is whether the latter intermediate decomposes (a) to give cobalt(I) salen and a radical,

$$[RCo(II)\ salen]^-\ \longrightarrow\ [Co(I)\ salen]^-\ +\ R\cdot$$

or (b) to give cobalt(II) salen and a carbanion:

$$[RCo(II)\ salen]^-\ \longrightarrow\ Co(II)\ salen\ +\ R^-$$

CATALYTIC REDUCTION OF IODOETHANE

Cyclic voltammograms obtained with a glassy carbon electrode in DMF containing 0.10 M TBABF$_4$ for the reversible reduction of cobalt(II) salen (curve A) and for solutions containing both cobalt(II) salen and iodoethane (curves B and C) are depicted in Figure 1, along with a cyclic voltammogram (curve D) for the direct reduction of iodoethane. For curve A, the cathodic and anodic peak potentials are -0.56 and -0.47 V, respectively. When iodoethane is introduced (curves B and C), the cathodic peak originally at −0.56 V shifts to −0.43 V (due to the rapid reaction between electrogenerated cobalt(I) salen and iodoethane, which causes the anodic peak at −0.47 V to disappear) and a second cathodic peak (caused by reduction of ethylcobalt(III) salen) is seen at −1.10 V. When the concentration of iodoethane is increased, the size of the first peak at −0.43 V does not change, but the height of the second peak at −1.10 V does increase.

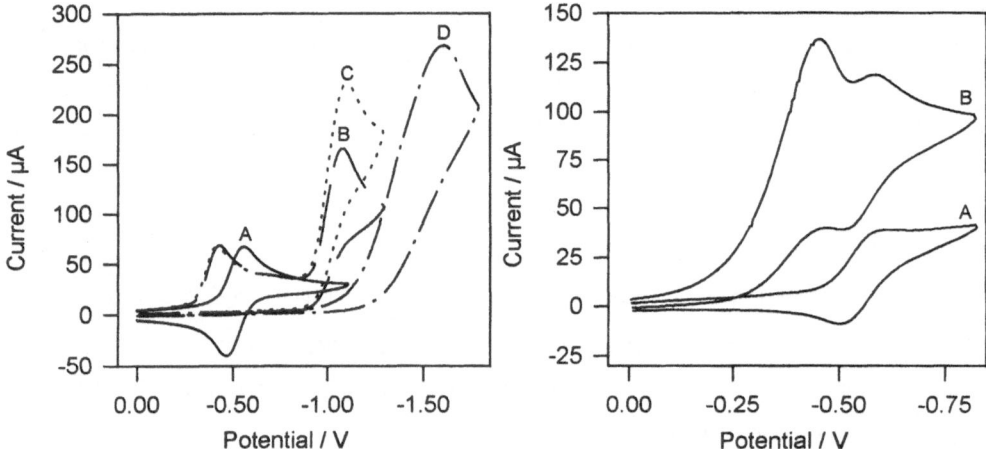

Figure 1. Cyclic voltammograms recorded with a glassy carbon electrode at 100 mV s^{-1} in DMF containing 0.10 M TBABF$_4$ and (A) 5 mM cobalt(II) salen, (B) 5 mM cobalt(II) salen + 7.5 mM iodoethane, (C) 5 mM cobalt(II) salen + 15 mM iodoethane, and (D) 15 mM iodoethane.

Figure 2. Cyclic voltammograms recorded with a glassy carbon electrode at 100 mV s^{-1} in CH$_3$CN containing 0.050 M TMABF$_4$ and (A) 1 mM cobalt(II) salen and (B) 1 mM cobalt(II) salen + 5 mM 3-chloro-2,4-pentanedione.

Table 1 shows results for macroscale catalytic reductions of iodoethane by cobalt(I) salen electrogenerated at reticulated vitreous carbon electrodes held at −1.10 V. An n value of 1 is obtained, indicating that each molecule of iodoethane accepts a single electron. Although the yields of volatile n-butane, ethane, and ethylene do not total 100%, no other products have been detected.

To gain insight into the behavior of ethylcobalt(III) salen, we performed a controlled-potential reduction of this species. A solution containing cobalt(II) salen was electrolyzed at −0.70 V to produce green-colored cobalt(I) salen. Then an equimolar quantity of iodoethane was injected into the cell, and orange-colored ethylcobalt(III) salen formed immediately. Next, the resulting solution was electrolyzed at −1.10 V; the n value was 1, and green-colored cobalt(I) salen reappeared. This experiment confirms that ethylcobalt(III) salen accepts one electron to form cobalt(I) salen and an ethyl radical:

$$H_5C_2Co(III) \text{ salen} + e^- \longrightarrow Co(I) \text{ salen} + \cdot C_2H_5$$

Table 1.   Coulometric Data and Product Distributions for Catalytic Reduction of 20 mM Iodoethane in CH₃CN and DMF Containing 0.050 M TMABF₄ and 0.50 mM Cobalt(II) Salen

solvent	n	Product Distribution, %			
		n-butane	ethane	ethylene	total
CH₃CN	1.06	49	17	8	74
CD₃CN	1.01	59	13[a]	7	79
DMF	1.01	21	38	6	67

[a]Consists of 7% C_2H_6 and 6% C_2H_5D.

Ethyl radicals undergo (a) coupling to form n-butane, (b) disproportionation to give ethane and ethylene, and (c) abstraction of a hydrogen atom from the solvent. Although disproportionation of ethyl radicals produces equimolar amounts of ethane and ethylene, Table 1 reveals that the yield of ethane is larger than that of ethylene, which can be attributed to hydrogen atom abstraction from solvent.   When a catalytic reduction of iodoethane was performed in CD_3CN (Table 1), the yields of C_2H_6 and C_2H_4 were identical; moreover, the yields of C_2H_5D and C_2H_6 were nearly the same, indicating that, in acetonitrile, ethane is formed approximately equally via disproportionation and hydrogen atom abstraction.

CATALYTIC REDUCTION OF 3-CHLORO-2,4-PENTANEDIONE

As shown in curve B, Figure 2, a cyclic voltammogram for reduction of cobalt(II) salen in the presence of excess 3-chloro-2,4-pentanedione at a carbon electrode in acetonitrile containing 0.050 M TMABF₄ exhibits two cathodic peaks, and an anodic peak is associated with the second cathodic peak. We attribute the first cathodic peak to a combination of processes: (a) reaction between electrogenerated cobalt(I) salen and 3-chloro-2,4-pentanedione to form 2,4-pentanedion-3-ylcobalt(III) salen and (b) reduction of the latter species to yield cobalt(II) salen and the 2,4-pentanedion-3-ate anion. Reduction of cobalt(II) salen to cobalt(I) salen is responsible for the second reversible process (curve A, Figure 2). Thus, catalytic reduction of 3-chloro-2,4-pentanedione differs from that for iodoethane in two ways: (a) 2,4-pentanedion-3-ylcobalt(III) salen is much easier to reduce than ethylcobalt(III) salen (and easier to reduce than cobalt(II) salen) and (b) 2,4-pentanedion-3-ylcobalt(III) salen undergoes reduction to cobalt(II) salen and the 2,4-pentanedion-3-ate anion, whereas ethylcobalt(III) salen is reduced to cobalt(I) salen and an ethyl radical.

Table 2 presents coulometric data and product distributions for the catalytic reduction of 3-chloro-2,4-pentanedione by cobalt(I) salen electrogenerated at −0.65 V at reticulated vitreous carbon cathodes in acetonitrile containing 0.050 M TMABF₄. Without an added proton donor, catalytic reduction of 3-

chloro-2,4-pentanedione is a two-electron process; 2,4-pentanedione is obtained in 53% yield, with the other product being tetramethylammonium 2,4-pentanedion-3-ate. However, addition of 100 mM water to an electrolyzed solution causes the yield of 2,4-pentanedione to increase to 85%. Performing an electrolysis in the presence of 100 mM HFIP causes no change in the n value, but the yield of 2,4-pentanedione increases to essentially 100%. In the presence of 100 mM $(CH_3)_2CD(OH)$, a deuterium atom donor, no deuterium incorporation into 2,4-pentanedione was detected; thus, the 2,4-pentanedion-3-yl radical is not produced by the catalytic process. To confirm the intermediacy of 2,4-pentanedion-3-ate, we added 100 mM iodoethane after the completion of an electrolysis, and noted both the formation of 3-ethyl-2,4-pentanedione in 84% yield and a drop in the yield of 2,4-pentanedione (11%).

Table 2. Coulometric Data and Product Distributions for Catalytic Reduction of 20 mM 3-Chloro-2,4-Pentanedione in CH₃CN Containing 0.050 M TMABF₄ and 1 mM Cobalt(II) Salen

		Product Distribution, %		
condition	n	2,4-pentanedione	3-ethyl-2,4-pentanedione	total[c]
	1.90	53	0	53
with 100 mM water[a]	1.86	85	0	85
with 100 mM HFIP[b]	2.01	105	0	105
with 100 mM C₂H₅I[a]	1.89	11	84	95

[a]Added after completion of the electrolysis.
[b]HFIP (1,1,1,3,3,3-hexafluoro-2-propanol) was present during the electrolysis.
[c]Remaining product is the tetramethylammonium salt of 2,4-pentanedion-3-ate.

In the absence of an added proton donor, the 2,4-pentanedion-3-ate anion is protonated by residual water in the solvent–electrolyte

$$^-C(COCH_3)_2 + H_2O \rightleftharpoons HC(COCH_3)_2 + OH^-$$

but the equilibrium is such that the reaction is incomplete. Deliberate addition of excess water after completion of an electrolysis leads to more 2,4-pentanedione, whereas the presence of HFIP during the entire course of an electrolysis leads to quantitative formation of the dione. When iodoethane is added after completion of an electrolysis, production of 3-ethyl-2,4-pentanedione

$$^-C(COCH_3)_2 + C_2H_5I \rightleftharpoons H_5C_2C(COCH_3)_2 + I^-$$

lowers the concentration of the 2,4-pentanedion-3-ate anion, thereby shifting the position of the preceding anion–water reaction toward the left and causing the yield of the dione to decrease.

PALLADIUM-CATALYZED CARBOXYLATION OF VINYL TRIFLATES.
ELECTROSYNTHESIS OF α,β-UNSATURATED CARBOXYLIC ACIDS

Anny JUTAND and Serge NEGRI

Ecole Normale Supérieure, Département de Chimie, CNRS URA 1679

24 Rue Lhomond 75231 Paris Cedex 5, France

ABSTRACT

The palladium-catalyzed electrocarboxylation of vinyl triflates affords α,β-unsaturated carboxylic acids. The reactivity of vinyl triflates has been reversed in the presence of an electron source, since they now react with electrophiles such as CO_2.

INTRODUCTION

Vinyl triflates usually undergo cross-coupling reactions with nucleophiles in the presence of a catalytic amount of palladium complexes. In the presence of an electron source, it is now possible to invert the reactivity of vinyl triflates and to bring them to react with electrophiles.

RESULTS AND DISCUSSION.

Vinyl triflates (easily synthesized from reactions of ketone enolates with triflate anhydride or from addition of triflic acid on alkynes) are electroactive compounds (Table 1). The direct reduction of vinyl triflates **1**, performed in the absence of any catalyst results, after hydrolysis, in the formation of the corresponding ketone **2** (eqn 1) by activation and cleavage of the O-S bond of the triflate group ($O-SO_2CF_3$). In the presence of a catalytic amount of $PdCl_2(PPh_3)_2$ **3**, the formation of the ketone is completely inhibited and a conjugated diene **4** is obtained in quantitative yield (eqn 2), showing that, under these conditions, the activation of the C-O bond of the vinyl triflate *via* a palladium complex, is highly favored.

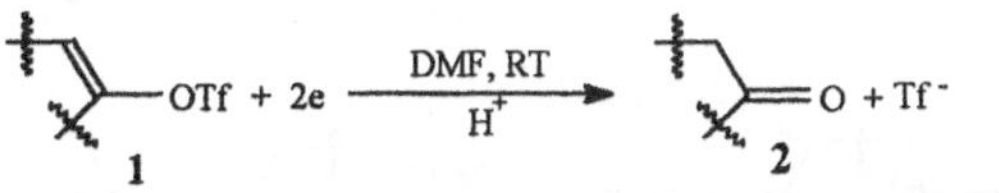

$$\text{(1)}$$

$$2 \overset{\hspace{0.5cm}}{\underset{\mathbf{1}}{\parallel}} \text{—OTf} + 2e \xrightarrow[\text{DMF, RT}]{\text{PdCl}_2(\text{PPh}_3)_2 \, \mathbf{3}\,,\,10\%} \overset{\hspace{0.5cm}}{\underset{\mathbf{4}}{\parallel}} + 2\,\text{TfO}^- \qquad (2)$$

In the presence of carbon dioxide as the electrophile, α,β-unsaturated carboxylic acids **5** are synthesized in good yields with $\text{PdCl}_2(\text{PPh}_3)_2$, as catalyst (eqn 3, Table 1).[1]

$$\overset{\hspace{0.5cm}}{\underset{\mathbf{1}}{\parallel}} \text{—OTf} + \text{CO}_2 + 2e \xrightarrow[\text{DMF, RT}]{\text{PdCl}_2(\text{PPh}_3)_2 \, \mathbf{3}\,,\,10\%} \overset{\hspace{0.5cm}}{\underset{\mathbf{5}}{\parallel}} \text{—CO}_2^- + \text{TfO}^- \qquad (3)$$

Table 1: Electrocarboxylation of Vinyl Triflates Catalyzed by $\text{PdCl}_2(\text{PPh}_3)_2$ in DMF at 20 °C (eqn 3).

Vinyl-OTf **1** [1 mmol]	Vinyl-OTf E^p_{red} (V vs SCE)	Vinyl-Pd-Cl(PPh$_3$)$_2$ **6** E^p_{red} (V vs SCE)	E^a	F/ mol	Vinyl-CO$_2$H **5**	%[b]
⬡—⬡—OTf	-2.95	-2.15	-2.0	2.4	⬡—⬡—CO$_2$H	85
(naphthalene) OTf	-2.16	-2.08	-1.6	2.0	(naphthalene) CO$_2$H	60
+—⬡—OTf	-2.91	-2.10	-2.0	2.2	+—⬡—CO$_2$H	80
⬡—OTf	-2.95	-1.75	-2.0	2.1	⬡—CO$_2$H	86
OTf (allylic)	-2.94	-2.20	-2.2	2.3	CO$_2$H (allylic)	70
(phenyl) OTf	-1.94	-1.92	-1.7	2.1	(phenyl) CO$_2$H	32

[a] Electrolysis potentials in Volt vs SCE. [b] Isolated yields are related to the vinyl triflate completely converted.

The palladium-catalyzed electrocarboxylation of vinyl triflates proceeds *via* a double activation: firstly, activation of the C-O bond of the vinyl triflate by a palladium(0) complex by oxidative addition and secondly, activation by electron transfer, of the vinylpalladium(II) complex **6** resulting from the oxidative addition, the vinylpalladium(II) complex **6** being more easily reduced than the vinyl triflate **1** (Table 1).

$$\text{Pd(0)(PPh}_3)_2\text{Cl}^- + \overset{}{\underset{\mathbf{1}}{\parallel}}\text{—OTf} \xrightarrow{-\text{TfO}^-} \overset{}{\underset{\mathbf{6}}{\parallel}}\text{—PdCl(PPh}_3)_2 \xrightarrow{+2e\;+\text{CO}_2} \overset{}{\underset{\mathbf{5}}{\parallel}}\text{—CO}_2^- + \text{Pd(0)(PPh}_3)_2\text{Cl}^-$$

Reference
1) A. Jutand, and S. Négri, *Synlett*, **1997**, 719.

INVESTIGATION OF THE MECHANISM OF PALLADIUM-CATALYZED REACTIONS BY ELECTROCHEMISTRY

Christian AMATORE, Emmanuelle CARRE, Anny JUTAND, Amine M'BARKI and Gilbert MEYER

Ecole Normale Supérieure, Département de Chimie, CNRS URA 1679

24 Rue Lhomond 75231 Paris Cedex 5, France

ABSTRACT

A new mechanism for Heck reactions catalyzed by a mixture of $Pd(OAc)_2$ and triarylphosphines has been established by means of electrochemical techniques. The complex $ArPd(OAc)L_2$ is found to be a key intermediate which reacts with the olefin. These results emphasize the role of the acetate ions present in $Pd(OAc)_2$, precursor of the palladium(0) catalyst.

INTRODUCTION

Since 25 years, Heck reactions have been widely developed and are key steps in the synthesis of organic compounds.[1]

$$\text{R} + \text{ArX} + \text{NEt}_3 \xrightarrow{\text{Pd}} \text{Ar-R} + \text{Et}_3\text{NH}^+ \text{X}^-$$

The postulated mechanism is described in Scheme 1. It involves a palladium(0) complex as a catalyst.

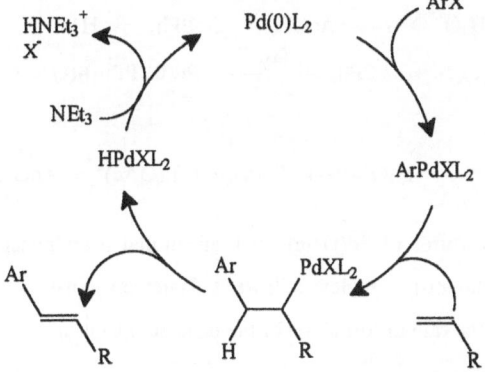

Scheme 1: postulated mechanism

However, Heck reactions are very often catalyzed by mixtures of a palladium(II) salt: $Pd(OAc)_2$ and triarylphosphines. Only palladium(0) complexes can activate aryl halides by an oxidative addition as shown in the first step of the postulated catalytic cycle (Scheme 1). Thus, the first problem to solve was to characterize, among all the reagents, the reductant able to reduce the palladium(II) to palladium(0).

RESULTS AND DISCUSSION.

It has been shown by cyclic voltammetry and ^{31}P NMR spectroscopy that a complex $Pd(OAc)_2(PPh_3)_2$ is formed in DMF, from mixtures of $Pd(OAc)_2 + nPPh_3$ (n ≥ 2). A palladium(0) complex is then generated *in situ* by an intramolecular reduction of the palladium(II) complex by the phosphine which is oxidized to phosphine oxide.[2,3] The resulting palladium(0) complex is an anionic complex ligated by one acetate ion. The kinetics of this inner sphere reduction has been monitored by amperometry performed at a rotating disk electrode and the following mechanism has been established for the spontaneous formation of the palladium(0) complex.

$$Pd(OAc)_2 + 2\ PPh_3 \xrightarrow{\text{fast}} Pd(OAc)_2(PPh_3)_2$$

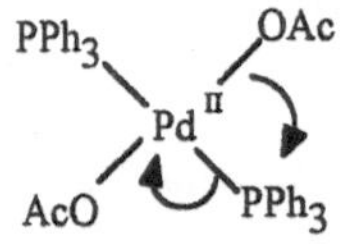

$$\xrightarrow[k = 4\times10^{-4}\ s^{-1}]{\text{rds}} \text{"Pd(0)(PPh}_3\text{)(OAc)}^-\text{"} + AcO\text{-}PPh_3^+$$

$$AcO\text{-}PPh_3^+ + H_2O \longrightarrow AcOH + O\text{=}PPh_3 + H^+$$

$$\text{"Pd(0)(PPh}_3\text{)(OAc)}^-\text{"} + 2\ PPh_3 \xrightarrow{\text{fast}} Pd(0)(PPh_3)_3(OAc)^-$$

with the overall reaction:

$$Pd(OAc)_2 + 4\ PPh_3 + H_2O \longrightarrow Pd(0)(PPh_3)_3(OAc)^- + AcOH + O\text{=}PPh_3 + H^+$$

When generated from mixtures of $Pd(OAc)_2$ and substituted triarylphosphines, $(p\text{-}Z\text{-}C_6H_4)_3P$, the formation of the palladium(0) complex follows a Hammett correlation with a positive slope (ρ = +2.4) showing that the rate of formation of the palladium(0) is higher when the phosphine is less electron rich.

The kinetics of the oxidative addition of phenyl iodide with the palladium(0) complex generated *in situ* from mixture of $Pd(OAc)_2 + nPPh_3$ (n ≥ 3) has been monitored by amperometry. The reactive complex is the less ligated complex $Pd^0(PPh_3)_2(OAc)^-$ involved in an equilibrium with the ligand and a saturated palladium(0) complex:[3]

$$Pd(0)(PPh_3)_3(OAc)^- \; \rightleftharpoons \; Pd(0)(PPh_3)_2(OAc)^- + PPh_3 \qquad K_0 = 2 \times 10^{-3} \, M$$

The complex resulting from the oxidative addition with PhI is not $PhPdI(PPh_3)_2$, as expected (Scheme 1), but a new complex $PhPd(OAc)(PPh_3)_2$, generated *via* an anionic pentacoordinated complex where the phenylpalladium(II) complex is ligated both by the acetate and the iodide anions.[4] $PhPd(OAc)(PPh_3)_2$ has been synthesized independently by reacting $PhPdI(PPh_3)_2$ with AgOAc.

$$Pd(0)(PPh_3)_2(OAc)^- + PhI \xrightarrow{61 \, M^{-1}s^{-1}} \left[Ph-\underset{\underset{L}{|}}{\overset{\overset{L}{|}}{Pd}} \overset{\cdots I}{\underset{OAc}{}} \right]^- \xrightarrow{3 \times 10^{-2} \, s^{-1}} PhPd(OAc)(PPh_3)_2 + I^-$$

$$\xrightarrow{\quad \times \quad} PhPdI(PPh_3)_2$$

The cyclic voltammetry of the complex $PhPd(OAc)(PPh_3)_2$ in solution in DMF exhibits two reduction peaks. The first reduction peak is assigned to the reduction of the cationic complex $PhPd(PPh_3)_2^+$ by comparison with an authentic sample (the later has been synthesized by reacting $PhPdI(PPh_3)_2$ with $AgBF_4$). The second reduction peak corresponds to the reduction of $PhPd(OAc)(PPh_3)_2$. The cationic and the neutral complexes are involved in an equilibrium and the equilibrium constant has been determined by chronoamperometry.[5]

$$PhPd(PPh_3)_2^+ + AcO^- \; \rightleftharpoons \; PhPd(OAc)(PPh_3)_2 \qquad K_{OAc} = 0.75 \times 10^3 \, mol^{-1}dm^3$$

$PhPd(OAc)(PPh_3)_2$ is found to be more reactive with an olefin (such as styrene which is transformed to stilbene) than the cationic complex. $PhPdI(PPh_3)_2$ does not react with styrene but reaction occurs in the presence of AcO^- *via* the formation of $PhPd(OAc)(PPh_3)_2$ according to the following equilibrium:

$$PhPdX(PPh_3)_2 + AcO^- \; \rightleftharpoons \; PhPd(OAc)(PPh_3)_2 + X^- \qquad K = 0.3$$

These results establish that $ArPd(OAc)L_2$ complexes are key intermediates in the Heck reaction.

The Heck reaction is always performed in the presence of a base (amines, acetate, carbonate) which is supposed to recycle an intermediate palladium(II) complex to palladium(0) (last step of the postulated catalytic cycle, Scheme 1). The role of the base has been investigated. In the presence of

NEt$_3$, the oxidative addition is slower but the reaction of ArPd(OAc)L$_2$ with the olefin is faster. Thus, the role of the base is more complicated than postulated. As indicated in Scheme 2, due to acidobasic reaction of acetate ions and protons, the equilibrium between ArPd(OAc)L$_2$ and the cationic complex ArPd(OAc)L$_2{}^+$ is shifted to the less reactive cationic complex. Protonation of the base results in a decay of the acetate ions concentration and consequently in a shift of the equilibrium towards ArPd(OAc)L$_2$ i.e. the more reactive species. On the basis of these results, a new mechanism is proposed for the Heck reaction (Scheme 2).

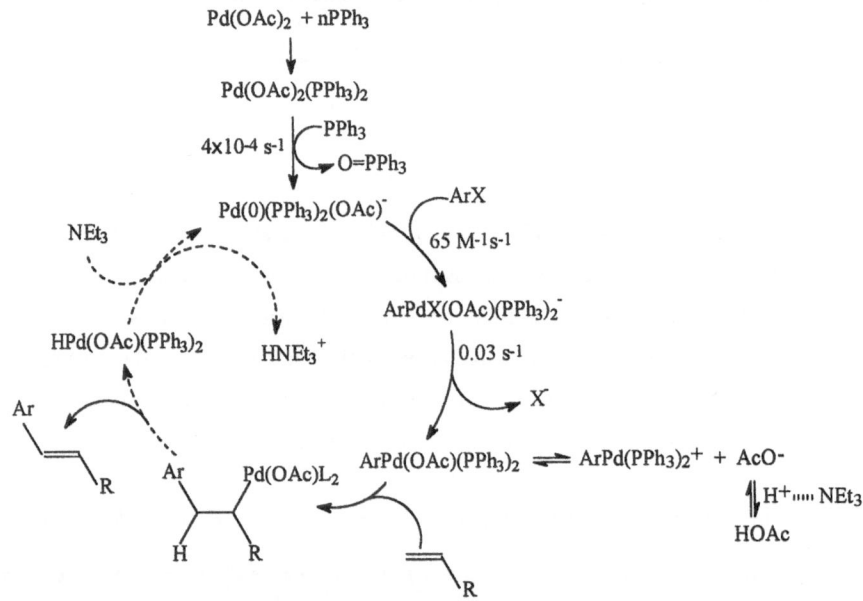

Scheme 2: new mechanism for the Heck reaction.

This new mechanism evidences the role of acetate ions coming from Pd(OAc)$_2$, precursor of the palladium(0) complex. Indeed, the acetate ion is a ligand of every intermediate palladium(0) and palladium(II) complexes of the catalytic cycle. In particular, it is involved in the rate determining step which is the reaction of ArPd(OAc)L$_2$ with the olefin.

References
1) A. de Meijere, and F. E. Meyer, *Angew. Chem. Int. Ed. Engl.* **1994**, *33*, 2379.
2) C. Amatore, A. Jutand, and A. M'Barki, *Organometallics*, **1992**, *11*, 3009.
3) C. Amatore, E. Carré, A. Jutand, and M. A. M'Barki, *Organometallics*, **1995**, *14*, 1818.
4) C. Amatore, E. Carré, A. Jutand, M. A. M'Barki, and G. Meyer, *Organometallics*, **1995**, *14*, 5605
5) C. Amatore, E. Carré, and A. Jutand, *Acta Chem. Scand.* **1998**. In press.

ELECTROCHEMICAL REDUCTION OF SOME
[(η^5-CYCLOPENTADIENYL) (η^6-ARENE) IRON(II)][PF$_6$] COMPLEXES BEARING AN IMINE OR A NITRONE FUNCTION IN BENZYLIC POSITION OF THE ARENE LIGAND

Fabrice PIERRE and Claude MOINET

Laboratoire :"Electrochimie et Organométalliques", UMR CNRS N°6509, Université de Rennes 1, Campus de Beaulieu, F-35042 Rennes Cedex, France

ABSTRACT

We report the selective electrochemical reduction of some mixtures of [(η^5-Cp) (η^6-benzophenone-N-(aryl)nitrone) Fe]$^+$ and [(η^5-Cp) (η^6-benzophenone anil) Fe]$^+$ complexes into the corresponding [(η^5-Cp) (η^6-α-(arylamino)diphenylmethane) Fe]$^+$ cations. In a similar manner, various [(η^5-Cp) (η^6-9-(arylamino)fluorene) Fe]$^+$ cations are synthesized from the corresponding [(η^5-Cp) (η^6-fluorenone anil) Fe]$^+$ derivatives. The electrochemical process allows the simultaneous reduction of both nitrone and imine functions with preservation of the organometallic moiety.

INTRODUCTION

[(η^5-Cp) (η^6-arene) Fe]$^+$ cations are an important class of organometallic sandwiches which has been extensively developed for macromolecular chemistry and organic synthesis. [1] However, owing to the reductible character of the organometallic moiety and the sensitivity of the arene ligand towards hydride addition, [2] their use in processes involving a reduction step may be potentially restricted. In this perspective, electrochemistry can be a powerful technique for selective reducing of the functions borne by the ligands.

In a precedent study, [3] we described the condensation reaction of various nitrosoarenes with (Cp Fe diphenylmethane)$^+$ and (Cp Fe fluorene)$^+$ complexes. This reaction afforded a very useful synthesis route to mixtures of (Cp Fe benzophenone-N-(aryl)nitrone)$^+$ and (Cp Fe benzophenone anil)$^+$ or to various (Cp Fe fluorenone anil)$^+$ cations. Within the context of our research on benzylic C-H activation applied to the formation of Carbon-Nitrogen bonds, we report herein a highly efficient electrochemical reduction of these nitrone and imine cations

RESULTS AND DISCUSSION

The controlled potential reduction of nitrone-imine mixtures **1a-d** + **1'a-d** at a mercury cathode, in ammonium buffer ((NH$_4$)$_2$SO$_4$ 0.25 mol l^{-1} + NH$_3$,H$_2$O 0.5 mol l^{-1})-acetone (1:1,v:v) led to the corresponding amine complexes **2a-d** with good yields (Scheme 1). The nitrone functions were reduced by a four electron process and the imines by a two electron process. We measured coulometric data between two and four Faradays per mole of substrate, depending on the composition of the starting mixtures. After work-up, the new compounds **2a-d** were fully characterized by Mass Spectrometry (LSIMS), and NMR spectroscopy.

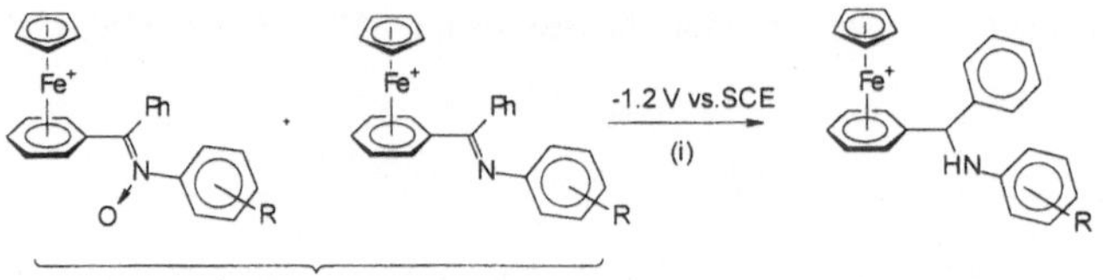

			mixtures		F.mol^{-1}		yield
R = H	**1a**	+	**1'a**	(90:10)	3.8	**2a**	90%
2'-Me	**1b**	+	**1'b**	(70:30)	3 4	**2b**	89%
3'-Me	**1c**	+	**1'c**	(80:20)	3.5	**2c**	85%
4'-Me	**1d**	+	**1'd**	(85:15)	3 6	**2d**	85%

(i) ammonium buffer ((NH$_4$)$_2$SO$_4$ 0 25 mol l^{-1} + NH$_3$,H$_2$O 0 5 mol l^{-1}) -acetone (1:1,v:v), Hg cathode

scheme 1

In a similar manner, (Cp Fe fluorenone anil)$^+$ derivatives **3a-h** and the imine-nitrone mixture **3i** + **3'i** were electrochemically reduced in acidic hydro-organic media, leading to the corresponding amine complexes **4a-i** with good yields (Scheme 2).

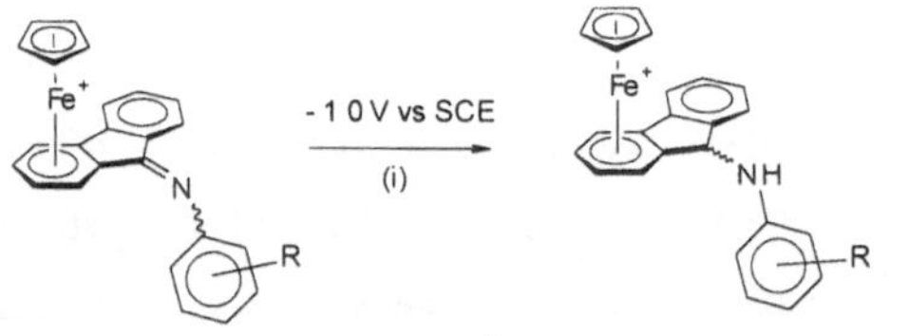

R =			F.mol^{-1}		yield	diastereomer ratio exo : endo
	H	**3a**	2.0	**4a**	88%	75 : 25
	2'-Me	**3b**	2.0	**4b**	84%	85 15
	3'-Me	**3c**	2.0	**4c**	81%	75 : 25
	4'-Me	**3d**	2.0	**4d**	87%	75 . 25
	2'-Cl	**3e**	1.9	**4e**	88%	80 20
	4'-Cl	**3f**	2.0	**4f**	79%	80 : 20
	2'-CH$_2$CO$_2$Me	**3g**	2.1	**4g**	70%	80 . 20
	4'-COMe	**3h**	2.1	**4h**	81%	70 : 30
	2'-CO$_2$Me	3i+nitrone 3'i(60 40)	2.9	**4i**	84%	45 : 55

(i) acetic buffer (CH$_3$CO$_2$Na 0.5 or 1 5 mol l^{-1} + CH$_3$CO$_2$H 0 5 mol l^{-1}) - acetone (1 1,v v), Hg cathode

scheme 2

The air-sensitive complexes **4a-i** were isolated in satisfactory purity under argon atmosphere and fully characterized by mass spectrometry (LSIMS), elemental analyses, ^{1}H and ^{13}C NMR spectroscopy. Two diastereomers *exo* and *endo* are formed during the electrolysis, as is evident from the NMR spectra which displays two signal sets of unequal intensity.

According to previous works on electrochemical reduction of (Cp Fe arene)$^+$ complexes in hydro-organic media, [4] one might attempt that upon electrolysis, the benzylic carbon protonation step takes place in *trans* to the metal, leading preferentially to the *endo* isomer. Surprisingly, it can be taken as evidence that, except for **4i**, the *exo* amino complex is present in the majority. This is based on an analysis of NMR protons chemical shifts of the R-C$_6$H$_4$-NH- fragment and the benzylic proton H-9, which undergo a downfield shift when they are located on the same side of the metal moiety. [5]

In order to explain these unexpected results, the amine **4b** (*exo:endo*-85:15) was submitted to recrystallization in CH$_2$Cl$_2$ at -25°C, leading to a mixture enriched with the *endo* isomer (*exo:endo*-55:45). By stirring it in acetic buffer-acetone at room temperature for one night, after work-up we could recover in 80% yield the initial mixture ratio (*exo:endo*-85:15). This result clearly demonstrates that in the acidic electrolysis media, the *endo* isomer is epimerized into the more stable *exo* isomer. According to precedent reports on highly stereoselective electrolyses of (Cp Fe arene)$^+$ cations, [4] we may assume that our reduction process is also stereoselective. However the experiments described above show that the reaction can be followed by a rearrangement which lead to the thermodynamic equilibrium (Scheme 3). The isomer ratio resulting from the electrosynthesis of **4i** (*exo:endo*-45:55) may be explained by a slower epimerization rate. This rate could be enhanced in DMSO, as is evident from an NMR spectrum in this solvent which displayed an *exo:endo* isomer ratio of 60:40 respectively.

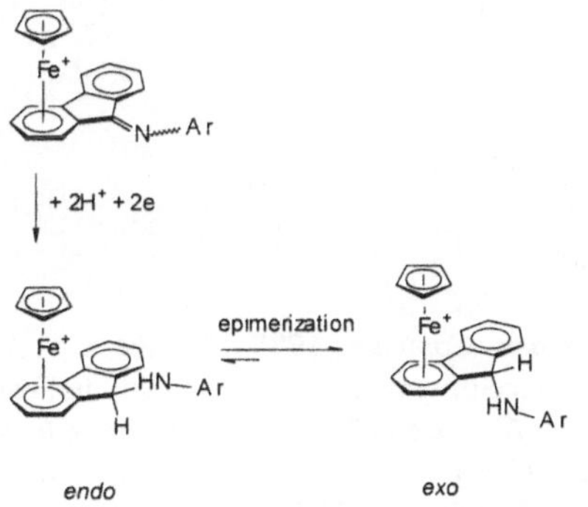

scheme 3

CONCLUSION

This work further demonstrates the efficiency of electrochemical processes. We have shown that highly chemio-selective reduction occurs on (Cp Fe arene)⁺ complexes, providing various (Cp Fe α-(aryl amino)arene)⁺ derivatives with good yield. In connection with the nitrosoarene condensation, [3] this methodology constitutes an efficient two-step introduction of a secondary arylamino group onto a benzylic chain (Scheme 4).

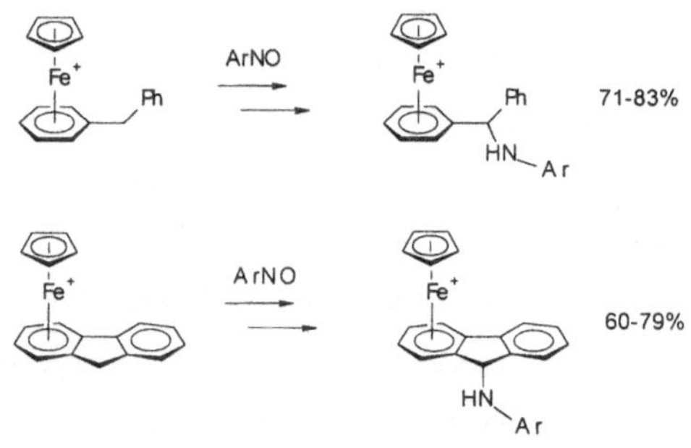

scheme 4

Otherwise, these amino complexes still possess a benzylic carbon liable to be deprotonated and to react with electrophiles. Additional studies are underway to use such complexes in further syntheses.

References

1) (a) V. Marvaud, D. Astruc, *Chem. Commun.*, **1997**, 773 ; (b) A. S. Abd-El-Aziz, C. R. De Denus, M. J. Zaworotko, and L. R. MacGillivray, *J. Chem. Soc. Dalton Trans.*, **1995**, 3375 ; (c) A. J. Pearson, A. M. Gelormini, M. A. Fox and D. Watkins, *J. Org. Chem.*, **1996**, *61*, 1297.

2) (a) D. Astruc, *Topics in Curr. Chem.*, **1991**, *160*, 47 ; (b) R.G. Sutherland, M. Iqbal and A. piorko, *J. Organometal. Chem.*, **1986**, *302*, 307.

3) F. Pierre, C. Moinet and L. Toupet, *J. Organometal. Chem.*, **1997**, *527*, 51.

4) (a) M. Le Rudulier et C. Moinet, *J. Organometal. Chem.*, **1988**, *352*, 337 ; (b) E. Roman, D. Astruc and A. Darchen, *J. Chem. Soc.,Chem. Commun.*, **1976**, 512 ; (c) E. Roman, D. Astruc and A. Darchen, *J. Organometal. Chem.*, **1981**, *219*, 221.

5) R.M. Moriarty, Y.-Y. Ku, U.S. Gill, R. Gilardi, R.E. Perrier, and M.J. McGlinchey, *Organometallics*, **1989**, *8*, 960.

STEREOSELECTIVITY IN THE ELECTROREDUCTIVE SYNTHESES WITH ARYL AND ALKENYL HALIDES CATALYZED BY NICKEL COMPLEXES

Muriel DURANDETTI, Sylvie CONDON-GUEUGNOT, Jacques PERICHON, and Jean-Yves NEDELEC

Laboratoire d'Electrochimie, Catalyse, et Synthèse Organique
2, rue Henri-Dunant, F-94320 Thiais, France

ABSTRACT

The nickel-catalyzed electroreductive coupling between aryl halides and 2-chloropropionic acid derivatives bearing chiral auxiliaries leads to chiral 2-arylpropionic acids with high enantiomeric excesses. The electrochemical alkenylation of activated olefins catalyzed by nickel complexes occurs with complete transfer of the streochemistry of the starting alkenyl halide onto the product.

ASYMMETRIC INDUCTION IN THE COUPLING BETWEEN ARYL HALIDES AND 2-CHLOROPROPIONIC ACID DERIVATIVES

Several 2-arylpropionic acids are known as important non-steroidal pharmaceuticals exhibiting anti-inflammatory activity and many methods of preparation of these acids have been developed. These methods lead generally to racemic compounds, but it has been shown that a higher activity is associated with the S configuration at the chiral center. Different approaches described so far to obtain the most active enantiomer include chemical resolution, microbial transformation, and various asymmetric syntheses.

We have previously described the electroreductive cross-coupling of α-halogenoesters with aryl halides catalyzed by nickel complexes in combination with the sacrificial anode process, and leading to α-arylpropionic esters in one operation in good to high yields.[1] We have now diclosed a very efficient remote asymmetric induction in this reaction using chiral auxiliaries.[2]

We first tried to obtain chiral products by using the commercially available chiral methyl (R)- or (S)-2-chloropropionates in the coupling with iodobenzene but we only obtained the racemic α-arylpropionic ester. We then attempted to induce the chirality remotely using chiral auxiliaries (Chart 1) attached to the carboxylic group. The best results (Table 1) were obtained from imidazolidinones derivatives **7** and **8** (ee > 92%). In addition to the high asymmetric induction obtained, the easy access to both enantiomers **7** and **8** of the imidazolidinone from (-)- or (+)-ephedrine respectively makes this process efficient to obtain either isomer of the desired α-arylpropionic acid, R from **7** and S from **8**.

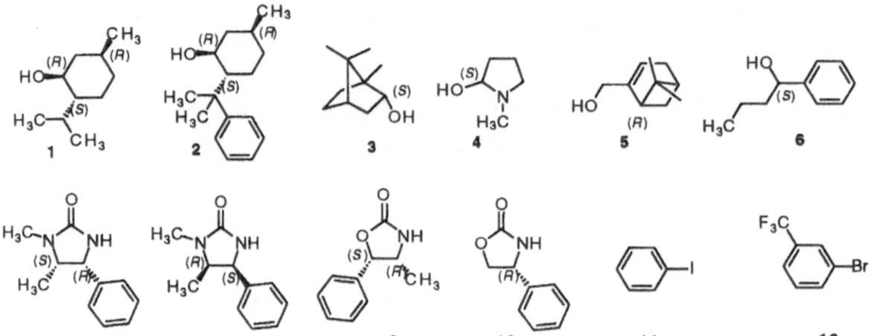

1 2 3 4 5 6

7 8 9 10 11 12

Chart 1

Table 1. Nickel-catalyzed electroreductive coupling between α-chloropropionic acid derivatives bearing chiral auxiliaries and iodobenzene (**11**) or 3-bromo-α,α,α-trifluorotoluene (**12**)

auxiliary	ArX	yield %	ee (de)	major configuration
1	**11**	60%	19% (30%)	*S*
2	**12**	70%	45% (50%)	*S*
3	**11**	50%	5% (10%)	*S*
4	**11**	41%	1% (6%)	*R*
5	**12**	50%	13%	*R*
6	**11**	60%	6% (6%)	*R*
7	**11**	57%	90% (96%)	*R*
7	**12**	51%	87% (92%)	*R*
8	**12**	51%	80% (92%)	*S*
9	**11**	64%	17% (20%)	*R*
10	**11**	50%	52% (63%)	*R*

The reaction was applied to the preparation of two commonly used drugs, (*S*)-Naproxen **13** and (*S*)-Flurbiprofen **14** (Chart 2). These compounds were obtained in ca 60% isolated yields and high diastereomeric excesses (93 and 82% respectively) at room temperature using **8** as chiral auxiliary.

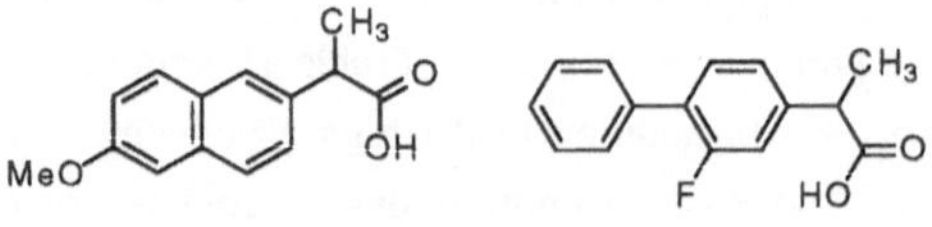

13 14

Chart 2

E/Z-STEREOSELECTIVITY IN COUPLING AND ADDITION REACTIONS INVOLVING ALKENYL HALIDES

E/Z-Stereoselectivity has been examined in three nickel catalyzed reactions, i.e. dimerisation, cross-coupling with aryl halides, and addition to activated olefins, which are efficiently carried out in the presence of nickel complexes as catalysts.

The nickel catalyzed dimerisation ofalkenyl halides occured with partial isomerisation to give more or less of the thermodynamic product.

$$C_5H_{11}CH=CHBr \xrightarrow[70\%]{NiBr_2bpy, e} (C_5H_{11}CH=CH)_2$$

Z   70%   Z,Z : 47.5; E,Z : 43.5; E,E :9

E   E,E : 100

$$C_5H_{11}CH=CHI \xrightarrow[73\%]{NiBr_2bpy, e} (C_5H_{11}=CH)_2$$

Z   73%   Z,Z : 64.5; E,Z : 28; E,E : 7.5

E   E,E : 98; E,Z : 2

$$C_6H_5CH=CHBr \xrightarrow[56\%]{NiBr_2bpy, e} (C_6H_5CH=CH)_2$$

Z   Z,Z : 83; E,Z : 17

$$CH_3CH=CH(CH_3)Br \xrightarrow[70\%]{NiBr_2bpy, e} (CH_3CH=CH(CH_3))_2$$

E/Z : 1/1   70%   Z,Z :9; E,Z : 4; E,E : 87

In the cross-coupling reactions partial isomerisation was also observed:

$$\underset{Z}{\overset{CH_3 \quad Br}{\underset{H \quad CH_3}{\diagup\!\!\diagdown}}} + ClCH_2CO_2CH_3 \xrightarrow[65\%]{NiBr_2bpy, e} CH_3CH=C(CH_3)CH_2CO_2Me$$

Z/E = 80/20

$$F\text{-}\langle\rangle\text{-}Br + Br\text{-}CH=CH\text{-}CH_3 \xrightarrow[66\%]{NiBr_2bpy, e} F\text{-}\langle\rangle\text{-}CH=CH\text{-}CH_3$$

E/Z =1   E/Z = 4

On the contrary, in the alkenylation of activated olefins catalyzed by NiBr$_2$ in DMF-acetonitrile the stereochemistry of the starting halide is transferred stereospecifically to the product (Table 2).[3]

$$\underset{1}{\overset{R^1 \quad R^3}{\underset{R^2 \quad X}{\diagup\!\!\diagdown}}} + \underset{2}{\diagup\!\!\diagdown_{EWG}} \xrightarrow[\substack{DMF/AN: 1/1 \\ 60-80°C}]{NiBr_2.3H_2O\ 10\%,\ e^-} \underset{3}{\overset{R^1 \quad R^3}{\underset{R^2}{\diagup\!\!\diagdown}}}\diagdown_{EWG}$$

The reaction has been applied to the synthesis of of *(Z)*-5-undecen-2-one (Chart 3) , a pheromone from the pedal gland of the Bontebok (Damaliscus dorcas dorcas).

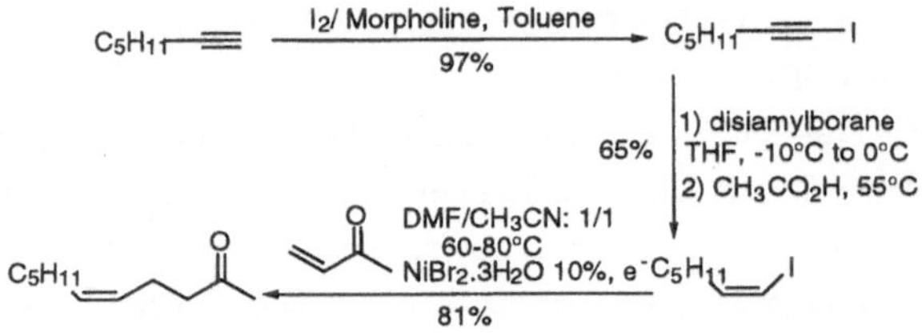

Chart 3

Table 2. Stereoselective Electrochemical Alkenylation of Electron-Deficient Olefins.

1	X	2 EWG	Product	Z/E purity %	Yield (%) of 3
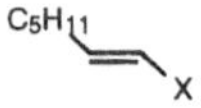 C₅H₁₁ ⌇X	Cl	COCH₃	3a	0/100	49
	Br	-		0/100	77
	I	-		0/100	80
 C₅H₁₁ ⌇X	Br	COCH₃	3b	100/0	77
	I	-		100/0	81
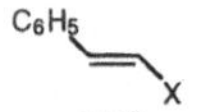 C₆H₅ ⌇X	Br	COCH₃	3c	7/93	73
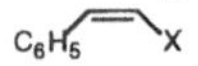 C₆H₅ ⌇X	Br	COCH₃	3d	97/3	84
		CN	3e	98/2	68
		CO₂Et	3f	95/5	76
⌇C₆H₅	Br	COCH₃	3g	-	61
		CO₂Et	3h	-	60
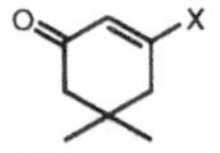	Cl	COCH₃	3i	-	60
	Br	-		-	75
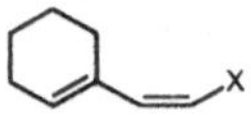	I	COCH₃	3j	94/6	50
		-			
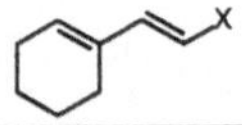	I	COCH₃	3k	0/100	71

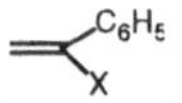

References

1) M. Durandetti, S.Sibille, J.Y. Nédélec, J. Périchon, *J. Org. Chem.* **1996**, *61*, 1748.

2) M. Durandetti, J. Périchon, J.Y. Nédélec, *J.Org.Chem.* , submitted.

3) S. Condon- Gueugnot, D. Dupré, J. Y. Nédélec, J. Périchon, *Synthesis*, in press.

NICKEL-CATALYZED ELECTROCHEMICAL REDUCTIVE CLEAVAGE OF THE OXYGEN-CARBON BOND OF ALLYL ETHERS : SYNTHETIC APPLICATIONS

Delphine FRANCO, Sandra OLIVERO, Jean-Paul ROLLAND and
Elisabet DUNACH
Laboratoire de Chimie Moléculaire, CNRS, Université de Nice-Sophia
Antipolis, Parc Valrose, 06108 NICE Cedex 2, France

ABSTRACT

Allyl aryl ethers can be selectively and catalytically cleaved to the corresponding phenols by electrogenerated Ni°-bipyridine species. The reaction constitutes a new method of alcohol deprotection. The intramolecular allyl transfer of allyl aryl ethers possessing a carbonyl function takes place regioselectively.

INTRODUCTION

In the field of organic electrosynthesis using organometallic catalysis, we have been interested by the reactivity of electrogenerated, low valent Ni(I) and Ni(0) complexes, generated by electrochemical reduction from stable and easily available Ni(II) compounds.

We have recently examined the electrochemical reactivity of allyl aryl ethers such as **1**, possessing an halogen function into the *ortho* position. Such derivatives present the possibility to undergo an electrochemical, nickel-catalyzed intramolecular cyclization[1-3]. The influence of the nature of the ligands associated to the nickel center can strongly determine the chemoselectivity of this reaction[4]. Thus, in the presence of Ni(II) with 2,2'-bipyridine (bipy) complexes, no cyclization of **1** occurs, and the cleavage of the O-C(allyl) bond takes place selectively (Scheme 1).

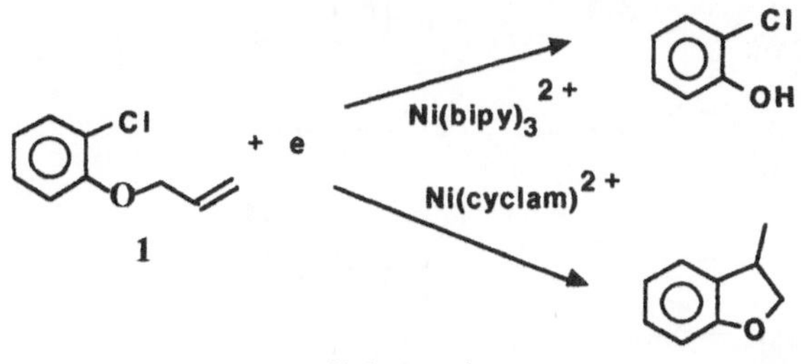

Scheme 1

RESULTS AND DISCUSSION

Ni(bipy)3(BF4)2 was shown to be a good catalyst for the electrochemical reductive cleavage of a series of allyl ether derivatives, affording the parent alcohols or phenols in good yields. Reactions were carried out in DMF, in single compartment cells fitted with a consumable magnesium anode. Some examples are presented in Table 1. The reductive cleavage process constitutes a new electrochemical method for the selective deprotection of allyl ethers, under simple and mild conditions[5]. The method enables the presence of several functional groups and avoids the use of strong acidic or basic media.

The O-C(allyl) cleavage reaction proceeds through the formation of a π-allyl Ni(II) complex, via the oxidative addition of electrogenerated Ni(0)(bipy)2 and 1.

Table 1 : Ni(bipy)3(BF4)2-catalysed electrochemical cleavage of allyl ethers (Mg/stainless steel electrodes).

Substrate	F/mol	Products	% Yield
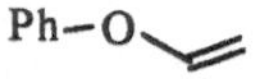 Ph–O⟍⟍	2.2	PhOH	99%
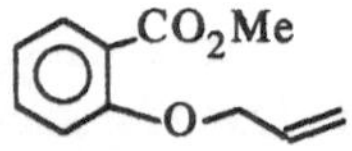 CO2Me ... O⟍⟍	2.2	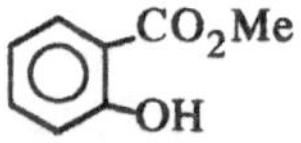 CO2Me ... OH	86%
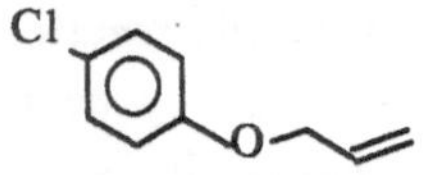 Cl ... O⟍⟍	2.2	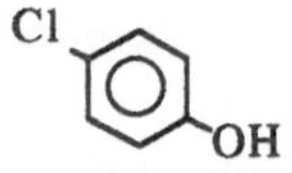 Cl ... OH	80%
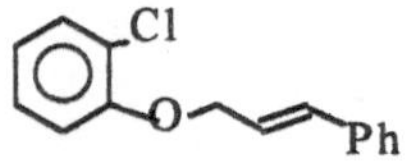 Cl ... O⟍⟍Ph	3.4	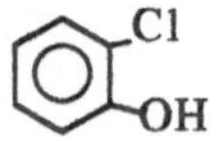 Cl ... OH	86%
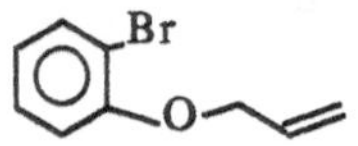 Br ... O⟍⟍	4.2	PhOH	99%
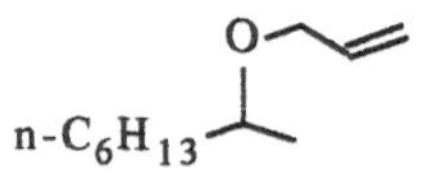 O⟍⟍ ... n-C6H13	2.8	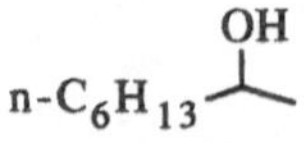 OH ... n-C6H13	55%

Experiments were carried out to quench the allyl moiety in the presence of carbonyl compounds in inter- and intramolecular reactions. The addition of the allyl unit to carbonyl compounds constitutes an important synthetic process, the homoallylic alcohols being useful synthetic intermediates. Although the electrochemical allylation reaction has been described[6], to the best of our knowledge, our studies constitute the first example of allylation of carbonyl compounds involving the cleavage of allyl ethers and the first example of electrochemical intramolecular allylation.

A series of *o*-(allyloxy)benzaldehydes and *o*-(allyloxy)acetophenones such as **2a**, possessing an allyl ether function and a carbonyl group in an *ortho* position of an aromatic ring have been used for the electrosynthesis of homoallylic alcohols of type **3** (eq. 1).

$$\text{2a} \quad + \quad 2e^- \quad \xrightarrow[\substack{\text{supporting electrolyte} \\ \text{DMF, 20 °C}}]{\substack{\text{Ni(bipy)}_3^{2+},\ 2BF_4^- \\ \text{Mg anode}}} \quad \text{3a} \tag{1}$$

The expected homoallylic alcohols **3**, issued from intramolecular allyl transfer reaction have been obtained in yields ranging from 45 to 88%[7].

The presence of the nickel-bipyridine catalytic system is essential to control the selectivity. The results indicate that, in the presence of the catalyst, the direct electroreduction of the carbonyl group does not occur and an exclusive activation of the allyl ether group involving the cleavage of the O-C allyl bond is obtained.

The regioselectivity of the allyl transfer reaction to aldehydes has been examined with unsymetrically substituted allyl ether units, the branched (B) homoallylic alcohols being preponderant over the linear (L) isomers. Regioselectivities up to 99% of (B) could be attained. The results on the regioselectivity in Ni-catalyzed allylation reactions are interesting, due to the poor reactivity of such complexes and to the little literature data dealing with the regioselectivity in π-allyl nickel additions.

From a mechanistic point of view, the overall transformation from **2** to **3** (eq. 1) involves a two-electron reduction process. A Ni(II) starting complex is the catalyst precursor, used in 10% molar ratio with respect to the substrate. The electroreduction of the Ni(II)-bipy system to form Ni(0)(bipy)$_2$ species in solution takes place at -1.2 V vs SCE and has already been described[8]. In the one-compartment cell electrolyses, the reactions at the electrodes are, at the anode, the oxidation of the magnesium rod into Mg^{2+} ions in solution, and at the cathode, the reduction of the Ni(II) into Ni(0) complexes. Electrogenerated Ni(0) complexes are the active catalytic species, able to react with the allyl ether part of **2** to form a π-allyl-type nickel complex, in analogy to the nickel-catalyzed allylation of carbonyl compounds with allyl chlorides[6b]. The intramolecular transfer of the allyl group to the carbonyl function leads to a nickel(II) alcoholate phenolate intermediate which, in the presence of the magnesium ions, undergoes a nickel-magnesium exchange reaction that forms a Mg^{2+} alcoholate-phenolate corresponding to **3**. This step liberates the Ni(II) complex, which can be further recycled. Ni^{2+} / Mg^{2+} exchange reactions in electrochemical processes have already been reported[9]. The alcoholate-phenolate is a stable species which accumulates during the electrolysis. It is simply hydrolyzed at the end of the reaction.

Intramolecular cyclization reactions of the phenol-alcohol products **3** under several experimental conditions (acid catalyzed, iodolactonization conditions) can be carried out. These several cyclizations allow the preparation of functionalyzed bicyclic ethers, structurally related to chromenes, hydroxy-chromanes, chromones or/and flavones, of interest in perfum industry and pharmacy.

References
1) S. Ozaki, I. Horiguchi, H. Matsushita, and H. Ohmori, *Tetrahedron Lett.*, **1994**, *35*, 725.
2) S. Olivero, J. C. Clinet, and E. Duñach, *Tetrahedron Lett.*, **1995**, *36*, 4429.
3) J. C. Clinet, and E. Duñach, *J. Organomet. Chem.*, **1995**, *503*, C48.
4) S. Olivero, and E. Duñach, SYNLETT, **1994**, 531.
5) S. Olivero, and E. Duñach, *J. Chem. Soc., Chem. Commun.*, **1995**, 2497.
6) (a) S. Torii, K. Uneyama, and H. Matsuda, *Tetrahedron Lett.*, **1994**, *25*, 6017; (b) S. Durandetti, S. Sibille, and J. Périchon, *J. Org. Chem.*, **1989**, *54*, 2198; (c) M. Tokuda, S. Satoh, and H. Suginome, *J. Org. Chem.*, **1989**, *54*, 5608.
7) D. Franco, S. Olivero, and E. Duñach, *Electrochimica Acta*, **1997**, *42*, 2159.
8) (a) L. Garnier, Y. Rollin, and J. Périchon, *New J. Chem.*, **1989**, *13*, 53; (b) S. Daniele, P. Ugo, G. Bontempelli, and M. Fiorani, *J. Electroanal. Chem. Interfacial Electrochem.*, **1987**, *219*, 259; (c) A. Misono, Y. Uchida, T. Yamagishi, and H. Kageyama, *Bull Chem. Soc. Jpn.*, **1972**, *45*, 1438.
9) (a) J. Chaussard, J.C. Folest, J.Y. Nédélec, J. Périchon, S. Sibille, and M. Troupel, *Synthesis*, **1990**, *5*, 369; (b) S. Dérien, E. Duñach, and J. Périchon, *J. Am. Chem. Soc.*, **1991**, *113*, 8447.

EXPLORING Co-MEDIATED ORGANIC REACTIONS. MODELLING MOLECULAR RECOGNITION IN VITAMIN B$_{12}$ DEPENDENT REACTIONS

T. Darbre, V. Siljegovic, A. Amolins, T. Otten, R. Keese, L. Abrantes and J. P. Correia

Dept. of Chemistry and Biochemistry, University of Bern, Freiestrasse 3, 3012 Bern, Switzerland and Faculty of Sciences, University of Lisboa, Rua Ernesto Vasconcelos, Ed. 5, 1700 Lisboa, Portugal

ABSTRACT

An *in vitro* catalytic cycle under electro-photochemical control has been set up for the coenzyme B$_{12}$ dependent methylmalonyl-succinyl rearrangement. For molecular recognition side chains for hydrophobic interactions and A-T-base pairing have been introduced into the vitamin B$_{12}$ derived catalyst and the corresponding substrates. Increased rearrangement has been observed in the hydrophobic model. Conservation of chirality in this catalytic reaction is low. According to CV the Co(II) - Co(I) reversible reduction of a thymine bearing corrin is low in acetonitrile, but complete in methanol. Anodic electropolymerisation has been observed with a pyrrol bearing corrin.

INTRODUCTION

Enzyme-coenzyme complexes are naturally occuring catalysts with active sites for selective recognition of substrates. Hydrogen bonds are often used for binding the substrates in a stereoelectronically optimal way. Enantioselective catalysis, often performed in aprotic solvents, depends on catalytically active transition metal centers and sterically demanding substituents restricting structural

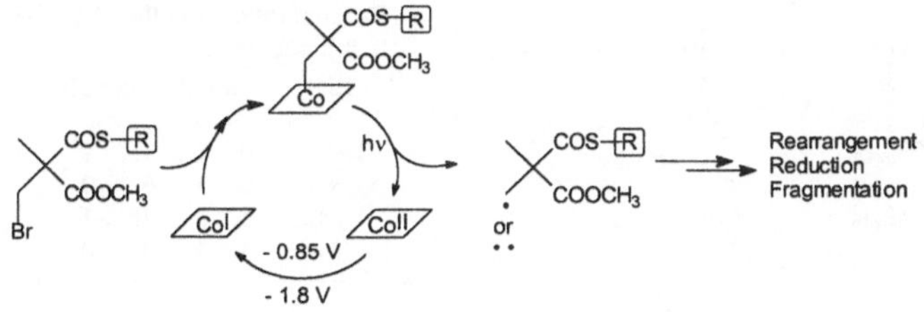

Scheme 1. General catalytic cycle for the methylmalonyl-succinyl rearrangement

options of substrates. Enzyme models are useful for understanding the mechanisms and control of stereoselectivities. They provide the stimulus for design of efficient catalysts where the substrate- and stereoselectivity might by controllable by highly selective recognition between substrates and catalyst in the periphery of the reactive centers.

395

An intriguing example is nature's coenzyme B_{12}, which is used *in vivo* as a multipurpose catalyst for a variety of important rearrangements and methyl transfer reactions. The potential of B_{12} derivatives as *in vitro* catalysts has been demonstrated by several groups. Here we present some results in our efforts to develop *in vitro* models for the coenzyme B_{12} dependent methylmalonyl- succinyl rearrangement under electrophotochemical control (Scheme 1).

RESULTS AND DISCUSSION

In the *hydrophobic Model* the methylmalonyl substrate as well as the Vitamin B_{12} derived catalyst habe been modified in the periphery by aliphatic groups of various chain length (Scheme 2). Variation of the chain length in the catalyst, the solvent and of the substrates shown, that the yield of the rearrangement product **4a** increases, when both the catalyst (**2**) and the substrate **3a** bear a $C_{18}H_{37}$ side chain and when the electro-photochemical reaction is run in aqueous methanol (Table 1a,b).

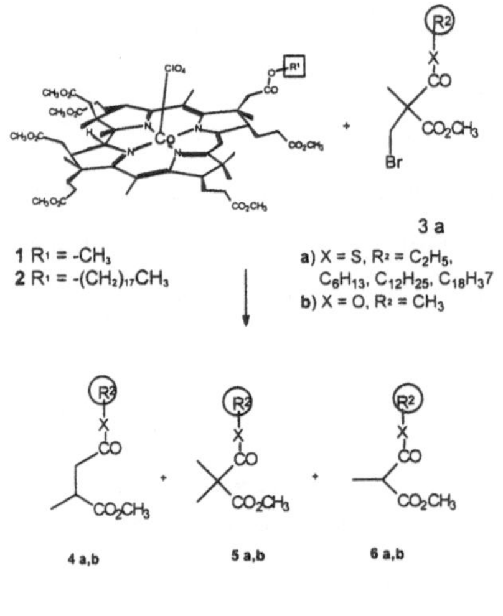

a) Product distribution, the chain length in the catalyst **1** , **2** and the solvent

Catalyst	Solvent	**4a**	**5a**	**3a**
none	CH_3OH-H_2O	-	trace	rec.
1		5.6	10	50
2		22	28	38
	CH_3OH	-	trace	-
	CH_3CN-H_2O	3	32	26
	CH_3CN	3	12	24

b) Product ratios and the chain length in the substrate **2a**

3a	**4a : 4b : 5a : 5b**
C_2H_5	0 : 2.3 :1.4 : 1
C_6H_13	0 : 1 :5 : 0
C_12H_25	1 : 2 :18 : 1
C_18H_37	1 : 1.3 .0 : 0

1 R_1 = -CH_3
2 R_1 = -(CH_2)_{17}CH_3

a) X = S, R_2 = C_2H_5, C_6H_13, C_12H_25, C_18H_37
b) X = O, R_2 = CH_3

3 a

4 a,b   5 a,b   6 a,b

Scheme 2. The electro-photocatalytic cycle for the methylmalonyl - succinyl rearrangement.

Table 1a,b. Product distribution for the catalytic reaction

With homochiral substrate (-)-(S)-**3a** extensively racemized succinate (S)-**4a** has been observed. In these experiments the potential was set at -0.85 C allowing the reduction of Co(II) to Co(I), but not the reductive cleavage of the Co-alkyl complex, formed as an intermediate. When the potential was lowered to -1.8 V, reductive cleavage of the intermediate Co-alkyl complex occurs and increased rearrangement is observed.

In order to explore the impact of H bond association between the catalyst and the substrate in apolar solvents, the *A-T base pairing model* has been developed (Scheme 3). Contrary to our expectations the cathodic CV of the catalyst in acetonitrile shows rather low reversibility. It can be improved by addition of methanol and is complete in pure methanol. Thus, it was not surprising, that a catalytic reaction could not be run under conditions, where H bonding is feasable. However, alkylation of the catalyst by the substrate has been achieved, when the catalyst was electrochemically reduced in methanol and treated with the substrate as shown. Photolysis in $CH_3CN/CHCl_3$ (1:1) gives the rearrangement and reduction product in a ratio of 2:1.

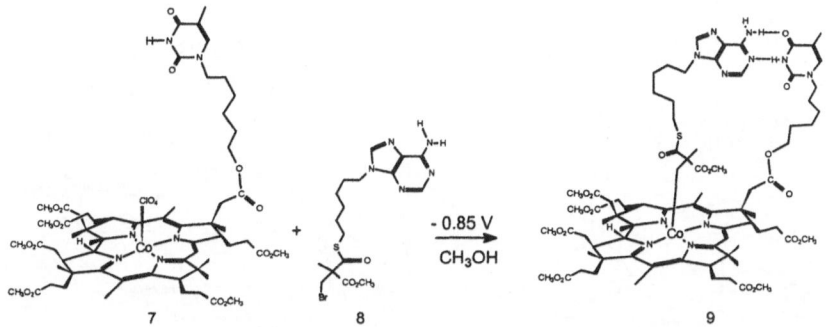

Scheme 3. A-T base pairing in the methylmalonyl-succinyl rearrangement.

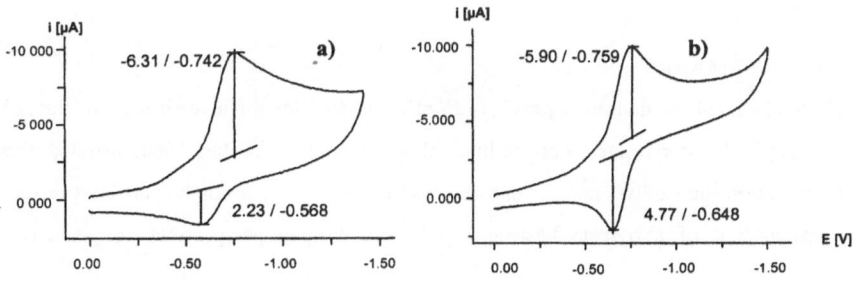

Scheme 4. CV for the solvent dependency of the thymine bearing B_{12} catalyst 7; a) in acetonitrile, b) in acetonnitril - methanol

For mechanistic studies and enhanced stereoselective recognition, fixation of the catalyst at the surface of an electrode was envisaged. As a first step towards this goal, the *electropolymerisation* of the corrin ring bearing a pyrrol head group in the periphery was studied.

Supported by complementary CV-investigations, the cyclic voltammmogram of **10a** being rather complex, is suggestive of polymerization. After the reduction Co(II) → Co(I) the complex undergoes oxidation to Co(II) and Co(III) and probably of the pyrrole group. The two anodic peaks, the decrease of the cathodic peak and the presence of a shoulder at low cathodic potential of the 2nd scan are in support of polymerisation. The increase of the current at the cathode in the 6th scan is taken as further evidence for polymer formation and thickening at the glassy carbon electrode.

The pyrrol derivative **10b** is rather resistent to polymerization and reacts at the anode only after reductive cleavage of the axial ligands.

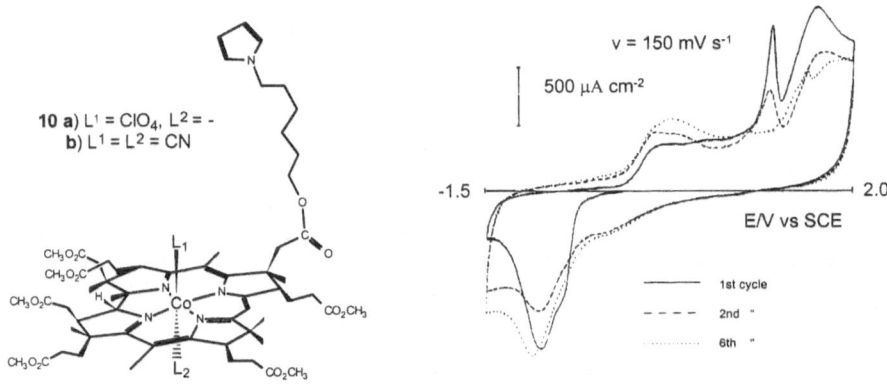

Scheme 5. Electropolymerization of a pyrrole bearing derivative of vitamin B$_{12}$.

CONCLUDING REMARKS

Electrochemical methodology provides an efficient tool for for investigation of the redox behaviour of vitamin B$_{12}$ derivatives. For the hydrophobic model it has been demonstrated, that recognition sites both in the catalyst and the substrat lead to increased rearrangement. First results in the electropolymerisation of a pyrrole bearing B$_{12}$ derivative open perspectives for selective redox catalysis.

References

1) Y. Murakami, Y. Hisaeda, T. Ohno, *J. chem Soc., Perkin 2*, **1991**, 405.
2) Y. Hisaeda, J. Takenaka, Y. Murakami, *Electrochimica Acta*, **1997**, *42*, 2165.
3) a) T. Darbre, A. Wolleb, R. Keese, V. Šiljegovic, *Helv. Chim. Acta* , **1996**, *79*, 2100; b) T. Darbre, R. Keese, V. Šiljegovic, *J. Chem. Soc. Chem. Commun.*, **1996**, 561.
4) P. Bonhôte, R. Scheffold, *Helv. Chim. Acta*, **1994**, *74*, 1425.

ELECTROCHEMICAL 1,2-MIGRATION REACTIONS MEDIATED BY VITAMIN B$_{12}$ DERIVATIVES

Yoshio HISAEDA
Department of Chemistry and Biochemistry, Graduate School of Engineering, Kyushu University, Hakozaki, Higashi-ku, Fukuoka 812-81, Japan

ABSTRACT

Electrolyses of an alkyl halide having two ester groups and one phenyl group on the β-carbon atom have been carried out under various conditions in the presence of a catalytic amount of the hydrophobic vitamin B$_{12}$. The migration of the pheny group was observed under the conditions forming a radical intermediate. The ester-migrated product was detected under the conditions forming an anionic intermediate. This is the first successful example in selecting a migrating group by electrolysis potential. A novel vitamin B$_{12}$ derivative, which has phenylalanine residues on the peripheral side chains, catalyzed effectively 1,2-migration of the carboxylic ester in 1-bromo-2,2-bis(ethoxycarbonyl)propane at −1.0 V vs. Ag/AgCl under irradiation conditions.

INTRODUCTION

Metalloenzymes are ingeniously designed catalysts existing in nature. Vitamin B$_{12}$-dependent enzymes, involving the cobalt species as a catalytic center, mediate various isomerization reactions leading to the intramolecular exchange of a functional group (X) and a hydrogen atom between neighboring carbon atoms as shown in Eq.1. These reactions have attracted much attention because of their novel nature from the viewpoints of organic and organometallic chemistry. Carbon-skeleton rearrangement reaction, mediated by methylmalonyl-CoA mutase, is shown in Eq. 2. Even though the real reaction mechanisms involved in the carbon-skeleton rearrangements have not been clarified up to the present time, radical mechanisms are considered to be the most plausible ones on the basis of ESR studies. In order to simulate the catalytic functions of vitamin B$_{12}$ as excerted in the hydrophobic active sites of enzymes concerned, we have been dealing with hydrophobic vitamin B$_{12}$ derivatives which have ester groups in place of the peripheral amide moieties of the naturally occurring vitamin B$_{12}$.[1] The present study demonstrates regioselective 1,2-migration of functional groups catalyzed by a hydrophobic vitamin B$_{12}$ under electrochemical conditions. We have also prepared a novel vitamin B$_{12}$ derivative, [Cob(II)7Phe(OBzl)]ClO$_4$, having phenylalanine residues on the peripheral side chains in order to enhance the catalytic efficiency in 1,2-migration reactions.

$$-\overset{\overset{\displaystyle X}{|}}{\underset{\underset{\displaystyle |}{|}}{C^1}}-\overset{\overset{\displaystyle H}{|}}{\underset{\underset{\displaystyle |}{|}}{C^2}}- \quad\Longleftrightarrow\quad -\overset{\overset{\displaystyle H}{|}}{\underset{\underset{\displaystyle |}{|}}{C^1}}-\overset{\overset{\displaystyle X}{|}}{\underset{\underset{\displaystyle |}{|}}{C^2}}- \qquad (1)$$

$$\begin{array}{c} CO_2H \\ | \\ H_3C-C-H \\ | \\ COS\text{-}CoA \end{array} \quad\xrightarrow{\text{methylmalonyl–CoA mutase}}\quad \begin{array}{c} CO_2H \\ | \\ H_2C-CH_2 \\ | \\ COS\text{-}CoA \end{array} \qquad (2)$$

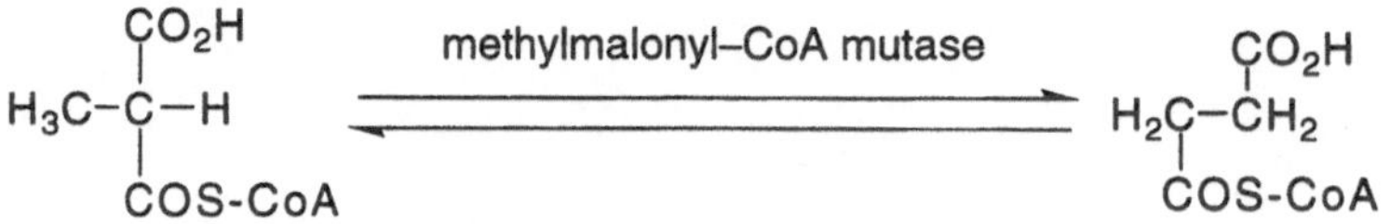

RESULT AND DISCUSSION

Redox Behavior of Hydrophobic Vitamin B12

We have previously reported on the redox chemistry of heptamethyl cobyrinate perchlorate, [Cob(II)7C$_1$ester]ClO$_4$, and pointed out that this complex is readily reduced to the univalent cobalt species of highly nucleophilic character by electrochemical means in nonaqueous media.[2] The redox potential for the Co(II)/Co(I) couple of [Cob(II)7C$_1$ester]ClO$_4$ in DMF was observed at −0.58 V vs. Ag/AgCl (−0.61 V vs. SCE).[3] As for [(CH$_3$)(H$_2$O)Cob(III)7C$_1$ester]ClO$_4$, an irreversible reduction peak was observed at −1.3 V vs. Ag/AgCl and assigned to the formation of the one-electron reduction intermediate.[4] The redox behavior of [(CH$_3$)(H$_2$O)Cob(III)7C$_1$ester]ClO$_4$ indicated that the cobalt-carbon bond is cleaved by electrochemical reduction. On the basis of the information, catalytic cycles were established as shown in Figure 1. An alkylated complex, generated by the reaction of a univalent cobalt complex and an alkyl halide, is generally decomposed by photolysis or electrolysis to afford reduction and/or rearrangement products.[5]

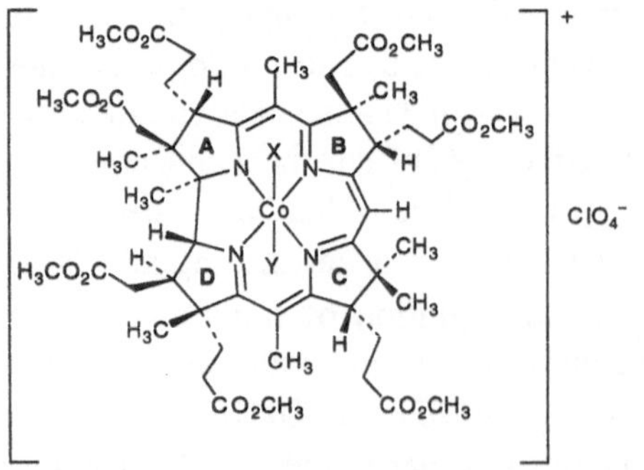

X=Y=None : [Cob(II)7C$_1$ester]ClO$_4$
X=CH$_3$, Y=H$_2$O :
    [(CH$_3$)(H$_2$O)Cob(III)7C$_1$ester]ClO$_4$

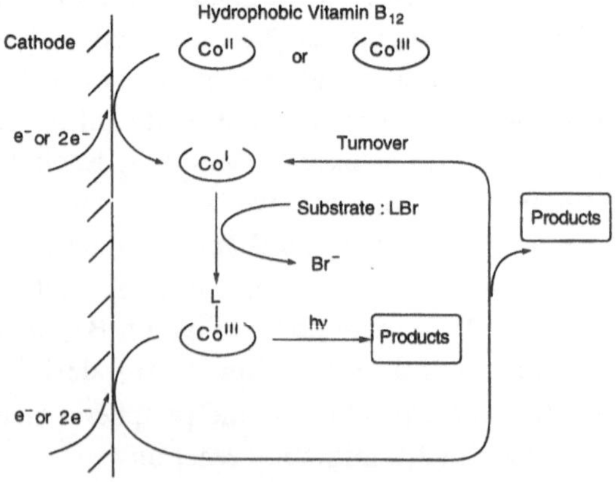

Fig. 1 Electroorganic reactions mediated by hydrophobic vitamin B$_{12}$

Regioselective 1,2-Migration of Functional Groups Catalyzed by Simple Hydrophobic Vitamin B12

A vitamin B$_{12}$ derivative is reduced to the Co(I) state in N,N-dimethylformamide (DMF) by the controlled potential electrolysis, and reacts with an alkyl halide to form the corresponding alkylated complex with a Co–C bond. This alkylated complex gives products after cleavage of the Co–C bond under irradiation with visible light. We have carried out the electrolysis of an alkyl halide having two ester groups and one phenyl group on the β–carbon atom in the presence of a catalytic amount of the hydrophobic vitamin B$_{12}$, [Cob(II)7C$_1$ester]ClO$_4$, in DMF. Products were analyzed by GLC and ^{1}H-NMR as shown below. The migration of the pheny group was observed under the conditions forming a radical intermediate; runs 1 and 2. The ester-migrated product was detected under the conditions forming an anionic intermediate, run 3. The electrolysis mechanism was investigated by various spectroscopic methods such as electronic spectroscopy and ESR spin-trapping experiments, and we proposed the electrolysis mechanism as shown in Scheme 1. This is the first successful example in selecting a migrating group by the electrolysis potential.

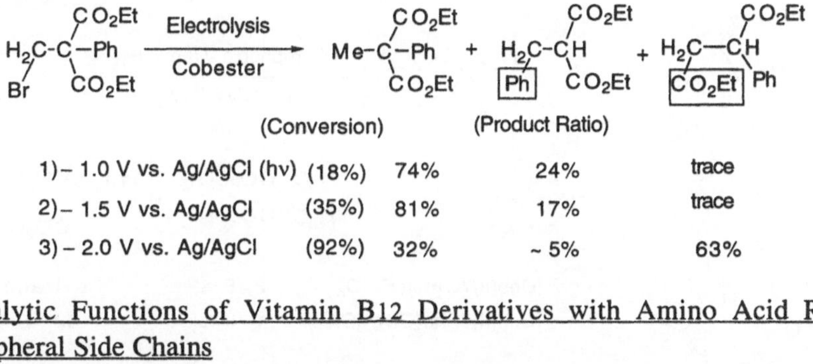

	(Conversion)		(Product Ratio)	
1) – 1.0 V vs. Ag/AgCl (hν)	(18%)	74%	24%	trace
2) – 1.5 V vs. Ag/AgCl	(35%)	81%	17%	trace
3) – 2.0 V vs. Ag/AgCl	(92%)	32%	~ 5%	63%

Catalytic Functions of Vitamin B12 Derivatives with Amino Acid Residues on Peripheral Side Chains

In order to enhance the catalytic efficiency in 1,2-migration reactions, we have prepared a vitamin B$_{12}$ derivative having phenylalanine residues on peripheral side chains. The electrolysis of 1-bromo-2,2-bis(ethoxycarbonyl)propane was carried out in the presence of vitamin B$_{12}$ derivatives in DMF. On using a vitamin B$_{12}$ derivative having phenylalanine residues, [Cob(II)7Phe(OBzl)]ClO$_4$, as catalyst, the ester migrated product was obtained as a major product at −1.0 V vs Ag/AgCl under irradiation conditions. On the other hand, when simple hydrophobic vitamin B$_{12}$, [Cob(II)7C$_1$ester]ClO$_4$, was used as catalyst, the product was only simple reduced one. It became clear from ESR measurements that these reactions proceed via radical intermediates. Peripheral amino acid residues may effectively assist 1,2-migration of the carboxylic ester.

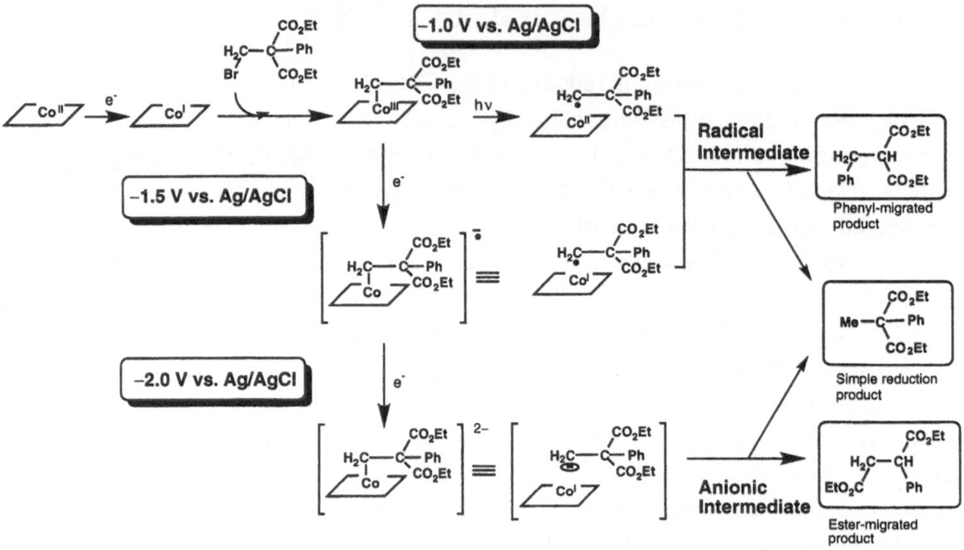

Scheme 1. Proposed Mechanism for the Electrolysis Catalyzed by Hydrophobic Vitamin B12

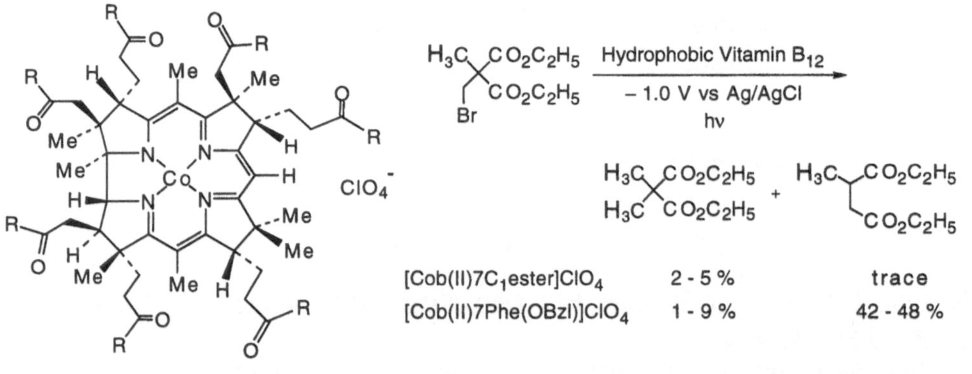

[Cob(II)7C₁ester]ClO₄	2 - 5 %	trace
[Cob(II)7Phe(OBzl)]ClO₄	1 - 9 %	42 - 48 %

R = L-Phe(OBzl) : [Cob(II)7Phe(OBzl)]ClO₄

References

1) Y. Hisaeda, J. Takenaka, and Y. Murakami, *Electrochimica Acta*, **1997**, *42*, 2165; Y. Murakami, Y. Hisaeda, T. Ohno, H. Kohno, T. Nishioka, *J. Chem. Soc., Perkin Trans. 2*, **1995**, 1175.

2) Y. Murakami, Y. Hisaeda, T. Ozaki, T. Tashiro, T. Ohno, Y. Tani, and Y. Matsuda, *Bull. Chem. Soc. Jpn.*, **1987**, *60*, 311.

3) Y. Murakami, Y. Hisaeda, A. Kajihara, and T. Ohno, *Bull. Chem. Soc. Jpn.*, **1984**, *57*, 405.

4) Y. Murakami, Y. Hisaeda, T. Tashiro, and Y. Matsuda, *Chem. Lett.*, **1985**, 1813 .

5) Y. Murakami and Y. Hisaeda, *Pure & Appl. Chem.*, **1988**, *60*, 1363; Y. Murakami, Y. Hisaeda, T. Tashiro, and Y. Matsuda, *Chem. Lett.*, **1986**, 555.

CATALYTIC ACTIVITY OF VITAMIN B_{12} DERIVATIVE HAVING AMINO ACID RESIDUES

Yoshio HISAEDA, Yoshiki INOUE, Kenshi ASADA, and Takuya NISHIOKA

Department of Chemistry and Biochemistry, Graduate School of Engineering, Kyushu University, Hakozaki, Higashi-ku, Fukuoka 812-81, Japan

ABSTRACT

A novel vitamin B_{12} derivative, [Cob(II)7Phe(OBzl)]ClO$_4$, having phenylalanine residues on the peripheral side chains was prepared. [Cob(II)7Phe(OBzl)]ClO$_4$ catalyzed effectively 1,2-migration of the carboxylic ester in 1-bromo-2,2-bis(ethoxycarbonyl)propane at -1.0 V vs. Ag/AgCl under irradiation conditions.

INTRODUCTION

Vitamin B_{12}-dependent enzymes, involving the cobalt species as a reaction site, catalyze various isomerization reactions leading to the intramolecular exchange of a functional group and a hydrogen atom between neighboring carbon atoms, for example methylmalonyl-CoA mutase as shown below. In order to simulate the catalytic functions of vitamin B_{12} as excerted in the hydrophobic active sites of enzymes concerned, we have been dealing with a hydrophobic vitamin B_{12}, [Cob(II)7C$_1$ester]ClO$_4$, which have ester groups in place of the peripheral amide moieties of the naturally occurring vitamin B_{12}.[1] In this work, we have prepared a novel vitamin B_{12} derivative, [Cob(II)7Phe(OBzl)]ClO$_4$, having phenylalanine residues on the peripheral side chains in order to enhance the catalytic efficiency in 1,2-migration reactions.

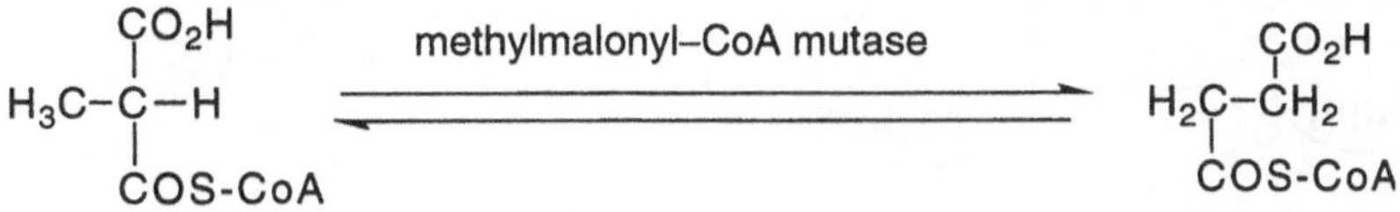

$$\underset{\text{COS-CoA}}{\overset{\text{CO}_2\text{H}}{\text{H}_3\text{C}-\text{C}-\text{H}}} \quad \xrightleftharpoons{\text{methylmalonyl-CoA mutase}} \quad \underset{\text{COS-CoA}}{\overset{\text{CO}_2\text{H}}{\text{H}_2\text{C}-\text{CH}_2}}$$

RESULT AND DISCUSSION

(CN)$_2$Cob(III)7Phe(OBzl) was prepared from (CN)$_2$Cob(III)7C$_1$ester[2] as shown in the below scheme. The compounds were analyzed by ^{1}H-NMR, IR, CD, UV-VIS, FAB-MS spectra as well as elemental analysis: MS (FAB), 2624 [M-CN]$^+$; Found: C, 70.80; H, 6.27; N, 6.56%. Calcd for C$_{159}$H$_{164}$CoN$_{13}$O$_{21}$•2H$_2$O; C, 71.04; H, 6.30; N, 6.77%. (CN)$_2$Cob(III)7Phe(OBzl) was transformed to [Cob(II)7Phe(OBzl)]ClO$_4$.

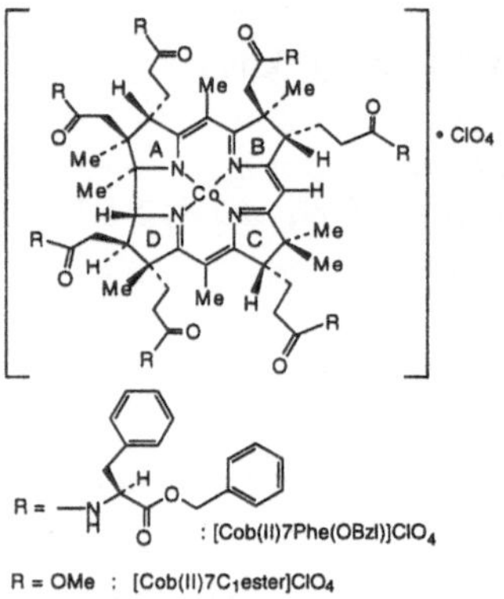

R = —N—... : [Cob(II)7Phe(OBzl)]ClO4

R = OMe : [Cob(II)7C1ester]ClO4

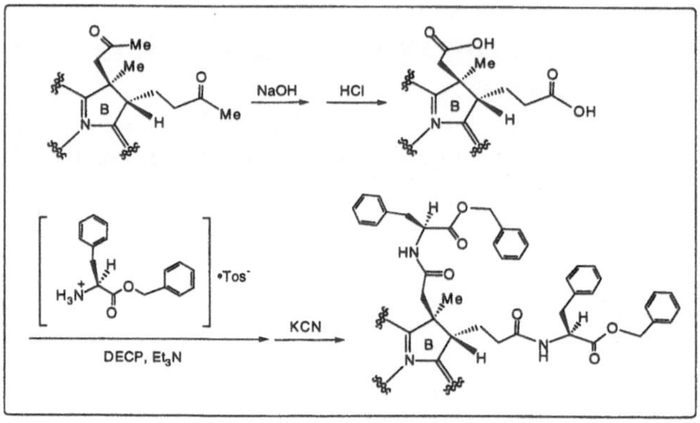

The electrolysis of 1-bromo-2,2-bis(ethoxycarbonyl)propane was carried out in the presence of vitamin B$_{12}$ derivatives in DMF. On using [Cob(II)7Phe(OBzl)]ClO$_4$ as catalyst, the ester migrated product was obtained as a major product at -1.0 V vs. Ag/AgCl under irradiation conditions. On the other hand, when [Cob(II)7C$_1$ester]ClO$_4$ was used, the product was only simple reduced one. ESR measurements indicate that these reactions proceed via radical intermediate. We propose that the reaction proceed via mechanism as shown below.

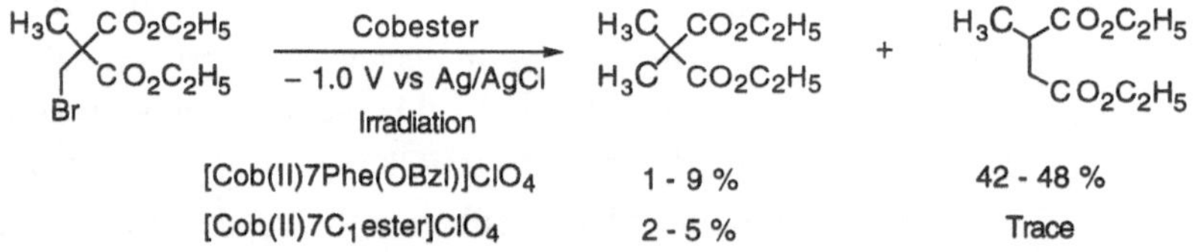

	1 - 9 %	42 - 48 %
[Cob(II)7Phe(OBzl)]ClO$_4$	1 - 9 %	42 - 48 %
[Cob(II)7C$_1$ester]ClO$_4$	2 - 5 %	Trace

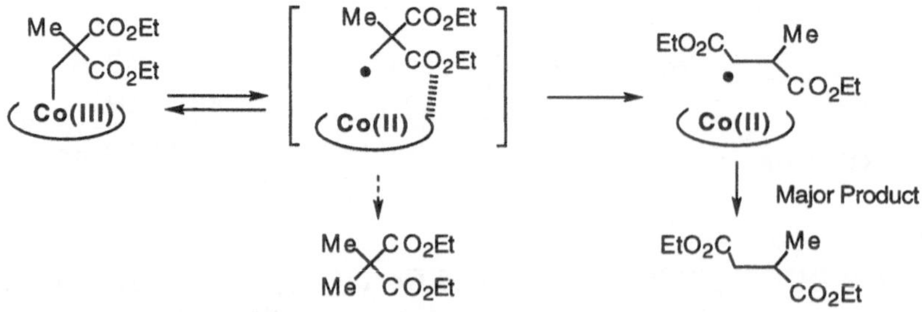

References
1) For example, Y. Hisaeda, J. Takenaka, and Y. Murakami, *Electrochimica Acta*, **1997**, *42*, 2165.
2) L. Werthemann, R. Keese, and A. Eschenmoser, unpublished results; see L. Werthemann, Dissertation, ETH Zürich (Nr. 4097), Zürich (1968); Y. Murakami, Y. Hisaeda, and A. Kajihara, *Bull. Chem. Soc. Jpn.*, **1983**, *56*, 3642.

ELECTRON TRANSFER REACTIONS OF SEMI-ARTIFICIAL BIOMOLECULES

Isao TANIGUCHI

Department of Applied Chemistry and Biochemistry, Faculty of Engineering, Kumamoto University, 2–39–1, Kurokami, Kumamoto 860, Japan

ABSTRACT

Using functional electrodes electrochemical responses of semi-artificially modified myoglobin and ferredoxin molecules have been examined to understand biological functions in more detail. Some reconstituted myoglobin molecules of which redox center was changed with artificially designed porphyrins showed that for determining the redox potential axial ligands are important when the porphyrin is incorporated inside the apomyoglobin. Change in metal ion of the redox center gave significant change in the electron transfer rate. For ferredoxin some evolutionary-conserved amino acid residues were changed to understand the roles of the amino acid ion biological functions using mutated molecules, and some particular amino acid residues were found to have distinguished roles in biological functions such as controlling the redox potential and as the binding site with enzymes.

INTRODUCTION

Recently, electron transfer reactions of metalloproteins at functional electrodes have become important in bioelectrochemistry not only for understanding biofunctions and bioelectrochemical reactions of metalloproteins but also for their applications [1, 2]. Various functional electrodes for metalloprotein electrochemistry are now available. Recently, surface functions of such electrodes have became more and more clear at the molecular level by using single crystal surfaces [3]. In addition, semi-artificial biomolecules are now easily prepared and are very useful to understand various biological functions. In the present paper, electron transfer reactions of semi-artificially modified myoglobins (Mb) and ferredoxins (Fd) have been demonstrated.

EXPERIMENTAL

The surfaces of In_2O_3 electrodes (ca. 5 x 5 mm, from Kinoene Optics Co., Japan) were cleaned up by ultrasonication in a New Vista (the anionic surfactant, AIC Corp.) solution followed by washing in ethanol and then in water. For both horse heart and sperm whale myoglobins, a stable redox waves at a highly hydrophilic surface of an In_2O_3 electrode was observed. The surface hydrophilicity of an In_2O_3 electrode affected very much the peak separation of the voltammogram of myoglobin, and only at fully hydrophilic surfaces (> 70 dyn/cm) of In_2O_3 electrodes were useful. The water layer of the electrode surface would suppress the strong adsorption of myoglobin to exclude denaturation of the protein. On the other hand, for ferredoxin electrochemistry, the positively charged surface of the electrode are effective. A poly-lysine modified electrode was prepared by immersing the In_2O_3 electrode into a 10 mM tris-HCl buffer solution (pH 7.2) containing 0.33 M NaCl and 1 mg/ml polypeptide (from Sigma at molar masses of ca. 25,000) of interest. Also, chemical modification of the In_2O_3 electrode surface with aminosilanes was also carried out by dipping the electrode into a 10 mM ethanol solution of the modifier of interest to give effective modified electrodes. These electrodes were applicable for various plant ferredoxins from different origins (e.g., spinach, maize and chlorella), but not for bacterial ferredoxins. Electrochemical measurements were carried out by using a BAS-100A Electrochemical Analyzer.

RESULTS AND DISCUSSION

For metMb, at the highly hydrophilic surface of an In2O3 electrode (surface tension of >70 dyn cm^{-1}) well-defined voltammograms for Mb were observed with the heterogeneous electron transfer rate constant, $k^{0'}$, of ca. 1-3 x 10^{-4} cm s^{-1} at pH 6.5 [4]. Similarly, for semi-artificially reconstituted Mbs showed well-defined voltammograms at the In2O3 electrode [5]. When nitrogen atom was introduced to hemin backbone to give mono-azahemin (see Figure 1), its formal redox potential, $E^{0'}$, shifted positively compared to hemin: e.g., the redox potential of monoazahemin in methanol shifted positively by ca. 70 mV and 100 mV compared to hemin and mesoheme XIII (see Figure 1). However, on the contrary to the positive shift in the redox potential of azahemins, the redox potentials of mono- and di-azahemin reconstituted Mbs shifted negatively than those expected from the redox potential of azahemins: e.g., the formal redox potential, $E^{0'}$, was evaluated to be -0.18 V vs. Ag/AgCl (Sat. KCl), which is more negative than that of native Mb and mesoheme reconstituted Mb by ca. 40 mV and 10 mV, respectively. These results suggest that the heme environment must be taken into account to consider its redox potential, when azahemins are reconstituted inside apoMb [5].

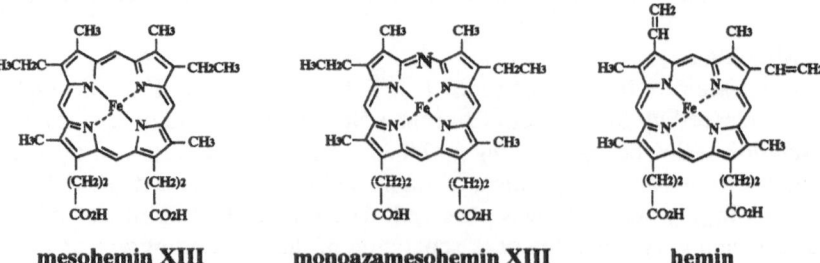

mesohemin XIII   monoazamesohemin XIII   hemin

Figure 1. Structures of artificially-designed and native (hemin) redox centers reconstituted into apomyoglobin.

For β, δ-diazamesoporphyrin III (diazahemin) reconstituted myoglobin (diazaMb), the distal histidine is the sixth ligand of the heme (so-called hemichrome type), and the observed voltammogram showed that the reaction was quasi-reversible and the ko' value of ca. 2×10^{-3} cm s^{-1} was obtained, which is one-order of magnitude larger than that of native and monoazahemin reconstituted Mbs.

Moreover, in the present study, for manganese reconstituted myoglobin, where no water molecule is coordinated to Mn (III) ion, a well-defined redox potential (- 0.32 V vs. Ag/AgCl) was obtained using a spectroelectrochemical technique, but the ko' value was found to be very small [6]. Also, tetramethyl- octamethyl- and ethio-hemin reconstituted myoglobins, where the structures of heme ligand were changed to give different stability of the heme moiety in myoglobin molecule, were also prepared [7]. The observed results on the $E^{0'}$ and ko' values showed that the heme environment is very important for the electron transfer reaction and their biological functions.

Maize ferredoxins from photosynthetic (FdI) and non-photosynthetic (FdIII) organs and those of which particular amino acid residues had been modified by site-directed mutagenesis were used. All mutated ferredoxin prepared showed very similar circular dichroism spectra to that of wild type (WT), and showed well-defined cyclic voltammograms at the modified electrodes [8]. Electrochemical study showed that particular evolutionary conserved amino acid residues have distinguished roles in biological functions. For example, when Ser-46 of FdIII or Ser-45 of FdI was modified to glycine (Gly), (S46G-FdIII or S45G-FdI), the redox potential showed a large positive-shift by ca. 180 mV compared to those of WT [9] as shown in Figure 2.

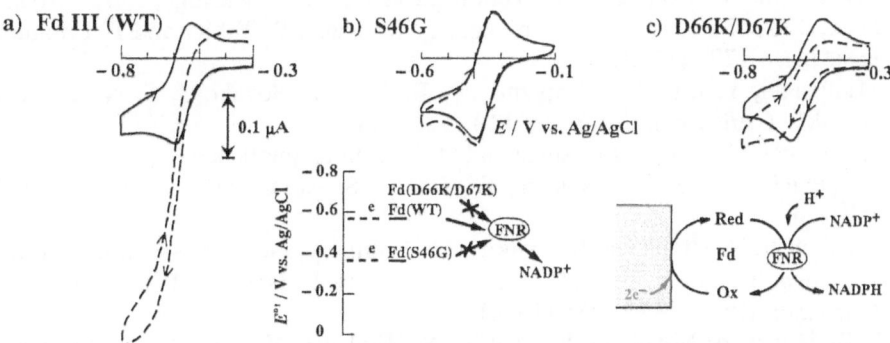

Figure 2. Cyclic voltammograms of a) native (WT), and b) ; c) mutated ferredoxins in the absence (solid curves) and presence (broken curves) of 0.05 U/ml FNR and 0.25 mM NADP$^+$. Scan rate: 2 mV/s. S46G: Serine (S)-46 of meize ferredoxin (FdIII) was changed to Glycine (G). D66K/D67K: Negatively charged Aspartic acid (D) at both 66 and 67 positions were changed to positively charged Lysine (K)

The redox potentials of D66K/D67K and D66N/D67N, where negatively charged aspartic acid (D) was converted to positively charged lysine (K) or neutral asparagine (N), did not change at all (see Figure 2). However, using the ferredoxin/ ferredoxin-NADP$^+$-reductase (FNR)/ NADP$^+$ system, no catalytic current for the reduction of NADP$^+$ to give NADPH was observed, suggesting that D66 and/or D67 are the binding sites with FNR to form the ferredoxin-FNR complex [9].

For bioelectroorganic reactions, thermostable (up to 60 °C) chlorella ferredoxin [10] is useful. With the aid of enzyme reactions, where NADPH is the co-factor, various bioorganic reactions can be designed [11]: e.g., L-malic acid was obtained from pyruvic acid with carbon dioxide fixation. L-glutamic acid was also obtained from oxoglutaric acid by introducing of ammonia.

These results suggests that for biological studies, electrochemical measurements are very useful to understand the biological functions of native and artificially modified metalloproteins, and electron transfer reactions of semi-artificial biomolecules are very promising for further developments in basic bioelectrochemistry and its applications.

References
1) I. Taniguchi, "Redox Mechanisms and Interfacial Properties of Molecules of Biological Importance," Eds. by F.A. Schultz and I. Taniguchi, Electrochemical Society, Inc., p. 9, Pennington, (1993), and references cited therein.
2) F. M. Hawkridge and I. Taniguchi, *Comments on Inorg. Chem.*, **17**, 163 (1995).
3) I. Taniguchi, S. Yoshimoto and K. Nishiyama, *Chem. Lett.*, 353 (1997).
4) I. Taniguchi, K. Watanabe, M. Tominaga and F.M. Hawkridge, *J. Electroanal. Chem.*, **333**, 331 (1992); M. Tominaga, T. Kumagai, S. Takita and I. Taniguchi, *Chem. Lett.*, 1771 (1993).
5) I. Taniguchi, Y. Mie, K. Nishiyama, V. Brabec, O. Novakova, S. Neya and N. Funasaki, *J. Electroanal. Chem.*, **420**, 5 (1997).
6) I. Taniguchi, C.-Z. Lee, M. Ishida and Q. Yao, to be published.
7) I. Taniguchi, Y. Mie, K. Sonoda, E. Krestyn S. Neya and N. Funasaki, to be published.
8) I. Taniguchi, Y. Hirakawa, K. Iwakiri, M. Tominaga and K. Nishiyama, *J. Chem. Soc., Chem. Commun.*, 953 (1994); K. Nishiyama, H. Ishida and I. Taniguchi, *J. Electroanal. Chem.*, **373**, 255 (1994).
9) I. Taniguchi, A. Miyahara, K. Iwakiri, Y. Hirakawa, K. Hayashi, K. Nishiyama, T. Akashi and T. Hase, *Chem. Lett.*, 929 (1997).
10) I. Taniguchi, K. Iwakiri, K. Nishiyama and T. Matsubayashi, *Denki Kagaku (J. Electrochem. Soc. Jpn.)*, **63**, 1191 (1995).
11) I. Taniguchi, "Novel Trends in Electroorganic Synthesis," Kodansha-Elsevier, p.393 (1995); *Ext. Abst. of the 185th Meeting of Electrochem. Soc.*, **94-1**, 1135 (1994).

ULTRAMICROELECTRODES:
THEIR USE IN SEMI-ARTIFICIAL SYNAPSES

Christian **AMATORE**,* Stéphane **ARBAULT**, Neso **SOJIC**
and Monique **VUILLAUME**
Ecole Normale Supérieure, Département de Chimie. URA CNRS 1679
24 rue Lhomond. 75231 Paris Cedex 05. France

ABSTRACT

Among their numerous other important properties, ultramicroelectrodes allow fundamental biological events to be monitored in real time at the single cell level. This is explained and discussed here based on the presentation of the first kinetic characterization of oxidative stress response from living cells.

INTRODUCTION

Among other mechanisms, communication between cells occurs through the release of particular molecules followed by their specific detection by the target cells. These messengers can be released into a restricted volume (*e.g.*, a synapse) to avoid diffusion of information to other cells, or into biological fluids (*e.g.*, in the blood circulation) or even in the extra-body environment (as, *e.g.*, with pheromones). Despite such important differences, a common factor is that these chemical molecules must be released by the emitting cell outside its own cytoplasm. From a *physico-chemical point of view*, two main release mechanisms can be considered. When the chemical messengers already exist in the cell before their emission (*e.g.*, neurotransmitters) the emission procedure is termed exocytosis; it is conditioned in part by the physico-chemical processes allowing the transfer of the messenger through the cell membrane. Alternatively, the messenger may be created at the moment of its emission by the cell directly into the extracellular fluid, using enzymes or proteins that are present in / on its membrane. Here, we wish to focus on this second aspect, in the context of the so called *oxidative stress*.

PRINCIPLE AND ROLE OF SEMI-ARTIFICIAL SYNAPSES

Because an ultramicroelectrode may only probe a volume that is comparable to its own size, it behaves like a flux-microscope observing concentration changes in its vicinity, as an optical-microscope displays pictures corresponding only to the focused region. The information provided by an ultramicroelectrode is thus essentially dynamic since it derives from variations in the nature and intensities of fluxes of chemical messengers in the space region where it is « focused ».

Positioning an ultramicroelectrode near a cell amounts to create a semi-artificial synapse (Figure 1). Variations of electrochemical responses at the ultramicroelectrode (the artificial synaptic side) reflect the variations of fluxes emitted by the cell (the natural synaptic side) in the extracellular fluid range limited by the cell and the active surface of the ultramicroelectrode. Cell emissions correspond to extremely small amounts of molecules: if ten million molecules is an extremely small number by chemical standards (*ca.* 20 attomoles!), it is extremely huge by biological ones. Precise electrochemical measurements of such quantities are nevertheless possible because of the considerable restriction of the volume in which these small quantities are released. For example, release of ten million molecules in an artificial synaptic cleft of *ca.* $10\mu m \times 10\mu m \times 1\mu m$ (*viz.*, Figure 1 for a 1μm electrode-membrane distance with an electrode tip and a cell diameter of 10 μm) corresponds to an average concentration rise of 0.2 mM, a variation which can be measured with extreme accuracy by most electrochemical methods [1]. This further supports the above analogy with biological synapses; *Nature* uses the same « trick » in synapses: a vanishingly small number of molecules released in an adequately small volume ensures a reliable and accurate detection because this results in a large concentration pulse.

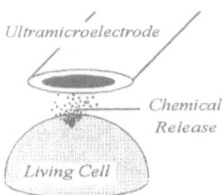

Figure 1. Schematic representation of a semi-artificial synapse. The ultramicroelectrode is positioned above a living cell maintained in its aerobic environment. Nota Bene: the respective size of the cell and of the ultramicroelectrode are roughly conserved on the drawing.

MONITORING AN OXIDATIVE STRESS BURST.

A. *The Biological Problem.* Aerobic cells derive their energy from controlled combustion catalyzed by transition metal based enzymes. Unfortunately, the same enzymes are also good reducing agents, prone reduce dioxygen (6-8% yield) [2]. At physiological pH, $O_2^{\cdot-}$ disproportionates spontaneously or through catalysis by superoxide dismutase (*SOD*) into dioxygen and hydrogen peroxide (Scheme 1). H_2O_2 is extremely hazardous since its life-time is sufficient to allow its diffusion to almost any cellular compartment, where it acts as a source of hydroxyl radicals (Fenton reaction).

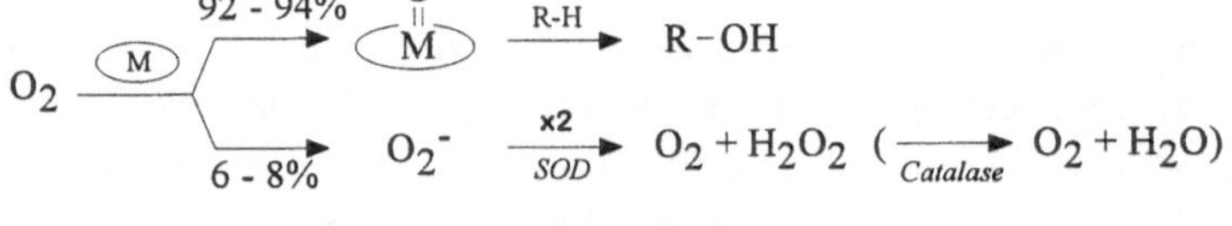

$$O_2 \xrightarrow{\;M\;} \begin{cases} \xrightarrow{92-94\%} \overset{O}{\overset{\|}{M}} \xrightarrow{R-H} R-OH \\ \xrightarrow[6-8\%]{} O_2^{-} \xrightarrow[SOD]{x2} O_2 + H_2O_2 \; (\xrightarrow[Catalase]{} O_2 + H_2O) \end{cases}$$

Scheme 1

Being among the best hydrogen atom acceptors, hydroxyl radicals are prone to induce a large variety of biological mutations, and to disrupt the cohesion of cellular membranes *via* radical catalyzed peroxidation of cell bilipids (Haber-Weiss reaction) [2]. In living cells these lethal routes are prevented by catalase which catalyzes the disproportion of H_2O_2 into dioxygen and water. However, catalase is inhibited by an excess of its substrate, so that it can perform efficiently only within the range of steady physiological concentrations of H_2O_2 (*viz.*, nM or less). To be protected against sudden rises of H_2O_2 levels, a cell must then express more efficient H_2O_2 scavengers [3]. This occurs *via* genetic expression of proteins and requires therefore time scales of 10-30 min at least. During this period the cell remains unprotected. This period of latency is used by cells to « fight » (*e.g.*, phagocytosis) or to « defend » themselves.

Indeed, a cell may change its normal dioxygen metabolism so as to produce spontaneously important quantities of superoxide [4] which provokes eventually a sudden rise of hydrogen peroxide levels into its extracellular environment. This important and ubiquitous phenomenon (the *oxidative stress* or *oxidative burst*) has been first anticipated and then extensively investigated by biologists but, it must be stressed, based only upon observation of cell metabolites. However, there was no indication about the intensity or about the time scale of the cellular response. Moreover, based on clinic observations and metabolites, a dysregulation of this mechanism is suspected to be involved directly or through apoptosis [5] in several human pathologies (aging, cancers, Parkinson and Alzheimer diseases, auto-immune pathologies, arthritis, *etc.*) [2-5], and has been shown more recently to play a crucial (yet still to be elucidated) role in the T cell depletion observed in AIDS phatogenesis [6].

B. Monitoring an Oxidative Stress response at a Single Living Cell. Mechanically induced oxidative stress at fibroblasts and lymphocytes has been investigated using the above semi-artificial synapses (Figure 1), the electrode tips having diameters ranging from *ca.* 20 μm to 2 μm as a function of the cell dimension [7]. The oxidative stress response is stimulated by intruding into the cell membrane with a sealed patch-clamp micropipette (0.2 μm ∅). This does not provoke lethal cell damages, being comparable to experiments in which genetic material is delivered into cells using identical micropipettes.

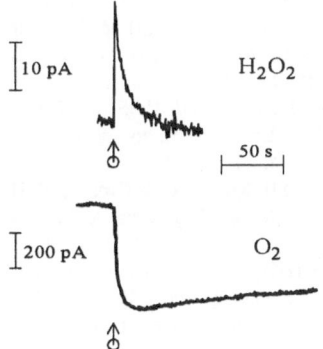

Figure 2. Oxidative stress responses stimulated in a human fibroblast by the intrusion of a patch-clamp micropipette into its membrane. The responses are monitored through an ultramicroelectrode positioned ca. 1 μm above the point of puncture as schematized in Figure 1. Top: response observed when the electrode potential is poised on the hydrogen peroxide oxidation wave (0.6 V vs Ag/AgCl). Bottom: Simultaneous variation of the reduction current of dioxygen. In both cases the moment of puncture is indicated by the vertical arrow.

Less than 0.1 s after penetrating the cell membrane, an intense burst of hydrogen peroxide is detected by the ultramicroelectrode (Figure 2, top). In human fibroblasts, this intense signal continues for several tens of seconds and, overall, corresponds to the release of *ca.* one hundred million H_2O_2 molecules, resulting in the instant building up of a few millimolar concentration of H_2O_2 near the cell surface. Connection of this oxidative burst with a sudden change of the cell dioxygen metabolism is indicated by the fact that (i) this intense burst disappears when the cell is exposed to anaerobic conditions during several minutes prior to its stimulation, or (ii) that the steady state concentration of dioxygen around the cell (monitored with a different electrode, Figure 2, bottom) rapidly drops off in phase with the sudden rise of H_2O_2, indicating a concomitant acceleration of the cell respiration.

The above time-resolved experiments [7] established for the first time that the oxidative burst is an intense and immediate response (less than 0.1 s) and lasts for several tens of seconds. Such a rapidity of the response rules out any involvement of protein expression mechanisms since this would have required several tens of minutes at least before a significant H_2O_2 release could occur. In other words, this and the simultaneous intense consumption of dioxygen strongly support the involvement of a pre-existing specialized enzyme in the cell external membrane.

ACKNOWLEDGMENTS

This work was supported in part by CNRS (URA 1679 and Ultimatech), ENS and the European Community (BioMed 2). We wish to thank P. Pantano whose strong resolution and great experimental skills were crucial in the initial stages of this research.

REFERENCES

1) C. Amatore, *Electrochemistry at Ultramicroelectrodes*, in *Physical Electrochemistry*, Rubinstein I. Ed., Marcel Dekker, New York. **1995**. pp. 131-208.

2) (a) B. Halliwell, J. M. C. Gutteridge, *Free Radicals in Biology and Medecine*, Clarendon Press, Oxford; (b) B. Chance, in *Annual Review of Biophysics and Biophysical Chemistry*, Vol. 15, Chap. 3, **1991**, pp. 267-353; (c) D. Engleman, Ed., *Annual Reviews*, Palo Alto, **1989**; (d) H. Sies, *Metabolic Compartimentation*, Academic Press, New York, **1982**; (e) W. Denk, J. H. Strickler and W.W. Webb, *Science*, **1990**, *248*, 73; (f) P.A. Cerutti, *Science*, **1985**, *227*, 375; (g) M. Vuillaume, *Mutat. Res.*, **1987**, *189*, 43; (h) J. S. Scandalios, Ed., *Molecular Biology of Free Radical Scavenging Systems*, Cold Spring Harbor Press, Cold Spring Harbor, **1992**.

3) (a) M.F. Christman, R.W. Morgan, F.S. Jacobson, and B.N. Ames, *Cell*, **1985**, *41*, 735; (b) D. Darr and I. Fridovich, *Free Radicals Med.*, **1995**, *18*, 195; (c) W. Löntz, A. Sirsjö; W. Liu, M. Lindberg, O. Rollman and H. Törmä, *Free Radicals Biol. Med.*, **1995**, *18*, 349.

4) (a) A. W. Segal and O. T. Jones, *Nature*, **1978**, *276*, 515; (b) B. Meier, A. R. Cross, J.T. Hancock, F. J. Kaup and O. T. Jones, *Biochem. J.*, **1991**, *275*, 245; (c) B. Meier, A. Jesaitis, A. Emmendorffer, J. Roesler and M.T. Quinn, *Biochem. J.*, **1993**, *289*, 481.

5) For a survey and discussion of the relationship between H_2O_2 and apoptosis see *e.g.*: J.F. Torres-Rocca, H. Lecoeur, C. Amatore and M.-L. Gougeon, *Cell Death and Differ.*, **1995**, *2*, 309.

6) See *e.g.*: (a) M.-L. Gougeon, R. Olivier, S. Garcia, D. Guetard, T. Dragic, C. Dauguet and L. Montagnier, *C.R. Acad. Sci. Paris*, **1991**, *312*, 529; (b) L. Meyaard, S.A. Otto, R.R. Jonker, M.J. Mijnster, R. Keet and F. Miedema, *Science*, **1992**, *257*, 217; (c) M.L. Gougeon, *Cell Death and Differ.*, **1995**, *2*, 1.

7) (a) S. Arbault, P. Pantano, J.A. Jankowski, M. Vuillaume and C. Amatore, *Anal. Chem.*, **1995**, *67*, 3382; (b) S. Arbault, M. Edeas, S. Legrand-Poëls, N. Sojic, C. Amatore, J. Piette, M. Best-Belpomme, A. Lindenbaum, and M. Vuillaume, *AIDS Sciences*, **1997**, *4*, 17. (c) S. Arbault, P. Pantano, N. Sojic, C. Amatore, M. Best-Belpomme, A. Sarasin, M. Vuillaume. *Carcinogenesis*, **1997**, *18*, 569. (d) For a more extensive review on this topic as well as on other biological applications related to neurotransmitters release see: C. Amatore. *C.R. Acad. Sci. Paris, Ser. II b*, **1996**, *323*, 757.

POLYPEPTIDES ENDOWED WITH ARTIFICIAL REDOX FUNCTIONS

Masahiko SISIDO, Teruhiko MATSUBARA, and Hiroaki SHINOHARA
*Department of Bioscience and Biotechnology, Faculty of Engineering,
Okayama University, 3-1-1 Tsushima-naka, Okayama 700, Japan*

ABSTRACT
Polypeptides carrying 2-anthraquinonyl groups or ferrocenyl groups were synthesized. Poly(L-2-anthraquinonylalanine) was found to form left-handed α-helical main chain and the anthraquinonyl side groups were arranged helically along the main chain. The anthraquinonyl polypeptide showed a unique redox property that was attributed to fast electron migrations along the anthraquinonyl side groups. Poly(L-ferrocenylalanine) was also synthesized. The latter polypeptide was found to take randomly-coiled conformations.

INTRODUCTION
Incorporation of nonnatural amino acids into peptides and proteins is a promising approach to introduce artificial functions into biomolecules.[1,2] Amino acids carrying redox functional groups have been synthesized and they have been incorporated into peptides and proteins through chemical and biochemical techniques. In this paper, synthesis and electrochemical properties of polypeptides carrying redox side groups, i.e., poly(L-2-anthraquinonylalanine) and poly(L-ferrocenylalanine) are described.

SYNTHESIS AND CONFORMATION OF POLY(L-2-ANTHRAQUINONYL-ALANINE)[3]
N-Acetyl-L-2-anthraquinonylalanine was synthesized from 2-chloromethyl-anthraquinone and optically resolved with acylase to give the corresponding free amino acid.   The amino acid was converted to the corresponding N-carboxyanhydride (NCA) by using phosgene dimer. The NCA was polymerized with aminoethyl poly(ethylene glycol) to produce a block copolymer containing a poly(L-2-anthraquinonylalanine) block (I). The average number of anthraquinonylalanine unit per polymer was 10.

H-(NHCHCO)$_m$—NHCH$_2$CH$_2$(OCH$_2$CH$_2$)$_n$OCH$_3$ (I)
CH$_2$

$n \approx 113$, $m \approx 10$

CD spectrum of the polypeptide in trimethyl phosphate (TMP) showed strong positive and negative peaks at the amide and anthraquinone absorption bands, indicating some regular conformation both in the main chain and in the side chain. The CD profile at the amide band was different from that of right-handed α-helix or β-sheet. Conformational energy calculations and theoretical CD computations suggested that the polypeptide takes a left-handed α-helical main chain with the side-chain anthraquinonyl groups being regularly arranged along the helix. The computer-predicted conformation is illustrated in Figure 1. The center-to-center distance between the adjacent anthraquinonyl units (7.9 Å) is close enough to achieve fast electron hopping among the pendant anthraquinonyl groups.

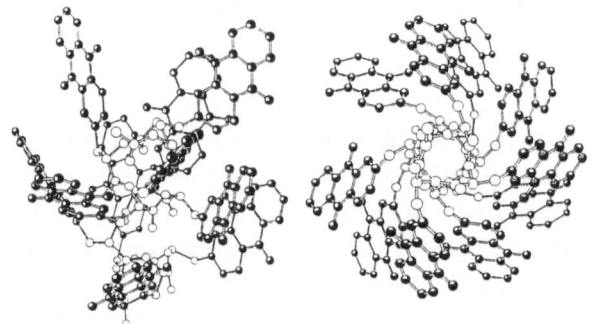

Fig.1. Computer-predicted conformation of poly(L-2-anqAla)

ELECTROCHEMICAL PROPERTIES OF POLY(L-2-ANTHRAQUINONYL-ALANINE)[4]

Electrochemical properties of the polypeptide was studied with cyclic voltammetry (CV) and differential pulse voltammetry (DPV) in 0.1M Bu_4NBF_4/DMF. The DPV diagram (Figure 2) showed the first reduction peak at about the same position (-1.11 V vs Ag/Ag$^+$) with about the same width as the monomeric model compound (Ac-anqAla-OMe). Contrary to the first reduction peak, no clear peak was observed for the polypeptide around the second reduction peak of the monomer model.

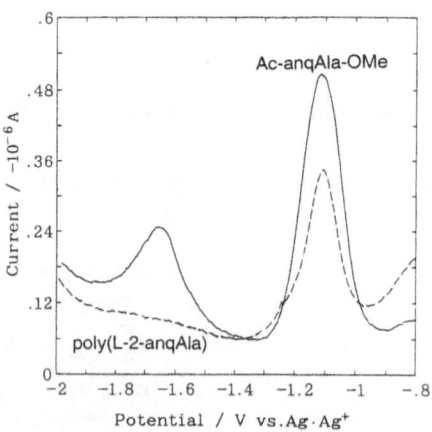

Fig.2. DPV diagrams of Ac-anqAla-OMe and poly(L-2-anqAla) in DMF/0.1M Bu_4NBF_4

Absorption spectra during the electro-chemical reduction of the monomer model polypeptide were measured using an optically transparent electrochemical cell. The monomer model shows spectrum of monoanions (AQ⁻·) after the electrochemical reduction under -1.0 V. On the other hand, the polypeptide showed spectrum of dianions (AQ²⁻) under the same potential. The results of DPV and absorption spectroscopy indicate that dianions are formed on the polypeptide under conditions where monoanions are formed in the monomer model. This may be interpreted in terms of a fast electron migration among the anthraquinonyl side groups of the polypeptide, followed by an efficient disproportionation to form dianions.

$$AQ^{-\cdot} + AQ \longrightarrow AQ + AQ^{-\cdot} \quad \text{(electron migration)}$$
$$AQ^{-\cdot} + AQ^{-\cdot} \longrightarrow AQ^{2-} + AQ \quad \text{(disproportionation)}$$

The efficient disproportionation indicates that an electron that was added to the polypeptide from electrode rapidly migrates among the helically-arranged anthraquinonyl groups. The helical peptide is, therefore, a promising candidate for a "**molecular wire**" that may transport electrons between electrodes and, for example, proteins.

The conformation during the redox process was examined with CD spectroscopy using the optically transparent electrochemical cell. The CD profiles before and after the electrochemical reduction are shown in Figure 3. The CD peaks at the anthraquinonyl absorption band shifted by the formation of dianions and returned to the original spectrum when neutral form was recovered. CD spectrum at the amide absorption band showed no change during the redox process, indicating the helical conformation is retained in the dianion form of the anthraquinonyl side groups.

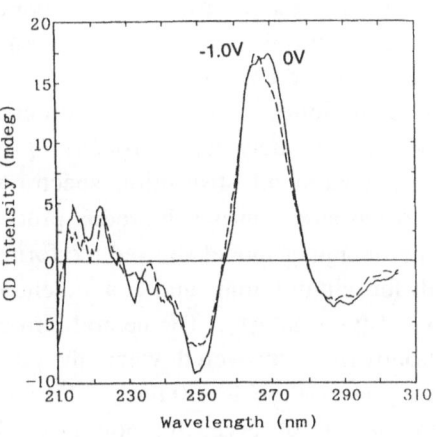

Fig.3 Electrochemical CD spectra of poly-(L-2-anqAla) in AN/0.1M Bu4NBF4 under 0V and -1V

SYNTHESIS, CONFORMATION, AND ELECTROCHEMICAL PROPERTIES OF POLY(L-FERROCENYLALANINE)

A novel nonnatural amino acid carrying a ferrocenyl group, L-ferrocenyl-alanine, was synthesized from ferrocenetrimethylammonium iodide[5] and converted to the corresponding NCA derivative. The latter was polymerized in the form of a block copolymer with poly(ethylene glycol) (II) The degree of polymerization of the poly(L-ferrocenylalanine) [poly(L-ferAla)] block was 8 and 21. .CD spectra of poly(L-ferAla) (n-21) in TMP showed a profile similar to right-handed α-helical conformation, but the intensity was much weaker than that of 100% helical conformation. The random-rich conformation is presumably due to the steric repulsion between bulky ferrocenyl side groups.

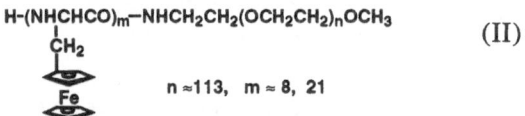

(II)

The DPV diagrams of poly(L-ferAla) and Ac-L-ferAla as the monomer model compound in TMP/0.1M Bu$_4$NBF$_4$ are shown in Figure 4. The polypeptides show reversible redox processes at about the same potential as that of the monomer model. In the case of n=21, the DPV peak is broadened, indicating electronic interactions between ferrocenyl groups. Electrochemical absorption spectra were also measured during the redox processes. The monomer model showed the formation of ferricinium ions under a potential of 0.6 V (vs Ag/Ag$^+$). The neutral ferrocenyl groups were recovered when the potential was set to -0.6V for 10 min. Similar be-

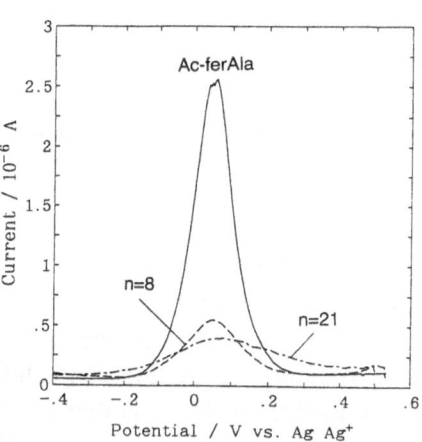

Fig.4. DPV diagrams of Ac-L-ferAla, and poly(L-ferAla) (n=8, 21) in TMP/0.1M Bu$_4$NBF$_4$

havior was observed for poly(L-ferAla) with lower yield of the ferricinium ions, probably due to slower diffusion rate of the polypeptide.

References
1) M. Sisido, *Prog. Polym. Sci.*, **1992**, *17*, 669.
2) M. Sisido, *Adv. Photochem.*, **1997**, *22*, 197.
3) T. Matsubara, H. Shinohara, and M. Sisido, *Macromolecules*, **1997**, *30*, 2651.
4) H. Shinohara, T. Matsubara, and M. Sisido, *Macromolecules*, **1997**, *30*, 2657.
5) M. Kira, T. Matsubara, H. Shinohara, and M. Sisido, *Chem. Lett.*, **1997**, 89.

SOME ASPECTS OF DEVELOPMENT IN DIRECT CATHODIC DEPROTONATION

Vladimir A. PETROSYAN

*N.D. Zelinsky Institute of Organic Chemistry, Russian Academy of Sciences,
Leninsky prospect 47, Moscow 117913, Russia*

ABSTRACT

Data on direct deprotonation of different types of organic acids were observed. It was noted that from the formal view point direct deprotonation was good in amperostatic electrolysis of weak acids reduced at the discharge potential of a supporting electrolyte where their latent indirect deprotonation occured. Such processes were called "Cathodic Deprotonation in Simple Systems" (CDSS). Characteristic features of CDSS processes were discussed and new facts of their realization were given.

INTRODUCTION

Carb- and heteroanions are fundamental particles in organic chemistry. This is the reason of a permanent interest to development of new approaches including electrochemical ones for the anion generation.

In the beginning of 70ies it was shown that some anionic particles generated at the cathode in nonaqueous media could be used as electrogenerated base. Such processes of organic acid indirect deprotonation play an important role in the modern electroorganic chemistry. It should be noted that at the mentioned period the first data about organic acid direct deprotonation under reduction in nonaqueous media at metal cathodes were published (for example [1-3]). Although individual communications appeared in literature up to the present, systematically such investigations were started in 80ies by two research groups in Russia.

In contrast to electrogenerated base method the processes of direct cathodic deprotonation were not widely discussed. At the same time it was not difficult to demonstrate synthetic attraction of such processes and development of a direct method gave perspectives of its use as a universal approach.

DISCUSSION and METHODS

First of all the question comes: "Is the direct cathodic deprotonation really direct and corresponds to the scheme (1)?"

(1)

This mechanism is in agreement with data of many experiments. Generalization of data obtained permits to indicate some characteristic features. So, the controlled potential electrolysis usually requires ~1F·mol^{-1} and the yield of a corresponding acid anion is rather high. The synthetic results are in agreement with voltammetric data. Usually one-electron cathodic wave of an acid reduction and anodic wave of corresponding anion oxidation were observed. These regularities are characteristic for electrochemical behaviour of primary nitramines (Pt, Bu$_4$NClO$_4$, AN) as we showed it recently [4]; for behaviour of nitrobenzylcyanide (Pt, Et$_4$NClO$_4$, AN) [1]; succinimide (Pt, Bu$_4$NF, DMF) [2]; 1,3-propanedithiol (Au, MeNClO$_4$, DMF) [3] and for many other acids under study. From the first publication quantitative hydrogen evolution as another product of the reaction (1) was noted (for example [2, 5—7]).

Not less characteristic feature of reviewed processes is the fact that their realization requires cathodes of definite metals. So during the electrolysis of primary nitramines at Hg-cathode in DMF a mechanism of self-protonation is accomplished while on their reduction at Pt-cathode the mechanism changed and after passing 1F·mol^{-1} corresponding N-anion formed as a result of N-H bond cleavage:

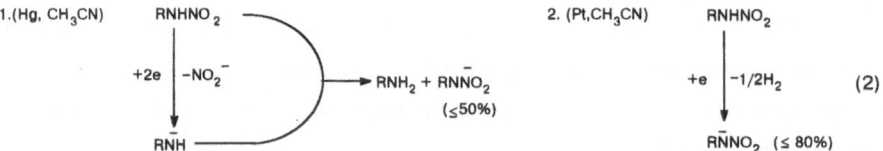

(2)

According to the existing data (for example the reduction of methylene-active compounds [5], mercaptans [6] and carbamides [7]) cathodic deprotonation of organic acids is the most effective at the cathodes of transient metals (Pt, Fe, Ni, Cu, Cr) which are capable to chemisorption of organic substances, whereas for example at Pb or Sn-cathodes such process can not be realized or it occurs with a minimal efficiency. Probably the effect of "electrode assistance" takes place during acid reduction at the cathodes with low hydrogen overvoltage, but this problem was not studied specially. Nevertheless the combination of available data permitted to conclude that potentiostatic electrolysis of C-H, S-H, O-H, P-H and N-H organic acids of different structure realized at a potential ≤2V (vs Ag/0,1NAg+) really corresponded to direct deprotonation (1).

Generation of carb- and heteroanions by this method led to effective processes of electrosynthesis [8] (scheme 3), for example the realization of electrochemical version of Michaelis-Becker, Wittig-Horner, Perkin reactions, base-catalysed chain reactions etc.

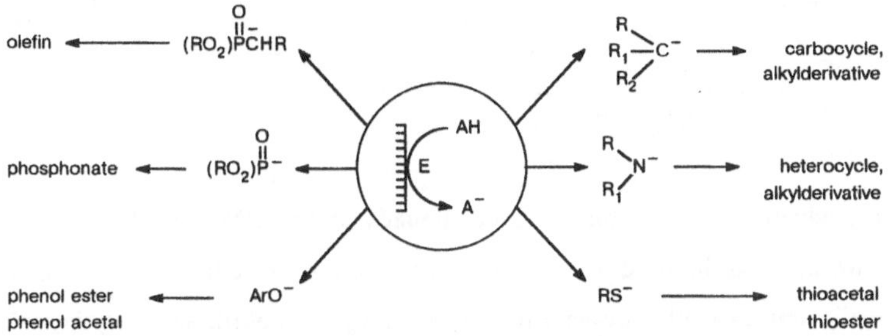

(3)

It should be noted that among the successfully occured processes were examples of amperostatic electrolysis with deprotonation of weak organic acids pK>20 which often reduced at the discharge potential of supporting electrolyte or at more negative potential [8]. It being known that this phenomenon is general, so it should be observed separately.

Actually in this case along with really direct deprotonation of organic acid a latent indirect deprotonation can take place and products of supporting electrolyte reduction can play a role of deprotonating agents. At the same time synthetic attraction of such processes is indubitable and, apart from details of mechanism, we called them "Cathodic Deprotonation in Simple Systems" (CDSS).

There are three characteristic features of CDSS-processes that should be distinguished:

– electrogeneration of anions under potentio- or amperostatic electrolysis occurs in a very simple system: organic acid/supporting electrolyte;

– in contrast to usual processes of indirect deprotonation CDSS processes do not require special probase and utilization of a reduced form of this probase after its protonation is not necessary;

– CDSS processes have universal character for different type of organic acids (pK≤30).

For example, electrochemical version of Wittig-Horner reaction with azobenzene [9a] and phenazine [9b] as a probase was described. However olefin formation by this reaction under CDSS conditions appeared not less effective [8], and recently it was found that alkylation of stereohomogeneous dialkyl phosphite occured with retention of configuration of hydrophosphoryl center and afforded stereoindividual phosphonate with ~70% yield [10].

Simple method of an anionic particle generation under CDSS conditions offers the possibilities for electroshemical carboxilation. At present the first results on carboxilation of chloroform, methylene chloride and cyanamide (Pt cathode, Mg anode, Bu4NBr, DMF) were obtained (scheme 4).

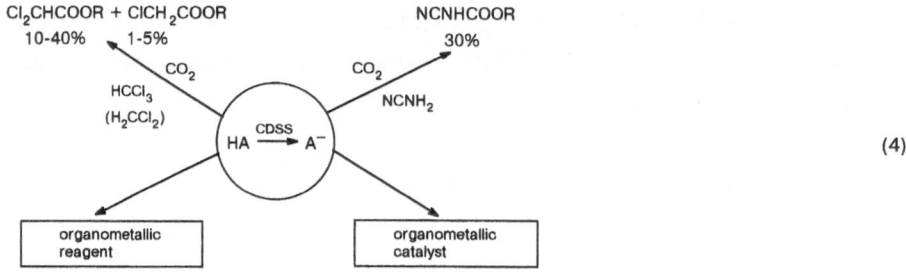

(4)

Another direction is a production of organometallic compounds under CDSS conditions with sacrificial anode (4). Some of them can be used as a reagent and others — as effective catalysts in organic synthesis. For example cuprous methyl cyanate that obtained by acetonitrile electrolysis with quantitative yield, effectively reacted with allyl bromide giving the target product with 70 % yield. The similar chemical process was more complicated and realized under extreme conditions (-70°) [11].

Replacement of cuprous chloride as catalyst by cuprous 3,5-dimethylpyrazolate extremely increased effectivity 1,2- (1,4-) addition of gem-polychloroderivatives to unsaturated compounds. The concentration of the insoluble cuprous azolate was usually 1000 times less than cuprous chloride and azolate could be easily filtered from reaction mixture and repeatedly used.

So the realization of different classes organic acid transformation under CDSS conditions makes possible future progress of this method in electroorganic synthesis.

References
1) D.E.Batrac, and M.D.Hawley, *J. Am. Chem. Soc.*, **1972**, *94*, 640.
2) W.M.Moore, M.Finkelstein, and S.D.Ross, *Tetrahedron*, **1980**, 727.
3) J.K.Howie, J.J.Hotus, and D.T.Sawyer, *J. Am. Chem. Soc.*, **1977**, *99*, 6323.
4) V.A.Petrosyan, V.A.Frolowsky, and D.A.Sadilenko, *Elektrokhimiya* (in press).
5) A.V.Bukhtiarov, V.N.Golyshin, and A.P.Tomilov, *Zn. Obshch. Khim.*, **1988**, *58*, 157.
6) A.V.Bukhtiarov, V.V.Mikheev, A.V.Lebedev, A.P.Tomilov, and O.V.Kuz'min, *Zn. Obshch. Khim.*, **1988**, *58*, 684.
7) A.V.Bukhtiarov, V.V.Mikheev, A.V.Lebedev, A.P.Tomilov, and O.V.Kuz'min, *Zn. Obshch. Khim.*, **1989**, *59*, 421.
8) V.A.Petrosyan, *Russ. Chem. Bull.*, **1995**, *44*, 1353 (Engl. Transl.).
9) (a) J.C.Le Menn, J.Sarrazin, and A.Tallec, *Can. J. Chem.*, **1989**, *67*, 1332; (b) M.Kimura, T.Kurata, and Y.Sawaki, in *Studies in Organic Chemistry Recent Advances in Electroorganic Synthesis* (Edited by S.Torii) Kodansha Tokyo, Elsevier Amsterdam, **1987**, *30*, 269.
10) V.A.Petrosyan, E.E.Nifant'ev, M.E.Niyazimbetov, R.K.Magdeeva, N.S.Magomedova, and V.K.Bel'sky, *Russ. Chem. Bull.*, **1996**, *45*, 427 (Engl. Transl.).
11) E.J.Corey, and Isao Kuwajima, *Tetrahedron Lett.*, **1972**, 487.

ELECTROCHEMICAL REACTIONS OF ORGANIC MATERIALS ON DIAMOND ELECTRODES

Akira FUJISHIMA, Elena POPA, Zhongyun WU and Tata N. RAO
Department of Applied Chemistry, Faculty of Engineering
The University of Tokyo, Tokyo 113, Japan

ABSTRACT

The electrochemistry of diamond has gained much importance due to its stability, low background current and wide potential window in aqueous electrolytes. We have used high quality diamond films both from fundamental and applied viewpoints. The detection of biologically important compounds such as dopamine was found to be possible at very low concentrations even in the presence of ascorbic acid, which interferes frequently in biological systems. Similarly, oxidation of nicotinamide dinucleotide (NADH) was observed, with a well defined cyclic voltammogram even in the micromolar range. The electrochemical reduction of C_{60} at diamond electrode was studied in acetonitrile-toluene mixture, and six reduction peaks were observed at room temperature.

INTRODUCTION

Diamond has become an interesting material from the viewpoint of applications, due to its outstanding properties, such as exceptional hardness, high thermal conductivity, good optical transparency and stability in corrosive chemical environments (1, 2). Apart from the applications of diamond in electronic devices, it also has potential applications in electrochemistry due to its chemical inertness, low background current and wide electrochemical potential window (3, 4). Highly doped conducting diamond electrodes have been tested by several researchers for demanding electrochemical processes (3-5). We have been exploring the use of diamond in electrochemistry from both fundamental and practical viewpoints. In the present paper we briefly summarize our recent work on diamond electrochemistry.

EXPERIMENTAL

The boron-doped diamond films were grown on Si (100) substrates, using a microwave plasma-assisted chemical vapor deposition (CVD) system (AsTeX Corp., Woburn, MA). A 9:1 (v/v) mixture of acetone and methanol was used as the carbon source. B_2O_3 was used as the boron source. The B_2O_3 was dissolved in the acetone-methanol mixture so that the B/C weight ratio in the feedstock was approximately 10^4ppm for the preparation of conducting diamond. Ohmic contacts were made on the front (i.e., last deposited) side of the films. Electrochemical oxidation of dopamine and ascorbic acid was carried out in 0.1 M $HClO_4$ solution. Electrochemical oxidation of NADH was carried out in a phosphate buffer (pH = 7.2). Electrochemical reduction of C_{60} was carried out in a mixture of acetonitrile and toluene.

RESULTS AND DISCUSSION

Electrochemical oxidation of dopamine and ascorbic acid

Dopamine (DA) is a biologically important chemical which plays an important role in neurotransmission. The electrochemical detection of DA is studied at platinum and various carbon electrodes (6). Ascorbic acid coexists with DA in the central nervous system and its oxidation peak is often overlapped with that of DA. Hence, the selectivity of DA in the presence of ascorbic acid is an important aspect. In the present work, we have made use of diamond for the detection of DA in the presence of ascorbic acid in acidic solution (Fig. 1a). At the diamond electrode, the cyclic voltammogram showed two well-defined anodic peaks, Such peak separation was not observed on the glassy carbon electrode in the same electrolyte. The peak height for DA increased linearly with concentration in the micromolar range without being affected by the presence of ascorbic acid. Another interesting point is the detection limit of DA, which approaches 2 μM, even in the presence of ascorbic acid.

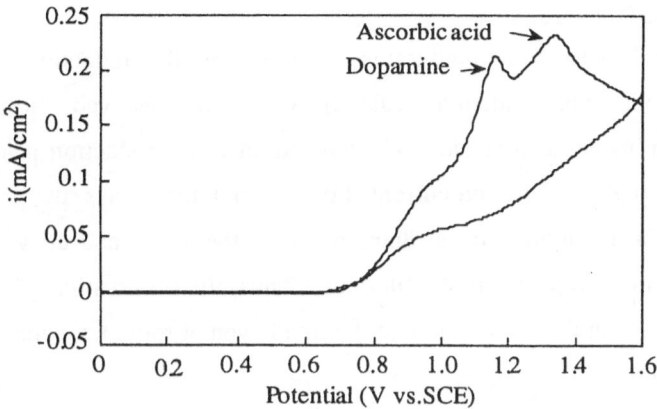

Fig. 1. Cyclic voltammogram for a mixture of 0.1 mM dopamine and 2 mM ascorbic acid in 0.1 M HClO$_4$ on diamond electrode; scan rate, 50 mV/s

Electrochemical oxidation of NADH

The electrochemical oxidation of β-nicotinamide adenine dinucleotide (NADH) has gained much attention due to its potential use in biosensors where NAD$^+$-dependent dehydrogenases are usually used (7). Although glassy carbon and Pt electrodes are used for the oxidation of NADH, well-defined cyclic voltammograms could not be obtained on these electrodes, especially at low concentrations. As in the case of dopamine, ascorbic also interferes with NADH. We have made use of bare diamond electrode to study the electrochemical oxidation of NADH. Well-defined cyclic voltammograms were obtained even in the low concentration region (1 μM to 50 μM) in contrast to those obtained on glassy carbon electrodes. The presence of the ascorbic acid could be detected in a mixture containing equimolar concentrations of NADH and ascorbic acid. The calibration curve obtained for NADH in the presence of ascorbic acid is close to that obtained for NADH alone. Thus the interference effects due to ascorbic acid could be largely minimized.

Electrochemical reduction of C_{60}

Differential pulse voltammograms obtained for the reduction of C_{60} on a diamond electrode. Six reduction peaks of C_{60} were observed. In contrast, the voltammogram obtained at a Pt electrode showed only five reduction peaks (Fig. 2b), probably due to high background current. For the first five peaks, the peak potentials measured both on platinum and on diamond were the same, and all values were in agreement with those reported in the literature. Thus, the observation of six reduction peaks for C_{60} was found to be possible at diamond even at room temperature.

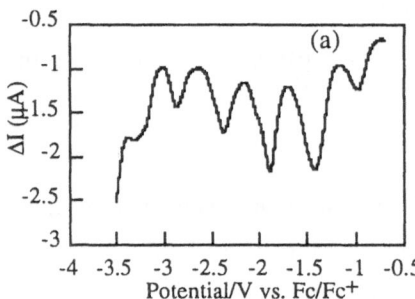 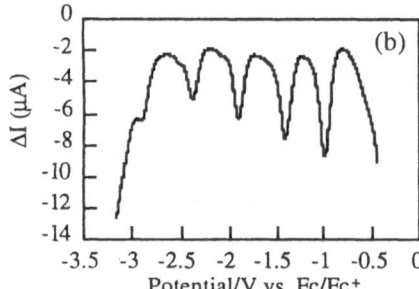

Fig. 2. Differential pulse voltammograms for the reduction of C_{60} at (a) diamond and (b) Pt electrodes; 80-mV pulse, 50-ms width, 200-ms pulse time; scan rate, 10 mV/s.

REFERENCES

1. J. C. Angus and C. C. Hayman, *Science*, **1988**, 241 913.
2. P. K. Bachmann and R. Messier, *Chem. Eng. News*, **1989**, May 15, 24.
3. G. M. Swain, *J. Electrochem. Soc.*, **1994**, 141 3382.
4. H. B. Martin, A. Argoitia, U. Landau, A. B. Anderson and J. C. Angus, *J. Electrochem. Soc.* **1996**, 143, L133.
5. R. Tenne, K. Patel, K. Hashimoto and A. Fujishima, *J. Electroanal. Chem.*, **1993**, 347, 409.
6. H. Kaneko, M. Yamada and K. Aoki, *Anal. Sci.*, **1990**, 6, 439 and references therein.
7. Q. Wu, M. Maskus, F. Pariente, F. Tobalina, V. M. Fernandez, E. Lorenzo and H. D. Abruna, *Anal. Chem.*, **1996**, 68, 3688.

CATALYSIS OF METAL IONS ON ELECTRON TRANSFER REACTIONS

Shunichi FUKUZUMI

Department of Applied Chemistry, Faculty of Engineering, Osaka University,
Suita, Osaka 565, Japan

The importance of electron transfer processes has been recognized in nearly every subdiscipline of chemistry, i.e., not only inorganic chemistry but also organic and organometallic chemistry. Numerous organic reactions, previously formulated by "movements of electron pairs" are now understood as processes in which an initial electron transfer from a nucleophile (reductant) to an electrophile (oxidant) produces a radical ion pair, which leads to the final products via the follow-up steps involving cleavage and formation of chemical bonds.[1-3] The follow-up steps are usually sufficiently rapid to render the initial electron transfer the rate-determining step in an overall irreversible transformation. In such cases, the catalysis on the electron transfer step is essential to control the overall redox reactions.

This study reports catalysis of metal ions on both thermal and photoinduced electron transfer by utilizing the strong Lewis acidities of metal ions introduced as an appropriate third component that can reduce the activation barrier of the electron transfer reactions. The comparison of the catalytic effects of metal metal ions on thermal electron transfer reactions with those on the redox reactions of the same substrates would provide a valuable quantitative basis to distinguish between conventional ionic or concerted mechanisms and electron transfer mechanisms of thermal redox reactions. More importantly both thermal and photochemical redox reactions which would otherwise be unlikely to occur are made possible to proceed efficiently by the catalysis of metal ions on the electron transfer steps. First the fundamental concepts of catalysis on electron transfer are presented and then the mechanistic viability is described by showing a number of examples of both thermal and photochemical reactions that involve catalysis of rare-earth metal ions on electron transfer processes as the rate-determining steps.

Fundamental Concepts of Catalysis on Electron Transfer. Electron transfer reactions are generally regarded as very fast processes which are in no need of catalysis to accelerate the rates of electron transfer. This is largely true for reversible electron transfer reactions in which electron transfer occurs only when the free energy change of electron transfer is negative, i.e., the electron transfer is exergonic. If the electron transfer is endergonic, no net electron transfer would occur. In the case of irreversible electron transfer when the electron transfer is endergonic but the follow-up reactions are highly exergonic to make the overall redox reactions proceed, however, the initial electron transfer step could be very slow, being the rate-determining step. In such a case the catalysis on electron transfer would play an essential role in reducing the activation barrier of the overall redox reaction. According to the Marcus theory,[4] rates of electron transfer are determined by the free energy change of electron transfer (ΔG^0_{et}) and the reorganization energy associated with the electron transfer (λ). As shown in Fig. 1, an electron

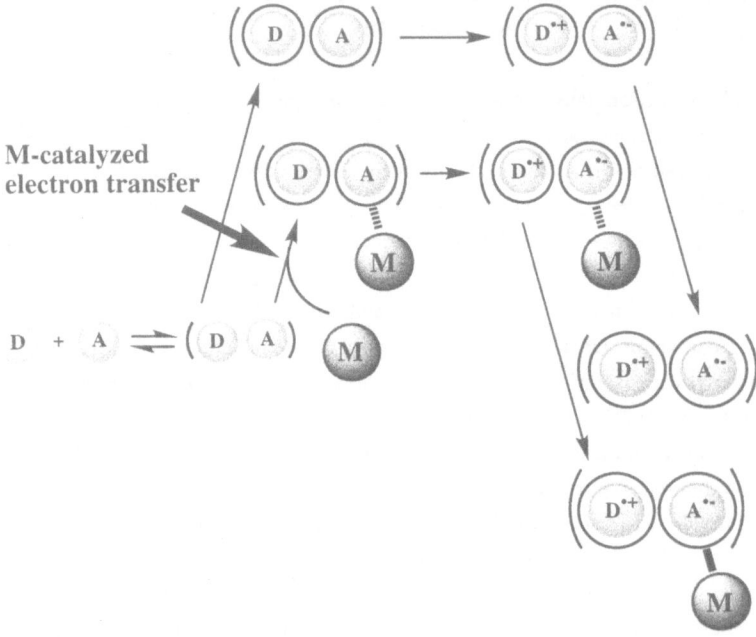

Fig. 1. General scheme for catalysis of metal ions on electron transfer.

is transferred from an electron donor (D) to an acceptor (A) instantaneously according to the Franck-Condon principle when the reactant pair is activated to reach the nuclear configurations which include the solvation, where the energy before and after the electron transfer is the same. If a third component (**M**) which can stabilize one of the products of electron transfer thermodynamically is introduced into the D-A system, the free energy change of electron transfer is shifted to the negative direction, when the activation barrier of electron transfer is reduced to accelerate the rates of electron transfer as shown in Fig. 1, where **M** forms a complex with $A^{\bullet-}$. It should be emphasized that there is no need to have an interaction of **M** with A and that the interaction with the reduced state ($A^{\bullet-}$) is sufficient to accelerate the rate of electron transfer. This contrasts well with the catalysis on conventional ionic or concerted reactions, in which the catalyst needs to interact with a reactant to accelerate the reactions. Since most organic compounds such as π acceptors in particular have small reorganization energies, the change of redox potentials by the interaction of the corresponding radical anions with **M** may be the main factor to accelerate the rates of electron transfer. Thus, any material **M** that can stabilize the radical anions thermodynamically by the complexation may act as an efficient catalyst to accelerate the rates of electron transfer. The stronger the interaction of **M** with the radical anions is, the faster will the rates of electron transfer be as the free energy change of electron transfer decreases. This means that electron is transferred instantaneously according to the Franck-Condon principle when the reactant pair is activated by the unfavorable interaction with **M** to reach the nuclear configurations where the energy before and after the electron transfer is the same. Once an electron is transferred, the interaction of **M** with the radical anion becomes energetically favorable to give the thermodynamically more stable products as shown in Fig. 1.

There may be the case when not only one but also two **M** molecules or ions can interact with A$^{\cdot-}$ as shown in Eqs. 1 and 2. In such a case the one-electron reduction potential of A (E_{red}) may be shifted to

$$A^{\cdot-} + M \underset{}{\overset{K_1}{\rightleftharpoons}} A\text{-}M^{\cdot-} \tag{1}$$

$$A\text{-}M^{\cdot-} + M \underset{}{\overset{K_2}{\rightleftharpoons}} A\text{-}2M^{\cdot-} \tag{2}$$

the positive direction with an increase in the concentration of M according to Eq. 3, which is derived from the Nernst equation of the one-electron reduction potential in the presence of **M**, where E^0_{red} is the

$$E_{red} = E^0_{red} + (2.3RT/F)\log (1 + K_1[M] + K_1K_2[M]^2) \tag{3}$$

one-electron reduction potential of A in the absence of **M**, K_1 and K_2 are the formation constants of the 1:1 and 1:2 complexes of A$^{\cdot-}$ with **M**, and F is the Faraday constant.

If one assumes that **M** has no effect on the oxidation potential of D, the free energy change of electron transfer from D to A in the presence of **M** (ΔG_{et}) can be expressed by Eq. 4, where ΔG^0_{et} is the

$$\Delta G_{et} = \Delta G^0_{et} - (2.3RT)\log(1 + K_1[M] + K_1K_2[M]^2) \tag{4}$$

free energy change in the absence of **M**. Thus, electron transfer from D to A becomes more favorable energetically with an increase in the concentration of M. If such a change in the energetics is directly reflected in the transition state of electron transfer, the dependence of the rate constant of **M**-catalyzed electron transfer (k_{et}) on [**M**] may be given by Eq. 5, where k^0_{et} is the rate constant in the absence of **M**.

$$k_{et} = k^0_{et} (1 + K_1[M] + K_1K_2[M]^2) \tag{5}$$

Thus, the dependence of the rate constant of **M**-catalyzed electron transfer on [**M**] is expected to change from the first-order to the second-order with respect to [**M**] at the concentration region where A$^{\cdot-}$ can form the 1:2 complex with **M**. The validity of Eq. 5 is confirmed in a number of examples of **M**-catalyzed electron transfer reactions (*vide infra*).

Rates of electron transfer from CoTPP to p-benzoquinones (Q) are significantly accelerated by the presence of rate-earth metal ions, particularly by Sc^{3+}. The dependence of the rate constant of electron transfer obeyed Eq. 5, showing the first order dependence on $[Sc^{3+}]$ at low concentrations and the second-order dependence at higher concentrations. The formation of the 1:2 complex between $Q^{\cdot-}$ and Sc^{3+} has been confirmed by the observation of the ESR spectrum of $Q^{\cdot-}$-Sc^{3+} which shows the super hyperfine due to the interaction of $Q^{\cdot-}$ with two equivalent Sc nuclei. Similarly the formation of 1:1 complex between $O_2^{\cdot-}$ and Sc^{3+} was confirmed by the ESR spectrum in solution.

Mechanistic Insight of Catalysis of Metal Ions on Electron Transfer. If catalysis of metal ions on electron transfer plays an essential role in some reactions, formulated previously as ionic or concerted reactions, the dependence of the rate constants on the metal ion concentration would be the same as that observed in the electron transfer reduction of the same substrate. Thus, the comparison of the catalytic function of metal ions in electron transfer reactions with that in apparently polar or concerted reactions of the same substrate would provide valuable insight into the mechanistic viability of electron transfer.

Diels-Alder reactions are usually regarded as concerted processes where the interaction between HOMOs of dienes and LUMOs of dienophiles determines the reactivity. The interaction of metal ions with radical anions of dienophiles may enhance the electron affinity of the dienophiles so that electron transfer may provide a favorable reaction pathway for such dienophiles. In fact, the [4+2] cycloaddition of 9,10-dimethylanthracene (DMA) with p-benzoquinone (Q) occurs efficiently in the presence of Sc^{3+} to yield the adduct selectively (Eq. 6), although p-benzoquinone derivatives have been regarded as inert or weak dienophiles. The [4+2] cycloaddition of anthracene and 9-methylanthracene with other p-benzoquinone derivatives (X-Q; X = 2,5-Cl$_2$ and 2,5-Me$_2$) also occurs efficiently in the presence of $Mg(ClO_4)_2$ to yield the corresponding adducts. The observed second-order rate constant (k_{obs}) increases

with an increase in $[Sc^{3+}]$ to exhibit first-order dependence on $[Sc^{3+}]$ at low concentrations, changing to second-order dependence at high concentrations, being consistent with Eq. 5.[5] In fact, there is a striking similarity with respect to the dependence of k_{obs} on $[Sc^{3+}]$ between the electron transfer reaction and the Diels-Alder reaction, despite the large difference in their reactivities. Thus, the catalysis of Sc^{3+} in the Diels-Alder reactions of DMA with X-Q is essentially the same as that in the electron transfer reduction of X-Q. The Sc^{3+}-catalyzed electron transfer from anthracenes to X-Q is thereby a rate-determining step of the Diels-Alder reactions.

The hydride transfer reaction from 10-methyl-9,10-dihydroacridine (AcrH$_2$) to p-benzoquinone (Q) is catalyzed by various metal ions such as Mg^{2+} and rare-earth metal ions in acetonitrile (MeCN).[3] Among triflate salts of metal ions examined, Sc(OTf)$_3$ was the most effective. The observed second-order rate constant increases with an increase in [Sc^{3+}] at low concentrations, changing to second order dependence at high concentrations. Essentially the same catalytic behavior of Sc^{3+} was observed in the electron transfer reduction of Q by CoTPP (TPP = tetraphenylporphyrin dianion) in MeCN. The second-order dependence of the rate constant with respect to [Sc^{3+}] is ascribed to formation of a 1:2 complex between Q$^{\bullet-}$ and Sc^{3+}. Thus, the hydride transfer from AcrH$_2$ to Q may also proceed via Sc^{3+}-catalyzed electron transfer from AcrH$_2$ to Q.

When AcrH$_2$ is replaced by 1-benzyl-1,4-dihydronicotinamide (BNAH) which is a stronger electron donor than AcrH$_2$, benzaldehyde and its derivatives are reduced by BNAH efficiently in the presence of Sc^{3+} in MeCN at room temperature to yield the corresponding alcohols (Eq. 7). The Sc^{3+}

$$(7)$$

also catalyzes photoinduced electron transfer from the excited state of Ru(bpy)$_3$$^{2+}$ to benzaldehydes in MeCN. There is a linear correlation between the observed rate constants of Sc^{3+}-catalyzed reduction of benzaldehydes by BNAH and those of the Sc^{3+}-catalyzed electron transfer reactions. The apparent catalytic reactivity of Sc^{3+} in the model system is comparable with that of the enzyme, and even higher if the effect of complexation of Sc^{3+} with BNAH is taken into account.

Complexation of metal ions with substrates can not only enhance the oxidizing ability of the substrate, but can also change the excited state property. Aromatic carbonyl compounds with lowest n-π^* singlet states are generally nonfluorescent, possessing large π-π^* triplet formation quantum yields (ca. 0.7) via the fast intersystem crossing. However, irradiation of the absorption band due to Mg^{2+} or Sc^{3+} complex of 1-naphthaldehyde (1-NA) or 2-naphthaldehyde (2-NA) causes strong fluorescence at 430-440 nm.[6] The change in the excited state properties by the complex formation with Mg^{2+} or Sc^{3+} may be attributed to increased energy of the n-π^* singlet state relative to the fluorescent π-π^* singlet excited state as reported for similar change in the photophysical and photochemical properties of carbonyl compounds by the complex formation with Lewis acids. The remarkable positive shifts (ca. 1.5 V) of the E^0_{red} values of the singlet excited states of the Sc^{3+}-carbonyl complexes as compared with those of the triplet excited states of uncomplexed carbonyl compounds result in a significant increase in the reactivity of the Sc^{3+} complexes vs. noncomplexed carbonyl compounds in the photoinduced electron transfer reactions.[6]

Such enhancement of the reactivities of the excited states of carbonyl compounds by the complexation with Sc^{3+} enables us to perform the photochemical reactions of the carbonyl compounds with various electron donors via Sc^{3+}-catalyzed photoinduced electron transfer.[7] For example, irradiation of the absorption band due to the Sc^{3+}-carbonyl complex in deaerated acetonitrile containing 2-NA, Me_4Sn and $Sc(OTf)_3$ with monochromatized light of 360 nm from a xenon lamp gives the methyl adduct via the Sc^{3+}-catalyzed photoinduced electron transfer from Me_4Sn to 2-NA as shown in Eq. 8. No photochemical reaction of 2-NA with Me_4Sn has occurred in the absence of Sc^{3+} or in the presence of Mg^{2+}.

$$(8)$$

Conclusion

As demonstrated above, both thermal and photoinduced electron transfer reactions are accelerated by rare-earth metal ions as catalysts when the products of electron transfer from complexes with the catalysts. Such catalysis on electron transfer processes is particularly important to control the redox reactions in which the electron transfer processes are involved as the rate-determining steps followed by facile follow-up steps involving cleavage and formation of chemical bonds. Once the thermodynamic properties of the complexation of rare-earth metal ions are obtained, we can predict the kinetic formulation on the catalytic activity. The scope and the application of catalysis of rare-earth metal ions on electron transfer are expected to expand much further in near future.

References

1. S. Fukuzumi, *Bull. Chem. Soc. Jpn.*, **70**, 1 (1997).

2. L. E. Everson, "Electron Transfer Reactions in Organic Chemistry; Reactivity and Structure, " Vol. 25, Springler, Heidelberg, 1987.;

3. S. Fukuzumi, in "Advances in Electron Transfer Chemistry," ed by P. S. Mariano, JAI press, CT (1992), Vol. 2, pp. 67-175.

4. R. A. Marcus, *Ann. Rev. Phys. Chem.*, **15**, 155 (1964).

5. S. Fukuzumi and T. Okamoto, *J. Am. Chem. Soc.*, **115**, 11600 (1993).

6. S. Fukuzumi, T. Okamoto, and J. Otera, *J. Am. Chem. Soc.*, **116**, 5503 (1994).

7. Recent our examples for photoinduced electron transfer reactions, see: K. Mikami, S. Matsumoto, A. Ishida, S. Takamuku, T. Suenobu, and S. Fukuzumi, *J. Am. Chem. Soc.*, **117**, 11134 (1995); M. Fujita, A. Ishida, S. Takamuku, and S. Fukuzumi, *J. Am. Chem. Soc.*, **118**, 8566 (1996).

SYNTHETIC USE OF ELECTROINDUCED HYDROGEN ABSTRACTION

Niels Them KJÆR and Henning LUND

Department of Organic Chemistry, University of Aarhus, DK-8000 Århus C, Denmark

The coupling of radicals generated by reduction of alkyl halides (RX) with aromatic radical anions (A$^{\cdot-}$) has been studied by the Aarhus group.[1] The proposed mechanism is the following:

Scheme 1

A + e$^-$	$\rightleftarrows$	A$^{\cdot-}$	(E_A°)
A$^{\cdot-}$ + RX	$\xrightarrow{k_1}$	A + R$^\cdot$ + X$^-$	(1)
A$^{\cdot-}$ + R$^\cdot$	$\xrightarrow{k_2}$	R - A$^-$	(2)
A$^{\cdot-}$ + R$^\cdot$	$\xrightarrow{k_3}$	A + R$^-$	(3)

The product determining competition between reactions (2) and (3) is dependent upon the standard potentials of the mediators and the reduction potentials of the alkyl radicals and the competion has been expressed by the parameter $q = k_3/(k_2 + k_3)$. By using mediators with lower and lower potentials, the mechanism will shift from reaction (2) to reaction (3). The potential at which the mechanism shifts and $q = \frac{1}{2}$, $E_{\frac{1}{2}}^q$, the reduction potential of the radical, has been determined for several short-lived radicals such as alkyl, allyl, benzyl and acyl radicals.[1] From the reduction potential the standard potential of the radical may be estimated.

It seemed of interest to measure the reduction potential of other radicals than carbon-centered radicals, and it was attempted to investigate oxygen-centered radicals in a way similar to that used for carbon-centered radicals. In a recent investigation of alkoxy radicals,[2] indirect electrochemical reductions of derivatives of *tert*-butyl hydroperoxide (*t*-BuOOR) by aromatic mediators were performed, and a plot of the *q*-values *vs* the standard potentials of the aromatic radical anions used in the experiment gave an S-shaped curve as expected. The reduction potential was, however, much lower than calculated from thermodynamic data; when the reaction was performed in dry DMF containing Bu$_4$NBF$_4$ as a supporting electrolyte with an electron donor which had q = 0 and thus should give a coupling product, the isolated product did not have a *tert*-butoxy group; when the mediator was anthraquinone, the isolated coupling product was 10-*N,N*-dimethylcarbamoyl-10-hydroxy-9(10 H)-anthracenone (**1**).

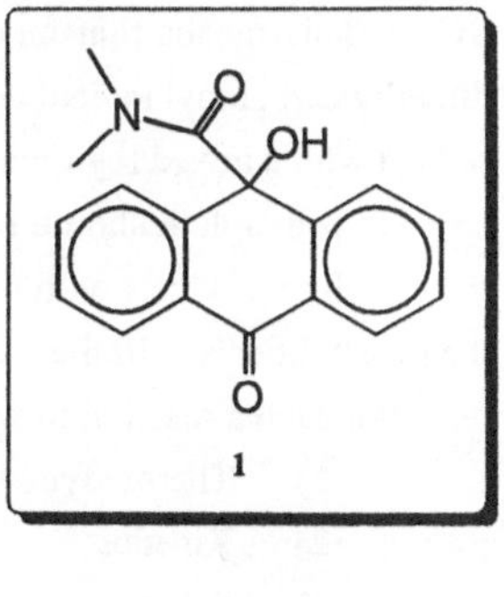

In Scheme 2 the electrogenerated *tert*-butoxy radical abstracts a hydrogen atom from the solvent DMF generating an *N,N*-dimethylcarbamoyl radical which can, like other radicals, either couple with or be reduced by the aromatic radical anion. The same coupling product (1) was isolated in a similar reduction where iodobenzene was used in place of *tert*-BuOOR. The electrogenerated phenyl radical is a good hydrogen atom abstractor and a reaction analogous to eq. (7) takes place to form benzene and an *N,N*-dimethylcarbamoyl radical.

Scheme 2

$A + e^-$	$\rightleftharpoons$	$A^{\cdot-}$	E_A°
$A^{\cdot-} + t\text{-BuOOR}$	$\xrightarrow{k_4}$	$A + t\text{-BuO}^\cdot + RO^-$	(4)
$A^{\cdot-} + t\text{-BuO}^\cdot$	$\xrightarrow{k_5}$	$t\text{-BuO} - A^-$	(5)
$A^{\cdot-} + t\text{-BuO}^\cdot$	$\xrightarrow{k_6}$	$A + t\text{-BuO}^-$	(6)
$t\text{-BuO}^\cdot + HCON(CH_3)_2$	$\xrightarrow{k_7}$	$t\text{-BuOH} + (CH_3)_2NCO^\cdot$	(7)
$A^{\cdot-} + (CH_3)_2NCO^\cdot$	$\xrightarrow{k_8}$	$(CH_3)_2NCO - A^-$	(8)
$A^{\cdot-} + (CH_3)_2NCO^\cdot$	$\xrightarrow{k_9}$	$A + (CH_3)_2NCO^-$	(9)

In the solution the radical anion of the mediator reduces the peroxide in a dissociative electron transfer. The formed *tert*-butoxyl radical can react with another $A^{\cdot-}$ in two ways: The *tert*-butoxyl radical can either couple with the radical anion or be reduced by it. Alternatively the *tert*-butoxyl radical may act as a hydrogen atom abstractor and react with the solvent, DMF. The formed *N,N*-dimethylcarbamoyl radical can then react with the radical anion of the mediator as shown in reaction (8) and (9).

If the concentration of the aromatic radical anion is low (~2 mM) reaction (7) will outrun reactions (5) and (6). The competition between the reactions (8) and (9) was followed and expressed by the parameter $q = k_9/(k_8 + k_9)$. Measurement of q-values for different aromatic mediators was used to estimate the reduction potential of the *N,N*-dimethylcarbamoyl radical which is equal to -1.62 V vs. SCE.[2] This means that mediators with standard potentials above -1.4 V will couple with the *N,N*-dimethylcarbamoyl radical and mediators with standard potentials below -1.9 V (vs. SCE) will reduce the radical with a mixed region in between.

In principle it should be possible to employ other hydrogen atom donors than DMF and to isolate the coupling products with the radicals generated by hydrogen abstraction. This paper describes some of the possibilities and limitations of this procedure in organic synthesis.

For such a reaction to work several factors should be considered, among them:

1) Thermodynamics
2) Kinetics
3) Selectivity
4) Solvent properties
5) Redox potential of mediator and radicals

Thermodynamics. Hydrogen bonds to silicon, tin, phosphorous and sulfur are relatively weak compared to the bonds to carbon, nitrogen and oxygen. For saturated hydrocarbons the bond strengths follow RCH_2-H > RR'CH-H > RR'R''C-H; benzylic and allylic C-H bonds are relatively weak. Generally speaking, compounds that form relatively stable radicals on loss of a hydrogen atom are good candidates for the proposed reaction, and by introduction of suitable substituents good hydrogen atom donors may be obtained.

Kinetics. Hydrogen abstracting radicals may roughly be divided into three groups, fast, medium fast, and relatively slow hydrogen abstractors. In the first group are radicals as F·, Cl·, and HO· which abstracts hydrogen too fast to be of interest for the proposed reaction. In the third group are radicals as alkyl and benzyl, which react too slowly. The medium fast hydrogen abstractors include phenyl and *tert*-butoxyl radicals. In Table 1 are shown rate constants for hydrogen abstraction by phenyl, *tert*-butoxyl, and methoxyl radicals from some hydrogen atom donors. When considering using $Me_3CO·$ as abstractor its decomposition to Me· and acetone ($k = 6.3 \times 10^5$ s^{-1} in MeCN)[3] should be considered.

Table 1. Rate constants for the reaction between phenyl, *tert*-butoxyl, and methoxyl radicals and some hydrogen atom donors.

	H-SnBu$_3$	H-CH$_2$Ph	H-CH$_2$OH	H-CH$_3$
Ph·	$5,7 \times 10^8$	$1,7 \times 10^6$	5×10^5	
t-BuO·	$2,2 \times 10^8$	$2,3 \times 10^5$	$2,9 \times 10^5$	
CH$_3$O·			$2,6 \times 10^5$	5,4

Selectivity. The selectivity of hydrogen abstraction from hydrogen donors follows generally the relative bond strength, but there are exceptions. The selectivity of phenyl and *tert*-butoxyl radicals towards RR'R''C-H is about 45 compared to RCH_2-H whereas the selectivity towards abstraction of benzylic hydrogens compared to RCH_2-H is lower (about 10) although the $PhCH_2$-H bond is weaker than the R_3C-H bond. The selectivity towards bridgehead hydrogens depends on the ring size; thus whereas the % of bromination at the bridgehead positions by $BrCCl_3$ is 9 % for bicyclo[2.2.2]octane, it is 86 % for adamantane and 100 % for bicyclo[3,3,1]nonane. The selectivity of the fast reacting radicals as Cl· is much lower.

Solvent. In Table 2 are given rate constants of hydrogen abstrcation from polar, aprotic solvents by some radicals. Table 2 shows that THF and DMF are relatively good hydrogen donors whereas abstraction from MeCN and DMSO is slower. Hydrogen abstraction from ammonia is slow, but the low solubility of unpolar compounds in ammonia limits its applicability.

Table 2. Rate constants for reactions between Ph·, *tert*-BuO· and 1-Naph· and some solvents at 25 ºC.

	Ph·	tBuO·	1-Naph·
DMF		$5 \times 10^6 \, M^{-1}s^{-1}$	$8 \times 10^6 \, M^{-1}s^{-1}$
CH$_3$CN	$1{,}0 \times 10^5 \, M^{-1}s^{-1}$	$2{,}8 \times 10^3 \, M^{-1}s^{-1}$	$2{,}5 \times 10^5 \, M^{-1}s^{-1}$
THF	$4{,}8 \times 10^6 \, M^{-1}s^{-1}$	$8{,}3 \times 10^6 \, M^{-1}s^{-1}$	
DMSO			$3 \times 10^5 \, M^{-1}s^{-1}$

Potentials. With the knowledge of the reduction potentials of the alkyl radicals from the previous work[1] it seemed possible to predict which radicals would give coupling products between mediators and radicals and which would be reduced. The mediators must have standard potentials higher than the reduction potentials of the radicals formed from the hydrogen atom donor. It should be kept in mind that although the introduction of a substituent may increase the selectivity by lowering the bond strength of the C-H bond the substituent may change the reduction potential of the radical so it will be reduced rather than couple with the mediator.

Using MeCN as a solvent, 1,4-dicyanobenzene as electron donor, iodobenzene as a source of phenyl radicals and a number of hydrogen donors, substituted benzonitriles were obtained. Thus with cyclohexene as hydrogen donor 4-(1-cyclohexenyl)benzonitrile **2** was the main product, with 4-(2-phenylcyclohexyl)benzonitrile as a side product and traces of 4-phenylbenzonitrile and 4-(3-cyclohexenyl)benzonitrile **3**; if some phenol is present during the reduction the latter is the main product rather than **2** which probably is obtained from the initially formed **3** through the influence of the electrogenerated base[4].

Similarly, with cyclooctane as hydrogen donor mainly 4-cyclooctylbenzonitrile is formed together with minor amounts of 2-cyclooctyl-1,4-dicyanobenzene.[4] In this case there is a competition between the substitution at the cyano group and an addition to the aromatic ring. Reactions with other types of hydrogen atom donors will be discussed.

References

1) R. Fuhlendorff, D. Occhialini, S. U.Pedersen, and H. Lund, *Acta Chem. Scand.*, **1989**,*43*, 803; ; D. Occhialini, S. U. Pedersen, and H. Lund, *Acta Chem. Scand.*, **1990**, *44*, 715; D. Occhialini, J. S. Kristensen, K. Daasbjerg, H. and Lund, *Acta Chem. Scand.*, **1992**, *46*, 474; H. Lund, K. Daasbjerg, D. Occhialini, and S. U. Pedersen, *Elektrokhimiya*, **1995**, *31*, 939.
2) N. T. Kjær, and H. Lund, *Acta Chem. Scand.*, **1995**, *49*, 848.
3) D. V. Avila, C. E. Brown, K. U. Ingold, and J. Lusztyk, *J. Am. Chem. Soc.*, **1993**, *115*, 466.
4) N. T. Kjær, and H. Lund, *Electrochim. Acta* **1997**, *42*, 2141.

π–π CHARGE-TRANSFER INTERACTION OF THE TETRACYANOETHYLENE π-DIANION GENERATED AT SEQUENTIAL ELECTROREDUCTION STEPS

Noriko OKUMURA, Kunimasa SETO and Bunji UNO

Gifu Pharmaceutical University, Mitahora-higashi, Gifu 502, Japan

INTRODUCTION

Increased attention has been paid to the π–dianions of organic molecules generated at sequential electroreduction steps, as a noticeably active electron-donor. The π–π stacking interaction of the dianion plays an important role in organic chemistry and biochemistry. In a series of our works the analyzing methods on the interaction of active species generated in electrode process has been proposed, and n-σ type charge-transfer interaction of the organic π-dianion has also been discussed.[1] In this paper, the π–π complexes of tetracyanoethylene (TCNE) π-dianion with hexamethylbenzene (HMB) and biphenylene (BP) were investigated with the aid of electrochemical techniques and molecular orbital calculations in order to clarify the π–π type charge-transfer interaction of the organic π–dianions.

RESULTS AND DISCUSSION

TCNE as typical π-electron acceptor forms the 1:1 charge-transfer complex with HMB ((4n+2)π system) and BP (4nπ system) in CH_2Cl_2, and is easily reduced, showing two reversible waves generating the corresponding anion radical (TCNE⁻) and π-dianion (TCNE⁼) in the cyclic voltammograms. Non-linear regression analysis based on our equations[1] have made it possible to evaluate the formation constants ($K^=$) of the TCNE⁼ π–π complexes. Temperature dependence of $K^=$ allows to yield the π–π formation enthalpy (ΔH^0) and entropy (ΔS^0) of the TCNE⁼–HMB and TCNE⁼–BP systems as collected in Table 1.

Table 1. Thermodynamic Data of the Charge-Transfer Complexes of the TCNE–HMB system and the TCNE⁼–BP system in CH_2Cl_2.

Compound	$K^{*)}$ /dm³ mol⁻¹	ΔH^0 / kcal mol⁻¹	Compound	$K^{*)}$ /dm³ mol⁻¹	ΔH^0 / kcal mol⁻¹
TCNE-HMB	3.35[**]	-4.32[**]	TCNE-BP	0.41[**]	-2.16[**]
TCNE⁼-HMB	0.77	-2.57	TCNE⁼-BP	15.8	-8.56

[*] Observed at 25°C.
[**] Values obtained from the spectroscopic method.

435

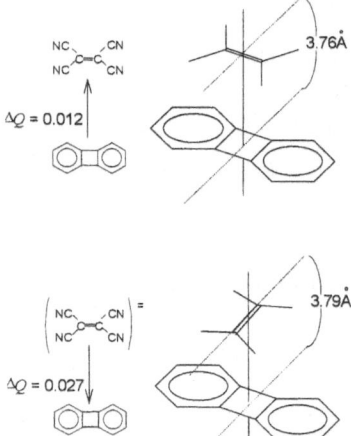

Fig. 1. Optimized Structures and charge migration (ΔQ) in the π–π Complexes of TCNE (a) and the dianion (b) with BP, calculated by the HF/4-31G energy gradient method.

The large value of the formation constants of the TCNE⁼–BP system indicates that TCNE⁼ strongly interacts with the weak electron-acceptor such as hydrocarbons. The TCNE⁼–BP complex may arise from the charge transfer from TCNE⁼ to BP. Postulating that the TCNE⁼ is a closed-shell singlet species, the energy gradient method was adopted to geometry optimization of the charge-transfer complexes within the framework of the HF calculation using 4-31G basis sets, the results being illustrated in Fig. 1. It is clearly found that TCNE is an electron-acceptor in the TCNE–BP complex, but TCNE⁼ is an electron-donor in the TCNE⁼–BP complex. These calculations have revealed that the difference in the optimized geometries of the TCNE–BP complex and TCNE⁼–BP complex arises from symmetries of the active molecular orbitals concerning the complex formation as shown in Fig. 2. Maximum overlap between the molecular orbitals of the electron-donor and acceptor plays an important role to form charge-transfer complexes. Further evidence of the TCNE⁼–BP charge-transfer complex formation has been donated by the direct observation for the CT spectra of the TCNE⁼–BP system.

In conclusion, we have explicitly shown that TCNE⁼ possesses strong electron-donating nature to the 4nπ BP. This might well explain that molecular switching in solution with potential sweep is owing to change in the complex geometry as shown in Fig. 2.

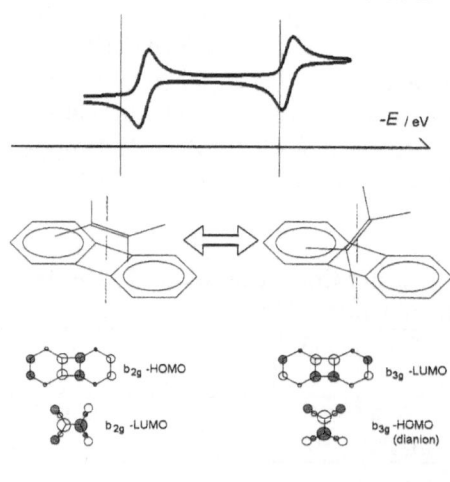

Fig. 2. Active Mo's in the TCNE-BP and TCNE⁼-BP complexes, and schematic illustration of the molecular switch with potential sweep.

References

1)K. Kano et al., *J. Am. Chem. Soc.*, **112**, 8645 (1990); B. Uno et al., *Chem. Lett.*, **1992**, 1017; N. Okumura et al, "Novel Trends in Electroorganic Synthesis, "S. TORII, ed., Kodansha, Tokyo (1995), pp. 31-32.

PHYSICAL-ENERGETICAL REACTION CONTROL IN ELECTROORGANIC PROCESSES

Tsutomu NONAKA

Departmemt of Electronic Chemistry, Interdisciplinary Graduate School of Science and Engineering, Tokyo Institute of Technology, 4259 Nagatsuta, Midori-ku, Yokohama, 226-8502, Japan

Unigue reaction control of electroorganic processes could be achieved by physical (hydrophobilization of electrodes/two-phase electrolytes/fluidization of electrolytes) and energetical (irradiation of lights / acoustic waves / mechanical fields) modifications of electrode interfaces.

Since electron transfer takes place in an electrode interface(the surface and vicinity of electrode), the electrode interface modification should be a useful method for controlling electrochemical processes. Hence, the chemical modification of electrodes has been applied to a variety of electroorganic processes so far. In this work, the physical-energetical modification of electrode interfaces is proposed as a new concept for the methodology of reaction control.

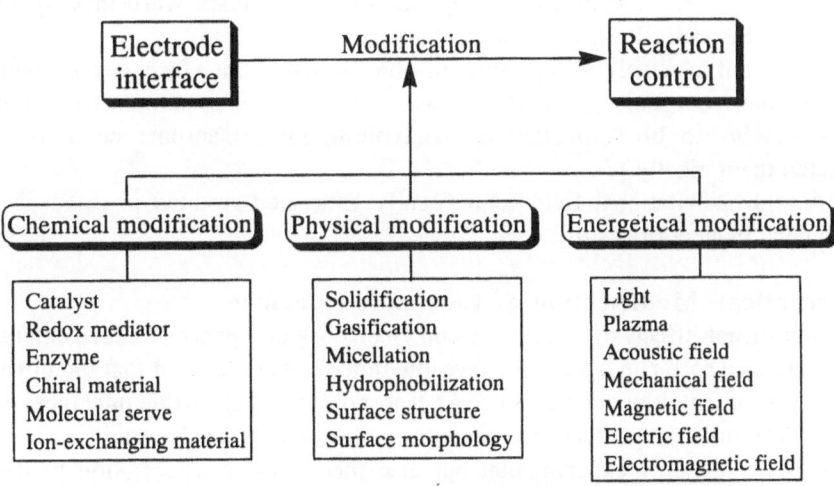

Electrode interface modifications for reaction control

Physical Modification of Electrode Interfaces

By making an electrode interface hydrophobic, the affiniy of the electrode with organic substrate molecules should increase, and consequently oxygen and hydrogen evolution by the electrolysis of water is suppressed in an aqueous medium while an efficiency for the desired electroorganic process is increased. In order to verify this idea, hydrophobic electrodes were prepared by composite-plating metals with fine particles of hydrophobic materials such as poly (tetrafluoroehylene)(Teflon, PTFE), fluorographite and fluorinated pitch. These hydrophobic electrodes resulted in significant increases in current efficiencies and/or selectivities for the desired electroorganic processes in homogeneous and hetergeneous (emulsion) systems [1-4]. A binary type composite plated nickel electrode with PTFE and silica gel as hydrophobic and hydrophilic components, respectively, was prepared and applied to electroreduction of aldehydes [5]. The electrode gave high current efficiencies and selectivities for the corresponding alcohols. The hydrophobic electrodes exhibited some remarkable effects as follows : Concentrating effect of organic molecules, filming effect of organic droplets, promoting effect of dehydration, reactivity-controlling effect of organic radical intermediate species, and decreasing effect of oxygen and hydrogen overvoltages.

Two-phase electrolysis is practically important for electrochemical processes of water-insoluble organic substrates. In this work, basic mechanisms of electron and mass transfers in emulsion and suspension electrolyses were investigated in detail form theoretical and experimental aspects [6].
Fluidization of electrolytic solutions in the electrode interfaces were found to influence electroorganic processes. It was indicated that reaction selectivities for the processes can be controlled by controlling mass transport conditions and predicted theoretically [7].

A high gravitational field centrifugally generated resulted in some peculier phenomena in electrolysis [8].

Energetical Modification of Electrode Interfaces

Significant effects of ultrasounds on electroorganic processes have extensively been examined so far in a series of investigation. It was clarified that the ultrasonic effects are caused by promotion of mass transport in the electrode interface, which is also promoted particularly by cavitation phenomena of ultrasonic waves (kHz level), from not only experimental but also theoretical consideraition [9,10]. In addition, not only ultrasonic irradiation to electrodes but also oscillation from electrodes was found efficient from energetics [11]. Ultrasounds with high frequencies (MHz level) gave quite different effects on electroorganic processes from kHz waves.

Product selectivity for the electrochemical reduction of aromatic ketones and benzyl halides was found to be controlled on semiconducting electrodes. It was concluded that the excitation of both the semiconducting electrode and the substrate molecules are inevitably required for the selectivity control [12].

It has been so far recognized that electrochemical processes are not sensitive to temperature. However, in this work it was found that product-selectivities in the reduction of carbonyl compounds were controlled by temperature [13].

The new concept of physical-energetical electrode interface modification is spreading to application of ultra-high pressure, microwave irradiation, high magnetic field, etc.

References

1) Y. Kunugi, T. Nonaka, Y.-G. Chong and N. Watanabe, *J. Electroanal. Chem.*,**1993**, *353*, 209.

2) Y. Kunugi, T. Nonaka, Y.-G. Chong and N. Watanabe, *J. Electroanal. Chem.*, **1993**, *356*, 163.

3) Y. Kunugi, P.-C. Chen,T. Nonaka, Y.-G. Chong and N.Watanabe, *J. Electrochem. Soc. Jpn.*, **1993**, *140*, 2833.

4) P.-C. Chen, P.-C. Cheng, Y. Kunugi and T. Nonaka, *J. Electrochem. Soc. Jpn.*,**1993**, *61*, 866.

5) Y. Ono, Y. Nishiki and T. Nonaka, *Chem. Lett.*,**1994** ,1623.

6) P.-C. Chen and T. Nonaka, *J. Electrochem. Soc. Jpn.*, **1995**, *63*, 395.

7) P.-C. Chen and T. Nonaka, *Bull. Chem. Soc. Jpn.*,**1995**, *68*, 378.

8) Unpublished.

9) M. Atobe, K. Matsuda and T. Nonaka, *Electroanalysis*, **1996**, *8*, 784.

10) M. Atobe and T. Nonaka, *Ultrasonics*, **1997**, *4*, 17.

11) M. Atobe and T. Nonaka, *J. Electroanal. Chem.*, **1997**, *425* , 161.

12) K. Tsunashima and T. Nonaka, *Electrochem. Acta*, **1997**, *42*, 2081.

13) K. Tsunashima and T. Nonaka, *Chem. Lett.*, **1995**, 862.

METHOD OF PRODUCING CHOLINE OF A HIGH PURITY

Shunji Hyoda

Sakaide Research Laboratory,   Japan Hydrazine Company Inc.
2-2-14 Irifune-cho, Sakaide 762, Japan

A choline is a strongly basic substance and is used as a photoresist-developing solution. It is, however, difficult to remove by-products formed during the production of a choline. So far, therefore, the choline has been used in the presence of by-products thereof.

The choline **1** is usually prepared in compliance with the following reaction formula (1),

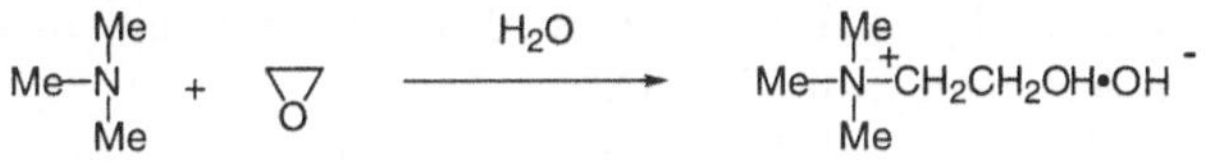

$$
\underset{\underset{Me}{|}}{\overset{\overset{Me}{|}}{Me-N}} \; + \; \underset{O}{\triangledown} \quad \xrightarrow{\; H_2O \;} \quad \underset{\underset{Me}{|}}{\overset{\overset{Me}{|}}{Me-\overset{+}{N}-CH_2CH_2OH \cdot OH^-}} \qquad \text{------ (1)}
$$

$$\mathbf{1}$$

This is the addition reaction in which an ethylene oxide is added to a trimethylamine, and it is estimated that the compounds of the following formula (2) are formed as by-products **2**,

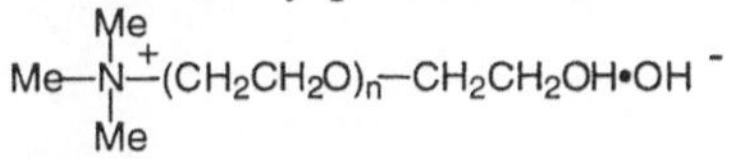

$$
\underset{\underset{Me}{|}}{\overset{\overset{Me}{|}}{Me-\overset{+}{N}-(CH_2CH_2O)_n-CH_2CH_2OH \cdot OH^-}} \qquad \text{------ (2)}
$$

$$\mathbf{2}$$

These by-products are the ones in which the ethylene oxides are successively added to the choline and, wherein n is chiefly 1 or 2. A conventional aqueous solution of the choline contains the above-mentioned by-products in amounts of from about 2 to 3 % by weight with respect to the choline.

It is virtually difficult to suppress the formation of, or to remove, by-products contained in the choline relying upon a generally employed chemical or physical treatment such as improving the reaction conditions for the choline, distillation, recrystallization, adsorption or column chromatography. If a highly pure choline could be produced without containing by-products, a further extended application is expected as a photoresist-developing solution and, besides, new applications can be explored in the fields of medicine, agricultural chemicals and functional materials.

We have already proposed a method by which a quaternary ammonium hydroxide and/or a salt thereof forms a molecule-inclusion complex with particular phenols.[1] The host compound that is used is particularly a 1,1'-bis-β-naphthol **3**. By using a molecule-inclusion complex of the choline, we first obtained the choline of a high purity without containing by-products.

1671 Grams (5.84 mols) of 1,1'-bis-β-naphthol and 1635 g of an aqueous solution containing 47.84 % by weight of the choline (6.45 mols) were added to 7326 g of pure water, and were heated and dissolved. The amount of by-products in the aqueous solution of the choline was 1.85 % by weight with respect to the choline. After heated and dissolved, crystals precipitated immediately. The aqueous solution of the choline was heated and stirred at 50 ℃. for two hours and was then cooled down to 1 ℃., subjected to the centrifugation, followed by washing with pure water. The obtained crystals were further washed with pure water under stirred and centrifuged once. After drying, there was obtained 1,1'-bis-β-naphthol / choline inclusion complex **4** in an amount of 1858 g (4.77 mols). The yield was 81.8 mol%. By the NMR and the elemental analysis, the molecular ratio of the inclusion complex was 1,1'-bis-β-naphthol / choline=1 / 1. The choline in the inclusion complex was subjected to the ionic chromatography, and it was learned that the amounts of by-products were smaller than a detectable limit (Fig.1).

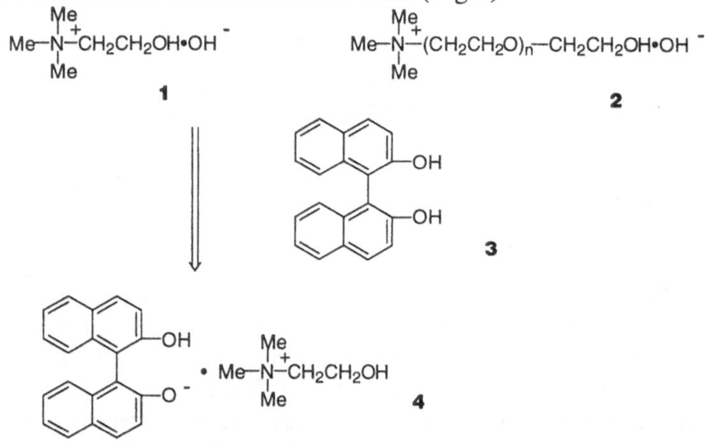

Fig.1

When a given compound takes in another compound to form a complex, this complex is called molecule-inclusion compound, the former compound is called host molecule and the latter compound is called guest molecule. In the host-guest inclusion lattice, host and guest molecules recognize each other and the inclusion crystal formation can be used for compound purification and separation of isomers. In this case, the choline only in an aqueous solution of the choline forms a molecule-inclusion complex with host molecule such as of 1,1'-bis-β-naphthol but does not form a molecule-inclusion complex with by-products.

That is, the choline taken in by the choline-inclusion complex reacts with the carbon dioxide gas to form a water-soluble choline salt which is then dissolved in water to obtain the choline carbonate **5** and / or the choline bicarbonate **6** of a high purity(Fig.2).

1858 Grams (4.77 mols) of the obtained inclusion complex without containing by-products was added to 10400 g of pure water, and 3000 g of carbon dioxide gas was blown thereto with stirring at 10 ℃. for four hours.

The obtained solution was centrifuged and washed with pure water and insoluble crystals were obtained as a residue. 7060 g of an aqueous solution containing 6.29 % by weight of the choline carbonate and / or the choline bicarbonate of a high purity (444 g as the choline bicarbonate) was obtained. The insoluble crystals were further washed with pure water under stirred and centrifuged once. Further, 13800 g of an aqueous solution containing 2.08 % by weight of the choline carbonate and / or the choline bicarbonate of a high purity (287 g as the choline bicarbonate) was obtained. The total yield was 92.8 mol%.[2]

After drying, the insoluble crystals were recovered in an amount of 1351 g (4.72 mols). By the IR analysis, it was confirmed that the insoluble crystals were 1,1′-bis-β-naphthol. Its recovery rate was nearly quantitative and the recovered 1,1′-bis-β-naphthol was reusable.

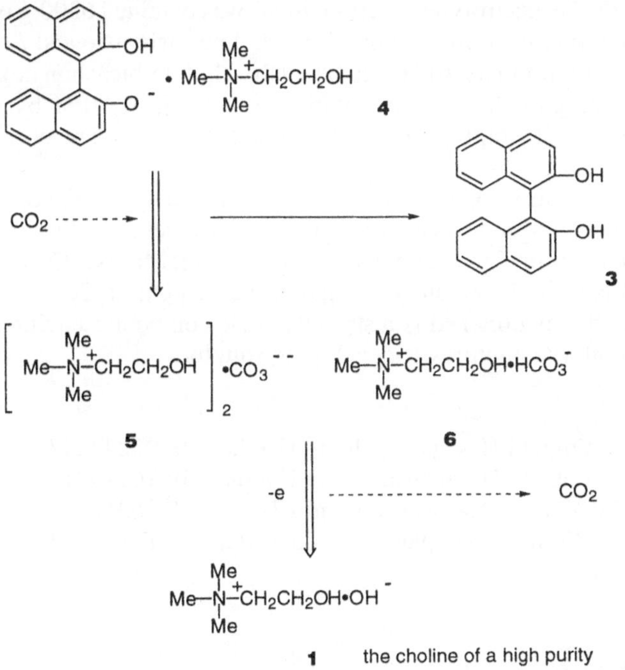

Fig.2

The thus obtained aqueous solution of the choline carbonate and / or the choline bicarbonate of a high purity is subjected to the electrolysis using a cation-exchange membrane as a diaphragm in order to obtain the choline of a high purity in the form of an aqueous solution (Fig.2).

There was used an electrolytic cell having a pair of anode and cathode each with an effective area of 1.8 dm². The electrolytic cell was sectionalized into an anode chamber and a cathode chamber using Nafion 966 (produced by Du Pont Co.) which was the cation-exchange membrane of the type of fluorine-containing resin. In the anode chamber of the electrolytic cell was provided an anode composed of DSE and in the cathode chamber was provided a cathode composed of Ni.

Into the anode chamber, was fed 7000 g of the aqueous solution containing 6.29 % by weight of the choline carbonate and / or the choline bicarbonate of a high purity without containing by-products obtained as described above (2.64 mols as the choline bicarbonate), and into the cathode chamber was fed 700 g of ultra-pure water.

The electrolytic reaction was carried out at a temperature of from room temperature to 35 ℃., at a constant-current density of 20 A / dm² for 2.5 hours and an electrolyzing voltage of from 7.1 to 21.0 V.

As a result of the electrolytic reaction, there was obtained 6200 g of the aqueous solution containing 0.56 % by weight of the choline carbonate and / or the choline bicarbonate of a high purity (0.209 mols as the choline bicarbonate) in the anode chamber and 1322 g of the aqueous solution containing 21.80 % by weight of the choline (2.38 mols) of a high purity in the cathode chamber. The current efficiency was 70.8 %.[3, 4]

The aqueous solution of the choline of a high purity obtained in the cathode chamber contained by-products at concentrations that were smaller than a detectable limit, and contained smaller than 5 ppb of Ag, Al, Mg, Mn, Ni, Cr, and Zn, smaller than 10 ppb of Na, Ca, K, Fe and Cu, smaller than 1 ppm of Cl.

The choline that is obtained is a strongly basic compound having a high purity and is very useful as a photoresist-developing solution.

References
1) Fumio Toda, Youichi Hasegawa, Shunji Hyoda,   JP 95247247
2) Fumio Toda, Youichi Hasegawa, Shunji Hyoda,   JP 96119911
3) Shunji Hyoda, Youichi Hasegawa, Fumio Toda,   JP 96218191
4) Shunji Hyoda, Youichi Hasegawa, Fumio Toda,   USP 5,618,978

THE STUDY OF NICKEL HYDROXIDE CATALYST SURFACE DURING ORGANIC ELECTROSYNTHESIS BY ATOMIC FORCE MICROSCOPY

Andrzej KOWAL

Institute of Catalysis and Surface Chemistry, Polish Academy of Sciences, ul. Niezapominajek 1, 30-239 Kraków, Poland

INTRODUCTION

Oxidation of alcohols on nickel hydroxide electrodes has been studied for twenty years. The mechanism of the oxidation reaction was proposed by Pletcher and Fleishmann [1] and extended by Schafer[2]. Recently Kowal and coworkers used the modern in-situ methods for understanding the mechanism of the oxidation reaction [3] and the stability of nickel hydroxide electrodes [4,5].

In this work, in-situ AFM is used for monitoring the changes in the surface topography of the electrode during the electrochemical oxidation of ethanol.

EXPERIMENTAL SCHEME :

Ni/NiO.H$_2$O   Natural electrode surface after

                                                          polishing in air

                    Cyclic voltammetry in 1M KOH

                    -0.1V ↔ 0.6 V vs. Hg/HgO ; 15 minutes, stop at -0.1 V

Ni/Ni(OH)$_2$ ⟵——— in - situ AFM examination of electrode surface : P-1

                    Potential step from -0.1 V to 0.44 V

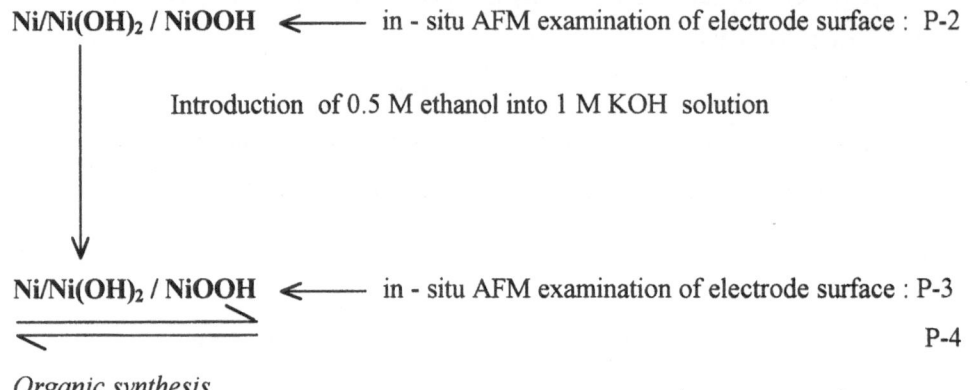

Organic synthesis

During the organic synthesis NiOOH reacts with hydrogen in α position in alcohol molecule and is reduced to $Ni(OH)_2$ then the potentiostat oxides the $Ni(OH)_2$ to NiOOH and again NiOOH can react with α hydrogen from alcohol molecule.

CONCLUSIONS

It was found that during the electrooxidation of $Ni(OH)_2$ to NiOOH the size of hydroxide grains changes in the horizontal and vertical directions. The decreasing of the thickness of $Ni(OH)_2$ grains during the electrochemical oxidation to NiOOH in 1M KOH was also shown in a separate experiment [4]. When ethanol was added to the solution the corrosion of the electrode surface was observed, especially after the prolonged oxidation. This fact was confirmed by the in - situ experiment carried out with the method of Potential Modulated Reflectance Spectroscopy [5].

REFERENCES
1a) M. Fleishmann, K. Korinek, D. Pletcher, J. Electroanal. Chem., **1972**, *31*, 39.
1b) M. Fleishmann, K. Korinek, D. Pletcher, J. Chem. Soc., Perkin Trans. II, **1972**, 1396.
2) H.-J. Schafer, Topics in Current Chemistry, Vol. 142, p. 101, Springer-Verlag, Berlin Heidelberg **1987**.
3) A. Kowal, S. N. Port, R. J. Nichols, Catalysis Today, in press.
4) A. Kowal, R. Niewiara, B. Peroñczyk, J. Haber, Langmuir **1996**, *12*, 2332.
5) A. Kowal, C. Gutierrez, J. Electroanal. Chem., **1995**, *396*, 234.

SANDWICH ELECTRODES AND THEIR POSSIBLE APPLICATION IN ORGANIC ELECTROSYNTHESIS STUDIED BY ATOMIC FORCE MICROSCOPY

Andrzej KOWAL [a], Joanna KOWAL [b] and Karl DOBLHOFER [c]

a. *Institute of Catalysis and Surface Chemistry, Polish Academy of Science, Kraków, Poland*
b. *Faculty of Chemistry, Jagiellonnian University, Kraków, Poland*
c. *Fritz-Haber-Institute of the Max-Planck-Society, Berlin-Dahlem, Germany*

The concept of sandwich electrodes and their application in organic electrosynthesis will be discussed. Sandwich electrode (SE) (see e.g. [1]) consists of at least 3 layers: flat current collector, thin catalyst layer and special organic membrane.

Polished glassy carbon (GC), highly oriented pyrolitic graphite (HOPG) or a thin monocrystalline layer from Au (111) is used as a current collector. As an example of catalyst, the thin layer of SnO_2 doped with Sb will be discussed. The SnO_2 layer doped with Sb is prepared by reactive ion sputtering or spray pyrolytic method [2]. Catalysts such as SnO_2 corrode in water solution. The progress of the corrosion process was monitored by the AFM topographic images recorded on a fresh catalyst surface and after 18 h oxidation [2] of formaldehyde solution. One of the authors studied the properties of membranes on electrode surfaces [3, 4] and proposed to cover the catalyst surface with a polymer membrane. In this work we used polyvinylpyridine or a thin layer of isotactic polystyrene as a membrane. The process of photooxidation and photodegradation of these polymers as well as their UV spectral properties were studied in detail [5].

There are three main benefits of using SE in organic electrosynthesis. First, the polymer layer can protect the catalyst layer from surface corrosion. Second, the polymer can be used to accumulate an electrode reactant to increase the rate of the electrochemical reaction (see diagram below). Third, the access of an undesirable reactant to the electrocatalytic surface may be hindered.

SANDWICH ELECTRODE:

CURRENT COLLECTOR (Au, GC, HOPG)	CATALYST	ORGANIC MEMBRANE + (anionic exchanger)	SOLUTION

EXAMPLE OF REACTANT ACCUMULATION (concentration profile):

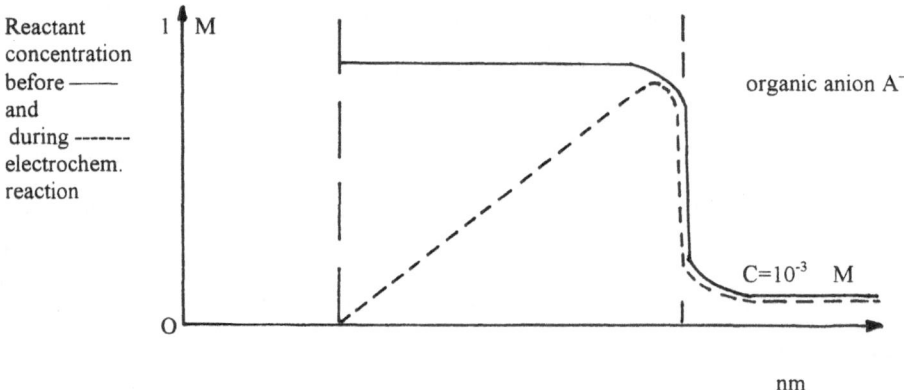

Reactant concentration before —— and during ------ electrochem. reaction

References

1) A. Kowal, K. Doblhofer, S. Krause, G. Weinberg, J. Appl. Electrochem., **1987**, *33*, 1246.

2) (a) A. Kowal, J. Kowal, J. Haber, ''Electrocatalytic Oxidation of Urotropin, Trioxane and Formaldehyde from Waste Water'', 4th Intern. Forum on Electrolysis, Fort Lauderdale 1992, lecture nr. 28; (b) A. Kowal, M. Janik-Czachor, A. Wołowik, ''Preparation and Characterization of Highly Doped SnO_2 Polycrystalline Layer by AFM and AES'', Proc. 18th Intern. Seminar on Surface Physics, Polanica Zdrój, Poland 1996; (c) H. Cachet, J. Stoch, A. Kowal, "Characterization of Highly Doped SnO_2 Layer by AFM, XPS and AES", EuropaCat-3, Kraków 1997.

3) K. Doblhofer, in ''Electrochemistry of Novel Materials'', J. Lipkowski and P. N. Ross, eds., VCH, New York 1994, pp 141-205.

4) H. Braun, F. Decker, K. Doblhofer, H. Sotobayashi, Ber. Bunsenges. Phys. Chem., **1984,** *88*, 345.

5) (a) J. Kowal, M. Nowakowska, Polymer, **1982**, *23*, 281; (b) J. Kowal, Polymer Degrad. Stabil., **1984**, *7*, 175; (c) J. Kowal, D. Jamróz, Makromol. Chem., **1991**, *192*, 461; (d) J. Kowal, Polymer, **1997**, *20*, 5059.

MICROSTRUCTURAL ANALYSIS OF CARBON FIBER FOR LITHIUM RECHARGEABLE BATTERY

Chang-Ho Sung, Ju-Seung Kim, Hal-Bon Gu
Dept. of Electrical Eng., Chonnam National University
300 Yongbong-Dong, Bukgu, Kwangju 500-757, Korea

ABSTRACT

Electrochemical properties of carbon fiber was investigated in order to find the relation between structure of carbons and performances as a negative electrode of lithium rechargeable batteries. In this study, we have studied microstructure of lithium intercalated carbn fiber using scanning electron microscope and x-ray diffractommetry. From SEM results of carbon fiber, film formed on the surface of carbon fiber appeared after discharging and disappeared after charging. X-ray diffractommetry of carbon fiber showed same results.

INTRODUCTION

Recently, rechargeable lithium batteries have become more attractive for application in power source for stationary electric power storage and electric vehicles because of their high energy density[1]. However, it is recognized that nonaqueous lithium rechargeable battery using lithium metal as the negative electrode exhibited such significiant problems as the short cycle life and unsafe operating characteristics. The rechargeability of this battery was lower than those of other batteries. It was unstable that lithium ion changed lithium metals during charge/discharge cycle[2]. To solve this problem, a rocking-chair battery that used an intercalation compound as the anode for a rechargeable lithium battery has been proposed[3].

EXPERIMENTAL

KCF 3000 was purchased from petca Co. Working electrode were fabricated by mixing carbon with 5wt% binder(Aldrich, PVDF). The paste-like mixture of 40mg was spread thinly onto a Cu foil and pressed. It was vacuum-dried for 12h at 110℃. Lithium metal was used as a counter electrode and reference electrode. The electrolyte was a 1M $LiPF_6$/EC-DEC.
All operations on the cell assembly were carried out in argon gas filled with glove box. The morphology of carbon fiber was investigated with scanning electron microscope(JEOL JSM5400), working at an accelerating potential of 25KV. The structural changes in carbon fiber were followed by x-ray diffraction measurements using a Philips diffractometer PW-1830 with CuKα radiations(1.5405Å).

RESULTS AND DISCUSSION

Fig. 1 represents the morphology of carbon fiber used lithiation studies. SEM photograph of carbon fiber charging up to the amount of $Li_{0.6}C_6$ showed film on the surface of carbon fiber. After second charging, this film was disappeared. And, after second discharging, this film was again appeared. The film formed after 2nd discharing was smaller than that of $Li_{0.6}C_6$. Also, we considered that lithium ion doped on film during intercalation of lithium ion to the carbon layer, according to the results of SEM photograph. Fig. 2 represents structure analysis result of carbon fiber as a function of cycle. After first discharging, x-ray diffraction patterns of carbon fiber showed new diffraction peak at the diffraction angle at 2θ =25.25°, corresponding to the value of d-space of 3.524Å. This peak disappeared after second charging and appeared after second discharge. The change in the diffraction patterns with the cycle revealed that the process was quite reverse. These results was found to intercalate Li ion into texture of carbon fiber reversibly.

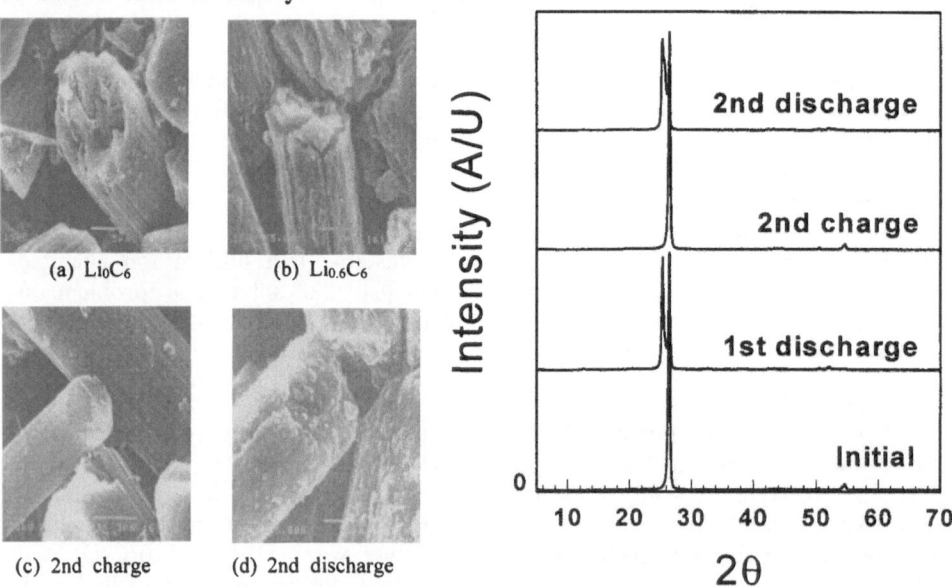

(a) Li_0C_6 (b) $Li_{0.6}C_6$

(c) 2nd charge   (d) 2nd discharge

Fig. 1 SEM photograph of carbon fiber.   Fig. 2 XRD patterns of carbon fiber.

REFERENCES

1) M. Morita, et al., J. of Electrochem. Soc., Vol. 143, pp. L26-L28, **1996**
2) R. Kanno, and Y. Kawamoto, J. of Electrochem. Soc., Vol. 139, pp. 3397-3404, **1992**
3) T. Iijima, K. Suzuki, and Y. Matsuda, DENKI KAGAKU, Vol. 61, pp.1383-1389, **1993**

APPLICATION OF SYDNONES TO LITHIUM BATTERY ELECTROLYTES

Minoru HANDA, Kan SEKINE, and Yukio SASAKI

Department of Industrial Chemistry, Faculty of Engineering,
Tokyo Institute of Polytechnics, Kanagawa 243-02, Japan

The sydnone compounds are well known as typical mesoionic compounds. A number of sydnones have been synthesized, and their structures, reactions and various physicochemical properties have been mainly studied in the field of medicine and pharmacology because they have varied types of biological activities. Some alkylsydnones are liquids at room temperature, and have high dielectric constants and large dipole moments. This paper focuses on the characteristics of the 3-alkylsydnones to lithium battery electrolytes. Figure 1 shows the conventional structure of alkylsydnones.

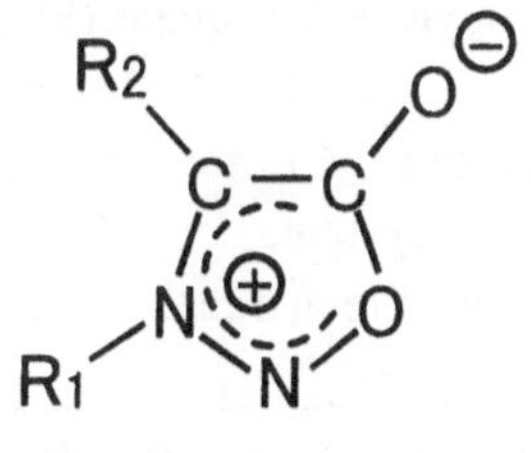

	R_1	R_2
3-MSD	CH_3	H
3-ESD	CH_3CH_2	H
3-PSD	$CH_3CH_2CH_2$	H
3-i-PSD	$(CH_3)_2CH$	H
3-BSD	$CH_3CH_2CH_2CH_2$	H
3-s-BSD	$(CH_3)_2CHCH_2$	H
3-PMSD	$CH_3CH_2CH_2$	CH_3
3-PESD	$CH_3CH_2CH_2$	CH_3CH_2
3-BMSD	$CH_3CH_2CH_2CH_2$	CH_3

Fig. 1   Structure of 3-alkylsydnones.

In Table 1, physical properties of 3-alkylsydnones at various temperatures are investigated by dielectric constant(ε), viscosity(η) and density(ρ) measurements. These 3-alkylsydnones have high dielectric constants and large viscosities compared with those of other polar solvents. Figure 2 shows the variation in specific conductivity(κ) of 3-alkylsydnone $-$ 1,2-dimethoxyethane (DME) binary mixed solutions. The higher dielectric constant solvent exhibits the higher conductivity in the solution. The 3-alkylsydnone-based binary systems are preferable to 3-alkylsydnones alone because of their high conductivities. The enhancement of specific conductivity upon addition of DME is due to the lower viscosity of the mixture, which increases the ionic mobility. With a larger mole fraction of DME, the conductivities decrease mainly due to the increase in ion pair formation upon addition of DME with low dielectric constant.

451

Table 1. Physical properties of 3-alkylsydnones.

		3-MSD	3-ESD	3-PSD	3-i-PSD	3-BSD	3-s-BSD	3-PMSD	3-PESD	3-BMSD
ε [a]	25℃	-------	102.3	93.0	-------	77.9	76.6	67.7	64.6	59.3
	40℃	141.8	95.6	87.1	-------	71.6	71.5	64.2	60.7	55.1
	60℃	126.4	87.5	79.2	65.9	64.6	65.1	60.9	55.9	51.0
η /cP	25℃	-------	6.6	10.3	-------	11.3	16.1	11.8	11.2	14.8
	40℃	4.6	3.9	6.2	-------	6.3	8.1	6.4	6.3	7.6
	60℃	2.6	2.3	3.5	2.9	3.3	3.8	3.3	3.2	3.7
ρ /10^3kgm^{-3}	25℃	-------	1.225	1.162	-------	1.115	1.119	1.121	1.068	1.085
	40℃	1.306	1.212	1.149	-------	1.103	1.107	1.107	1.054	1.073
	60℃	1.287	1.194	1.132	1.132	1.088	1.091	1.089	1.035	1.057

a) Measured at a frequency of 1 MHz.

The variation in charge-discharge coulombic cycling efficiency of the metal Li anode in the electrolytes was estimated by a galvanostatic plating-stripping method [1]. Figure 3 shows the variation in Li cycling efficiencies of 3-alkylsydnone−DME binary systems by reference to the commercially available propylene carbonate (PC)−DME and ethylene carbonate (EC)−DME systems. The 3-PSD−DME system exhibits about 70 % good efficiencies in higher cycle numbers and the efficiencies are larger and more stable than those of PC−DME and EC−DME systems. This means that it is possible to use sydnone compounds as electrolytes in Li rechargeable cells.

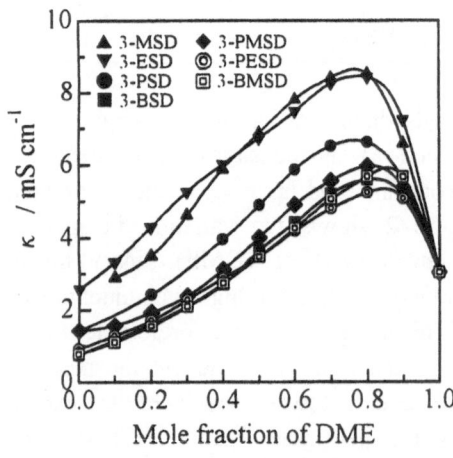

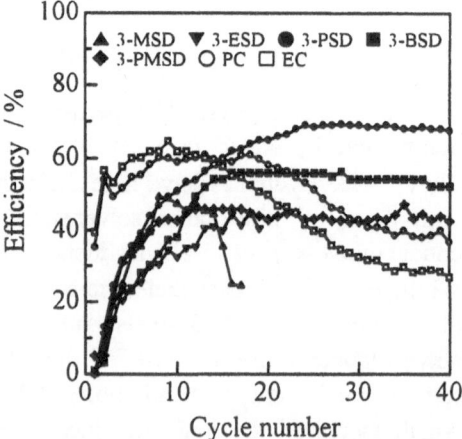

Fig. 2 Variation in specific conductivity (κ) of 3-alkylsydnone − DME binary solutions containing 0.5 mol dm^{-3} LiClO$_4$ at 25℃.

Fig. 3 Variation in Li cycling efficiencies of binary solutions (mol fraction of DME; 0.5) containing 0.5 or 1.0 mol dm^{-3} LiClO$_4$ at 25℃.

Reference
1) V. R. Koch, S. B. Brummer, *Electrochim. Acta*, **1978**, *23*, 55.

INTERFACIAL REACTION AND CHARGE/DISCHARGE PROPERTIES OF LiMn$_2$O$_4$ WITH ORGANIC ELETROLYTE

In-Seong Jeong, Jong-Uk Kim, Gye-Choon Park [*] , Hal-Bon Gu
Dept. of Electrical Eng., Chonnam National University
300 Yongbong-Dong, Bukgu, Kwangju 500-757, Korea
[*] *Dept. of Electrical Eng., Mokpo National University*
Cheongkye-myun, Muan-kun, Chonnam 534-729, Korea

ABSTRACT

The time dependents of discharge property of LiMn$_2$O$_4$ cathode active material and interfacial reaction property of LiMn$_2$O$_4$/electrolyte interface are studied. LiMn$_2$O$_4$ cathode active material heated at 800 ℃ for 36h exhibits excellent discharge property and shows significant advantage at reversibility. Use of super-s-black as conductive agent in LiMn$_2$O$_4$ cathode, charge/discharge property is best because interfacial reaction of LiMn$_2$O$_4$ cathode with organic electrolyte is more stable than other conductive agent.

INTRODUCTION

In recent years, use of lithium ion battery have been increasing because of high energy density and anti-pollution. LiMn$_2$O$_4$ have been using as cathode active material of Lithium ion battery because Mn is low cost and plentiful sources[1-3]. However, manganese dissolution reaction takes place in LiMn$_2$O$_4$/Electrolyte interface. It affects to charge/discharge property. Therefore, it is important to understand interfacial reaction property of LiMn$_2$O$_4$/electrolyte interface.

In this study, we report the results on variation of charge/discharge property by kind of conductive agent involved in LiMn$_2$O$_4$ cathode and heating time of LiMn$_2$O$_4$ cathode active material.

EXPERIMENTAL

LiMn$_2$O$_4$ was prepared by reacting stoichiometric ratio and heating at 800 ℃ for variable time, respectively. Then, LiMn$_2$O$_4$ analyzed by X-ray diffraction for crystal structure analysis. For electrochemical measurement, LiMn$_2$O$_4$/Li cell was assembled in glove box. The cut-off voltage for charge and discharge limit were fixed at 4.3 and 3.0V. Current density was 0.1mA/cm^2.

RESULTS AND DISCUSSION

Fig. 1 shows the time dependents of discharge property of LiMn$_2$O$_4$ obtained from the heating at 800. The results demonstrate that when heating at 800 for 36h, discharge property show remarkably excellent. Discharge capacity of first cycle was and after 10cycle, discharge capacity was stabilized about 110mAh/g. Fig. 2 shows charge/discharge properties by kind of conductive agent involved in $LiMn_2O_4$ cathode. Charge/discharge properties of $LiMn_2O_4$ cathode mixed with super-s-black as conductive agent are excellent than acetylene black or that mixed super-s-black and acetylene black.

CONCLUSION

Charge/discharge property of $LiMn_2O_4$ cathode active material heated at 800℃ for 36h was excellent. Charge/discharge property of $LiMn_2O_4$ cathode mixed with super-s-black as conductive agent was excellent than acetylene black or that mixed super-s-black and acetylene black. Because interfacial reaction of $LiMn_2O_4$ cathode with organic electrolyte is more stable than other conductive agent.

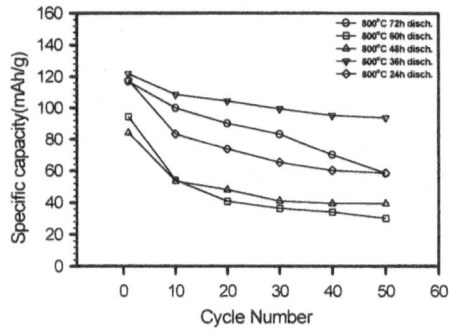

Fig. 1. Discharge capacity of cathode active material prepared at 800℃ for various time.

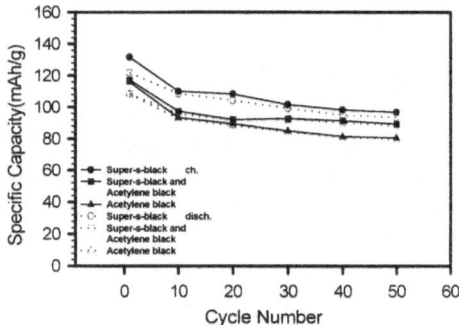

Fig. 2. Discharge capacity of cathode by kind of conductive agent

REFERENCES
1) J. N. Reimers, J. R. Dahn and U. von Sacken, *J. Electrochem. Soc.*, **1993**, 140, 10, 2752.
2) J. R. Dahn, U. von Sacken, M. W. Juzkow and H. Al-Janaby, *J. Electrochem. Soc.*, **1991**, 138, 8, 2207
3) A. Momchilov, V. Manev, A. Nassalevska and A. Kozawa, *J. Power Sources*, **1993**, 41, 305

Stereochemistry and Electrochemistry, Synthetical and Analytical Implications in the Synthesis of Molecules of Defined Handedness

P. M. Bersier*, L. Carisson and J. Bersier

ElectroCell AB, Täby, Sweden:
* Cagliostrostrasse 38, 4125 Riehen, Switzerland

1. Optically pure substances, why?

Inactive, even though innocuous enantiomers are "isomeric ballast" [1]. Clinical studies of absorption and metabolism of racemic drugs lead according to Ariens [1] to "highly sophisticated scientific nonsense", as individual stereo isomers can have different pharmacological, toxicological, metabolic and pharmacokinetic properties.

In 1991 more than 500 drugs were currently marketed as racemic mixtures with negligible relevant information available on the properties of the individual stereoisomers.[3] Enantiomeric drugs are growing at the expense of racemates, as shown by the (predicted) dollar sales for 1991 to 1996.[2]

For industrial production, asymmetric synthesis of enantiomerically pure intermediates or compounds now requires synthetic techniques that take into account not just cost, but also environmental impact, in terms of process integrated environmental protection, as well as by-products and degradation products (false stereoisomers and enantiomers).[4]

2. Analytical methods

Central to the art and practice of stereoselective synthesis is the analysis of the outcome of the reaction [5] necessitating an extensive armory of analytical methods in terms of selectivity and sensitivity (<< 1%). Ultimately, the ratio of isomers and the configuration of each new stereocenter should be determined. Two types of analysis exist: (i) those that separate: chromatographic processes: TLC, GC, HPLC; CSCF (supercritical capillary electrophoresis); (ii) those that do not separate: polarimetry (outdated), NMR, electrochemical techniques: polarography (assay of axial and equatorial isomers, e.g. determination of β-fluorenon impurity in α-fluorenon; of organic mediators, etc.), voltammetry, HPLC with electrochemical detection, tensammetry.

455

3. Routes/basic methods for the synthesis of optically pure enantiomers
3.1. Conventional routes:
(1) Isolation from natural sources (still the most important source of optically pure compounds on a technical scale).

(2) Asymmetric synthesis: (i) Chiral pool: use of readily available, inexpensive optically active natural products and synthetic products; use of synthetic building blocks ("new pool"), e.g. technical scale synthesis of 3-hydroxyethyl-4-acetoxyazetidin-2-one, a key intermediate for a host of a new type of β-lactam antibiotics (penems, and carbapenems); (ii) stochiometric and catalytic methods; (iii) enzyme-mediated processes.

3.2. Electrochemical routes [6,7,8] include:
(i) electrolysis of chiral starting materials to make new chiral products; (ii) asymmetric induction by alkaloids in electrochemical reduction/oxidation; (iii) asymmetric synthesis using supporting electrolytes; (iv) indirect synthesis with chiral redox mediators; (v) electrolysis of nonchiral starting materials at chiral electrodes; (vi) stereoselective electro-reductive/oxidative cyclization; (vii) synthesis of chiral and prochiral strting materials using enzymatic methods; (viii) enantioselective membrane-based electrosynthesis; (ix) electrochemical stereoisomerization.

Selected examples taken from the literature[9] are presented.

Pros and cons, scope and limitations and versatility of electrochemical processes as compared to conventional processes are discussed.

4. Conclusion
Compared to the totality of examples of the manufacture of optically active materials, especially at the industrial level, electrochemical stereochemistry/syntheses, are still fairly few in number.

5. References

(1) Ariens, E.J., Eur.J.Clin.Pharmacol 26(1984)663; (2) Stinson , St.S., C & EN , May 16,1994, p.10; (3) Cayen, M.N., Chirality 3(1991)94; (4) Sedelmeier, G., in: Organic Synthesis, Highlights II, VCH, Weinheim, 1995, p.277; (5) Gawley, R.E. and J.Aubé, Principles of Asymmetric Synthesis, Tetrahed. Org.Chem. Ser.Vol. 14, 1996; (6) Nonaka, T., in: H.Lund and M.Baizer (Eds.), Organic Electrochemistry, 3rd. Ed., Marcel Dekker, 1991, 1131; (7) Tilborg, van W.J.M., C.J. Smit, Rec.Trav.Chim.Pays-Bas, 97(1978)89; (8) Tallec, A., Bull.Soc.Chim.France, 1985,743; (9) Schäfer, Steckhan, Torii, Weinberg, et al.   FlKu97/2

REVERSIBLE ELECTRON TRANSFER IN ELECTROORGANIC CHEMISTRY

Vitalii V. YANILKIN,[a] Vakhid A. MAMEDOV,[a] and Eugene A. BERDNIKOV[b]

[a]A.E. Arbuzov Institute of Organic and Physical Chemistry,
Russian Academy of Sciences, Arbuzov str.,8, Kazan, 420088, Russia
[b]Department of Organic Chemistry, Kazan State University,
18 Kremlevskaya Str., Kazan, 420008, Russia

Electron transfer reactions in condensed medium are accompanied by the reorganization of the nuclear system of a molecule (ΔG°_r) and solvation sheath (ΔG°_{solv}). In electrochemical reactions of organic and elementorganic compounds the spatial reorganisation of a molecule might happen ranging from minimal structural changes to the rapture (or formation) of σ-bonds. The extreme case corresponds to the dissotiative (concert) mechanism of electron transfer, all other cases corresponds to the formation of anion-(or cation-) radical. The total energy of the reorganization ($\Delta G^\circ_{chem}=\Delta G^\circ_r+\Delta G^\circ_{solv}$) is represented by the energy of the chemical component of electrochemical reaction and it determines kinetics of electron transfer reactions under conditions of standard redox-potential E°

$$k_s = k_0 \exp(-\rho \Delta G^\circ_{chem}/RT) \quad (1)$$

The kinetics of activation electron transfer at other potentials could be described by the equation

$$k_E = k_0 \exp(-\rho \Delta G^\circ_{chem}/RT) \exp[\alpha nF(E-E^\circ)/RT] \quad (2)$$

Electrode reactions of other spheric adiabatic electron transfer ($B_{solv} - e = B^{+}_{a\ solv}$) could be described by the following thermochemical cycle

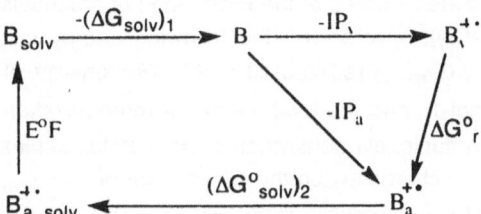

The total energy of the reorganization and its components could be found experimental values of potentials of oxidation and from energies of vertical (IP_v) and adiabatic (IP_a) ionization.

$$\Delta G^\circ_{chem} = IP_v - E^\circ F \quad IP_v - IP_a = \Delta G_r^\circ$$
$$\Delta G^\circ_{solv} = (\Delta G^\circ_{solv})_2 - (\Delta G^\circ_{solv})_1 = IP_a - E^\circ F$$

457

Therefore kineticks of electrochemical reactions of oxidation could be described by the equation

$$k_E = k_o \exp[-\rho n F(IP_v - E^o)/RT] \exp[\alpha n F(E - E^o)/RT] \quad (3)$$

The dependency for reactions of electrochemical reduction is analogous, only the ionization potential has to be changed for electron affinity.

Since the energy barrier depends on the chemical component of electrochemical reaction, we suggested the independence of position of transition state on potential. This led us to conclusion about the constancy of the transfer coefficient $\alpha=0.5$ for the reactions of activation electron transfer at the wide ranges of potentials beginning with barrierless ($\alpha=1$), coming to activationless ($\alpha=0$) processes, which is in full agreement with numerous evidences from literature.[1]

The values $k_o = 9.3 \text{cm}^{-1}$ and $r = 0.033$ were evaluated on the basis of experimental data (k_s, E^o, IP_v, a) of the kinetics of electroreduction of K^+ and Li^+. To convert E^o to the absolute scale we used the latest data of Trasatti[2] on the values of the potential of hydrogen electrode ($E^o_{NHE} = 4.48$ V).

We used equation (3) to calculate

1) potentials of transition of activation process into barrierless and activationless processes for the reaction of hydrogen escape; 2) standard rate constants (k_s) of electron transfer for several redox-systems. Our calculated values are in satisfactory agreement with the experimental data from literature.[1] This empirical equation has diverse consequences. Here we will discuss only the problem of reversibility. In polarographic methods processes of electron transfer are reversible at the reaction rate of $k_s \geq 2.4.10^{-2}$ cm s^{-1} and $\Delta G^o_{chem} \leq 650$ kJ mol^{-1}. Our experiments on electrochemical oxidation of Nitrogen-, Phosphorus-, Sulphur- and Oxygen - containing organic compounds,[3] which are oxidated at the wide range of potentials (from -0.37 to +4.00 V vs Ag/0.01 M AgNO$_3$ in MeCN), have shown the following energy values: $\Delta G^o_{chem} \leq 250$ kJ mol^{-1}, $\Delta G^o_r \leq 90$ kJ mol^{-1}, $\Delta G^o_{solv} \leq 130-200$ kJ mol^{-1}. The energy of dissociation of single bonds is less than 400 kJ mol^{-1}. and the total energy of reorganization for all the reactions of reduction and oxidation of organic, element organic and metalocomplex compounds, including dissociate electron transfer, is characterised the lower value of ΔG^o_{chem} and therefore it meets the criteria of reversibility. The irreversibility of the resultant electrode reactions a whole is due to the irreversibility of the following chemical and electrochemical stages. Experimental registration of the retarded electron transfer is possible in cases of essential reorganization of the molecule and presence of the following rapid chemical and electrochemical reactions.

1. Krischtalik L.I. Electrodnie reactii, Mechanism elementarnogo akta. M. : Nauka, 1979. 2. Trasatti S. In: Comprehensive treaties of electrochemistry. Ed. by Bockris J.O'M., Conway B.E., Yeager E.V. 1.New York: Plenum Press. 45 (1980). 3. Zverev V.V., Yanilkin V.V., Gromakov V.S., Mironov V.F. Abstr. of XI Inter. Confer. on Chemistry of Phosphorus Compounds. Kazan, Russia. 216 (1996).

Subject Index

Numbers refer to the pages on which a presentator's paper begins.